The Petroleum Engineering Handbook

The Petroleum Engineering Handbook
Sustainable Operations

The Petroleum Engineering Handbook: Sustainable Operations

Gulf Publishing Company
2 Greenway Plaza, Suite 1020
Houston, TX 77046

Library of Congress Cataloging-in-Publication Data

Khan, M. Ibrahim, 1968–
    The petroleum engineering handbook : sustainable operations / M. Ibrahim Khan and M. R. Islam.
      p. cm.
    Includes bibliographical references and index.
    ISBN 978-1-933762-12-8 (alk. paper)
  1. Petroleum engineering–Handbooks, manuals, etc. I. Islam, M. R. (M. Rafiq), 1959– II. Title.
    TN870.K475 2007
    665.5–dc22

                                   2007020540

Transferred to Digital Printing in 2013

Printed on acid-free paper. ∞

Contents

Foreword

Sustainable Operations Handbook for Petroleum Engineers is a timely book, which not only provides a summary of various sustainable practices in petroleum engineering operations, but does it in a manner that raises a greater awareness of environmental sustainability.

It is not the traditional "technical handbook" one relies upon to get technical information for finding solutions to an "engineering" problem. It is a book that provides a comprehensive coverage of the systems with a focus on managerial decision-making. In doing so, it shifts the focus from finding "technical solutions", to "managerial decisions" leading towards sustainable practices. It provides step-by-step guidelines towards such practices in Chapter 2.

Subsequent chapters carry forward a similar unifying theme in combining sustainable management decisions and operational practices to various segments of the petroleum sector including exploration, drilling and production, reservoir engineering, enhanced oil recovery, transportation and refining and waste management practices.

Overall, this book examines current practices of an important industrial segment through the lenses of "environmental sustainability" and provides a valuable resource to those who will have to steer the petroleum industry from its current practices down the long road towards achieving true sustainability, not only from an environmental angle but also from an economic one.

Professor Amit Chakma, Ph.D., P.Eng.
Vice President Academic and Provost
University of Waterloo
Canada

Preface

Nature is perfect, both in space and time. To understand this perfection, one must use the science of intangibles. The premise underlying such a concept is that Nature operates and Humanity detects the tangible effects and formulates a response. However, the response has to take into consideration the role of processes within Nature that have up to now remained inaccessible to our capabilities of detection or measurement and hence "intangible" – either because of where we are looking, or because we become fixated on only certain tangible aspects that can maximize a financial return in the shortest possible time. The *hubris* of the contemporary scientific and engineering enterprise resides in its bedrock belief that we have dealt with everything of importance and done all the heavy lifting necessary once we have identified all tangible features. However, a little reflection discloses that the tangible aspects of anything do not go beyond very small elements in space, i.e. Δs approaching zero; and even a smaller element in time, i.e. $\Delta t = 0$ (meaning, time = "right now"). Hence, the urgency for Humanity to elaborate the science of intangibles has never been greater. From the moment water, air, soil, and fire were identified as ingredients essential for the sustainability of human life until very recently, human beings did not face, and were not compelled to reckon with, a crisis of unsustainability. The time-honored principle that "nature is infinite" is now being challenged because, all of a sudden we discover that natural resources are depleting despite a decline in population in the industrialized world. We discover natural resources are not enough to sustain human civilization. Is this a perception or reality?

Some blame seemingly infinite corporate greed for the mess that we are in today. Not that corporate greed cannot be damaging, but, even if corporate greed is infinite, so is the Universe, and we are thus still left with no answer to the question: why *this* sustainability crisis? There is enough water, air, soil and fire to go around, each element being regenerated through nature's ecosystem. Each element is recycled and in these processes of recycling, each element enriches itself to make it more suitable for some portion or aspect of natural existence, all of which eventually contributes to the welfare of mankind. This beneficial endpoint derives not from humans being some "superior species", or "on top of the food chain", but only from the condition, and to the extent, that humans have the ability to think (*Homo sapiens* means "thinking man") and make use of natural processes. This act of thinking, if driven by conscience (science of intangibles), should help us avoid harmful natural products. This awareness should at the same time invoke processes that enhance the natural processes, in order to achieve greater quality of life for all. Innovating science along this line, and engineering solutions to problems accordingly, opens some exciting and compelling prospects. Many of the obstacles built into present-day corporate arrangements could be countered and even shed in the most industrialized countries. Humanity generally would be enabled to counter and shed many other obstacles built into present-day systems of political and economic governance found throughout all countries on this planet. Let all those who remain skeptical about or lack confidence in this overwhelming power realise that this is "an idea whose time has come": from grasping reliable knowledge of scientific truth, consider what happened in the wake of Galileo's insistence 350 years ago that the Earth revolves about the Sun. What everyone accepts today as science would not have come into existence without this affirmation – and yet, no one, least of all Galileo himself, actually "saw" the Earth moving around the Sun. Even with modern-day space exploration, no one has been able yet to record the Earth's actual 365-day transit around the Sun, but without accepting Galileo's irrefutable conclusion, there would be no modern-day space exploration.

Nature is infinite within a closed system. It is infinite as well because it is a closed, i.e. complete, system. Because of this infinite dimension, Nature is also perfect (balanced). So, what is the origin of the imbalance and unsustainability that seems to manifest itself so ubiquitously, in the atmosphere, the soil and the oceans? As the "most intelligent creation of nature", men were expected to at least stay out of the natural ecosystem. Einstein might have had doubts about human intelligence or the infinite nature of the Universe (as evidenced in his often-quoted remark that "there are two things that are infinite, human stupidity and the Universe, and I am not so sure about the Universe"), but human history tells us human beings always managed to go with the infinite nature of Nature. From central American Mayans to Egyptian Pharaohs, from Chinese Hans to the Mannaeans of Persia, the Edomites of the Petra Valley to the Indus Valley civilization of the Asian subcontinent, all managed to remain in harmony with nature. They were not necessarily "righteous" people nor were they free from practices that we would no longer countenance (Pharaohs sacrificed humans to accompany the dead royal for the resurrection day), but they did not produce a single gram of inherently anti-nature product, such as DDT. In the modern age, we managed to give a Nobel Prize (in medicine) for that "invention". What becomes clear is this: whatever it was that our ancestors did in terms of technology remains something to be desired today.

Consider the marvels of the people who carved rocks in the crystal valley of Petra. What did these people use to cut rock? It surely was not lasers, or nucleo-thermal devices, or even TNT. What did the builders of pyramids use to calculate precisely the shapes that defies today's mathematicians, computer designers, and architectures combined? It surely was not linear algebra, finite elements, or even number-crunching supercomputers. What did the makers of the Taj Mahal use to ensure continuous waterjets flowing through fountains, air conditioning inside the building, and the evergreen lushness of the trees? It was not electric pumps, freon, or synthetic fertilizers. What did the chemical engineers of Egypt use to preserve the mummies for thousands of years? It was not formalin, bezoate, and numerous other toxins that we call "preservatives".

Today, we brag about how we do things better, faster, and cheaper. Yet, we took longer to carve out four faces in Mount Rushmore than the stone-carvers of the Petra Valley took in making those stunning crystal valleys out of solid rocks. We took longer to carve out the monument of Crazy Horse than did the makers of the Taj Mahal. Not only did we take longer, we made an immeasurable mess by using TNT and other inherently anti-nature explosives. Today, we brag about a quantum leap in all branches of sciences, yet we only recently discovered our knowledge is nowhere close to what our ancestors had many years ago. We have to ponder what was the basis for Harrapan mathematics, Jain and Tamil mathematics, or Babylonian and Sumerian mathematics. Only recently we discovered Islamic scholars were doing mathematics some 1,000 years ago of the same order that we think we discovered in the 1970s[1] – the difference being that our mathematics can only track symmetry, something that does not exist in nature. Recently, a three-dimensional PET-scan of a relic known as the Antikythera Mechanism has demonstrated that it was actually a universal navigational computing device – with the difference being that our current-day versions rely on GPS, tracked and maintained by satellite.[2] We would also be shocked to find out what Ibn Sina (Avicenna) said regarding nature being the source of all cures still holds true[3] – with the proviso that not a single quality given by nature in the originating source material of, for example, some of the most advanced pharmaceuticals used to "treat" cancer remains intact after being subject to mass production and accordingly stripped of its powers actually to cure and not merely "treat", i.e. delay, the onset or progress of symptoms. What are we missing?

This book recognizes that civilization is driven by energy needs and uses the modern-day supplier of energy needs, *viz.*, petroleum engineering, as the case study. The book challenges readers with the pointed question, "If we have progressed as a human race, why has our efficiency in sustaining human civilization regressed?" For every phase of petroleum operations, ranging from exploration to refining, the authors investigate the root cause of the failure in sustainability. Once the cause is identified, it becomes quite simple to recommend practices that are sustainable. Once sustainable practices are in place, never again should petroleum operations be synonymous with polluting the environment. This book could be a textbook on fundamentals of sustainable energy management, yet it is called a "handbook". It is so because it gets beyond the smokescreen of "blue sky", i.e. fundamental, science, tackling the justifications for various engineering practices to show exactly which practices are responsible for which effects and thus how simple it would be to remedy those practices to come up with solutions that are starkly different from the ones previously being practiced.

This book is not meant to frighten the reader. It does not lecture; it does not indoctrinate. It elucidates some of the fundamental principles of sustainability that made it possible for nature to continuously improve the environment, while making comfort available to all. The book shows comfort in lifestyle doesn't have to come at the cost of long-term unsustainability. In fact, the book argues the best lifestyle even in the short-term can only be assured with a long-term approach. It is heartening to see the authors, with very distinct track records in developing sustainable technologies, have taken up this task of "greening petroleum operations". A back-to-nature approach is long overdue. The authors propose that approach in a convincing manner. They start with the definition of sustainability. With this definition, zero-waste production strategies are in place. Such schemes are inherently sustainable. However, with their definition, it is also necessary that every practice and additive also meet the sustainability requirement. There lies the recipe for reversing global warming. Overall, this book represents what can be considered as the cookbook for evergreen petroleum operations. They do that with fundamental science but without the rhetoric of scientists. They introduce the first premise, "Nature is perfect", without the rhetoric of philosophy or even religious dogma. Who could argue with that?

Hans Vaziri (BP America), Houston, USA
Gary Zatzman (EEC Research Org.), Halifax, Canada
M. Rafiqul Islam (Dalhousie University, on sabbatical in Sultan Qaboos University, Muscat, Oman)

Notes

1. Lu, P.J. and Steinhardt, P.J. (2007) Decagonal and Quasicrystalline Tilings in Medieval Islamic Architecture, *Science* **315** [27 Feb], 1106.

2. Freeth, T. et al. (2006) Decoding the ancient Greek astronomical calculator known as the Antikythera Mechanism, *Nature* **444** [30 Nov], 587–91; and also John Noble Wilford, (2006) An Ancient Computer Surprises Scientists, *The New York Times* [29 Nov], which discusses some interesting aspects of this technology's likely context. There are not a few among those who have been more than ready to grant the wisdom of the ancients but who also nevertheless persist in believing that some ancients, especially in Europe, just had to be smarter than the ancients in, say, Muslim regions of central and west Asia. Discussion sparked around the results of the PET-scan of the Antikythera Mechanism has renewed questions about precisely this long-assumed hierarchy and sequence of ancient genius. In this regard, Wilford notes the remarks of François Charette, from the University of München, in a separate article elsewhere in the same edition of *Nature*, that "more than 1,000 years elapsed before instruments of such complexity are known to have re-emerged. A few artifacts and some Arabic texts suggest that simpler geared calendrical devices had existed, particularly in Baghdad around A.D. 900. It seems clear . . . that 'much of the mind-boggling technological sophistication available in some parts of the Hellenistic and Greco-Roman world was simply not transmitted further,' [and that] 'the gear-wheel, in this case, had to be re-invented.' "

3. Steenhuysen, J. (2007) Mother Nature Still a Rich Source of New Drugs, *Reuters* [20 Mar].

Acknowledgements

This book has been in the works for quite a few years. The initial work started as early as 1999, when R. Islam was inspired by the mission statement of Canada's then NRCan Minister, Hon. Ralph Goodale, who often talked about developing technologies that are innovative, economically attractive, environmentally appealing and socially responsible. Not too long ago that statement would be considered to be absurd, worthy of a mention within "blue sky" category. This statement formed the basis of our research group for last seven years and this book personifies that statement in the topic of petroleum engineering.

The book is a result of a number of government/industry funded research grants, worth some $4 million over the last seven years. During this time we also received invaluable advices from many researchers and industry personnel. Dr. Hans Vaziri of BP America always kept in touch and provided useful comments in numerous occasions throughout the research period of the book, spanning over six years. Dr. Scott Wellington of Shell was truly an inspiration during the early period of the writing of this book. Maj. Gen. Parvez Akmal, the former Managing Director of Oil and Gas Development Corporation (OGDC) of Pakistan mentored a number of ideas that pursued in this research. Dr. Jadoon, Chief Engineer of OGDC, was a true believer of the science that has been included in this book. His comments and suggestions were most helpful. Professor Lakhal of the University of Moncton gave us the idea that "greening" of any operation is possible, including the most difficult one, namely, petroleum engineering operations. Gary Zatzman of EEC Research Organization has been most helpful in providing critical comments on many fundamental topics, forming the core of this book. Professor Mysore Satish made many useful comments and gave many valuable tips for proposing techniques that would eventually render oil production operations sustainable. Professor Farouq Ali continued to mentor the progress of our research group and played a vital role by visiting us in several occasions and sending his colleague, Dr. Sara Thomas, who herself was very helpful. David Prior of Veridity Environmental Technology helped us develop numerous ideas into usable tools. Our research group also benefited from researchers from Canada and around the world. Dr. Omar Chaalal, Mr. Ronal Moberg, Ms. Serperi Sevgur, Mr. Frank Proto, Dr. David Bernard, Dr. Amit Chakma, and many others made a difference in the line of thinking that was needed to write such a book.

The entire research group that had at times nearly 40 members contributed to this endeavor. In particular the contributions of M. E. Hossain, A. B. Chhetri, Dr. Ketata, Dr. Agha, Y. Mehedi, S. Rahman, E. Smit, Dr. Belhaj, Dr. Basu, Dr. Tango, Dr. Satish, and Dr. Butt are noteworthy.

Nomenclature

C_p : Heat capacity
C_g : Gas compressibility
C_o : Oil compressibility
C_r : Rock compressibility
C_t : Total (rock + fluid) compressibility
C_w : Water compressibility
D : Well diameter
E_e : Effective total energy
E_r : Effective total energy
H : Thickness
h : Enthalpy
I_s : Instability number
K : Permeability
K_r : Relative permeability
k : Thermal diffusivity
L_w : Well depth or length
M : Mobility ratio
N : Rotary speed
p : Pressure
p_c : Capillary pressure
p_g : Pressure in the gas phase
p_w : Pressure in the water phase
p_o : Pressure in the oil phase

q : flow rate
q_{laser} : laser source energy
R : Rate of rock penetration
R_{wi} : Relative weight of an indicator, j
r : radius

S_{SL} : Solid/liquid energy transfer
S_{LV} : Liquid/vapor energy transfer
S_{wi} : Initial water saturation
S_{or} : Residual oil saturation
Δt : Time interval
Δt_F : Time interval in fluid
Δt_s : Time interval in solid
T : Temperature
T_m : Mean temperature
T_{sat} : Saturation temperature
t_d : drilling time
V_F : Total pore volume
V_{EP} : Effective pore volume

V_{NP} : Non-effective pore volume
V_p : Total pore volume
V_s : Solid volume
V_T : Bulk volume
W   : Weight of a drilling bit
x,y,z   : Coordinates

Greek
δ : Thermal penetration depth
ϕ : Porosity
γ : interfacial tension
μ : dynamic viscosity
η_E : Mechanical energy efficiency
ρ : density
σ : surface tension
σ_e : pseudo-effective surface tension

Subscripts:
D   : Dimensionless
g   : gas
o   : oil
w   : water

Chapter 1

Introduction

The evolution of human civilization is synonymous with how it meets its energy needs. Yet, for the first time in human history, an energy crisis has seized the entire globe and the very sustainability of civilization itself has suddenly come into question. If there is any truth to the claim that Humanity has progressed as a species, it must exhibit, as part of its basis, some evidence that overall efficiency in energy consumption has improved. In terms of energy consumption, this would mean that less energy is required per capita to sustain life today than, say, 50 years earlier. Unfortunately, exactly the opposite has happened.

We used to assume that resources were infinite and human needs finite. After all, it takes relatively little to sustain an individual human life. However, things have changed and today we are told, repeatedly: resources are finite and human needs infinite. What is going on?

Some Nobel Laureates (e.g., Robert Curl) or environmental activists (e.g., David Suzuki) have blamed the entire technology development regime. Others have blamed petroleum operations and the petroleum industry. Of course, in the context of increasing oil prices, the petroleum sector becomes a particularly easy target. Then, there is US President George W. Bush, talking about "oil addiction". Even his most ardent detractors embrace this comment as some sign of deep thinking. Numerous alternate fuel projects have been launched – but what do they propose? The same inefficient and contaminated process that got us into trouble with fossil fuel! Albert Einstein famously stated, "The thinking that got you into the problem, is not going to get you out." As Enron collapsed, everyone seemed even more occupied with trying to recoup by using the same (mis-)management scheme that led to its demise. In this book, a new management approach is proposed. It addresses the problem of petroleum resources from the root. It proposes a solution that is inherently sustainable. With this management scheme, we would not only cover up the "oil addition" but would cure it. As such, the healthy lifestyle will not be compromised.

Chapter 2 introduces the guidelines for a management practice that can truly say: Enron never again! It provides step-by-step guidelines toward achieving sustainability by breaking out of the management practice that can only be characterized as "managing through fear". A manager of a service company once said, "We excel by doing things faster, safer, and cheaper." Doing things faster, of course, does not mean anything particularly beneficial if it also entails violating a characteristic time (who would want chickens that were hatched in only a few hours?). Even a nuclear bomb might appear to be safe, if the duration of safety (Δt) is small enough. Anything is cheap if the real costs are not considered. The manager later admitted that this safer, faster, and cheaper was also a recipe for continuing a short-term approach, the likes of which had just collapsed (at Enron). For this approach, however, he placed the blame on the lawyers and business managers who had "taken over petroleum management". Chapter 2 recognizes that criticizing, especially this blame-shifting kind, is easy, but neither necessary nor sufficient. This chapter does not blame anyone. However, it corrects the management style and introduces the notion of and proposes the content of management practices based on the long term. This can be characterized as the approach of obliquity, which is well-known for curing both long-term and short-term problems. It stands 180 degrees opposite to the conventional band-aid approach, which prevailed in the Enron decades.

This book promises the "greening" of every practice in the petroleum industry, from management style to upstream to downstream. In the past, petroleum engineers only focused on drilling, production and transportation, and reservoir engineering. This book starts with management and continues through exploration all the way up to refining and gas processing. In the past, exploration meant increasing the chances of production. Reminiscent of the GDP *vs* spending graph that ignores, or otherwise neglects to clarify, whether spending is for wasting or for reconstruction, petroleum exploration practices were for a long time looked at only as a means of increasing tangible benefits, either in physical oil production, or financial gain, or both. In the end, many of these cases ended up costing Humanity much more in the form of environmental damages, which, as Chapter 2 points out, were never part of the management equation.

Chapter 3 shows how current practices of exploration can be rendered environmentally acceptable. It discusses the long-term impacts of some current practices and provides guidelines for exploration practices that will allow us to operate even in the most sensitive parts of the Earth. It shows how to work with the communities that would be most affected by the exploration practices and talks about how to humanize the environment that includes all species.

Chapter 4, discusses drilling and production as they are practiced today. It also includes the topics of waterjet drilling, laser drilling, and production operations that use a no-chemical approach. There is also a discussion of zero-waste oil production. These are progressive topics that show clearly how replacing current practices will benefit both the short and the long term. With this approach, the cost-benefit analysis does not have to be adjusted every quarter and a manager can plan well ahead of time with few, if any, surprises in the future.

Chapter 5 discusses waste management practices – and challenges the implicit assumption that "waste is waste" by boldly proposing the conversion of waste liabilities into assets as a general principle of waste management overall, and not just when or where it might produce maximum profits in the short term. Inspired by Nature, this waste management practice truly converts every waste item into something useful for others. As a result, the time-tested, yet recently touted as an absurd concept, zero-waste approach is implemented.

Chapter 6 addresses reservoir engineering at its fundamental theoretical level, proposing methods for solving non-linear problems without first linearizing their conditions. Anyone reading through this chapter will then know all about current practices in reservoir engineering and see all the shortcomings of the current practices, from both theoretical and practical standpoints. There is no need to despair at the current conditions, as solutions are proposed for all problems, ranging from wellbore monitoring to well testing. Step-by-step guidelines are provided for addressing some of the most difficult problems encountered in reservoir engineering practices.

Chapter 7 elaborates on rendering enhanced oil recovery (EOR) schemes sustainable, using waste to recover more oil. This approach offers a triple dividend: converting waste into an asset, enhancing the environment by engineering an inherently pro-Nature approach, and using locally-available materials. As Chapter 2 points out, such a bottom-up management approach has long-been considered the right thing to do, but until now, has never been converted into engineering practices. This chapter explains why so many EOR projects have failed, both economically and technically. It goes on to show how the recovery schemes can be implemented to drastically increase the probability of success. It provides a 15-point procedure so any company can find the optimum EOR solutions, uniquely suitable to their needs, while humanizing the environment at the same time. Real life examples are given to demonstrate how it is possible for anyone to increase efficiency and productivity while drastically decreasing environmental liabilities.

Chapter 8 examines the fields of transportation, refining and processing. The long-standing assumption that these functions must be as highly-capitalized and as technologically complex as possible, in order to be profitable, is relaxed. Such approaches as tackling the hydrate problem in pipelines by employing bacteria instead of ethanol, refining by employing solar heating in place of thermal (and chemically highly toxic) "cracking", and processing by means of natural absorbents are taken up. Of course, all these solutions are offered after pointing out the long-term implications of current practices. It is shown that petroleum products are not the culprit (not even plastic), but the processes that are responsible. Alternatives to numerous toxic chemicals that are used are proposed in order to truly clean up the current mess. This is fitting considering the discovery that there are some 4000 toxins, most from "oil addiction", recent being released into the environment.

In Chapter 9, the thorny issues of offshore-rig and production decommissioning are reassessed. Far from having to remain closeted as one of the petroleum industry's dirty little secrets – very much like the waste disposal problem confronting managers of nuclear power plants – the same inherent-sustainability criteria informing the rest of the book are applied to develop and elaborate a number of pro-Nature practices and possibilities.

Chapter 10 advances the essential conclusions: the pathways advanced to handle energy production and development can be either pro-Nature or anti-Nature. Any course of action based on short-term considerations must manipulate short-term surpluses or shortages in the natural environment, thereby disrupting the normal operation of the laws of conservation of mass and of energy. Consequently, intervening on this short-term basis will harm the environment and, with it, the possibilities and prospects for Humanity as a whole. On the other hand, any intervention planned on the basis of keeping the long term in mind must operate within the characteristic boundary conditions of the natural environment itself. Therefore, in the long term, there can be no such thing as surpluses or shortages in Nature, and therefore, costs and benefits planned around such boundary conditions can never violate the natural order, ensuring such intervention will be truly sustainable.

Chapter 11 provides a comprehensive list of references. Some 50 pages of complete references make sure the pro-Nature, pro-environment technology train that was set in motion in this book never stops.

Chapter 2

The New Management Guidelines

2.1   Introduction

Petroleum hydrocarbons are considered by some to be the lifeblood of modern society. The petroleum industry that took off from the golden era of the 1930s has never ceased to dominate all aspects of our society. Until now, there has been no suitable alternative to fossil fuel and trends indicate continued dominance by the petroleum industry in the foreseeable future. Even though petroleum operations have been based on solid scientific excellence and engineering marvels, only recently has it been discovered that many of the practices are not environmentally sustainable. Practically all life-cycle activities of hydrocarbon operations are accompanied by undesirable discharges of liquid, solid, and gaseous wastes, which have an enormous impact on the environment.

The life cycle of petroleum operations includes exploration and development, production, refining, marketing, transportation/distribution to the end-user, and final utilization. For example, during drilling, water-based drilling muds and cuttings are discharged overboard, whereas during production, the major discharge is produced water. In addition, treated sanitary and domestic wastewaters, deck drainage, and miscellaneous wastes, such as ballast, may be discharged at any point in the operation (Khan and Islam 2003b). Oily muds, produced sands, and trash and debris are also produced but cannot be discharged overboard. However, accidental releases of different wastes and toxic chemicals are reported (Cranford et al. 2003; Veil 1998, 2000). There are risks of spills from a blow-out during exploratory drilling, which are not significantly low (<1%).

The ecological impacts of these discharges, including habitat destruction and fragmentation in both terrestrial and aquatic environments, are now recognized as major concerns associated with petroleum and natural gas developments. Consequently, reducing environmental impact is the most pressing issue today and many environmentalist groups are calling for curtailing petroleum operations altogether. Thus there is clearly a need to develop a new management approach in hydrocarbon operations, which will have to be environmentally acceptable, economically profitable, and socially responsible. These problems might be solved by developing new technologies that guarantee sustainability. Recently, Khan et al. (2005a) and Khan and Islam (2005b) introduced a new approach by means of which it is possible to develop a truly sustainable technology. Under this approach, the temporal factor is considered the prime indicator in sustainable technology development.

This chapter discusses some implications of how the current management model for exploring, drilling, managing wastes, refining and transporting, and using the by-products of petroleum has been lacking in foresight, and suggests the beginnings of a new management approach. Sustainability or sustainable operations in the petroleum sector is rarely addressed in the present management regime. This chapter provides a framework for the new management guidelines with respect to sustainability of petroleum operations management. It introduces novel sustainability criterion and an integrated green supply chain model that is structured to achieve sustainability in this sector. In addition, the framework of truly "sustainable management" for practically all aspects of oil and gas operations is detailed.

Different applicable concepts, methods, and guidelines for achieving sustainability are addressed with respect to petroleum operations. It is demonstrated with detailed examples that using the new approach will be economically more beneficial than the conventional approach, even in the short term. This chapter proceeds by providing a brief overview of different technological phases and current practices of petroleum operations. There is a comprehensive description of the different tools to achieve sustainability and proposed

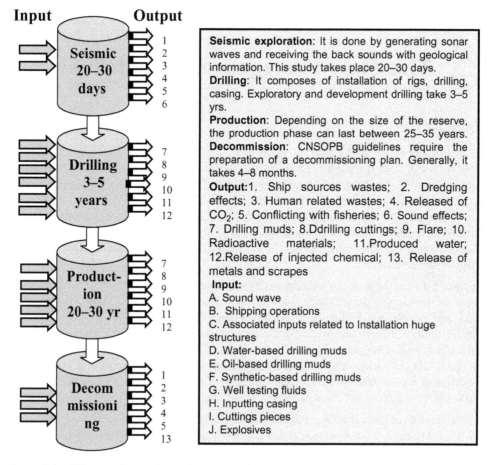

Figure 2-1   Different phases of petroleum operations which are seismic, drilling and production, transportation and processing, decommissioning and production, and their associated wastes generation and energy consumption.

measures to incorporate the green supply chain model at this stage. Finally, it analyzes the process by which the green supply chain model leads to environmental sustainability.

2.2   Current Practices of Petroleum Operations

In a very short time (relative to the history of the environment), the oil and gas industry has become one of the world's largest economic sectors, a powerful globalizing force with far reaching impacts on the entire planet that humans share with the rest of the natural world. Decades of continuous growth of oil and gas operations have transformed the natural environment and the way humans have traditionally organized themselves. The petroleum sectors draw huge public attention due to their environmental consequences of petroleum activities. All stages of oil and gas operations generate a variety of solids, liquids, and gaseous wastes (Currie and Isaacs 2005; Wenger et al. 2004; Khan and Islam 2003a; Veil 2002; Groot 1996; Wiese et al. 2001; Rezende et al. 2002; Holdway 2002). Different phases of petroleum operations and their associated problems are discussed in the following sections.

2.2.1 Petroleum operation phases

Petroleum operations are divided into four phases. These phases are: seismic exploration, drilling and production, transportation and processing, and decommissioning of the platforms and production facilities (Table 2-1). The technological details and scale involved for each of these phases are significantly different and generate different effluents and impacts, as shown in Table 2-2.

Table 2-1   Different technological phases of offshore oil and gas development

Seismic exploration	Drilling	Production	Decommissioning
When a license is issued, the proponent is given 5 years to explore the resources. The actual process may be 20–30 days.	Generally 3–5years, including onshore fabrication, installation, and commissioning.	Depending on the size of the reserve, the production phase can last between 25–35 years.	Proponents require preparing a decommissioning plan; however, no information on the time frame of decommissioning activities was found.

Table 2-2   Types of wastes generated in different offshore oil and gas development phases

Seismic exploration	Drilling and installation	Production	Decommissioning
• Sounds • Associated wastes • Human generated wastes: sanitary wastes, kitchen and food wastes, laundry wastes, sink and shower drainage, and trash	• Drilling muds • Drillings cuttings • Produced sands • Storage displacement water • Bilge and ballast water • Deck drainage • Well treatment fluids • Naturally occurring radioactive materials • Cooling water, desalination brine • Water for testing fire control • Other assorted wastes • Accidental discharges: air emission: oil spills, chemical spills, blowout • Human generated wastes: sanitary wastes, kitchen wastes, sink and shower drainage, laundry wastes, and trash • Other industrial wastes: cardboard, empty containers, scrap metal, wood pallets, used chemicals and paint, sandblasting grit and paint, and cooling water	• Produced water • Treatment and completion fluids • Deck drainage • Produced sand • Storage displacement water • Bilge and ballast water • Deck drainage • Well treatment fluids • Naturally occurring radioactive materials • Cooling water, desalination brine • Water for testing fire control • Accidental discharges: air emission: oil spills, chemical spills, blowout • Other assorted wastes • Human generates wastes: sanitary wastes, kitchen wastes, sink and shower drainage, laundry wastes, and trash • Other industrial wastes	• Abandoned structures • Cut pieces of oil structures • Scrap materials

Seismic exploration observes the behavior of induced sound waves to map sub-seafloor longitudinal sections (2D) or features (3D). A vessel tows an array of streamers, with several airguns attached, below the water surface. These release high-pressured air every 12 s, creating a sounding of generally 225 dB to produce seismic waves in the water. They are focused at the seafloor, reflected by sub-seafloor features, and received by hydrophones attached every 12.5 m along streamers up to 6 km long; typically run lines 100 m to 1 km apart (CEF, 1998). Detailed seismic operations are discussed in Chapter 3.

A drilling rig is installed to perform drilling of the subsurface area of the seabed, the tasks performed by the hosting, circulating, and rotation system, backed-up by the pressure-control equipment. Figure 2-2 represents drilling operations, showing the drill bit mounted at the end of the drill pipe, to cut through the hard substratum. The drill pipe is rotated mechanically to enable it to cut into the rocks. During drilling, drilling fluids are circulated from tanks through a standpipe into the drill pipe and drill collar to the bit.

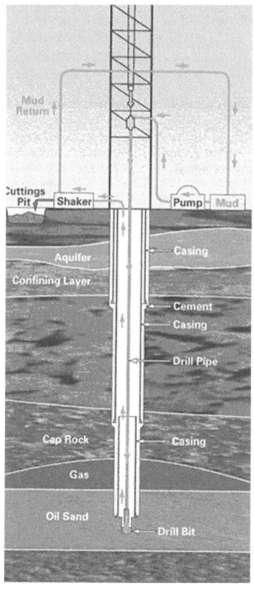

Figure 2-2   Pictorial view of drilling operations. Photo courtesy of DOE (2005).

The process of decommissioning involves removal of structures used during the operation of the well, together with any environmental remediation that is considered necessary. It involves clearing the project area of any material or equipment that could interfere with other commercial uses, preventing fluid from escaping from the well-bore, and cutting off all casing at a level below the seabed, so as not to be affected by ice scour. However, sometimes some parts are allowed to be left at the project site, if considered not to pose any threat in the future for example, during the decommissioning of the Cohasset site on the Scotian Shelf. There are instances when approval is granted to leave sub-sea pipelines in place, providing they pose no threat to navigation or to other sectors. Moreover, the removal of these parts may cause more environmental damage than leaving them in place.

2.3 Problems of Current Operations

In petroleum operations, different types of wastes are generated. They can be broadly categorized as drilling wastes, human-generated wastes, and other industrial wastes. A detailed list of wastes generated in different phases of petroleum operations is presented in Table 2-2. There are also accidental discharges, for example, via air emission, oil spills, chemical spills, and blowouts.

During the drilling of an exploratory well, several hundred tonnes of drilling muds and cuttings are commonly discharged into the marine environment. Though an exploratory activity, such as seismic exploration, does not release tangible wastes, it nevertheless has a potential negative impact (Cranford et al. 2003; Putin 1999). According to a report (SECL 2002; Putin 1999), seismic shooting kills plankton, including eggs, larvae of many fish, and shellfish species, and juveniles that are close to the airguns. The most important sublethal effect on adult organisms exposed to chronic waste discharges, from both the ecological and fisheries perspectives, is the impairment of growth and reproduction (GESAMP 1993; Putin 1999). Thus, damage to growth and reproduction are generally considered to be the most important sublethal effects of chronic contaminant exposure (Cranford et al. 2003). Seabirds aggregate around oil drilling platforms and rigs in above average numbers due to night lighting, flaring, food, and other visual cues. Bird mortality has been documented to be due to impact with the structure, oiling, and incineration by the flare (Wiese et al. 2001).

Khan and Islam (2005a) reported that the large quantity of water emissions taking place during production primarily result from the discharge of produced water, which include fluid injected during drilling and highly saline water. They also reported that produced water contains various contaminants, including trace elements and metals from formations through which the water passed during drilling, as well as additives and lubricants necessary for proper operation. This water is typically treated prior to discharge, although historically this was not always the case (Ahnell and O'Leary 1997).

Based on the geological formation of the well, different types of drilling fluids are used. The composition and toxicity of these is highly variable, depending on their formulation. Water is used as the base fluid for roughly 85% of drilling operations internationally, and the remaining 15% predominantly use oil (Reis 1996). Spills make up a proportionately small component of aquatic discharges (Liu 1993).

CO_2 emissions are one of the most pressing issues in the hydrocarbon sector. There are direct emissions from production sites through flaring and from burning fossil fuels. For example, during exploration and production, emissions take place due to control venting and/or flaring and the use of fuel. Based on 1994 British Petroleum figures, it is reported that emissions by mass were 25% volatile organic compounds (VOCs), 22% CH_4, 33% NO_x, 2% SO_x, 17% CO, and 1% particulate matter. Data on CO_2 are not provided (Ahnell and O'Leary 1997). Until now, flaring is considered as a production and refining technique which wastes huge amounts of valuable resources through burning.

The air emissions during petroleum processing are primarily due to uncontrolled volatilization and combustion of petroleum products in the modification of end products to meet consumer demand (Ahnell and O'Leary 1997). Oils, greases, sulfides, ammonia, phenols, suspended solids, and chemical oxygen demand (COD) are the common discharges to water during refining (Ahnell and O'Leary 1997). Natural gas processing generally involves the removal of natural gas liquid (NGLs), water vapor, inert gases, CO_2,

Figure 2-3   Flaring from an oil refinery.

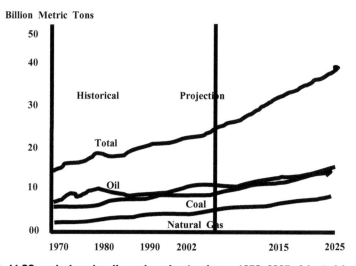

Figure 2-4   World CO$_2$ emissions by oil, coal, and natural gas, 1970–2025. Adapted from EIA (2004).

and hydrogen sulfide (H$_2$S). The by-products from processing include CO$_2$ and H$_2$S (Natural Resources Canada 2002a).

The oil sector contributes a major portion of CO$_2$ emission. Figure 2-4 presents the world historical and projection of CO$_2$ emissions from different sectors. About 29 billion tonnes of CO$_2$ are released into the air annually by human activities, with 23 billion tonnes coming from burning fossil fuels and industry (IPCC 2001; Jean-Baptiste and Ducroux 2003), which is why this sector is blamed for global warming. The question is, how might these problems best be solved? Is there any solution possible?

Throughout the life cycle of petroleum operations there are accidental discharges, for example, via air emission, oil spills, chemical spills, and blowouts. The spilled oil released to terrestrial and aquatic environment is different, with Figure 2-5 showing the annual total amount of oil released into the marine environment.

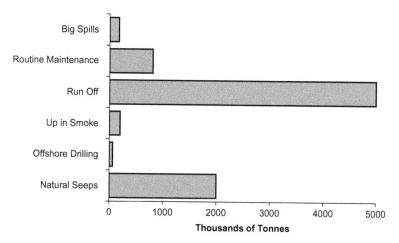

Figure 2-5   Annual total amount of oil release in the marine environment. Data source: Oil in the Sea (2003).

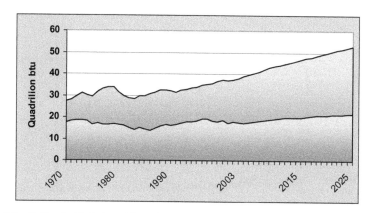

Figure 2-6   Estimated total world production and demand of oil. Data source: EIA (2003).

Crude oil is one of the major toxic elements released into this environment by the oil industry. On average, 15 million gallons of crude oil are released into the sea annually, from offshore oil and gas operations. There are in total 700 million gallons of oil discharges from other sources (United States Coast Guard 1990; Khan and Islam 2004). These other sources of oil release are from routine maintenance of shipping, domestic/urban runoff, up in smoke, and natural seepages.

The above are only a few examples of the current technology development mode in the petroleum sector. It is hard to find a single technology that does not have such problems. In addition to the use of technologies that are unsustainable, the corporate management process itself is based on a structure that resists sustainability. Generally, corporate policy is oriented toward gaining monetary benefits, while producing very little (Zatzman and Islam 2007). This model has imploded spectacularly in the aftermath of the fall of the world energy giant Enron, in December 2001 (Deakin and Konzelmann 2004; Zatzman and Islam 2006a). Post-Enron events, including the crisis that afflicted World Dot Com, indicate that practically all corporate structures are based on the Enron model (Zatzman and Islam 2006a).

It is clear from the above discussion, that there are enormous environmental impacts from current petroleum operations. Due to high market demand and technological advancement in exploration and development, petroleum operations have spread worldwide, even into more remote and deeper oceans (Wenger et al. 2004; Pinder 2001), as shown in Figure 2-6. Due to the limited supply of onshore oil and gas reserves, and the fact that these reserves have been declining due to exploitation over a long time, there is an increasing pressure to explore and exploit offshore reserves. As a result of these declining onshore reserves, offshore oil and gas operations have increased dramatically within last two decades (Pinder 2001). This phenomenon has already been evident in many parts of the world. For example, in the 1970s, the feasible gas reserves on the Scotian Shelf, Canada, were found to be economically attractive (Khan and Islam 2007). Figure 2-7 shows active leasing areas in the Nova Scotian area of eastern Canada. These figures show that all the available offshore areas have been allocated to different oil companies. At present, this area is capable of operation of oil and gas for more than 3000 meters.

Petroleum operations are not only ecologically risky, but also involve high capital investment. The petroleum sector is one of the largest global economic sectors. Currently, the industry is projected to spend $100 billion on deepwater oil and gas fields, pipelines, drilling rigs, and production platforms around the world. The volume of expenditure is increasing as exploration is moving toward ultra-deepwater offshore environments (Wenger et al. 2004; Pinder 2001; Khan and Islam 2003a). Operations in these environments are based on even higher levels of technology and monitoring. The large-scale investment at stake entails higher risks in terms of economic success and environmental well-being and safety (Hossain et al. 2006). The success of a high risk offshore oil and gas operation depends on the use of sustainable technology (Khan et al. 2005b; Khan and Islam 2005b) and on the improvement of sustainability in the operating company. Similar to other

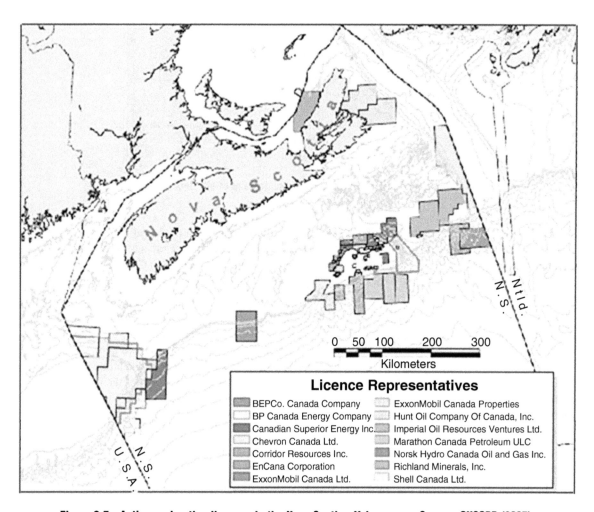

Figure 2-7   Active exploration licenses in the Nova Scotia offshore area. Source: CNSOPB (2005).

industrial sectors, the present hydrocarbon development process, which produces oil and gas, is developed with the goal of economic success in the short term.

Existing management regimes try to address these problems through developing waste management guidelines, such as the Canada-Nova Scotia Offshore Petroleum Board (CNSOPB) guidelines and the US Environment Protecting Agency (US EPA) guidelines. For example, the CNSOPB guidelines allow the release of produced water containing 10 mg/L oil or less (OWMGL 2000). However, other components of produced water, such as radioactive materials, are not taken into account. Moreover, the current regulations give the operator/industry the legal right to pollute the environment as long as a certain concentration level is not detected.

As already discussed, true sustainability cannot come without management attitude change as the corporate management process itself is based on a structure firmly resisting such sustainability. Corporate policy tends to be solely oriented toward monetary gain, with little concern for ethics or law, as observed in the aftermath of the fall of Enron 2001 (Deakin and Konzelmann 2004), which confirmed that practically all corporate structures are based on the Enron model (Zatzman and Islam 2006a). Moreover, management strategies in the oil and gas sector are not well adapted to natural resource management. Community-based management is found to be a more suitable approach in natural resource management, including renewable energy development (Khan et al. 2006a). In this process, stakeholders, or a community, play a role in a participatory management system and benefit from tangible and intangible values. Present petroleum operations management regimes operate in centrally-controlled and un-integrated ways. All their activities, such as planning, central engineering, upstream operations, processing, and supply and transportation groups are undertaken as complete, separated entities (Forest and Oetti 2003).

It is neither difficult nor sufficient to criticize the current *modus operandi*. The question we must ask is how to change this management and technology development style. Such a change cannot take place unless there is a change in the process itself. Albert Einstein said, "The thinking that got you into trouble will not get you out." Thus, a change in thinking is tantamount to a change in the process. The hypothesis proposed in this chapter is that Nature is perfect therefore the process of Nature should be emulated in order to develop technologies that are inherently sustainable. In the past, numerous claims have been made as to how we emulate Nature. However, (Islam 2005, 2006; Zatman and Lslam 2006). Let us review how the current management model and engineering practices compare with Nature. Table 2-3 elaborates the comparative scenarios of natural, engineering, and management processes. Engineering and management show that these processes diverge sternly from the process of Nature. If Nature is perfect and inherently sustainable, the

Table 2-3   Comparison of natural, engineered, and management processes. Adapted from Zatzman and Islam (2006b)

Nature	Engineering	Management
1. Multiple/flexible	1. Exact/rigid	1. Stochastic
2. Nonlinear	2. Linear	2. Geometric (Linear-regressive) (statistics-based)
3. Heterogeneous	3. Homogenous/uniform	3. Reducible to money-value
4. All-natural processes	4. Artificially-imposed processes	4. Artificial, ascribed to "human nature"
5. Characteristic life cycles	5. Disposable (one-time use)	5. Periodic
6. Infinite	6. Finite	6. Finite
7. Non-symmetric	7. Symmetric	7. Complementary
8. Productive design	8. Reproductive design	8. Extended reproductive design
9. Reversible	9. Irreversible	9. Threshold
10. Knowledge	10. Ignorance (anti-knowledge)	10. Differential "knowledge"
11. Sustainable	11. Unsustainable (aphenomenal)	11. Profit driven
12. Dynamic/chaotic	12. Static	12. Astable
13. No boundary	13. Based on boundary conditions	13. Path-dependent and bounded

currently practiced development model is inherently unsustainable. In this, both processes of technology development and management policy have to be addressed.

2.4   Sustainability in Petroleum Operations

It is time to recognize the unpleasant truth that the present natural resource management regime governing activities such as petroleum operations has failed to ensure environmental safety and ecosystem integrity. The main reason for this failure is that the existing management scheme is not sustainable (Khan and Islam 2007). Figure 2-8 shows the sustainable and unsustainable management techniques and their related costs. In fact, existing management approaches are formulated to reflect moving in directions that are the opposite from where we should be going. Under the present management approach, development activities are allowed so long as they promise economic benefit. Once that is likely, management guidelines are set to justify the project's acceptance.

Sustainable petroleum operations development requires a sustainable supply of clean and affordable energy resources that do not cause negative environmental, economic, and social consequences (Dincer and Rosen 2004, 2005). In addition, it should consider a holistic approach where the whole system will be considered instead of just one sector at a time (Mehedi et al. 2007a, 2007b). Figure 2-9 represents the coordination or integration of basic components such as environmental, technological, social, and economic variants of each operation. The concept of sustainability was brought into the core of social, economic, and environmental debates after the well-known definition of sustainable development created by the World Commission on Environment and Development (WCED 1987). However, sustainable development has remained an obscure term due to the absence of explicit guidelines for sustainability. Recently, Khan and Islam (2005b, 2006b) and Khan et al. (2005b) developed an innovative criterion for achieving true sustainability in technological development. This criterion can be applied effectively to offshore technological development. New technology should have the potential to be efficient and functional far into the future in order to ensure true sustainability. Sustainable development is seen as having four elements – economic, social, environmental, and technological. Delivery is the overarching concept that drives both implementation and further strategic development (Khan and Islam 2005b).

2.4.1   Concept of sustainability

The concepts expressed in the terms "sustainable", "sustainability", and "sustainable development" essentially lack clear definition even though we see them in various government documents, hear them in mainstream media, and read them in corporate newsletters and international agreements (Wright 2002). It is also apparent that even those few who may understand the true meaning of sustainability are not able to agree on a criterion (Judes 2000; Leal Filho 2000; Wright 2002). Table 2-4 shows popular definitions of sustainability.

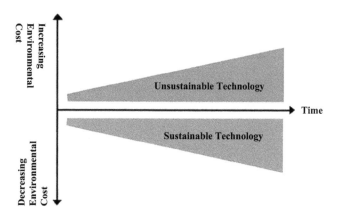

Figure 2-8   Comparative environmental cost of a sustainable and an unsustainable technology.

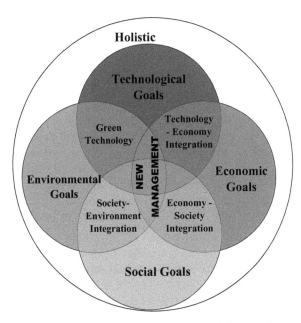

Figure 2-9 Conceptual framework of new management for petroleum operations.

Table 2-4   Meaning of "sustainable" management or development

Definition of sustainability	Source	Comments
"Development that meets the needs of the present without comprising the ability of future generations to meet their own needs."	WCED (1987)	Most popular definition, but lacks clear direction, i.e., what is the scale of needs? It can be treated as a "paper tiger".
"Management practices that will not degrade the exploited system or any adjacent system."	Lubchenco et al. (1991)	Generally a system approach. There is no specific direction about time.
"Development without throughput growth beyond environmental carrying capacity and which is socially sustainable."	Daly (1992)	Considers assimilative capacity of Nature in a spatial scale. A time direction is completely missing.
"Improvement in the quality of human life within the carrying capacity of supporting ecosystems."	Robinson (1993)	Tries to integrate the social and ecological context in spatial scale, but not in temporal.
"Sustainability is defined as minimizing the consumption of the world's resources by pursuing better environmental performance within product life cycles."	Donnelly et al. (2006)	Very weak definition; misguided sustainability.

In the international environmental context, the idea of sustainability is based on the notion that planetary resources are finite, a highly contentious assertion in itself. But essentially, the numerous mutually reinforcing intentional initiatives that have promoted the idea since the Second World War have been united by the need to foster a global understanding of the environment and to address how humanity could ameliorate the delegation of the biosphere.

Founded on the concern for ecological security expressed by the 1972 United Nations Conference on the Human Environment, it was the World Commission on Environment and Development (WCED) that popularized the concept, defining it as the obligation of "meeting the needs of the present without compromising the ability of future generations to meet their own needs" (WCED 1987). The intervening 15-year period was marked by a titanic battle of ideas in which British Prime Minister Margaret Thatcher, with her declaration that "there is no such thing as 'society'", played a leading role. She was backed by the Reagan Administration in the United States, which, by withholding more than US $1-billion in unpaid dues, brought pressure on the FAO and other UN agencies involved in supporting government-sponsored funding of economic development projects among developing countries. Meanwhile, the appeal of the environmental portion of the overall message – the portion that trumpeted "small is beautiful" and similarly idealistic sentiments relating to sustainability – corralled considerable support among younger academics and other young people. Thus it came about that sentiments similar to those of the Brundtland Report were echoed in the United Nations Conference on Environment and Development (UNCED), held in Rio de Janeiro in June 1992.

Criticisms of the Brundtland definition of sustainable development exist (Table 2-4), including those centered on the anthropocentric connotations of the wording; that it generates strategies that ignore the carrying capacity of the planet; that it contains vague wording that allows individuals to manipulate the definition; and that it apparently supports sustaining growth (Brown et al. 1991; Leal Filho 2000; Gibson 1991; Hawken 1992; Miller 1994; Nikiforuk 1990; Rees 1989; Wackenagel and Rees 1996; Welford 1995). Recently, Appleton (2006) criticized the Brundtland definition, "satisfying human need", but how much is a limit for a human? He asked at what level human needs should be satisfied. Is it the American per capita income level, the Chinese per capita income level, the Millennium Development Goals, or some similar bundle of minimum services (i.e., clean water, healthy sanitary conditions, an x calorie and nutrient a day diet, and heat levels in the winter)? He also asks questions, such as is air conditioning in the summer a human need; is the measurement of need ownership of 0.6 cars per capita, as in the United States; or 1 car per capita, the current worldwide ratio; or the current 0.08 cars per capita in China? However, this definition is very popular to date and has been used in many policy and government documents worldwide. Someone can argue that this definition is weak and based on imperfect direction (Figure 2-10).

To assess the sustainability of projects, technologies, as well as overall implications of different types of "sustainability frameworks", have been used (Labuschagne et al. 2005). Some of them are: Global Reporting Initiative (GRI 2002), United Nations Commission on Sustainability Development Framework (UNCSD 2001), Sustainability Metrics of the Institution of Chemical Engineers (IChemE 2002), and Wuppertal Sustainability Indicators (Spangenberg and Bonniot 1998).

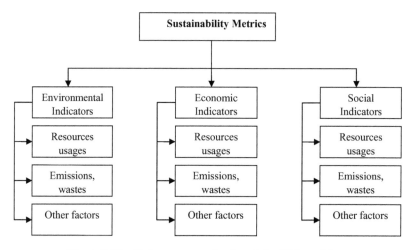

Figure 2-10   Basic components of sustainability metrics.

The GRI was launched in 1997. The United Nations Environment Program together with the United States NGO, Coalition for Environmentally Responsible Economics (CERES), developed GRI. The main goal of GRI is enhancing the quality, rigor, and utility of sustainability reporting (GRI, 2002). Reporting is, therefore, the focal point of these guidelines. The GRI uses a hierarchical framework in the three focus areas, namely: social, economic, and environmental.

The UNCSD constructed a sustainability indicator framework for the evaluation of governmental progress in achieving sustainable development goals. The hierarchical framework groups indicators into 38 subthemes and 15 main themes that are divided between the four aspects of sustainable development (UNCSD 2001). The main deference between this framework and the GRI, is that it addresses institutional aspects of sustainability.

The Wuppertal Institute proposed indicators for the four dimensions of sustainable development, as defined by the UNCSD, together with interlinkage indicators between these dimensions (Spangenberg and Bonniot 1998). These indicators are applicable at both macro- and micro-levels. The approach used to address business social sustainability is worth noting: the UNDP Human Development Index has been adapted to form a Corporate Human Development Index.

The (IChemE) published a set of sustainability indicators in 2002, to measure the sustainability of operations within the process industry (Figure 2-10). The IChemE provides standard reporting forms and conversion tables. This framework is less complex and impact oriented. However, the framework strongly favors environmental aspects, as well as quantifiable indicators that may not be practical in all operational practices, for example, in the early phases of a project's life cycle.

However, a review of the literature that has sprung up around the concept of sustainability indicates a lack of consistency in its interpretation. More important, while the all-encompassing nature of the concept gives it political strength, its current formulation by the mainstream of sustainable development thinking contains significant weaknesses (Wright 2002). Most of the sustainability assessment models mentioned above are based on broader socio-economic and environmental considerations. Overall, it is obvious that different matrix systems and indexes have been used to measure sustainability, there being no straightforward guidelines to achieve true/inherent sustainability.

2.4.2   Problems in technological development

The technologies promoted in the post-industrial revolution are based on the aphenomenal model (Islam 2005; Zatman and Islam 2006). This model is a gross linearization of Nature ("Nature" in this context includes Humanity in its social nature). This model assumes that whatever appears at $\Delta t = 0$ (or time = "right now") represents the actual phenomenon. This is clearly an absurdity. How can there be such a thing as a natural phenomenon without a characteristic duration and/or frequency? When it comes to what defines a phenomenon as truly natural, time, in one form or another, is of the essence.

The essence of the modern technological development scheme is the use of linearization or reduction of dimensions in all applications. Linearization has provided a powerful set of techniques for solving equations generated from mathematical representations of observed physical laws – physical laws that were adduced correctly, and whose mathematical representations, as symbolic algebra, have proven frequently illustrative, meaningful, and often highly suggestive. However, linearization has made the solutions inherently incorrect. This is because any solution for t = "right now" represents the image of the real solution, which is inherently the opposite from the original solution. Because this model does not have a real basis, any approach that focuses on the short term makes us travel on the wrong path. Unlike common perception, this path does not intersect the true path at any point in time, other than t = "right now". The divergence begins right from the outset. Any natural phenomenon or product always travels an irreversible pathway that is never emulated by the currently used aphenomenal model of technological development. Because, by definition, Nature is nonlinear and "chaotic" (Glieck 1987), any linearized model merely represents the image of Nature at a time, t = "right now", from which instant their pathways diverge. It is safe to state that all modern engineering solutions (all are linearized) are anti-Nature. Accordingly, the black box was created for every technology promoted (Figure 2-11).

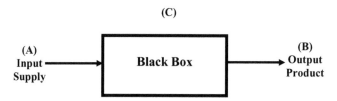

Figure 2-11   Classical "engineering" notion. Redrawn from Islam (2005a).

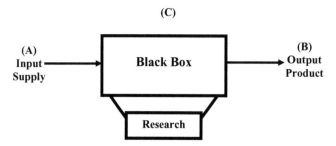

Figure 2-12   Research based on observing Nature intersects classical "engineering" notion. Redrawn from Islam (2005a).

This formulation of a black box helped keep "outsiders" ignorant of the linearization process that produced spurious solutions for every problem apparently solved, its crucial and least remarkable feature. The model itself has nothing to do with knowledge. In a typical repetitive mode, the output (B) is modified by adjusting input (A). The input itself is modified by redirecting (B). This is the essence of the so-called "feedback" mode that has become so popular in our day. Even in this mode, nonlinearity may arise as efforts are made to include a real object in the black box. This nonlinearity is expected. Even a machine will generate chaotic behavior, which becomes evident only if we have the means of detecting changes over the dominant frequency range of the operation.

What needs to be done is to improve our knowledge of the process? Before claiming to emulate Nature, we must implement a process that allows us to observe it (Figure 2-12). Research based on observing Nature is the only way to avoid spurious solutions due to linearization or elimination of a dimension.

Sustainable development is characterized by certain criteria. The "time" criterion is the main factor in achieving sustainability in technological development. However, in the present definition of sustainability, a clear time direction is missing (Table 2-4). To better understand sustainability, we can say that there is only one alternative to sustainability, viz., unsustainability. Unsustainability involves a time dimension: it rarely implies an immediate existential threat. Existence is threatened only in the distant future, perhaps too far away to be properly recognized. Even if a threat is understood, it may not cause much concern now, but will cumulatively work in its effect in the wider timescale. This problem is depicted in Figure 2-13.

In Figure 2-13, the impact of the wider timescale is shown where "A" and "B" are two different development activities that are undertaken in a certain time period. According to the conventional environmental impact assessment (EIA) or sustainability assessment process, each project has insignificant impacts on the environment in the short timescale. However, their cumulative impacts will be much higher and will continue under a longer timescale. The cumulative impacts of these two activities (A, B) are shown as a dark line.

Technology plays a vital role in modern society. The causes of present-day environmental and social problems are related to the use of unsustainable technology. The problem associated with the current technology

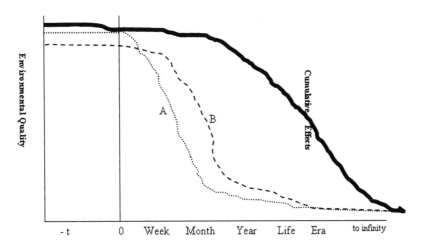

Figure 2-13   Cumulative effects of activities A and B within different temporal periods.

development mode is easily identified if we attempt to correlate between the increase of cancer, non-genetic diabetes, Alzheimer's disease, and others, with the increase in the usage of sugar ("refined" biofuel), gasoline ("refined" fossil fuel), and petrochemicals (ranging from fertilizers to ubiquitous plastics materials). This has recently been highlighted by Islam (2005c). Most modern technologies are developed on principles that focus on short-term economic benefits.

The most important feature of this technology development is the focus on tangibles. The overwhelming assumption behind this kind of "technology development" is that Nature is chaotic and needs fixing. The most important manifestation is the fascination with imitations, homogenization, and plastics. Today, we use some 100 million barrels of crude oil and produce 4 million tonnes of plastic every day. Yet, "Reduce, Reuse, and Recycle" remains the only slogan available. Disinformation is the essence of the current technology development scheme (Table 2-5).

The present-day technological development process focuses on turning maximum profit in minimum time at the expense of many environmental consequences. As human beings in a fast-moving modern society, our vision of time is extremely short term. For example, we commonly think in terms of, for example, a bi-weekly pay check. This two-week period is our standard in the civilized world and it has its reflection in development activities. Long-term planning may be considered within time frames ranging from hours to weeks, months or perhaps even years – but never in terms of, say, generations (25-year periods). This short-term focus is exactly the opposite from what is needed to ensure sustainability. Similar to other sectors, the technological development in the petroleum sector also follows the unsustainable path.

2.5   Tools Needed for Sustainable Petroleum Operations

Sustainability can be assessed only if technology emulates Nature. In Nature, all functions or techniques are inherently sustainable, efficient, and functional for an unlimited time period. In other words, as far as natural processes are concerned, "time tends to infinity". This can be expressed as t, or $\Delta t \rightarrow \infty$.

By following the same path as the functions inherent in Nature, an inherently sustainable technology can be developed (Khan and Islam 2005b). The "time criterion" is a defining factor in the sustainability and virtually infinite durability of natural functions. Figure 2.14 shows the direction of Nature-based, inherently sustainable technology, as contrasted with an unsustainable technology. The path of sustainable technology is its long-term durability and environmentally wholesome impact, while unsustainable technology is marked by Δt approaching to 0. Presently, the most commonly used theme in technological development is to select technologies that are good for t = "right now", or $\Delta t = 0$. In reality, such models are devoid of any real

Table 2-5   Analysis of "breakthrough" technologies

Product	Promise (knowledge at t = "right now")	Current knowledge (closer to reality)
Microwave oven	Instant cooking (bursting with nutrition)	97% of the nutrients destroyed; produces dioxin from baby bottles
Fluorescent light (white light)	Simulates the sunlight and can eliminate "cabin fever"	Used for torturing people, causes severe depression
Prozac (the wonder drug)	80% effective in reducing depression	Increases suicidal behavior
Anti-oxidants	Reduces aging symptoms	Gives lung cancer
Vioxx	Best drug for arthritic pain, no side effects	Increases the chance of heart attack
Coke	Refreshing, revitalizing	Dehydrates; used as a pesticide in India
Transfat	Should replace saturated fats, including high-fiber diets	Primary source of obesity and asthma
Simulated wood, plastic gloss	Improve the appearance of wood	Contains formaldehyde that causes Alzheimer's Disease
Cell phone	Empowers, keeps connected	Gives brain cancer, decreases sperm count among men.
Chemical hair colors	Keeps young, gives appeal	Gives skin cancer
Chemical fertilizer	Increases crop yield, makes soil fertile	Harmful crop; soil damaged
Chocolate and "refined" sweets	Increases human body volume, increasing appeal	Increases obesity epidemic and related diseases
Pesticides, MTBE	Improves performance	Damages the ecosystem
Desalination	Purifies water	Necessary minerals removed
Wood paint/varnish	Improves durability	Numerous toxic chemicals released
Leather technology	Will not wrinkle, more durable	Toxic chemicals
Freon, aerosol, etc.	Replaced ammonia that was "corrosive"	Global harms immeasurable and should be discarded

Source: Islam (2005a, 2005b)

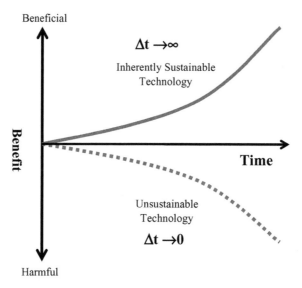

Figure 2-14   Direction of sustainable/green technology.

basis (termed "aphenomenal" by Khan et al. 2005c) and should not be applied in technological development if we seek sustainability for economic, social, and environmental purposes.

Considering pure time (or time tending to infinity) in terms of sustainable technology development raises thorny ethical questions. This "time tested" technology will be good for Nature and good for human beings. The main principle of this technology will be to work toward, rather than against, natural processes. It would not work against Nature or ecological functions. All natural ecological functions are truly sustainable in this long-term sense. We can take a simple example of an ecosystem technology (natural ecological function) to understand how it is time-tested (Figure 2-15).

In Nature, all plants produce glucose (organic energy) by using sunlight, CO_2, and soil nutrients. This organic energy is then transferred to the next highest level of organisms, which are small animals (zooplankton). The next highest (tropical) level organism (high predators) use that energy. After the death of all organisms, their body mass decomposes into soil nutrients, which again take plants to keep the organic energy loop alive (Figure 2-15). This natural production process never dysfunctions and remains for an infinite time. It can be defined as a time-tested technique.

This time-tested concept can be equally applied technological development. The new technology should be functional for an infinite time. This is the only way it can achieve true sustainability (Figure 2-16). This is the idea that informs the new assessment framework that is developed and shown in Figures 2-17 and 2-18. The triangular sign of sustainability in Figure 2-16 is considered as the most stable sign. In this, a triangle is formed by different criteria that represent a stable sustainability in technological development. This idea informs the new assessment framework that is developed and shown in Figure 2-17. Any new technology could be evaluated and assessed by using this model. There are two selection levels, one a primary level and the other a secondary level. A technology must fulfill primary selection criterion before being taken to the secondary level of selection. The primary selection criterion is "time".

For a simulation test, we imagine that a new technology is developed to produce a product name "Ever-Rigid". This product is non-corrosive, nondestructive, and highly durable. The "Ever-Rigid" technology can be tested using the proposed model to determine whether it is truly sustainable or not. The prime first step of the model is to find out if "Ever-Rigid" technology is "time-tested". If the technology is not durable over infinite time, it is rejected as an unsustainable technology and so not considered for follow-up testing. For, according to the model, time is the prime criterion for the selection of any technology.

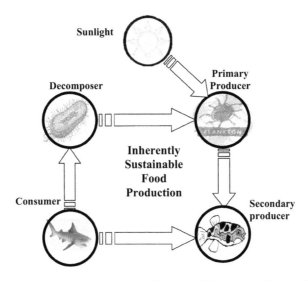

Figure 2-15   Inherently sustainable natural food production cycle.

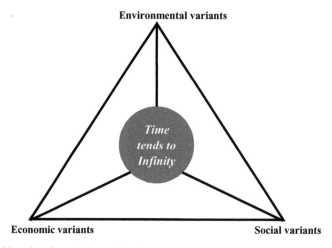

Figure 2-16   Pictorial view of the major elements of sustainability in technology development.

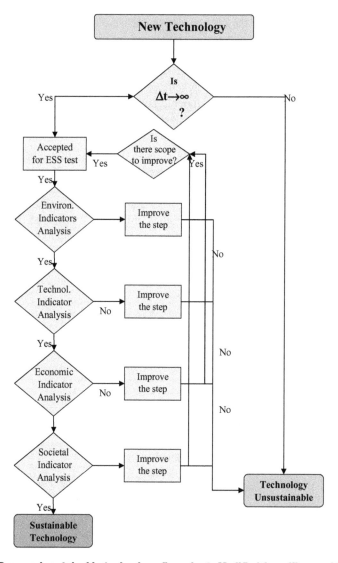

Figure 2-17   Proposed sustainable technology flow chart. Modified from Khan and Islam (2005a).

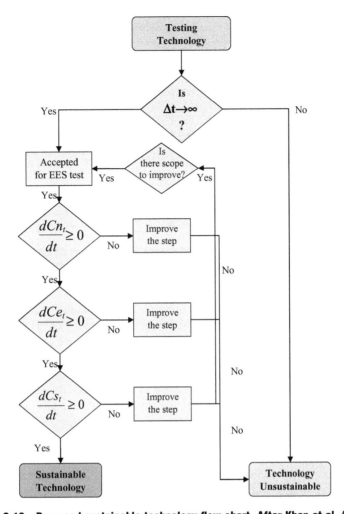

Figure 2-18   Proposed sustainable technology flow chart. After Khan et al. (2006a).

If the "Ever-Rigid" technology is acceptable with respect to this time criterion, then it may be taken through the next sorting process to be assessed according to a set of secondary criteria. The initial set of secondary criteria analyzes environmental variants. If it passes this stage, it goes to the next step. If this technology is not acceptable with respect to environmental factors, then it may be rejected, or further improvements may be suggested to its design. After environmental evaluation, the next two steps involve technological, economic, and societal variants analyses, each of which follows a pathway similar to that used to assess environmental suitability. At these stages also, either it will ask for improvements required to be made to the technology, or it might be rejected as unsustainable.

2.5.1   *Condition of sustainability*

In order to consider a technology, which is used in petroleum operations, an inherently sustainable method is needed for evaluation. This evaluation method should be based on the principle of true sustainability, which is defined and shown in Figure 2-18. Based on this newly developed method, a practical tool is proposed and presented Figure 2-19. In this evaluation method, for the sake of sustainability, the total critical natural resources should be conserved in the whole technological process. Also, waste produced in the process of using the technology should be within the assimilative capacity of a likely to be affected

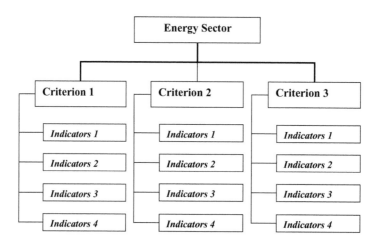

Figure 2-19   Hierarchical position of criteria and indicators of sustainability.

ecosystem. This means that an intra- and inter-generation ownership equity of natural resources, on which the technology depends, must be ascertained (Daly 1999; Costonza et al. 1997). Daly (1999) points out that all inputs to an economic process, such as the use of energy, water, air, etc., are from natural ecology and all the wastes of the economic process are sunk back into it. In other words, energy from the ecological system is used as throughput in the economic process, and emissions from the process are given back to ecology. In this sense, an economic system is a subsystem of ecology and, therefore, the total natural capital should be constant or increasing. Man-made capital and environmental capitals are complementary but are not substitutable. As such, any energy system should be considered sustainable only if it is socially responsible, economically attractive, and environmentally appealing (Islam 2005c).

To consider the petroleum operations system sustainable, it should fulfill basic criteria with respect to environment; social, economic, and technological (Pokharel et al. 2003, 2006; Khan and Islam 2005b, 2005c; Khan et al. 2005b, 2006a). In this study, the following criteria are taken into consideration:

Natural (Environment) Capital (Cn) + Economic Capital (Ce) + Social Capital (Cs) ≥ Constant for all Time Horizons

$$(C_n + C_e + C_c)_t \geq \text{constant for any time "t" provided that } \frac{dCn_t}{dt} \geq 0 ,\ \frac{dCe_t}{dt} \geq 0 ,\ \frac{dCs_t}{dt} \geq 0$$

These conditions are shown in Figure 2-18. In the proposed model, a technology is only "truly sustainable" if it fulfills the time criterion. Other important criteria that it must also fulfill are related to environmental, social, and economic factors (Figure 2-18).

2.5.2   Sustainability indicators

Indicators can be used to measure the sustainability state of petroleum operations. Sustainability or sustainable operations is accepted as a vision for managing the interaction between the natural environment and social and economic progress with respect to time. However, there is no suitable method to measure the sustainability of petroleum operations. Experts are still struggling with the practical problem of how to measure it. The Centre d'Estudis D'Informaci Ambiental (CEIA 2001) stated that "the move towards sustainability would entail minimizing the use of energy and resources by maximizing the use of information and knowledge." In effect, in order to develop sustainable technology and manage natural resources in a sustainable manner, decision- and policy-makers need to improve the application of knowledge gained from information. However, there is generally a large communication gap between the provision of data and the application of knowledge.

One method of providing information in a format that is usable by policy- and decision-makers is through the use of sustainability indicators (Figure 2.19). An indicator is a parameter that provides information about the environmental issue with a significance that extends beyond the parameter itself (OECD 1993, 1998). Indicators have been used for many years by social scientists and economists to explain economic trends, a typical example being the Gross National Product (GNP). Different NGOs, government agencies, and other organizations are using indicators for addressing sustainable development. Some of them are: the World Resources Institute; the World Conservation Union-IUCN; United Nations Environmental Program; the UN Commission on Sustainable Development; European Environmental Agency; the International Institute of Sustainable Development (IISD); and the World Bank (IChemE 2002).

Indicators for addressing sustainable development are widely accepted by development agencies at national and international level. For example, *Agenda 21* (Chapter 40) states that "indicators of sustainable development need to be developed to provide solid bases for decision-making at all levels and to contribute to the self-regulating sustainability of integrating environmental and development systems" (WCED 1987). This has led to the acceptance of sustainability indicators as basic tools for facilitating public choices and supporting policy implementation (Dewulf and Langenhove 2004; Adrianto et al. 2005). It is important to select suitable indicators, because they need to provide information on relevant issues, identify development-potential problems and perspective, analyse and interpret potential conflicts and synergies, and assist in assessing policy implementations and impacts.

Khan and Islam (2005a, 2007) developed sets of indicators for technology development for oil and gas operations. The hierarchy positions of criteria and indicators are presented in Figure 2-19. They developed indicators for environmental, societal, policy, community, and technological variants, which are shown in Figures 2-20–2-23. By analyzing these sets of indicators, they also evaluated the sustainable state of offshore operation and its technology.

2.5.3 Sustaining capacity

Sustaining capacity concept is newly introduced in this chapter to ensure sustainable petroleum operations. The view here is that energy projects should be selected based on the capacity of the environment or ecosystem to carry them, not as individual projects, but in consideration of the total and cumulative impact of

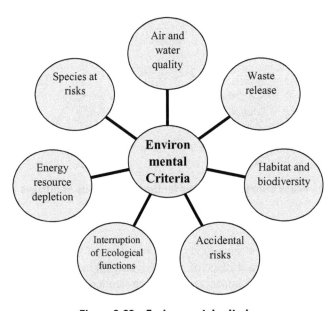

Figure 2-20 Environmental criteria.

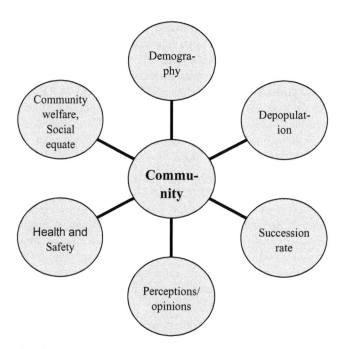

Figure 2-21   Criteria to consider for sustainability study of offshore oil and gas.

Figure 2-22   Policy criteria to consider for the sustainability of offshore oil and gas.

all the projects that would or could be supported by the resources of the environment in question. In other words, the proposal being put forward is based on the obvious truth that an ecosystem cannot accommodate or carry the environmentally degrading impacts of many petroleum or economic development projects. This is because every ecosystem has its own bearing or carrying capacity. If that limit is exceeded, then the ecological functions that it supports would become unsustainable.

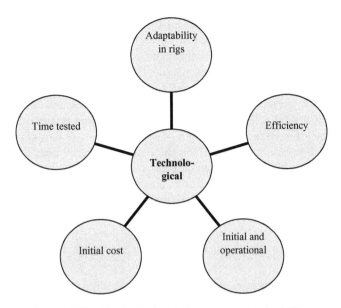

Figure 2-23   Technological indicators of sustainability.

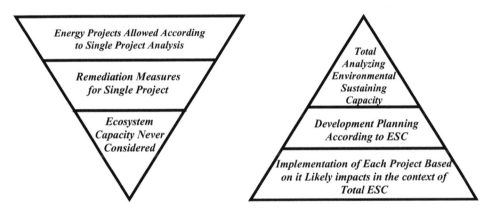

Figure 2-24   Existing management model (left diagram) and proposed sustaining capacity-based petroleum operations (right diagram).

The ecological term "carrying capacity" speaks of the capacity of an ecosystem being the size of the population or community that can be supported indefinitely upon the available resources and services of that ecosystem (Harding 1986, 1993). This proposed model applies to the carrying capacity principle in the petroleum operations management context. For purposes of this work, the adaptive concept is referred to as Sustaining Capacity (SC). In biological research, carrying capacity is determined by estimating the correlation between the sizes of the populations as balanced against the space and food that is available to support them in a particular ecosystem. In contrast, the SC is defined in terms of the degree or intensity of human use and development activities that an ecosystem can sustain at an environmentally acceptable level (Figure 2-24), considering the biological, physical, social, and resource characteristics of the ecosystem in issue.

Today's management approaches do not consider an SC perspective. Management policy is based on the "single project" approach. What this means is that each project is assessed then accepted or rejected in terms of its meeting established criteria for acceptability, irrespective of whether other competing and environmentally compromising projects are located in the same environment general area. The procedure that is followed to accept each project, as if the others do not exist, is to assess it in light of all the different environmental impact assessment (EIA) processes, and to examine the different remediation plans and environmental monitoring plans that its proponent put forward based on the nature of the project. The management regime is geared to consider project-based impact assessment, and to try to solve the problems generated by one project at a time. The limitation of this approach is that we do not know whether the ecosystem can actually bear the volume and weight of development activities approved within it, in terms of their adverse impacts. The overall result in that the current approach does not establish a sustainable relationship between human activities and their impacts on environmental carrying capacity to generate strategies for managing resources. The existing management approach also makes no room to quantify the carrying capacity of an ecosystem. The solution to this, as suggested, is to adopt an ESC-based management approach where the capacity of the environment/ecosystem will be estimated first, and then development activities may be allowed according to its total sustaining capacity in relation to all activities that are allowed. The upside down nature of the existing approach is diagrammatically represented in Figure 2-26.

Even though the SC-based management model might be an ideal tool to facilitate proper decisions for natural resource management, its application in petroleum management is a major challenge. This is because of the question of how to determine the SC of an ecosystem. The proposed SC model is to quantify the total SC of a specific region by taking all of its site specific activities (i.e., fisheries, aquaculture, tourism, oil and gas operations, urban developments, and waste releases) into account (Figure 2-25). One option would be to estimate SC of individual operations and then mathematically calculate the total SC of the whole ecosystem through quantification of the variables attached to the possible impacts associated with each site-specific activity. Figure 2-25 represents the different activities in an ecosystem, where the arbitrary values for different activities are shown. The outer boundary of the figure represents the threshold limit of each activity, except for fisheries, which is shown to be exceeded the threshold boundary. The inner boundary represents the status of the operations. This figure is based on arbitrary values.

It would also be necessary to consider the short-term, long-term, direct, indirect, cumulative, irreversible, and irretrievable consequences of these activities on temporal (past, present, and future) and spatial scales.

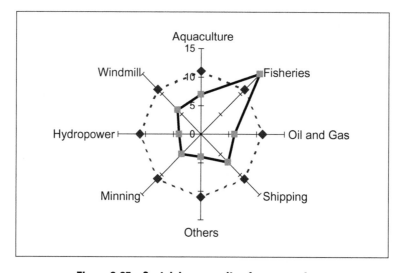

Figure 2-25   Sustaining capacity of an ecosystem.

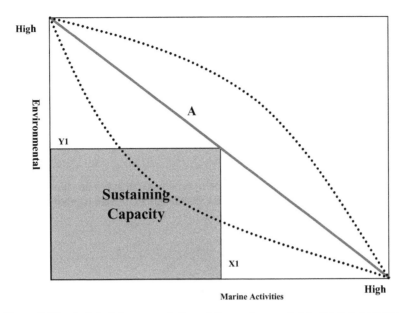

Figure 2-26   A pictorial representation of the environmental sustaining capacity.

The currently applicable management principle of petroleum operations is more focused on engineering solutions than on ecosystem and resource sustainability ones. This is why, for resources conservation and reasons of ecosystem maintenance, it is necessary to sustain the capacity-based management approach. The proposed model is pictorially represented in Figure 2-26. In this figure, hypothetical relationships of all activities in an ecosystem and their impacts are shown, where X1, Y1 represents the maximum sustaining capacity and line "A" the sum of all activities.

2.5.4   VECs for sustainability

The Environmental Impact Assessment (EIA) commonly measures the potential impacts of a project or energy project. However, the EIA never considers the ecosystem level impacts. On the other hand, the Valued Ecosystem Components (VEC) study can represent the ecosystem level impact. By determining the VEC status, the overall sustainability of petroleum operations can be determined. For studying the ecosystem level impacts of petroleum operations, the ecosystem is categorized into different components (Figure 2-27), which are known as Valued Ecosystem Components (VECs). The Valued Ecosystem Components are classified based on their importance and functions. Figure 2-27 presents the species at risk, fisheries, special habitat, and carbon as the VECs for an oil and gas development project. For analyzing the status and effects levels on each VEC, such as species at risk, different issues such as, its distributions, policy commitment, applicable monitoring protocol, and typical pre-mitigation, are discussed. Figure 2-28 shows what detailed issues are considered to study the VECs. Detailed VEC-based sustainability assessment is discussed in Chapter 4.

VECs are different in different ecosystem. For example, fisheries can be identified as an important VEC because in the offshore operations on the Scotian Shelf, it is locally important. The status of fisheries is critical at this time. There is a potential link to effects caused by oil and gas development project activities. Also, it has great social and economic value. Fishing is Nova Scotia and Canada's oldest industry sector, having been in existence for centuries (NHNS 2002).

In addition to fisheries, species at risk can be considered as a VEC. For example, three species at risk have been chosen from those listed for Atlantic Canada and Nova Scotian coastal areas (Table 2-6) for their

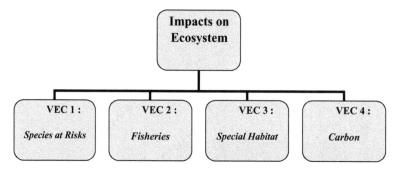

Figure 2-27   Pictorial representation of determination of total impacts based on individually-valued ecosystem components (VEC) in determining sustainable petroleum operations.

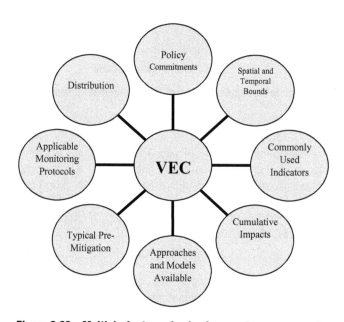

Figure 2-28   Multiple factors of valued ecosystem components.

combined representation of the various habitats that may be impacted by petroleum development. The endangered species selected are the leatherback turtle, *Dermochelys coriacea*, which is found in marine waters off Nova Scotia; the diadromous species Atlantic salmon, *Salmo salar*, which uses marine, coastal, and estuarine habitats; and the coastal-nesting, intertidal-feeding piping plover, *Charadrius melodus*, subspecies *melodus*.

Sable Island on the Scotian Shelf has been selected as an example of a special area valued ecosystem component. Numerous parcels of land have been purchased by oil companies for exploration on the Sable Island Bank, a series of relatively flat banks located on the outer margin of the Scotian Shelf. Sable Island is considered a "geomorphic oddity". It is composed entirely of unconsolidated sand and is the only exposed portion of the outer continental shelf in the northwest Atlantic. It has a diverse flora and fauna and is stabilized, primarily, by its vegetation cover and the oceanic current. The uniqueness of both the physical structure of the island and the wildlife thereon, and the large potential for future development of oil fields near the island make a strong case for the inclusion of Sable Island as a special area VEC in future environmental impact assessment.

Table 2-6   Species at risk in Nova Scotian marine and coastal environments. Committee on the Status of Endangered Wildlife in Canada [COSEWIC] website, March 13 2002

Status	Atlantic Canada marine species	Nova Scotia coastal species
Extirpated	Atlantic Walrus, Grey Whale	No listing
Endangered	Beluga Whale, Right Whale, Leatherback Turtle, Atlantic Salmon	Eskimo Curlew, Piping Plover Roseate Tern, Atlantic Salmon, Atlantic Whitefish
Threatened	Harbor Porpoise, Beluga Whale, Northern, and Spotted Wolffish	No listing
Special Concerns	Whales: Blue, Fin, Humpback, Northern Bottlenose, Sowerby's Beaked. Fish: Atlantic Cod, Atlantic Wolffish	Harlequin Duck, Barrow's Goldeneye

In studying VEC, different important factors of each VEC should also be studied. Figure 2-28 shows the different factors, such as distribution, policy commitment, temporal bounds, commonly used indicators, cumulative impacts, approaches and models available, typical mitigation, and applicable monitoring protocols.

2.6   How the Green Supply Chain Model Leads to Sustainability

Recently, the "green supply chain" approach has been introduced in achieving sustainability in petroleum operations (Khan et al. 2006c). A green supply chain provides guidelines to a company to implement the principles of sustainability in the areas of corporate governance, the environment, social responsibility, and economic contribution to the community (Sarkis 2003). However, the application of a green supply chain in the energy development sector is currently at an embryonic stage (Lakhal et al. 2005). Some work has been done on the petroleum supply chain by Lasschuit and Thijssen (2004), Neiro et al. (2004), and Colella et al. (2005). These works mainly describe business plans that include the supply and marketing of petroleum products, rather than focusing on considerations allied to oil extraction or development, or the technological problems involved in greening production. Lakhal et al. (2005) took the initiative by applying the green supply chain model to the oil refining process of a Canadian company.

The concepts of sustainability and sustainable management have also been applied to companies (Khan and Islam 2007; Khan et al. 2006b; Lakhal et al. 2005; Atkinson 2000; Huizing and Dekker 1992). On one hand, companies use capital, which is undesirable because capital is valuable and limited. On the other hand, companies' output, that is, the products and services they produce, is desirable. Companies thus need to optimize the way in which they use capital. To achieve sustainability in the petroleum company, the "Olympic" green supply chain model is proposed.

The supply chain of petroleum operations is developed on the basis of their life-cycle operations (Figure 2-1). Cylinders 1 to 7 indicate the process of operations. Seismic exploration is at the top level of the chain and decommissioning is at the bottom. The intermediate steps include drilling, production, and transportation. The solid arrows on the left-hand side represent input, while the blank arrows on the right-hand side represent emission.

In present operations, activities associated with the petroleum operations supply chain emit numerous wastes into the air, water, and sediments (Figure 2-29). The seismic process emits 10 different wastes while the drilling process is the source of the highest emissions within the chain. The main wastes of petroleum operations that impact the environment are drilling-waste fluids or muds, drilling-waste solids, produced water, and volatile organic compounds. Moreover, produced water constitutes the highest amount of waste released in the whole supply chain (Khan and Islam 2005a, 2007).

Analogous to the "Olympic" logo, the proposed supply chain for petroleum operations consists of five zeros (zero-wastes), which are:

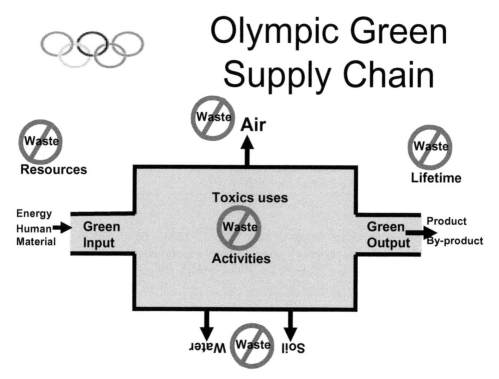

Figure 2-29   The proposed "Olympic" green supply chain for offshore oil and gas operations. Adopted from Khan et al. (2006c).

(i)   zero emissions

(ii) zero resource wastes

(iii) zero wastes in activities

(iv) zero use of toxics and

(v)   zero waste in product life cycle.

Figure 2-29 captures the concept of the Olympic supply chain for petroleum operations in a representative diagram. According to this model, the system is enabled to input green resources. In this figure, energy, human, and materials are shown as green inputs. The rectangular box illustrates a processing unit showing that toxic compounds cannot be used in the chain. Production and administrative activities can be wasted with regard to inefficiency in processing. The three arrows at the top indicate that the chain does not create any gaseous, solid, or liquid emissions or discharges. The outlet of the system also shows green products that generate a no product life cycle, in terms of transportation, use, and end-of-use of the product. This supply chain model can be used in petroleum operations to achieve sustainability. This is because such an approach is always the norm in Nature. Thus, the proposed "green" supply chain is a new approach for use in the petroleum (Bjorndalen et al. 2005; Lakhal et al. 2007) and renewable energy sectors (Khan et al. 2005b, 2006a).

2.6.1   *Zero emission of air, water, solid wastes, soil, toxic wastes, and hazardous wastes*

It is evident from the above discussion (Section 2.3) that petroleum operations generate huge amounts of solid, liquid, and gaseous wastes that are released into the air, water, and sediments. In the proposed "Olympic" green supply chain model (Figure 2.29), it has been demonstrated that it is possible to not release any wastes into air, water, or soil; a goal that present petroleum operations have not been able to achieve. As a consequence, current offshore operations have negative impacts on the environment (Holdway 2002; Khan and Islam 2003a, 2004, 2005a, 2005b, 2005c; Colella et al. 2005; Currie and Isaacs 2005).

To achieve sustainability in the petroleum operations sector, it is imperative to stop all emissions that the sector is currently producing. The challenge facing this industry is to devise ways to stop the emissions substantially so as to bring them down to zero levels in petroleum operations. Following a general evaluation of currently available technology, it has been concluded that no technology is currently available that could completely eliminate emissions (Khan and Islam 2006b). Therefore, the need to develop sustainable technology is extremely urgent, as proposed by Khan et al. (2005b).

Some technologies do exist that reduce discharges considerably. For example, solid wastes (drill cuttings and barites) can be reduced significantly by adding extra shale screen or centrifuger during the drilling process. Re-injection or onshore disposal is also considered as an effective way to achieve zero emission of solids, liquids (produced water), and gases (Malik and Islam 2000). In addition to the implementation of new cleaning processes, other technical and operational modifications at installation sites can contribute to lowering the levels of some specific wastes.

2.6.2 Zero waste of resources (energy, materials, and human)

The proposed supply chain model (Figure 2-29) requires zero waste generation from input materials, involvement of humans, and use of energy. Present petroleum operations lose many resources in these areas through flaring, blow-outs, and spills. Drilling fluids are generally re-used but large amounts are washed out with drilling cuttings, According to Khan and Islam (2006c), a single development well in deep water lost 217,491 kg of SBF and 71,772 kg of barites. This kind of loss can be significantly reduced by developing new eco-friendly technologies. As mentioned above, to reduce drill cutting wastes, more shale screen and cetrifuger may be used. However, the use of more screening sometimes turns the small sized particles into finer sizes, making it hard to separate and reduce the SBF from the fine solid particles.

An oil and gas development project is a high, energy-intensive project and there is still a lot of room to improve energy use, such as reducing the length of drilling periods. Drilling a well takes a couple of months to a year. This time can be significantly reduced by employing better technology. Besides drilling time, durations of seismic exploration, production, and decommission can also be reduced by use of improved technology.

Fatal human accidents are often reported in petroleum operations. These frequently involve lifting incidents on decks or drilling activities. Table 2-7 shows the fatality and injury figures per 100,000 employees. According to Powell (2004), although no fatalities were reported between April 2002 and March 2003, three deaths have occurred since. Recently, the oil firm Shell was fined £900,000 following the death of two workers on the Brent Bravo platform in the North Sea in September 2003 (BBC 2005). In 2007, BP admitted to "gross mismanagement" in the Texas decident.

2.6.3 Zero wastes in activities (administration, production)

The proposed "Olympic" green supply chain calls for zero activity wastes in regard to production and administration (Figure 2-29). It is vital for petroleum operations to achieve this goal because these opera-

Table 2-7   Total fatalities/injuries among 100,000 employees.
Source: Powell (2004)

Category	1999/2000	2000/2001	2001/2002	2002/2003 p
Fatalities	2 (10.5)	3 (12.9)	3 (12.9)	0 (0)
Major injuries	53 (227.2)	53 (227.2)	47 (202.5)	64 (287.5)
Over 3-day injuries	193 (1015.8)	177 (758.7)	187 (805.8)	118 (530.0)

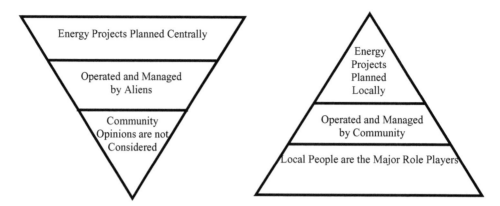

Figure 2-30   Upside-down, bottom-up approach and community-based energy development approach (Khan et al. 2005b).

tions involve high expenditure. Reduction in activity means reduction in monetary expenditure. Only large companies are involved in petroleum operations and their current management systems are centrally controlled. However, their operations are in remote areas and sometimes field geologists or engineers cannot take *in situ* decisions in response to the particular behavior of a well. It has been reported by Lojan (2005) that drilling activities had to be stopped in an Ecuador oil field to await a simple decision from the company head office in China. According to Lojan, this entails a lot of revenue loss and is an example of wasteful and costly administrative activities.

An alternative solution to the existing bottom-up approach in petroleum operations is the introduction of a community-based management regime. This approach is effective in securing administrative, production, and other environmental benefits (Khan et al. 2005b). Figure 2-30 indicates the major difference between existing management practices and the proposed community-based management arrangement. At present, most of the oil industry still does its planning, engineering, upstream operations, refining, supply, and transportation in a centralized, non-integrated way (Forest and Oetti 2003). This method opposes the proposed community-based model.

2.6.4   Zero use of toxics

In present practices of petroleum operations, many toxic compounds are used in drilling fluids, well testing chemicals, and corrosion inhibitors. Instead of using toxic compounds, natural non-toxic compounds can be used in this process. Al-Darbi et al (2002) found that natural vegetable components are highly efficient in controlling pipeline corrosion. They suggest replacing conventional toxic corrosion inhibitors with these natural components. The composition of drilling fluids has many toxic compounds, including carcinogens. Recently, many of these compounds were replaced by vegetable based compounds (EPA 2000). It is important to realise that long-term toxicity arises from the use of artificial chemicals (Chhetri et al. 2007; Zatzman and Islam 2006).

2.6.5   Zero waste in product life cycle

An "Olympic" green supply chain would stop wastes in the life cycle, including transportation, usages, and end-of-life of a product. In existing offshore operations, transportation of oil involves many accidents resulting in large oil spills. It has been reported that 62,000 tons of oil are released from offshore operations annually and that 172,000 tons are released into the marine environment (Khan and Islam 2004).

Decommissioning a platform also generates many waste products. The production installation is made up of large amounts of material whose dismantling is a serious environmental concern. For example, as reported by Khan et al. (2006d), a single energy project can use more than 900,000 m of pipe, 109 million kg of steel, 53,000 m of chrome tubing, and 160 million kg of concrete. Some innovative technologies have been suggested by Khan and Islam (2003b) to achieve zero waste in the product life cycle of petroleum operations, including rigs for use as artificial reefs as fish shelters and the platonic utilization of wastes.

2.6.6 How an Olympic green supply supports sustainability

The proposed "Olympic" supply chain for petroleum operations supports the major goals of sustainability through improved technology, economic well-being, environmental protection, and social benefits. As mentioned earlier, no current technology is suitable to completely achieve the zero-waste objective and there is a need to develop new technology to meet this goal. New technology must ensure sustainability, following the recently developed criteria of sustainability proposed by Khan et al. (2005b).

Economic well-being is improved when organizations are able to identify inefficiencies in processes, products, and services and find cost-saving solutions to these problems. For example, waste reduction improves efficiency and lowers costs. Zero waste generation will directly help to improve regulatory compliance and reduce project costs. Also, by maintaining good environmental standards, a company can acquire a good reputation and greater consumer acceptance, helping it to achieve long-term benefits.

Reduction of wastes to zero levels would significantly improve the environmental quality of areas where offshore oil and gas development is conducted. At present, petroleum operations are blamed for adverse environmental effects due to the wastes they release (Holdway 2002; Wells et al. 1995; Khan and Islam 2003a). Following compliance with the "Olympic" green supply chain model (Figure 2-29), social well-being can be achieved through reliance on the proposed new management technology, which would ensure that oil and gas resources are available to all. In addition, a more complete use of "wastes" will create jobs, such as logistics and reprocessing activities. For example, the Titanium Corporation, a company that produces valuable products from oil industry wastes, has become a profitable venture (Eliot 2005, personal communication).

2.7 Benefits of the New Model

Although sustainability is not an established reality in the petroleum sector, apparent benefits for companies, communities, and the natural environment can be identified. For example, choosing Nature-based technology leads to human health benefits, social benefits, environmental benefits, and business benefits. According to Bjorndalen et al. (2005), the no-flaring technique does not only improve the environmental benefit, but also brings the health benefit of reducing asthma in the locality. Sustainable petroleum operations ensure sustainable wealth generation that protects and preserves local communities, maintains ecosystem health in perpetuity, and provides new opportunities for businesses. Sustainable or "greening" businesses minimize the environmental impact and such a company tends to attract the best employees, helping the business make more money, and enhancing its reputation.

It is important to realize that oil producers can play a beneficial role in managing their environment practices, by exceeding governmental guidelines. For example, according to environmental guidelines, if oil content in the drill cuttings of 10 mg/L is allowable, then the company should maintain it at less than 5% or better. In this general point of view, it seems as if companies are spending more money to maintain higher environmental standards. However, in the actual calculation, its production costs will decrease through avoiding any sort of potential environmental compensation (Veil 1998; Ofiara 2002), saving management costs, avoiding environmental effects monitoring (EEM) programs, etc. It will also significantly decrease the processing time by the regulatory agencies. The truth is that if a company maintains better environmental standards than the mandatory guidelines require, this will minimize the costs and regulatory processing time. It also can gain an intangible value, through goodwill and improve reputation in the eyes of the general public.

The current practice is to barely meet the regulatory requirements, while claiming that the strict nature of environmental regulations are making the corporate oil and gas operating non-profitable. Consider the *modus*

Table 2-8   Costs and benefits of sustainable development

Costs	Benefits
• In the short term may be less economically hence costly, but it is also proved that it can bring even short-term/apparent benefit as well (discussed above) • Requires new line of research and planning • Requires development of markets/infrastructure • Requires compromise and communication • Requires mechanisms for education/training • May require new management approach, policy, and laws • May require conflict management mechanisms • Requires new nature-based technology\ development	• In the long it is the only solutions to sustain an individual sector • Involves local people and provides them with rewarding, immediate work, income, and education • Preserves functionality and diversity of ecosystem, while providing a wide range of economic benefits • Brings short-term and long-term company benefit and reputation and contribution to social progress • Provides a niche for indigenous peoples in modern, free market society, should they choose

operandi in this: oil companies invest in research and determine how inadequate current regulations are. This will lead to the development of techniques that are so beneficial in the long term that the initial research and development costs are paid off multifold in clearing environmental costs alone (Khan and Islam 2007).

2.7.1   Basic environmental benefits of sustainable management

Sustainable development is based on the understanding that quality of life and personal well-being are determined by many factors. In the initial phase of the journey toward sustainable development, the necessity to comply with environmental laws and regulations drives improvements in environmental performance. As the journey continues, business strategies for sustainable development tend to evolve beyond compliance to areas such as eco-efficiency, niche marketing, and a greater ability to meet client requirements.

Still, the final evolution of thinking includes the understanding that the health of the national (for example Canadian) economy and in fact, the global economy will eventually depend on how natural capital resources (including the petroleum sector) are managed.

There is an upward trend in greenhouse gases, such as carbon dioxide and methane, that leads to climate change, which might increase the frequency of storms and thus insurance costs. Sustainable technology development and management will reduce the production of greenhouse gases and so will slow the rate of climate change.

Unsustainable techniques are responsible for health problems. Cleaner air benefits everyone. Health care is a major component of our social system and it commands more tax dollars every year. Poor air quality is a major cause of asthma and other respiratory conditions. In fact, in 1990, the total cost of asthma in Canada was estimated at $504–648 million per year (SBOR 2003). According to Bjorndalen et al. (2005), by developing sustainable technology, such as the no-flaring technique, health problems can be reduced and so reduce health costs.

Zero-waste petroleum operations or converting waste into value added products creates jobs and brings economic benefit. Finally, efforts to better understand ecosystems will not only provide us with a healthier environment, but may help to prevent the collapse of resource-dependent industries such as fishing and forestry.

2.7.2   Economic and social benefits of sustainable management

As a global society, we can see that the growth of economic development continues to be a huge focus of our nations. New and complex initiatives are undertaken to ensure that our economic growth continues and that it can be sustained by the available human and natural resources. The sustainability concept is an integral part of the decision-making process on economic development matters. For that reason, sustainable development is not a stand-alone environmental or social program and it needs to be integrated into business practices. Sustainability of the world's largest business sector, the petroleum sector, must improve the environmental, social, and economic future.

Programs designed to resolve environmental problems and limit resource depletion will help maintain economically and environmentally sustainable growth. It has been estimated that only 6% of material ends up in a product. In fact, the ratio of waste to the durable it creates may well be closer to 100 to 1 (Chhetri 2006). Waste reduction, energy efficiency, and pollution prevention make economic sense in cost savings, increased efficiency, and better use of tax dollars.

According to SBOR (2003), sustainable management practice can form "niche" enterprises and product lines that provide more jobs and wealth, as well as assist in meeting new and evolving supplier requests.

Sustainable use of natural resources (including petroleum) maintains employment, trade, exports, and product development. Eco-efficiency results in benefits such as cost savings, effective risk management, and business expansion into new products, processes and services.

Sustainable development creates jobs within the community. For example, development of community-based energy creates more jobs than central-based energy development (Khan et al. 2006a). Sustainable operations also enhance research and technology development, truly sustainable innovation, and result in the development of export opportunities.

Our quality of life and the values of diversity and equity will increasingly link to the quality of our environment. As human beings, we are particularly concerned about the legacy we leave to future generations in terms of their natural heritage, their economic opportunities, and potential health risks.

In addition to the above-mention benefits, more environmental, economic, and social benefits can be achieved by implementing sustainable petroleum development (SBOR 2003). To gain these benefits, each petroleum operational phase (from exploration to decommissioning) should be managed in a sustainable way. To reach that goal, the chapters of this book discuss sustainable management of individual phases. In Chapter 3, sustainable petroleum explorations operation is detailed.

Exploration Operations

3.1 Introduction

The main goal of exploration operations is to obtain an image of the reservoir prior to drilling. By using exploration techniques, geological information data are collected on subsurface conditions to evaluate the potential for the presence of oil and natural gas. The collected information helps a decision to be made as to whether there is a hydrocarbon reserve or not and where more intensive exploration activities should take place. In a typical seismic exploration operation, a line of data receivers are laid out, called geophones for terrestrial programs or hydrophones for aquatic operations. Explosives or mechanical vibrators are commonly used on land and airguns are used in aquatic or offshore environments. Seismic surveys are used by the offshore oil and gas industry to help determine the location of oil and gas deposits beneath the seafloor.

The popular seismic exploration techniques are used for both terrestrial and offshore exploration. Among all of the exploration technologies, three-dimensional (3D) and two-dimensional (2D) techniques are the most popular (Diviacco 2005). Since the beginning of the petroleum industry in the 19th century, the large, shallow, and high-quality reservoirs of oil and gas have been explored and most of the "low-hanging fruit" of world oil and gas reservoirs has been picked. At present, oil and gas exploration is focused increasingly on finding resources located in more geologically complex, deeper, and lower-quality reservoirs (EPA 2005).

To face the new challenge of exploring deeper and more complex reservoirs, new exploration technologies are emerging. For example, four-dimensional (4D) techniques are drawing more attention. Lumley and Behrens (1998) reported that 4D techniques, which are also known as time-lapse seismic, which geophysicists often abbreviate as 4D seismic, have the ability to image fluid flow in the interwell volume by repeating a series of 3D seismic surveys over time. The 4D seismic also shows great potential in reservoir monitoring and management for mapping bypassed oil, monitoring fluid contacts and injection fronts, identifying pressure compartmentalization, and characterizing the fluid-flow properties of faults (Lumley and Behrens 1998).

All phases of petroleum operations have environmental impacts, with exploration operations bearing the least environmental consequences. Browning et al. (1996) studied environmental consequences of seismic survey. Unlike 2D surveys, where profile positioning is flexible, 3D surveys involve high density coverage over many square miles. Khan and Islam (2006c, 2006d) analyzed different seismic exploration activities and their associated environmental effects. Currently used exploration techniques are based on advanced technological or engineering solutions. In addition, this sector is currently one of the most governmentally regulated (Khan and Islam 2003a, 2003b; Thanyamanta 2004). The oil and gas activities are restricted by dozens of rules, regulations, and guidelines, such as the Canada Nova Scotia Offshore Petroleum Board (CNSOPB) and Environment Protection Agencies (EPA) guidelines. To meet these guidelines and regulatory requirements, appropriate technologies are required.

At present, technologies are selected from the best available technologies (BAT) using different evaluation methods. Recently, EPA (2000), Thanyamanta (2004), and Zatzman et al. (2006) management techniques have been evaluated for use in offshore oil and gas operations. These technologies were evaluated with regard to their technical feasibility, rig adoptability, cost-effectiveness, regulatory requirements and compatibility, and environmental impacts. In addition, Thompson and Nour (1998) studied environmental risk exploration activities and identified the hazards posed by seismic surveys and drilling operations and their potential environmental effects. However, considering the sustainability issue as a component of these studies is rare in technology evaluation in petroleum operations or in seismic operations. Therefore, new technologies are developed without considering the cost to the environment.

Recently, sustainability issues in energy development, especially in petroleum operations, have been introduced by Khan (2006a), Khan and Islam (2005b,c), and Khan et al. (2005b, 2006a, 2006b). Chapter 2 of this book elaborates the sustainable management framework for the petroleum operations. This study proposes a novel method that posits that a technology is sustainable if it assimilates the long-term functional efficiency of natural systems. As in Nature, any such technology must be functional for an infinite duration. This approach has been applied effectively in renewable (Khan et al. 2006b) and non-renewable energy development and technological evaluations (Khan and Islam 2005). To achieve overall sustainability in oil and gas operations, from technology development to company operations, the "Olympic" green supply chain model has been proposed by Khan et al. (2006c).

There are few studies that have been conducted to achieve sustainability in the seismic technological operations. This chapter initiates a new dimension of the technology development method for the exploration operations to achieve sustainability. It examines the context in which oil and gas exploration is regulated by government, evaluates seismic exploration activities in the terrestrial and offshore, evaluates these activities from the point of view of sustainability, and presents guidelines for sustainable exploration activity. In doing this, presently available and emerging technologies are compared with natural functions/processes, which might be applicable to seismic operations. This chapter also discusses an overview of sustainability and available seismic technologies, followed by methods for evaluation. The findings are presented and discussed to evaluate a technology that is more acceptable, based on its sustainability status. Finally, this chapter provides guidelines to achieve sustainability in exploration operations.

3.2   Current Practices of Exploration Operations

Petroleum exploration operation methods can be divided into two different categories. The first are land-based operations and the others are aquatic, which might be conducted in river and offshore environments. Figure 3-1 shows a pictorial view of seismic activities in offshore exploration operations. The objectives of both methods are similar, but operation methods are different. Therefore, these two methods are discussed separately.

3.2.1   Terrestrial seismic exploration

The terrestrial seismic exploration includes 2D and 3D surveys. Generally, dynamite or vibroseis equipment is used as an energy source to generate sounds and geophones are used to receive sound. Explosives or mechanical vibrators generate sounds and the resulting energy waves are then sent into the subsurface. The geophones collect the energy as seismic waves reflect back from the subsurface deposits, both unconsolidated and bedrock. The wave energy is converted into electrical impulses transmitted to a bank of computers in a recording vehicle. The data is processed and interpreted to provide an image of the subsurface geology.

Prior to the line operation, a survey is usually conducted on the ground and/or by aerial reconnaissance to identify the new or existing lines and access routes, site-specific terrain problems, and assess potential stream crossing sites. Corridors or lines are cleared for equipment access and deployment of geophones. A complete seismic survey of an area typically involves a series of seismic lines running parallel to each other, usually at a distance of 400 m or more between lines. After the survey, equipment is collected and removed from the area. Survey approaches to minimize environmental impacts include the implementation of low-impact or minimal-impact procedures and the use of heliportable seismic methods.

Seismic waves can be generated by using explosives or vibroseis. For explosives, shot holes are drilled at defined locations along the seismic line for the placement of dynamite charges. The dynamite charges are sequentially exploded and the reflected sound waves are recorded at the surface using the recording

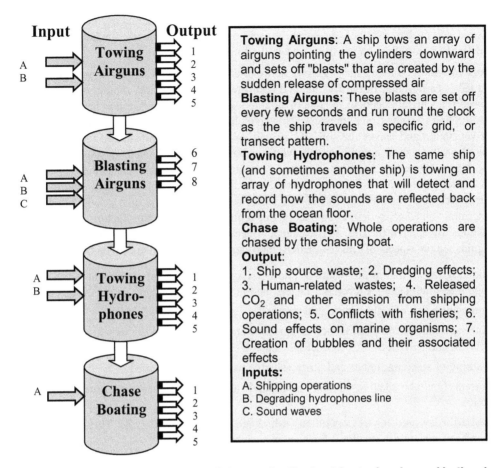

Figure 3-1   Schematic seismic activities in offshore exploration involving towing airguns, blasting airguns, towing hydrophones, and chase boat operations.

equipment. Because this approach allows placing the geophones after the shot holes have been drilled and charges set, lines can be narrow.

Vibroseis involves large vehicles with a large rectangular pad in the midsection that is lowered to the surface of the ground and then vibrates sending sound waves through the underground rock formations. Vibroseis buggies weigh about 60,000 lbs but have large tires or tracks to distribute the weight of the vehicle over a broad area. Normally four trucks travel in a line and operate in unison.

Vibrating pads are lowered to the ground and vibrate in one location for 10 to 14 s. The reflected sound waves are collected by the geophones. Vibroseis buggies replace the conventional use of shot holes with dynamite charges. Line widths of 6.0 to 8.0 m are required to allow the buggies access.

3.2.1.1   Line clearing approaches

Conventional Seismic: For the survey, bulldozers are used to create conventional seismic lines. Generally these lines are 6 to 8 meters wide to allow the bulldozers to operate (MVEIRB 2003). Generally these lines have long lines of sight and important vegetations/trees cut down to develop the line. Based on the size of the lease area, the lines can be several tens of kilometers (MVEIRB 2003). Global Positioning System (GPS) readings are used to maintain the direction of seismic lines during bulldozer operation. It helps if the line is less straight so as to avoid important sites, but still provide many kilometers of line of sight. The conventional seismic survey has relatively negative impacts on the environment compared to modern seismic methods (MVEIRB 2003) which are discussed in the following sections.

Low-Impact Seismic: To avoid the negative environmental impacts of seismic operations a Low-Impact Seismic (LIS) is carried out. In the LIS method comparatively narrow and continuously meandering lines are developed. This method reduces the line of sight to less than 200 meters, but in the conventional seismic operations the line is couple of kilometers. This short line of sight can avoid cutting of larger trees and important vegetation and leaves the soil and ground cover generally undisturbed (ASRD 2002; MVERB 2003). Based on the size of equipment, the width of line is also minimized. In the LIS the line width is reduced to 1.5 to 4.5 meters, but it is more than 6 meters in the conventional method. For safety reasons the width cannot be less than 1.5 meters. Instead of using heavy equipment, hand cutting or mechanical cutting are carried out. However, the line cutting varies with many factors, such as vegetation cover, density of forest, terrain, valued ecosystem components, line requirements, and other factors. At present, some specific conditions are applied for LIS. For example, Alberta Ministry of Sustainable Resource Development provides guidance on aspects of LIS (MVEIRB 2003) and some of them are given below:

- Considering the minimum line width to allow passage of mechanical equipment is 2.5 meters, the average line width cannot exceed 4.5 meters, and maximum line width cannot exceed 5.0 meters.
- The maximum line of sight can be 200 meters.
- For LIS, specialized equipment, such as gyromowers that are tracked units with blades that cut the underbrush and leave a trail of woodchips, may be used to clear the right-of-way.
- The cutting of standing timber and other valuable trees must be avoided.
- Measurers should be taken to maintain minimal disturbance of ground cover during the LIS operation.
- Special drills are used for LIS operations, which are know as enviro-drills. These are about 3.0 m long and can maintain a smaller 2.5 meter wide right-of-way.

Minimal Impact Seismic: Minimal impact seismic (MIS) is carried out in a site where there is high ecological and cutting of forest is restricted. The main objective of the MIS is to avoid environmental effects due to seismic operations. To achieve this objective small walk trails are created for foot access. Commonly, there is no cutting of standing trees and little if any cutting of shrubs (MVEIRB 2003). These small trail lines are used to connect heli-portable drill sites or for 3D receiver lines. In some cases, MIS lines are used for survey purposes (ASRD 2002). MIS operations not only avoid the cutting of important trees/vegetations, but also avoid the alteration of topsoil in the study sites.

Heliportable: Oil and gas exploration activities are not allowed in special environmental sites because of the negative environmental impacts (Khan and Islam, 2006a). In conventional seismic operations, construction of access roads for survey vehicles and the development of wider seismic lines are common. The main objective of heliportable seismic operations is to avoid the access of vehicles to the survey area. As a result, instead of developing wide access roads and lines only small trails are developed for geophone placement. The trail line is developed using small drills/heli-drills for shot holes. All equipment and personnel are brought to the site by helicopter. However, heliportable seismic programs still require a 1.5 m wide corridor along the seismic line and helicopter landing sites must be cleared every 2 km for safety reasons (MVEIRB 2003).

Heliportable seismic operation is expensive compared to conventional seismic operations, but it is environmentally more acceptable. It has minimal environmental impacts compared with other seismic operations.

3.2.2   Offshore seismic exploration

Seismic surveys are used by the offshore oil and gas industry to help determine the location of oil and gas deposits beneath the seafloor. Such surveys use large, specialized ships, which tow an array of powerful

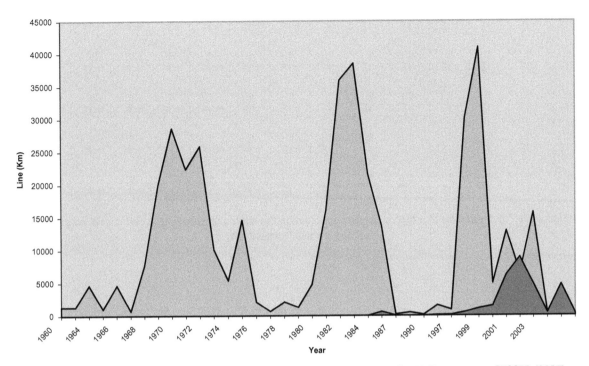

Figure 3-2   Amount of 2D and 3D seismic (line km) shooting on the Scotian Shelf. Data source: CNSOPB (2005).

airguns that generate sound waves by firing off explosive blasts of air. The sound waves are reflected off the seafloor and create a picture of underwater geological formations.

A typical seismic survey lasts 2–3 weeks and covers a range of about 300–600 miles. The intensity of sound waves produced by the firing of seismic airguns can reach up to 250 decibels (dB) near the source and can be as high as 117 dB over 20 miles away. The sound intensity produced by a jackhammer is around 120 dB, which can damage human ears in as little as 15 s.

Potential hydrocarbon reserves are extensively studied by seismic technology. The Scotian Shelf of Canada is one such example. On this shelf, the first seismic operation was initiated in 1960. By 2004, 400,034.33 km of 2D seismic tracks were completed. Since 1985, there have been 3D seismic explorations, and 29,511.86 km^2 has been completed (Figure 3-2). Marine seismic exploration uses the behavior of induced sound waves to map the sub-seafloor longitudinal sections (2D) or features (3D). Different seismic activities and their related effects are shown in Figure 3-1, the four cylinders representing the four major activities in this area, which are connected sequentially. They are towing airguns, blasting airguns, towing hydrophones, and chase boating. The arrows with numbers represent types of input and output or impact that the activities have.

The "towing airguns" activity is done by a boat towing an array of streamers with several airguns attached below the water surface. These release high-pressured air every 12 s, creating a sound pulse of generally 225 dB to produce seismic waves through the water. The airguns are focused at the seafloor, reflected by sub-seafloor features, and received by hydrophones attached every 12.5 m along streamers up to 6 km long. Typical run lines are 100 m to 1 km apart. The 3D survey needs more closely spaced lines that yield highly resolved spatial structures, with the need for interpolation. These operations are escorted by "chase boats" to protect the dragging unit from other boat traffic (Figure 3-3).

The following activities take place during a typical seismic exploration in an offshore operation:

1. A ship tows an array of airguns aiming the cylinders downward and sets off "blasts" that are created by the sudden release of compressed air.

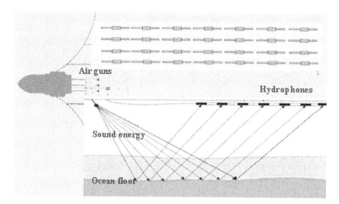

Figure 3.3   Pictorial view of a typical seismic operation in the marine environment, where a ship tows an array of airguns and hydrophones.

2. These blasts are set off every few seconds around the clock as the ship travels a specific grid.

3. The same ship tows an array of hydrophones. Sometimes another ship tows the hydrophones. Hydrophones detect and record how the sounds are reflected back from the ocean floor. Different types of rock and other substrate reflect sound differently.

4. Explosives were used when underwater seismic exploration first began, but they have been replaced by compressed air.

During the operation, air bubbles rising to the surface interfere with the received signal. To avoid this problem, multiple airguns of varying sizes are used. Generally, noises are created by other types of sound in the marine environment. Figure 3-4 shows the different possible types of sounds that can occur, as well as their intermittent and local effects. In this figure, the frequency of sound and their pressure density spectrum level are also shown. The seismic study involves complex technology, and every seismic project is different from another, depending on where it operates and the requirements of the survey.

3.2.3   Technology scale

Seismic surveys generally operate for a period of 5 to 20 days. During operations, activities run 24 hours/day. The area for each survey varies. Current leases in Nova Scotian waters range from 10,000 to 40,000 km^2, and survey areas may only cover a proportion of the lease area. A survey takes the form of a grid of lines, which may range from 20 to 1000 km in length, but typically about 100 km^2. A total seismic survey in the Scotian Shelf is presented in Figure 3-2. From 1960 to 2004, there were 400,034.33 km of 2D seismic tracking. The peak survey period in the Scotian Shelf was 1999 to 2000. Since the introduction of 3D seismic technology in 1985, there have been a total of 29,511.86 km^2 of surveying done on the Scotian Shelf. Van Daessel (1991) reported that the effects of 2D and 3D are similar and both physically damage, disturb, and even kill marine organisms (Jepson et al. 2003). The total cost of the seismic survey was about US $168 million between 1967 and 1997 (DFO 2003).

3.2.3.1 Underwater acoustics

In the exploration study, the noise/sound source is from an airgun, which generates a large amount of underwater noise traveling long distances through the water media. Marine animals are exposed to this noise/sound. Different organisms have different sound detecting capacities, with some having higher hearing capacity than others. Species with higher hearing capacity can detect the signals easily. There are many other

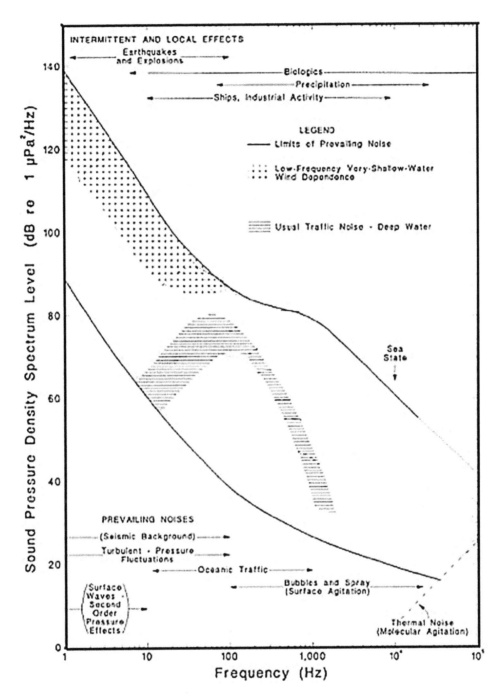

Figure 3-4 **Frequency level and sound pressure density spectrum of sea noise spectrum and different types of sounds in the marine environment. According to Wenz (1962); Kndudsen et al. (1944).**

underwater acoustics in the sea and detecting the airgun sign depends on the background noise. Figure 3-4 shows the frequency level and sound pressure density spectrum of sea noise and different types of sounds in the marine environment. Common background acoustic sources are ocean traffic, bubbles and spray, turbulent pressure, surface waves (second-order pressure effects), thermal noise (molecular agitation), earthquakes and explosions, precipitation, and shipping activities.

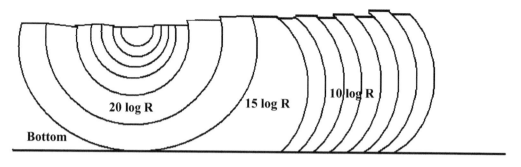

Figure 3-5   Spreading loss. Redrawn from CNSOPB (2005).

Humans hear sounds with a complex biological logarithmic process. The intensity of sound is denoting in decibels (dB), but in underwater acoustics, sound is expressed as Sound Pressure Level (SPL). SPL is equal to 20 log (P/P_0); P_0 is a reference level, usually 1 µPa. SPLs are usually expressed as received sound levels at the receiver location or the sound at the source. In the case of airguns, the source level is measured as the SPL at 1 m from the source. For example, it is indicated as 160 dB re 1 µPa at 1 m. The unit distance is necessary for comparing the source levels. Figure 3-5 shows the source of sound and its loss with distance.

For seismic exploration, an impulse is created by a burst of compressed air. It is composed of a positive pressure pulse followed by a negative pressure pulse. The difference in pressure between the highest positive pressure and the lowest negative pressure is the peak-to-peak pressure. The peak positive pressure, usually known as the peak or zero-to-peak pressure (O-P), is approximately half the peak-to-peak pressure. Thus, the difference between the two is approximately 6 dB. The average pressure recorded during the pressure pulse is described as the root mean square (RMS). In the case of seismic sounds, the RMS or average pressure is usually about 10 dB, which is lower than the peak pressure (Davis et al. 1998).

3.3   Exploration Techniques

At present, commonly used technologies for seismic surveys are 2D, 3D, and 4D. The major difference between these three surveys is the method of sampling. 3D data provide information about the sub-surface on a tight grid, usually 25 m in each direction, while 2D data provide information on every 25 m in one direction, but rarely less than every 1 km in the other, leaving it entirely to the imagination what happens between adjacent survey lines (Davis et al. 1998). Newly introduced 4D technologies consider the multiphase of the time factor in cases of water, oil, and gas movement in a reservoir. Details of these technologies are discussed in the following sections.

3.3.1   *Two-dimensional technology*

Two-dimensional seismic technology gathers only single directional data about the geological information of the studied area. This is because survey lines with airguns and hydrophones are placed about 1 km apart and laid out in a single direction. As a result, 2D survey data require further investigation by looking at other exploration techniques, such as drilling and 3D survey. The 2D technology is suitable for gathering preliminary information of large areas in a relatively inexpensive way. Figure 3-6 shows the increased preci-sion resulting from closer spacing of seismic lines. The 2D technology has less precision, because the space between two signals is much greater. However, 3D technology has greater precision.

3.3.2   Three-dimensional technology

Three-dimensional seismic imaging is a widely accepted technology. It provides more detailed geological information than is possible with 2D technology. By using 3D, extracted information gives a detailed "image" of the sub-surface to help make realistic estimates of the amount and distribution of hydrocarbons in a reservoir (Davis et al. 1998). Thus, 3D seismic imaging has replaced 2D imaging to improve the industry standard for mapping the extent and thickness of oil and gas reservoirs in three dimensions and to identify fluid-flow pathways and barriers within a reservoir (NETL 2005). Generally, a 3D survey is undertaken where hydrocarbons exist in economically viable quantities. 3D survey is expensive in comparison to 2D, but proves cost-effective to use prior to the design and construction of production facilities. It helps to develop better planning of the locations and numbers of production wells and platforms. Due to advances in offshore streamer technology, on-board processing and interpretation have improved the speed and reli-ability of data capture while reducing cost. At present, processing time has decreased dramatically, from 52 weeks to 4 weeks or less.

3.3.3   Four-dimensional technology

Four-dimensional seismic imaging is a newly introduced technology in seismic exploration that adds a time-lapse component to the process. It gives a snapshot offering a clearer picture of reservoir evolution and a better understanding of gas, oil, and water movement over time (Figure 3-6). The operational cost of this technology might be higher, but it decreases overall field development expenditures by allowing the operator to reduce the number of wells needed to access economically viable hydrocarbon deposits on various producible horizons. According to Lumley and Behrens (1998), soft unconsolidated sands or heavily fractured rocks are best for 4D seismic monitoring of fluids, whereas hard cemented sandstones and carbon-ates are difficult monitoring candidates. Figure 3-7 shows schematically that compressible, high-porosity rocks, like unconsolidated sandstone or heavily fractured rock, can be optimal for 4D seismic monitoring, whereas highly consolidated or cemented sandstones, or rigid carbonates, are difficult candidates for 4D seismic monitoring. In general, the rule of thumb is that the reservoir rocks must exhibit a minimum 4% impedance change during production, including all pressure, temperature, and saturation effects, to be con-sidered reasonable candidates for monitoring (Lumley and Behrens 1998).

Table 3-1 presents the relative cost of 4D seismic to well work. All values are in US dollars. The cost of 4D is higher than that of 3D, with 2D technology being the cheapest of all.

3.3.4   Mechanism of seismic technologies

In offshore seismic exploration, pressure waves (sound waves) are generated by using airguns. They simul-taneously and instantaneously release high-pressure air of 2000 psi. The rapid expulsion of compressed air causes the water around the airguns to accelerate rapidly outward, creating a pressure wave (sound wave), which serves as the seismic energy source.

A typical marine seismic airgun source (both 2D and 3D) has a volume of 3000–4000 m^3 in a peak-to-peak pressure output of ~262 dB – 1 μPa at 1 m (Davis et al. 1998). The array will typically be made up of 30–40 airguns which are towed by a ship. The airguns are set in such a way that they can release sound vertically rather than sideways. It is necessary to employ a source (i.e., sound) with this level of output in order to map sub-surface structures that may be 5000 m below the seafloor.

The released seismic sound waves travel both down through the Earth and then return, to be received by hydrophones. The layers of rock below the seafloor cause much of the energy to be absorbed. Thus a power-

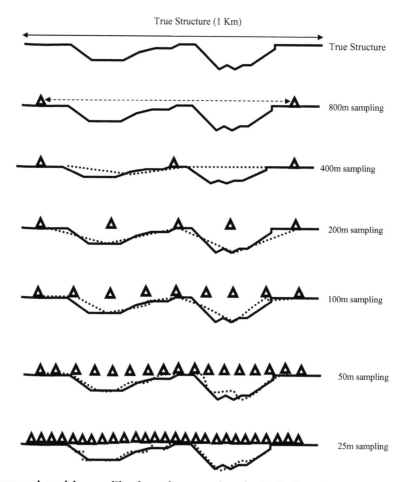

Figure 3-6   Increased precision resulting from closer spacing of seismic lines. By courtesy of CNSOPB (2005).

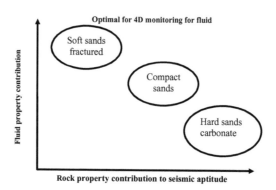

Figure 3.7   Seismic amplitudes are a coupled contribution of rock and fluid properties. Redrawn from Lumley and Behrens (1998).

Table 3-1   Relative cost (in thousands of US dollars) of 4D seismic to well work

Water	Indonesia	West Africa	North Sea	Gulf of Mexico shelf	Deep
Recompletion	50	1,500	600	500	–
Side Track	(70)	2,500	4,000	500	2,500
New Well	140	4,000	8,000	3,000	20,000
Seismic	400	1,000	1,000	1,000	1,000
Seismic acres	640	640*9	640*9	640*9	640*9
Number of wells per seismic area	80	36	20	36	5
Seismic cost per 10% well work	35	10	15	55	5

Source: Lumley and Behrens (1998)

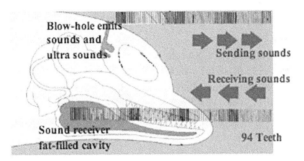

Figure 3-8   A pictorial view of dolphin's communication system that produces sounds in the ultrasonic frequency band, and received back by the fat-filled cavity.

ful source is necessary to ensure penetration to the desired depths. In addition, the towed hydrophones are in an environment with relatively high ambient noise levels, making it difficult to detect the desired geophysical information above the noise if the signal strength is too weak.

3.3.5   Nature-based dolphin's sonar technique

Dolphins live in dark or murky waters where visibility is poor. As a result, they rely on sound production and reception to navigate, communicate, hunt, and rescue. A dolphin's sonar communication technique has similar objects and functions to seismic surveys. In both cases, sound waves are used for geating an "image" of the target by producing, sending, receiving, and interpreting the sound waves. However, one is natural and the other is artificial. In this section, the detailed mechanism of dolphin communication is examined.

3.3.5.1   Generating ultrasound

Dolphins produce ultrasound by complex tissues situated in the nasal region. Figure 3-8 shows the blowhole responsible for the production of sounds and ultrasounds. The blowhole, the rounded region of a dolphin's forehead, consists of lipids. It also acts as an acoustical lens to focus these sound waves into a beam, which is projected forward into the water in front of the animal.

Sounds are generated by movements of air in the trachea and nasal sacs. Bottlenose dolphins produce clicks and sounds that resemble moans, trills, grunts, squeaks, and creaking doors. They also produce whistles. They can make these sounds at any time and at considerable depths. The sounds vary in volume, wavelength, frequency, and pattern.

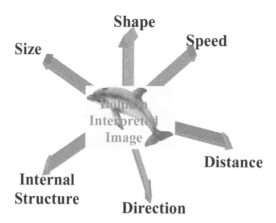

Figure 3-9   Six-directional sketch of ultrasound signals interpreted by a dolphin, where size, shape, speed, distance, direction, and internal structure of an object are determined.

The frequency of the sound produced by a bottlenose dolphin ranges from 0.25 to 200 kHz. The lower frequency vocalizations (about 0.25 to 50 kHz) are believed to be used in social communication. Social signals have their highest energy at frequencies less than 40 kHz. Higher frequency clicks (40 to 150 kHz) probably are used primarily in echolocation. The peak frequency of typical echolocation clicks is about 100 kHz, but frequency varies considerably with specific echolocation tasks.

3.3.5.2   Receiving ultrasound

The major areas of sound reception are the fat-filled cavities of the lower jaw bones. Sounds are received and conducted through the lower jaw to the middle ear, inner ear, and then to hearing centers in the brain via the auditory nerve. The brain receives the sound waves in the form of nerve impulses, which relay the messages of sound, and enable the dolphin to interpret the meaning.

3.3.5.3   Mechanism of sound interpretation

A dolphin has the ability to "see" with its ears by listening to echoes known as odontocetes. This process is echolocating by producing clicking sounds and then receiving and interpreting the resulting echo. The blowhole produces clicks (sound waves) in trains toward a targeted direction. Each click lasts about 50 to 128 microseconds (Au 1993). Sound waves travel through water at a speed of about 1.5 km/sec (0.9 mi/sec), which is 4.5 times faster than sound traveling through air. These sound waves bounce off objects in the water and return to the dolphin in the form of an echo.

The released signals reflect back to the dolphin and are received by the "receiver" fat-filled cavities of the lower jaw bones, and sent to the brain in the form of nerve impulses. The brain analyzes the signals to get a clear picture of the object. A dolphin interprets the received ultrasound signals in such a smart way that it can determine size, shape, speed, distance, direction, and even some of the internal structure of objects. However, many of the details of echolocation are not completely understood. Research on echolocation continues (Figure 3-9).

3.3.6   *Emerging Exploration Technologies*

Unconventional geological formations are often difficult to explore with existing technologies. For example, it is especially critical for "tight" (low-permeability), as naturally fractured gas reservoir exploration is playing an increasingly important role in the petroleum industry (NETL 2005). Detecting, predicting, and simulating these reservoirs' fracture systems are critical for realizing the potential of these tight gas reservoirs. It is necessary to find new technologies, which will enable the petroleum operator to produce a crisper, more coherent view of the subsurface. Such innovative imaging technologies will help operators to find hydrocarbons in complex geological settings and will help to produce more "stranded" oil and gas when optimizing infill drilling programs. In addition, new/emerging technologies are needed to improve drilling success, reduce risk, and to prevent many dry holes. Some of the emerging technologies for exploration operations are discussed.

3.3.6.1 Vibration

Airguns are the only practical source currently available for deep penetration marine seismic operations. Considering the detrimental impacts of high sound pressure, industries are looking for alternative technologies, especially alternative seismic sources. For very shallow site survey work, sparkers are also used. Some experimental work has been carried out using a marine vibrator, a source that delivers a similar amount of energy as an airgun array, but delivers the energy over a 5 or 6 s time window rather than instantaneously. This results in lower peak acoustic pressure and slower rate of change in pressure compared to airguns and, consequently, eliminates or, at least, reduces potential damage to marine life close to the source, such as fish larvae and juveniles (Davis et al. 1998).

3.3.6.2 Advanced seismic technologies

Advanced seismic technologies are required to obtain accurate information about subsurface geological information in order to avoid drilling risks and to increase resource recovery in oil and gas fields. They also help reduce economic and environmental costs. At present, exploration technologies are required to gather information from deeper and more complex reservoirs. There are also growing needs for more accurate, more durable, and more affordable seismic tools.

At present, 4D has replaced 3D and 2D for mapping the extent and thickness of oil and gas reservoirs, and for identifying fluid-flow and barriers within realtime. However, the seismic technology can be extended to include Fiber Optic MEMS-based seismic receivers for borehole applications. Moreover, improved algorithms for using seismic attributes, such as attenuation as a direct indicator of reservoir properties, can also be added. Figure 3.10 shows a packaged fiber optic MEMS pressure sensor.

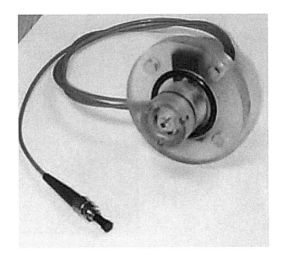

Figure 3-10   Packaged fiber optic MEMS pressure sensor (NETL 2005).

The existing seismic technology can be improved further for gathering more accurate information. For example, specialized tools, including multi-component seismic data acquisition and processing, seismic attribute analysis, wave equation migration, advanced AVO (amplitude variation with offset) analysis, borehole seismic imaging, and time-lapse imaging are being adopted in combination with advanced log analysis techniques to delineate oil and gas sweet spots, especially in complex reservoirs. The application of these advanced seismic tools and techniques has led to more effective targeting of wells in complex reservoirs, and to overall improvements in oil and gas recovery in existing fields.

Currently, it is reported that NETL (2005) have developed successful technologies for detecting fractured areas in tight gas reservoirs; for sub-salt imaging in oil fields; for reservoir characterization using seismic attenuation; and for improving seismic image quality and resolution in structurally complex oil and gas reservoirs to ensure a reliable, affordable, and environmentally sound supply of natural gas and oil to meet future demand. Advanced seismic data analysis offers a huge potential impact in helping to monitor reservoir production processes in a volumetric sense, which is crucial for optimal reservoir management (Lumley and Behrens 1998).

3.3.6.3 Natural gas production from tight gas accumulations

From different studies it is proved that natural gas production from tight gas has great potential in the many places, especially in the USA. It is mainly in the case of rocks with less than one tenth of a millidarcy permeability. However, the production of natural gas from tight gas is not well developed yet. Even, the accumulation of tight gas is little understood. To exploit this gas advanced technologies are needed, such as advanced detection and prediction of tight gas reservoir "sweet spots" (NETL 2005). Figure 3.11 shows microscopic sections of sandstone. The pores of tight gas sandstone (right) are irregularly distributed and poorly connected by very narrow capillaries. Because of this low connectivity (low permeability), gas trapped within tight gas sandstones is not easily produced. NETL (2005) has developed advanced remote sensing technologies such as, multi-component, multi-azimuth seismic surveys, which aim to detect fractures and quantify their density and orientation prior to drilling. Other techniques are advanced geomechanical analyses and computer based simulations.

3.3.6.4 Water detection and prediction

Identification of production sweet-spots in the basin-centered, low-permeability reservoirs does not guarantee the production of gas. Sometimes, these sweet-spots end up only with water. Therefore, it is necessary

Figure 3-11   Microscopic sections of sandstone. Conventional sandstone (left) has well-connected pores (dark blue). Photo courtesy of NETL (2005).

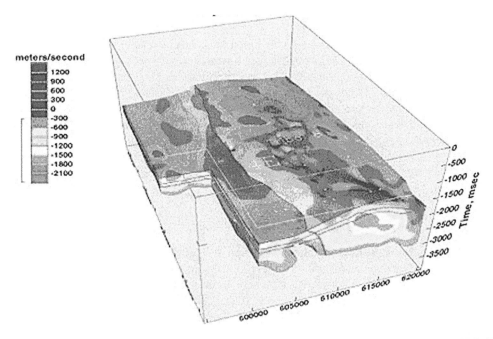

Figure 3-12   Anomalous Velocity Model, Riverton Dome Project top of frontier. Photo courtesy of NETL (2005).

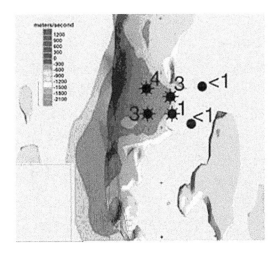

Figure 3-13   Anomalous Velocity Model, Riverton Dome Project top of frontier. Photo courtesy of NETL (2005).

to confirm whether the reservoir will produce gas or water. NETL (2005) recently reported a method for testing the validity of an exploration rationale for basin-centered gas that does not rely on fracture detection. It maps the 3D structure of the boundary between normal and anomalous pressures combined with prediction of areas of enhanced secondary porosity and permeability related to diagenetic or other events (NETL 2005). Figure 3-12 shows the Anomalous Velocity Model, Riverton Dome Project. The main objective of this is to determine whether these "sweet spots" are likely to contain gas or water. Two 3D seismic surveys from the Anomalous Velocity Model, Riverton Dome Project area were analyzed, and presented in Figures 3-12 and 3-13. From these the areas of anomalously-slow velocity were detected (NETL 2005). Comparing these two figures, the anomalies suggest where significant volumes of gas may be present.

3.3.6.5   Natural fracture detection

Current natural fracture detection technologies are based on advanced seismic data collection and analysis. The collected information helps to identify subsurface fractures prior to drilling. At present, obtaining a 3D subsurface is common in the exploration technique. In seismic exploration techniques, two types of returning waves are analyzed, which are know as pressure-waves and shear-waves. Pressure-waves or P-waves are the conventional data source where individual particles oscillate in the direction the wave is moving. Figure 3.14 presents the P-wave seismic velocity ratios. Shear waves or S-waves are expensive to collect, but represent more important information. In the S-waves particles oscillate perpendicular to the direction of wave motion. To overcome the current limitation and achieve better fracture detection NETL (2005) is developing improved multi-azimuth 3D seismic data collection and analysis techniques that may provide fracture detection capabilities to those without 3D seismic capability.

3.3.6.6   Natural fracture prediction

After successfully collecting natural fracture information it is necessary to discern the likely location, orientation, and relative aperture of natural fractures from the structural configuration of an area and the mechanical qualities of the reservoir rock (NETL 2005). Advanced technologies are needed to make a better prediction of natural fractures. To achieve this goal NETL (2005) proposed the "geomechanical approach" that can augment or be augmented by direct detection methods, or it can provide a cost-effective alternative where advanced seismic survey is not feasible. Figure 3-15 shows the geomechanical prediction of fault-related stress fracturing. The geomechanical prediction is better than the current conventional approach and its predictions are based on the mapping of fault systems within the reservoir from existing seismic data (NETL 2005).

3.3.6.7   Natural fracture simulation

Finally, better simulation technologies are required for analyzing natural fracture information. Recently, NETL (2005) developed a new fractured reservoir simulation model which is known as FRACGEN/

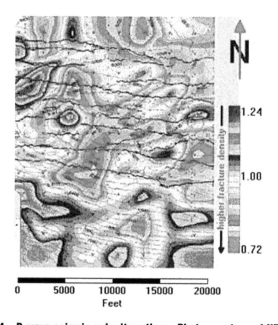

Figure 3-14   P-wave seismic velocity rations. Photo courtesy of NETL (2005).

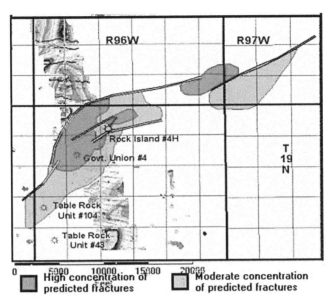

Figure 3-15 Geomechanical model prediction of fault-related stress fracturing. Photo courtesy of NTEL, 2005).

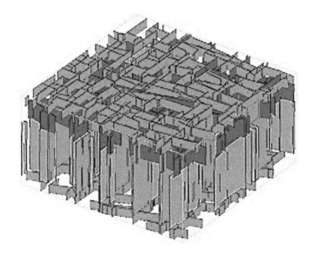

Figure 3-16 Image of multilayer fracture network. Photo courtesy of NTEL (2005).

NFFLOW. This newly developed model consists of two key components which are FRACGEN and NFFLOW. FRACGEN or fracture network generators provides detailed two-dimensional or three-dimensional representations of reservoir fracture networks. Figure 3-16 shows an image of a multilayer fracture network which is generated using FRACGEN model. Then, NFFLOW estimates the interaction of the fractures with the rock matrix and simulates the flow of gas through the fracture network to one or more boreholes. FRACGEN/NFFLOW significantly improves fracture simulation. Figure 3-17 is an image of drawdown in a fractured-reservoir produced by a horizontal well. This image is generated as a FRACGEN-NFFLOW model output.

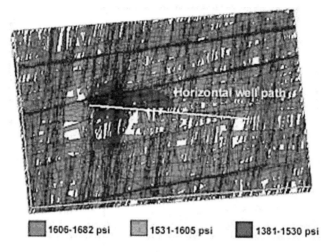

1606-1682 psi   1531-1605 psi   1381-1530 psi

Figure 3-17   Pressure drawdown in fractured reservior produced by a horizoned well.

3.4   Problems with Current Exploration Techniques

Seismic exploration is examined for the preliminary investigation of geological information in the study area and is considered to be the safest among all other activities in petroleum operations, with little or negligible negative impacts on the environment (Diviacco 2005; Davis et al. 1998). However, recent studies have shown that it may have adverse environmental impacts (Jepson et al. 2003; Khan and Islam 2007). Most of the negative effects are from the intense sound generated during the survey. Detailed life cycles of seismic exploration were studied and reported by Khan and Islam (2007) and Khan et al. (2006c). Figure 3-1 shows the different activities of seismic exploration and their associated impacts on the ecosystem.

Table 3-2 identifies the detailed seismic activities and their impact on wildlife, species at risk, special habitat, and other ecological components. Impacts of seismic exploration on organisms vary from species to species. For example, dolphins, whales, and seals use their sense of hearing to locate prey, avoid predators, choose migration routes, and to communicate over long distances. The noise associated with seismic surveys can affect the ability of these animals to detect natural underwater sounds, thereby disrupting these critical activities. Different research studies have observed signs of physical stress, such as startle responses in humpback whales, while seismic surveys were being conducted many miles away. Figure 3-18 shows impacts of seismic noise on whales at various distances.

Terrestrial seismic studies and related deforestation have major impacts on wildlife and clearing forest. Based on the operation areas, the impacts on wildlife can be very harmful. Table 3-2 shows the impact of exploration operations on wildlife as higher than in other ecosystem components. Forests are needed to be cut clear for the developing seismic operations. In addition, offshore exploration operations also have direct or indirect impacts on fisheries and important aquatic species. For understanding the effects of current seismic exploration operations, different aspects of their impacts are discussed in the following sections.

3.4.1   Effects on fish

Seismic operations and their impacts on fish is an issue of concern for fisherman and fisheries biologist. There is not enough information of direct impact of seismic operation on fisheries, but it is generally consider that the powerful sound waves generated by seismic surveys might have adverse impacts on fish. According to AMCC (2005) at close range, seismic surveys have been found to kill adult fish as well as larvae and fish eggs. Seismic sound can cause direct and indirect physical and behavioral effects on fish. Some studies

Table 3-2   Potential impacts of exploration operations on wildlife, species at risk, special habitat, forest, fisheries, and fish farming

Activities	Potential Impact Levels on VECs					
	Wildlife	Species at Risk	Special habitat	Forest	Fisheries	Aqua-culture
Land survey	2	2	2	1	0	1
Aerial survey	2	2	2	0	0	2
Seismic trail clearing	2	3	3	5	3	3
Seismic wave production	7	7	2	1	5	7
Clearings vegetation	5	7	7	6	6	3
Road Construction	4	5	6	3	3	1
Mobilization of truck/equip.	3	4	3	1	1	1
Site development	5	5	5	3	3	0
Drill pad construction	4	4	5	3	2	1
Accidents	7	10	10	7	8	9
Secondary recovery	5	5	7	5	3	3
Airstrip/helipads	3	3	4	2	1	2
Worker accommodations	4	4	4	1	1	1
Increase in local population	5	5	7	3	3	1

Source: Khan and Islam (2007)

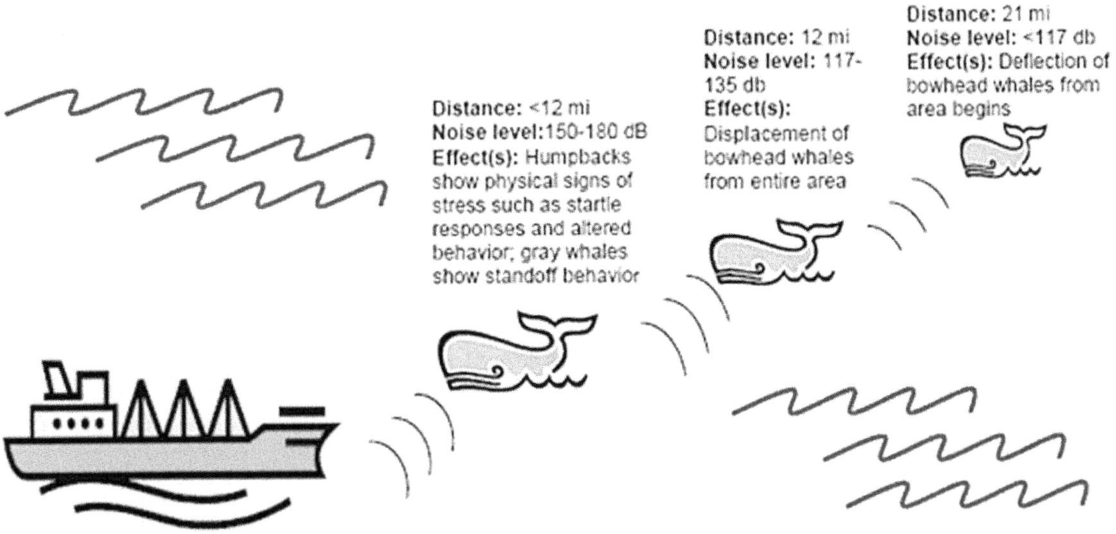

Figure 3-18   Observed impacts of seismic testing on whales at various distances and sound levels. Photo courtesy AMCC (2005).

show that exposure to seismic sound can cause a variety of sublethal impacts on fish, such as damaging orientation systems and reducing their ability to find food (AMCC 2005). Due to higher locomotive capacity, fish tend to avoid the seismic operation site. As a result, the migratory activities of fish can be affected severely in a way that will also directly affect the breeding activities. However, these problems can be overcome by avoiding seismic activity during breeding and migratory seasons.

Seismic surveys can cause direct physical damage to a fish. High pressure sound waves can damage the hearing system, swim bladders, and other tissues/systems. These effects might not directly kill the fish, but they may lead to reduced fitness, which increases their susceptibility to predation and decreases their ability to carry out important life processes (AMCC 2005). There might be indirect effects from seismic operations. If the seismic operation disturbs the food chain/web, then it will cause adverse impacts on fish and total fisheries. The physical and behavioral effects on fish from seismic operations are discussed in the following sections.

3.4.1.1   Physical effects

Although direct fish killings have been reported by oil spill accidents, there are no documented cases of fish mortality upon exposure to seismic sound under field operating conditions (DFO 2003b). However, it has also been argued that the efficiency of detecting fish kills by the follow-on vessels was not tested independently, so the possibility of undetected fish kills cannot be eliminated (DFO 2003b; CNSOPB 2004).

There are limited scientific studies that have been conducted in the case of exploration operations. Under laboratory conditions, one study reported that some subjects from three of four species tested suffered lethal effects from low-frequency (<500 Hz) tonal sounds, under exposure levels of 24 h at >170 dB (DFO 2003b). These experimental conditions differ greatly from field operating conditions of seismic surveys. Some of the researchers argued that the result indicates that risk of direct fish mortality from sounds, with some characteristics of seismic sound, cannot be discounted completely. However, exposure to seismic sound is considered unlikely to result in direct fish mortality (DFO2003a, 2003b).

Under laboratory conditions, sub-lethal and/or physiological effects, including effects on hearing, have sometimes been observed in fish exposed to an airgun. The experimental design made it impossible to determine, to the satisfaction of all experts, what intensity of sound was responsible for the damage to ear structures, nor the biological significance of the damage was observed (DFO 2003b). Simulated field experiments attempting to study such effects have been inconclusive. Currently, only inadequate information is available to evaluate the likelihood of sub-lethal or physiological effects of seismic sounds under field operating conditions (DFO 2003b).

3.4.1.2   Behavioral effects

According to DFO (2003b) there is high likelihood of obtaining the behavioral effects in some fish exposed to seismic sound. For example, startle response, change in swimming patterns (potentially including change in swimming speed and directional orientation), and change in vertical distribution are some of the effects. These effects are expected to be short-term, with duration of effect less than or equal to the duration of exposure, they are expected to vary between species and individuals, and be dependent on properties of received sound. The ecological significance of such effects is expected to be low, except where they influence reproductive activity.

Several studies have investigated other behavioral effects on fish during seismic surveys. Some have found the change in horizontal distribution of fish not closely associated with habitat structures such as a reefs or pinnacles. According to DFO (2003b) the change in catchability of fish is possibly related to changes in behavior. The duration of these effects may or may not extend beyond the duration of exposure, are expected to vary between species and individuals, and be dependant on the properties of received sound (DFO 2004a, 2004b).

3.4.2   *Effects on fishing efforts*

From the earlier discussion in the previous section it is evident that seismic operations have negative impacts on fish. Some fisheries experts and fisheries conservation councils assume that exploration operations can threaten commercial and subsistence fishing by harming fish resources, but also by interfering with fishing

Table 3-3   Reductions in fish catch rates as a result of seismic survey activity

Species	Gear type	Noise level seismic testing	Catch reduction	Source
Atlantic cod (*Gadus morhua*)	Trawl	250 decibels (dB)	46–69% lasting at least 5 days	Engas et al. 1993
Atlantic cod (*Gadus morhua*)	Longline	250 dB	17–45% lasting at least 5 days	Engas et al. 1993
Atlantic cod (*Gadus morhua*)	Longline	Undetermined, 9.32 miles from source	55–79% lasting at least 24 hours	Lokkeborg and Soldal 1993
Haddock (*Melanogrammus aeglefinus*)	Trawl	250 dB	70–72% lasting at least 5 days	Engas et al. 1993
Haddock (*Melanogrammus aeglefinus*)	Longline	250 dB	49–73% lasting at least 5 days	Engas et al. 1993
Rockfish (*Sebastes* spp.)	Longline	223 dB	52%–effect period not determined	Skalski et al. 1992

Source: FBB (2005)

operations and dramatically affecting catch rates (AMCC 2005). During seismic operations a long array of airguns and receivers are dragged in the survey site. Detailed pictures of offshore seismic operations and activities are pictorially presented in Figure 3-3. This line is longer than 1000 meters and there is a risk of tangling up with some fishing gear, such as crab pots, set nets and trawl nets (AMCC 2005). Table 3-3 shows the impacts of seismic survey on fishing. It shows that seismic survey reduces fishing.

3.4.3   Effects on planktonic organisms

Few studies of the effects of seismic sound on eggs and larvae or on zooplankton were found. However, data are generally insufficient to evaluate the potential damage to eggs and larvae of fish and shellfish and other planktonic organisms that might be caused by seismic sound under field operating conditions.

From the experiments reported to date, results do show that exposure to sound may arrest development of eggs, and cause developmental anomalies in a small proportion of exposed eggs and/or larvae; however these results occurred at numbers of exposures much higher than are likely to occur during field operation conditions, and at sound intensities that only occur within a few meters of the sound source. In general, the magnitude of mortality of eggs or larvae that could result from exposure to seismic sound predicted by models would be far below that which would be expected to affect populations. However, special life history characteristics such as extreme patchiness in distribution and timing of key life history events in relation to the duration and coverage of seismic surveys may require case by case assessment. Khan and Islam (2007) reported the impacts or potential impacts of offshore oil and gas operations on planktonic organisms, but there is very limited published information available to investigate the role of seismic sounds in recruitment variation of marine fish or invertebrates.

3.4.4   Effects on invertebrates

3.4.4.1   Physical effects

Most of invertebrates are slow moving organisms and it is considered that they are the main victims of the seismic operations. However, according to DFO (2003b) there are no documented cases of invertebrate mortality upon exposure to seismic sound under field operating conditions. A few studies reported some big invertebrates are affected during the seismic operation. For example, there is an uncorroborated report of beaching of giant squid on two occasions (DFO 2003b). To investigate the potential impacts of offshore seismic activities many laboratory experiments were carried out (DFO 2003a, 2003b). In controlled conditions some effects, such as lethal and/or sub-lethal effects, including effects on external structure, have sometimes been observed in invertebrates exposed close (less than 5 m) to an airgun. Considering the overall situation DFO (2003b) the report concluded that exposure to seismic sound is considered *unlikely* to result in direct invertebrate mortality.

3.4.4.2 Physiological effects

It is reported by DFO (2003a, 2003b) that many studies have been conducted to investigate the physiological effects of seismic operations. However, there is no clear evidence from these studies. In laboratory conditions, it was found that non-seismic sounds have direct impacts on the physiology of crustacean (DFO 2003b). However, in the case of gastropods, direct impacts are reported from seismic operations. The major effects on gastropods were on growth, reproduction, and behavioral changes. Direct physiological changes due to seismic sounds were reported in case of gastropods. However, in other invertebrates such physiological effects were rarely present, except for some sign of excitation of ensonified crabs compared to control crabs (DFO 2003b). There is no information available to evaluate the sub-lethal or lethal effects due to seismic operations.

3.4.4.3 Behavioral effects

Compared to other mammals or fish species the locomotive capacity of invertebrates is limited. As a result, the direct effects from seismic operations are higher than other organisms with higher locomotive capacity. For example, fish and dolphin can escape the operation area quickly. There is very limited data about the seismic operation on the behavioral effects on invertebrates. However, it is unlikely that some behavioral activities, such as startle response and change in swimming or movement patterns in some species can be changed. DFO (2003b) reported that both increases and decreases in catch rates of commercially exploited species have been documented, but changes do not occur consistently. They also mentioned that these effects are expected to be short-term, with duration of effect often less than the duration of exposure, and they are expected to vary between species and individuals, and with biological characteristics. In addition, the properties of the received sound plays an important role on the magnitude of effects.

3.4.5   Effects on marine turtles

The turtle is an endangered species all around the world. Any negative impact from seismic sound will further increase the risk level of this organism. In addition, the lifecycle of the turtle is very long and any adverse impacts on it will hamper the juvenile recruitment of this organism. This organism is a slow moving organism and cannot escape the hazard site immediately. As a result, there are many chances of adverse impacts on this species from seismic sound. According to DFO (2003b) sea turtles, specifically loggerhead and green turtles, are able to hear and respond to low frequency sound. Their activities, such as increased swimming speed, increased activity, and change in swimming direction, and avoidance are reported in other studies (DFO 2003a, 2003b; CNSOPB 2004).

3.4.6   Effects on marine mammals

Marine mammals depend on their hearing instead of visual systems. Marine mammals, such as dolphins, whales and seals utilize their sense of hearing to find prey, avoid predators, and find migratory paths. These activities can be hampered by seismic sounds. During offshore seismic operations the frequencies of emitted sound pulses range from 10 to 300 Hz and 215 to 250 dB. According to DFO (2003b) there are no documented cases of marine mammal mortality upon exposure to oil and gas exploration seismic surveys. This report also mentioned that under experimental conditions, sub-lethal, temporary elevations in hearing thresholds (TTS) have sometimes been observed in captive marine mammals exposed to pulsed sounds. Figure 3-

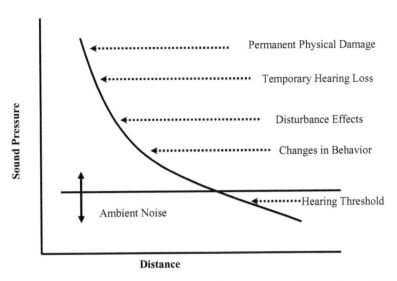

Figure 3-19   Pictorial view of different effect levels of sound pressure over distance traveled (Davis et al. 1998).

19 shows how different levels of sound pressure affect organisms. In this figure the hearing threshold is shown by the horizontal solid line. If the sound pressure crosses the hearing threshold, then different levels of effects might take place. These includes: changes in behavior, disturbance effects, temporary hearing loss, and permanent physical damage. The potential effects of seismic operations are elaborated in the following sections based on the DFO (2003b) study. The major effects are direct behavioral effects and functional consequences of physical and behavioral effects.

3.4.6.1 Direct behavioral effects

Displacement and migratory diversion: According to DFO (2003b) exposure to seismic sound can result in displacement and/or migratory diversion in some marine mammals, but this effect is species-, individual-, and contextually-related. The marine mammals avoid the seismic operation site during the operation period. The displacement and migratory diversion occurs during the operation period. If the operation period is longer, then the effects are higher.

Changes in dive and respiratory patterns: Some marine mammals, such as bowhead whales and harbour and grey seals show changes in dive and respiratory patterns if they are exposed to seismic sound. The effects might last longer than the exposure period. However, in an earlier section, it was observed that the diversion or displacement only happens during the operations period. The changes in dive and respiratory patterns might interfere with feeding as well as substantial loss of energy due to hyperactivity. According to DFO (2003b) changes in dive and respiratory patterns are expected to vary with species, individual, and context.

Changes in social behavior: Most marine mammals have important social behavior to sustain their lives. Common social behaviors of marine mammals are mating, cooperative feeding, play, aggressive interactions, and communication. According to DFO (2003b) the changes in social behavior due to exposure to seismic sound in marine mammals is unknown, but if it were to occur there are conditions under which the worst-case consequences of such changes could be highly significant.

Changes in vocalization patterns: Due to seismic operations, changes in vocalization patterns can happen to marine mammals. For example, the loss of contact between individuals or reduced ability to coordinate social behaviors. In many studies it is reported that exposure to seismic sound can result in changes in marine mammal vocal behavior, and when it occurs there are conditions under which the worst-case consequences could be highly significant (DFO 2003b). However, it is hard to quantify the degree of impact due to changes in vocalization patterns.

3.4.6.2   Functional consequences of physical and behavioral effects

Reduced communication efficiency: Marine mammals, such as dolphins communicate with each other by using ultrasound signals. In an earlier section and Figures 3-8 and 3-9 the mechanism of dolphin communication is described. If this communication is hampered, then there will be great impacts on their activities, such as feeding, breeding, parental care, predator avoidance, or maintenance of social groupings. According to DFO (2003b) there have been no published studies of the potential for seismic sound to reduce the efficiency of communication in marine mammals. As a result, the impacts on communication efficiency of marine mammals due to exposure to seismic sound are unknown.

Reduced echolocation efficiency: Marine mammals are capable of using echolocation in dark, turbid ocean water. This mechanism of echolocation is presented in Figure 3-8 and 3-9. According to DFO (2003b) there have been no direct studies of the potential for seismic sound to reduce the efficiency of echolocation in marine mammals. Therefore, it is *unknown* if exposure to seismic sound can result in reduced echolocation efficiency in marine mammals (DFO 2003b).

Hampered passive acoustic detection of predators and prey: Marine mammals, such as dolphins detect the presence of predators by their acoustic sensor system. The mechanism of dolphin acoustic detection is presented in Figures 3-8 and 3-9. Seismic sound might hamper this system. According to DFO (2003b) there have been no direct studies of the potential for seismic sound to hamper the passive acoustic detection of predators by marine mammals. Therefore, it is *unknown* whether exposure to seismic sound could have any adverse impacts on the passive acoustic predator detection of marine mammals.

The mechanism of prey detection is similar to predator detection (Figures 3-8 and 3-9). Similarly, we can conclude that the impact of seismic sound on prey detection of marine mammals is unknown.

Hampered avoidance of anthropogenic threats (such as ship strikes, net entanglement): Due to different human activities many accidents, such as ship strikes and net entanglements happen. According to DFO (2003b) other types of sounds interfere with the ability of individual whales to avoid anthropogenic threats such as ship strikes and net entanglements, but it is not known how widespread this response is.

Hampered parental care or bonding: Parental care is very important for higher level species such as mammals. If there are any adverse impacts on the parental care, then the population of that species will be decreased significantly. According to DFO (2003b) there have been no direct studies of the potential for seismic sound to hamper parental care or bonding in marine mammals.

3.5   Sustainability of Current Exploration Techniques

3.5.1   Analyzing sustainability

In this study, the sustainability status of seismic exploration technologies is evaluated based on the methods described by Khan et al. (2005d). The details of the steps are shown in Figure 3-20. The flow chart shows that a supposedly sustainable technology will be tested according to a set of inter-related criteria. According to the model, the first evaluative step is to determine whether a technology is time-tested. If this criterion is met, then the technology will be defined as potentially truly sustainable. If the technology is not suitable for an infinite time, then it cannot be accepted. According to the model, time is the prime criterion for the selection of any technology to the sustainable category (Khan et al. 2005c, 2006; Khan and Islam 2005d).

If the studied technology passes the "time" test, then it will go through the next sorting process, based on the secondary criteria. The first set of secondary criteria consists of environmental variants. If these

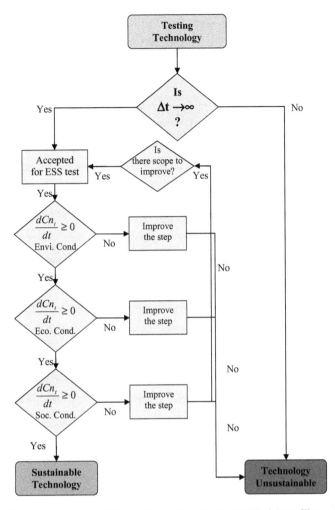

Figure 3-20 Proposed sustainability analysis flow chart. Modified from Khan et al. (2005c).

criteria are satisfied, there is another level of testing. However, if it fails to satisfy the environmental factors, it might be rejected, or suggestions may be made for its improvement. After the environmental evaluation, the next two steps involve technological feasibility, and regulatory and cost-effectiveness. Both of these steps follow a pathway similar to the environmental one. The same criteria for rejection or improvement apply.

3.5.2 Scoring and data analysis

At this stage, direct weight is assigned to all environmental, technical, cost-effectiveness, and regulatory indicators. Total weights of 100 points are explicitly distributed to all indicators based on their degree of importance. The scores of all indicators must satisfy the following requirements:

$$0 \leq S_{ji} \leq 100 \quad and \quad \sum S_{ji} = 100 \qquad (3.1)$$

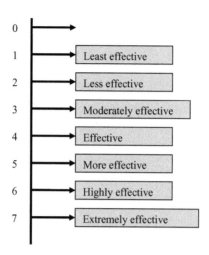

Figure 3-21   Semantic scale of degree of importance.

where j indicates criterion, i indicates indicator, and S_{ji} is the score for indicator i of j criterion. The sum of all indictors will be less than or equal to 100.

The selected sustainability indicators are evaluated in terms of their degree of importance, simply by ranking each indicator following a modified semantic scale (weight) of multi-criteria analysis (MCA) (Khan and Islam 2005d). This semantic scale employs a 7-point scale. Figure 3-21 shows the detailed weighting/ordering and their degree of importance. In the weighing scaling, most studies adopt a 5-point scale. In this study, a 7-point scale is adopted so as to get a more realistic result (Mendoza and Prabhu 2003; Khan and Islam 2005b; Adrianto et al., 2005). Based on this ranking, relative weight on an indicator can be estimated using the following formula (Mendoza and Prabhu 2003; Khan and Islam 2005b; Adrianto et al. 2005):

$$Rw_j = \frac{\bar{a}_j}{\sum RoP_j} \tag{3.2}$$

where Rw_j is the relative weight of indicator j, $\bar{a}_j$ is the average of responses of participants for indicator j, and $\Sigma SRoP_j$ is the summation of all of the Responses of Participants for indicators j.

Calculated scores are normalized according to Equation 3.3. The main purpose of normalizing data is to give a standard form.

$$Overall\ Score\ (in\ 0-7) = \frac{(Quantitative\ Datavalue - Minimum\ Value) \times 7}{Maximum\ Value - Minimum\ Value} \tag{3.3}$$

To determine the relative weight of each set of criteria, each technology is judged against the current condition relative to the perceived target or desired condition (Khan and Islam 2005c). MCA is also used for judging these indicators where a 5-point semantic scale is adopted, and where 1 means excrementally weak performance and strongly unfavorable; 2 means poor performance, unfavorable and major improvement needed; 3 means acceptable; 4 means very favorable performance; and 5 means state of the art performance, clearly understanding performance. In this scale, the numerical values of the score are shown with their related meaning. Following the Mandoza and Prabhu (2003) and Khan and Islam (2005c) methods, the sustainability index of criteria (SIC) is calculated by using following Equation 3.4, as follows:

$$SIC = \sum S_j Rw_j \tag{3.4}$$

SIC is a sustainability index of criteria i (environmental, technological, socio-economy, community, and policy). S_j is the score of indicator j and Rw_j is the relative weight of indicator j (Equation (3.1)).

Overall value of environmental, technological, regulatory, and cost criteria are estimated based on the following equations:

Environmental, $V_{ei} = \sum_{i=1}^{p} W_{ei}S_{ei}$, Technology, $V_{ti} = \sum_{i=1}^{q} W_{ei}S_{ei}$, Regulatory, $V_{ri} = \sum_{i=1}^{s} W_{ri}S_{ri}$, and Cost,

$V_{ci} = \sum_{i=1}^{s} W_{ci}S_{ci}$.

$$\text{Overall Value, } V(X) = V_{ci} + V_{ti} + V_{ri} + V_{ci}$$

$$= \sum_{i=1}^{p} W_{ei}S_{ei} + \sum_{i=1}^{q} W_{ei}S_{ei} + \sum_{i=1}^{s} W_{ri}S_{ri} + \sum_{i=1}^{s} W_{ci}S_{ci} \tag{3.5}$$

After scoring, all of the indicators' overall value for each criterion is estimated based on the Equation 3.4. Where V is overall value of a criteria; W is weight indicator i; and X is the score under i th criterion, n is the total number of criteria. This value is determined by the overall status of a technology in respect to environmental, technological and regulatory requirements, and cost effectiveness.

3.5.3 Technology evaluation

For this study, information about different seismic technologies was gathered by reviewing available published articles, reports, and guidelines from different sources. To evaluate the technological feasibility, cost effectiveness, and regulatory requirements of different technologies, a survey was undertaken. In the survey, experts such as petroleum engineers, geologists, lawyers, fishers, and environmentalists, were interviewed.

In this study, seven different technologies are evaluated. Table 3-4 shows the evaluated technologies and their brief overview. Out of seven, three are commonly used, namely 2D, 3D, and 4D. Among these, 2D and 3D have been accepted worldwide. The explosive technique (Table 3-4) is used in seismic exploration, but is not permitted by many countries because of its greater environmental impacts. The 4D technology is a newly introduced technology, which is reportedly used on a smaller scale. To compare with the presently available technologies, emerging technologies, such as vibration, are also studied. This technology is still at the developmental stage. Considering the same objective of seismic exploration, drilling technologies are also studied. This helps us to understand the qualitative and quantitative differences in respect to the sustainability of the technologies. The natural communication system of the dolphin is also studied for comparative

Table 3-4   Functional definitions of existing seismic technology

No	Name of Technology	Operational Definition
T1	Uses of explosive	It was a widely used technology, but at present it is not much used in view of its huge environmental impacts.
T2	2D	It is used for getting preliminary information in an inexpensive way. It does not give real picture of geological formation.
T3	3D	Technologically efficient, but very expensive to operate and data manipulation.
T4	4D	Cost effectiveness of the technology is studied.
T5	Drilling	Drilling is considered an exploratory technique to obtain geological information in the study area.
T6	Vibration	It is an emerging technology which, it is hoped, will reduce ecological effects
T7	Natural-dolphin	Dolphins use this method to explore and communicate in the aquatic environment

purpose to facilitate understanding the difference between artificial and natural technologies, as dolphins also use ultrasound to receive images similar to seismic technologies, with no adverse impact on Nature.

Sustainable technology development is a mechanism that paves the way for developing inherently sustainable technologies. This concept was recently introduced by Khan and Islam (2005c) and Khan et al. (2005c, 2006a). Sustainability analysis, especially in seismic explorations, is a particular type of process that can be determined by analyzing certain criteria of sustainability and evolutionary development, which can be clearly specified (Khan and Islam 2005b). However, limited research has been done in the case of offshore seismic technology evaluation. In this study, diverse sets of sustainability indicators were selected based on their importance judged by a group of stakeholders.

Based on the nature of operations, environmental consequences, and the importance of offshore oil and gas development in modern society, four major criteria were selected for each technology. They are environmental criterion, technological criterion, regulatory criterion, and cost criterion. Generally, most research in sustainability analysis considers another four major sets of variables, namely environmental, societal, economic, and institutional (Khan and Islam 2005). Instead, in this study, technological, regulatory and cost effectiveness criteria are selected for the seismic technological evaluation. The reason for selecting the other criteria is to evaluate a technology within a theoretical perspective and also to gain a practical/realistic overview.

Considering the success of the participatory approach in natural resource management (Matta and Alavala-pati 2005), the observations of various people are gathered in this study. To obtain a representative and standard weight to each indicator, a representative group of 15 people was selected from different backgrounds. The participants range from highly skilled petroleum engineers to fishermen. The selected indicators and variants for evaluating the technologies are shown in Table 3-5. The environmental criteria have 16 indicators. Technological criteria consist of 6 indicators. The regulatory and cost evaluation criteria have 4 and 2 indicators, respectively. At the beginning, each technology is described to participants and their potential consequences are also discussed. The functional definition of each indicator is also given to the participants.

Participants were asked to give a score to each indicator for the different technologies. Table 3-6 shows the average weight of each indicator for the different technologies. The values shown in the table are the average weight of participants' judgment, as well as direct assigned weights. Some studies only rely on participants' weight (such as Mendoza and Prabhu 2003; Khan and Islam 2005c; Adrianto et al. 2005) and some studies are based on direct weighting as argument (such as Thanyamanta 2004; Zatzman et al. 2005a). This study combines participants' opinions, and direct weight in order to create an overall better indicator set. Thus, this combination gives robustness to the weighting methods.

To determine the acceptability of a certain seismic technology, participants were asked to give their opinion as to whether that technology should be selected or rejected. The detailed results of interview processes are shown in Table 3-6. The 3D method and "dolphin" scored the highest regarding acceptability (16/16), while drilling scored the lowest. 2D and 4D got similar acceptability scores (Figure 3-22). Dolphin communication scored the second highest just after 3D. The reason for this was that some participants thought this technology (dolphin) is not functional or really achievable.

The total weight for each technology was calculated based on participants' weighting, and direct additive values. The total weight for each technology is shown in Figure 3-23. Dolphin communication scored the highest while the use of explosives got the lowest score. 2D, 3D, and 4D received similar weight values. Drilling got the second lowest score, and this may be due to its environmental effects and poor cost-effectiveness. Due to the ban on the use of explosives, it did not get any score in the regulatory criterion.

Overall values for each technology are estimated using Equation 3.5. These values are shown in Figure 3-24. According to these values, the best technologies for seismic exploration are dolphin communication, 2D, 3D, and 4D in that order. Drilling and explosive techniques received the lowest scores. Moreover, these two technologies got negative values in some criteria, such as environmental quality and cost. Participants

Table 3-5   Technology evaluation criteria and their values

Criteria	Total	Name of Technology						
Name Time tested	Value Y/N	Explosive	2D	3D	4D	Vibration	Drilling	Dolphin Y/N
Environmental	**35**	**18**	**18**	**18**	**18**	**18**	**17**	**35**
Associated wastes	*15*							
GHG	3	2.5	2.5	2.5	2.5	2.5	1	3
Non-GHG	3	2.5	2.5	2.5	2.5	2.5	1	3
Liquid	3	3	3	3	3	3	0	3
Solid	3	3	3	3	3	3	0	3
Sound/pressure	3	0	0	0	0	0	3	3
Impact levels	*10*							
Planktons	2.5	2	2	2	2	2	2	2.5
Invertebrates	2.5	2	2	2	2	2	1	2.5
Fishes	2.5	0	0	0	0	0	1	2.5
Mammals	2.5	0	0	0	0	0	1	2.5
Impact types	*10*							
Hearing	2	0	0	0	0	0	2	2
Dead	2	1	1	1	1	1	1	2
Activities	2	1	1	1	1	1	1	2
Reproduction	2	1	1	1	1	1	1	2
Displacement	2	0	0	0	0	0	2	2
Tech feasibility	*24*	*11*	*20*	*22*	*22.25*	*19*	*18*	*24*
Proven tech	4	2	4	4	3	3	4	4
Ease of operation	4	1	4	4	4	4	2	4
Installation	4	0	4	4	4	4	2	4
Maintenance	4	4	4	4	4	4	2	4
Performance	4	2	2	3	4	2	4	4
Accuracy	4	2	2	3	3.25	2	4	4
Regulatory	*20*	*0*	*20*	*20*	*20*	*8*	*15*	*20*
International	5	0	5	5	5	2	4	5
Federal	5	0	5	5	5	2	4	5
Provincial	5	0	5	5	5	2	4	5
Guidelines	5	0	5	5	5	2	3	5
Cost	*21*	*16*	*18*	*16*	*18*	*16*	*8*	*21*
Investment	10.5	8	9	8	9	8	4	10.5
Operational	10.5	8	9	8	9	8	4	10.5
Total	**100**	**45**	**76**	**76**	**78.25**	**61**	**58**	**100**

Table 3-6   Response rate, average weight and standard deviation for each seismic technology

No	Name of Technology	Responses of Participants																Total	Ave Wt	S.D. Wt
		1	2	3	4	5	6	7	8	9	10	11	12	13	14	15	16			
T1	Explosive	2	–	–	–	–	–	–	–	–	–	–	2	–	1	2	2	9	**1.5**	**0.45**
T2	2D			5	4	6	2	2	3	5	4	4	3	3	3	4	6	54	3.85	1.30
T3	3D	7	7	5	7	7	6	5	5	6	3	4	6	6	7	5	7	93	5.81	1.22
T4	4D	–	7	6	6	7	6	4	5	5	6	7	7	5	4	7	7	89	5.93	1.1
T5	Drilling	3	–	–	–	–	2	2	–	3	3	2	3	2	3	3	3	29	2.63	0.50
T6	Vibration	5	–	–	–	–	–	3	2	3	4	3	2	4	3	4	3	36	3.27	0.90
T7	Dolphin	7	–	7	7	6	7	6	7	5	–	4	7	6	6	5	7	87	6.21	0.97

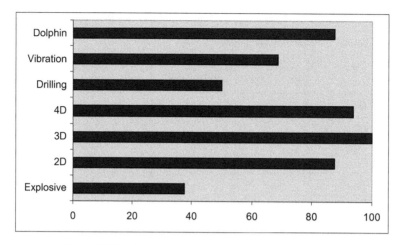

Figure 3-22   Acceptance rate of different technologies.

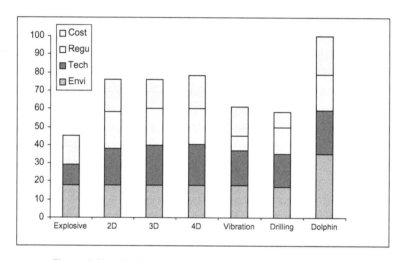

Figure 3-23   Total weight for different seismic technologies.

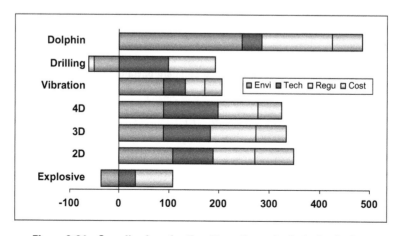

Figure 3-24   Overall values for the alternative seismic technologies.

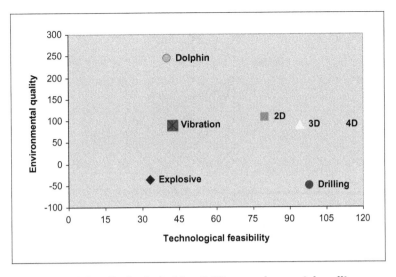

Figure 3-25 Technological feasibility vs environmental quality.

considered these two technologies to be more environmentally detrimental than other technology scoring the highest because, as a Nature-based technology, it has no negative impact environmentally. Moreover, it is technologically more efficient than presently available seismic technologies. The dolphin communication technique is capable of identifying the size, shape, direction, speed, and internal composition of an object. Using the same mechanism that dolphins use to communicate, pro-Nature technologies, which are efficient without any hazard to humans, can be developed both to be efficient and to end environmental problems.

The studied technologies are evaluated with respect to environmental quality, technological feasibility, regulatory requirement, and cost-effectiveness. This figure shows that drilling is not a good option to consider as a preliminary exploration technique. The 2D seismic technology, having the second highest score in quality, scores considerably lower than 3D and 4D technologies in technological feasibility. In other words, 3D and 4D are voted as technologically better than 2D, yet environmentally inferior to 2D.

In the part of the analysis that directly evaluates the "sustainability status" of each seismic technology, the "sustainability state" is determined based on the average weight, and relative weight sustainability indicators, which are calculated following Equations 3.1, 3.2, and 3.3. The detailed results are shown in Table 3-6. In addition to estimating the relative weights of indicators, the "sustainability state" is elaborated from the perceived targets or conditions judged by the stakeholders. This analysis also began with the judgments of the stakeholders as to what they perceive as the targets of each indicator and, finally, for each technology. Figure 3-26 shows the perceived target of each technology, which are used in the calculation of the sustainability index criteria (SIC) and, ultimately, the "sustainability status".

To estimate the "sustainability state" of each technology, the sustainability criteria (SCI) are estimated. The results are presented in Figure 3-27. These are based on the perceived target or scoring conditions given in Table 3-5. It is found that the sustainability index of dolphin technology is highest among all the technologies. Drilling got the lowest sustainability index score of 3.63. These mean that overall the sustainability status of dolphin communication is better than any other technology, while the sustainability status of drilling is the worst. As a result, the drilling technology needs more discretion when used in terms of its uses for preliminary exploration. The 3D, 4D, and vibration technologies had similar index scores of 4.38, 4.53, and 4.64, respectively. This implies that all three technologies have similar "sustainability status". Among all operational phases of offshore development, seismic exploration has negligible impacts (Davis et al. 1998; Khan and Islam 2007; Khan et al. 2006c). The overall values of all SCI scores for presently used

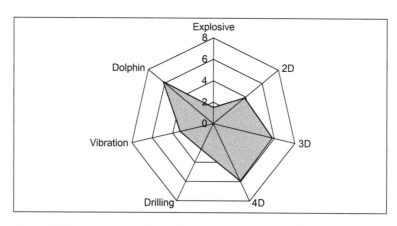

Figure 3-26   Average weight of different seismic exploration technologies.

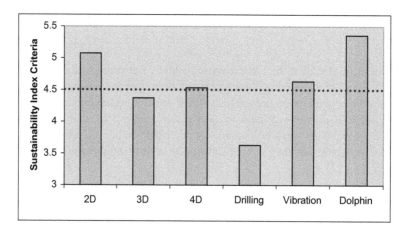

Figure 3-27   Sustainability index criteria.

technologies, such as 2D and 3D, support the argument of Davis et al. (1998) and Khan and Islam (2007) that these technologies are comparatively less detrimental to the environment. However, these technologies are not sustainable when compared to the dolphin communication technique.

3.6   Accuracy and Uncertainty of Explorations

Seismic survey has a huge impact on the exploration operations. Accurate seismic exploration can bring economic success by identifying hydrocarbon reserves. Accurate prediction of geological formation can help avoid unsuccessful drilling. Table 3-7 summarizes success rates of 2D *vs* 3D in a gas well, an oil well, and a drilled well. These results are based on the data collected in 1993 (Aylor Jr. 1999). However, since 1994, when the measurements were first made, both the number of exploration wells drilled with 3D seismic and the drilled success rate has increased enormously.

Uncertainty needed to be measured and the impact of different technologies on uncertainty appraised. In this way, the most effective technologies are nurtured to grow. For less-broad-spectrum technologies, use of confidence modeling can greatly improve the ability of the technology to predict outcomes. All these measures should be used more routinely to optimize technology application and development decisions.

Table 3-7   Comparative success rate of different exploration technologies

	Percent
Overall success rate, 1990–97, 2D *vs* 3D	13 *vs* 47
Gas-well success rate, 1990–97, 2D *vs* 3D	24 *vs* 54
Oilwell success rate, 1990–97, 2D *vs* 3D	3 *vs* 37
Drilled wells covered by 3D seismic, 1990 *vs* 1997	5 *vs* 97
Annual drilled success rate, 1990 *vs* 1997	10 *vs* 47

Source: Aylor Jr. (1999)

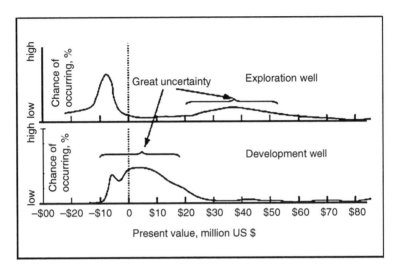

**Figure 3-28   Uncertainty associated with exploration prospects and development location.
Source: Aylor Jr. (1999).**

Uncertainty, although largely ignored in many of these measurements, is a parameter of great importance to both the operators and the management. In addition, reduction of uncertainty is ultimately a primary goal of management because, with more certain outcomes, it is much easier to decide quickly on the most prudent course of action. Figure 3-28 illustrates the interplay of technology and uncertainty in E&P work. This figure shows that both high risk and high uncertainty typically exist before technology is applied.

3.7   Special Measures for Sustainable Exploration Technology

Seismic surveys are an indispensable component of oil and gas exploration. From the above discussion, it is revealed that oil and gas exploration activities have significant impacts on the valued ecosystem. The environmental impact of seismic operations can be minimized. Special kinds of measures are needed to minimize these effects of exploration operations. The following guidelines should be abided by during exploration operations. These guidelines are taken from the work of ASRD (2002), MVEIRB (2003) and Khan and Islam (2007):

3.7.1   General

- Existing cutlines and access routes should be identified and evaluated for their suitability for usage both as access and survey lines.

- Screeners and developers should be aware of other land use activities in the area. For example, directly adjacent 3D programs should be avoided, or at least separated by a buffer.

- Heliportable operations are accepted as causing the least impact to the environment as a result of the smaller physical footprint on the ground. However, these operations involve highly specialized training for the use of helicopters, portable drills, and recording equipment.

- Parallel existing lines within 400 m or less of proposed new cut lines should be identified and used if they are suitable.

- When possible, small cats, tracked equipment, low-ground-pressure vehicles, and heliportable equipment should be used.

- Use existing clearings as much as possible for access and location of new seismic lines to minimize the amount of new clearing and surface disturbance.

- Practices show that source line widths of 5.5 m and receiver line widths of 3.0 m are usually sufficient for obtaining data. These widths should be used as much as possible.

- Ancillary facilities should not be located where they will interfere with natural drainage patterns, cause erosion, or seriously alter the hydrologic regime or water quality.

- Fuel tanks must have 110% secondary containment (double walled tanks are preferred) to prevent releases to the environment. Bladders are not acceptable for fuel storage.

- Precautions should be taken to prevent spills and releases and the improper disposal of fluid and solid wastes.

- If water wells are present near the seismic line, well water quality should be monitored to eliminate environmental liabilities.

- An environmental monitoring should accompany the operations crew to monitor adherence to protection procedures and provide environmental advice, as needed.

3.7.2   Public consultation

- Public consultation should be undertaken to notify local communities of the seismic operation and timing.

- Economic benefits accrue to the community through the use of local contractors, hotels, restaurants, shops, and helicopter services. The extent of local benefits depends on the nature of the seismic program. Generally 3D operations involve more work than 2D operations and, therefore, provide more economic benefits to communities. However, the potential of environmental hazards from 3D operations is greater.

- Despite the additional work created from clearing more and wider seismic lines, most communities prefer to see less cutting through the incorporation of low impact seismic methods.

- Traditional land uses (i.e., trapping, hunting, fishing, cabins, archaeological, or heritage sites) in the project area should be identified. Archeological sites or sites of high spiritual value may not be included on project maps but should be avoided by all development components.

- Hunting and trapping activities should be identified and seismic programs should be designed to minimize interference with these activities.

- Representatives from aboriginal organizations should be consulted to coordinate field activities and determine any traditional environmental knowledge requirements.

- Seismic operations should be planned in accordance with local or regional land use plans.

3.7.3 *Vegetation/timber resources*

- The proposed seismic route should be planned in order to avoid clearing merchantable timber and reforested areas as much as possible. Access routes and seismic lines must be scouted prior to any clearing.
- Vibroseis should not be used in areas with commercially viable timber stands.
- The estimated amount of merchantable timber to be cut should be determined.
- To eliminate unnecessary clearing of trees, survey stakes and flagging should be placed at short intervals to indicate access roads, lease boundaries, and detours (shooflies) required during clearing.
- Zero ground disturbance, through high blading and the use of mulchers, should be employed to promote fast recovery.
- Avoidance cutting techniques should be used to minimize clearing of tree cover. This type of cutting avoids larger standing trees, avoids merchantable timber, and leaves short lines-of-sight.
- Vegetation should be cut at a height of 10 cm on seismic lines to accommodate natural regeneration.
- All trees should be felled away from watercourses to reduce impacts to creek beds and adjacent trees.
- Windrowed slash should not be pushed into the adjacent timber. Breaks of 10 m in length should be made in the windrows at intervals of 60 m to allow movement of wildlife and reduce fire hazard.
- All wood debris and leaning trees should be slashed, limbed, and bucked to lie flat on the ground to allow them to decompose and to minimize the fire hazard.
- Low shrub and ground vegetation should be kept intact.
- Slash and debris should not be burned.
- Natural re-vegetation of rights-of-way should be promoted by avoiding disturbance to the root layer.
- Rather than being removed from the area or burned, slash and shrubby vegetation should be rolled back onto seismic lines to promote natural re-vegetation and to prevent access.
- Rollback on steep hills and areas of highly erodible soil should be used as a method of erosion control. Rollback is generally restricted to debris less than 15 cm in diameter.

3.7.4 *Wildlife*

- Soft start technique should be used for the protection of marine mammals.
- Critical wildlife habitat, den sites, calving areas, wintering areas, and mineral licks should be identified and avoided.
- All activities should be scheduled to avoid interference with endangered species in the project area and to protect these animals and their habitats.
- As late winter and spring are sensitive periods for caribou, work within caribou ranges should be scheduled for the early part of the winter season.
- In ranges where evidence or informed opinion suggests that caribou populations may be compromised by incremental linear development, 3.0 m (or less) low-impact seismic lines using enviro-drill technology should be used.
- Localized habitat features such as beaver ponds, nests, dens, or burrow sites should be avoided as much as possible.
- Lines should be as narrow as possible to reduce wildlife habitat fragmentation.

- Avoidance cutting techniques should be employed to render lines impassable by public vehicles, thus restricting public access and reducing hunting pressure on wildlife.

- Hand-cutting should be employed within 90 m of watercourse crossings to minimize disturbance to habitat within the riparian zone.

- The clearing of large diameter (>25 cm) snags offering nest sites for cavity nesters or raptors should be avoided wherever possible. Snags should be assessed for wildlife value while minimizing hazard potential.

- Lines should be doglegged at intersections with access roads and trails to eliminate continuous lines of sight along the line.

- Alternating windrow methods should be used to break up the continuity of potential fire fuels and not to impede wildlife movement.

- Localized habitat features such as beaver ponds, nests, dens, or burrow sites should be avoided as much as possible.

- Special attention should be paid to areas known to support species considered to be rare and/or endangered.

- Harassment of wildlife should be avoided.

- Where feasible, helicopters should maintain a ceiling of 500 m in areas of wildlife concentration (i.e., caribou herds) and 3000 m in goose staging areas.

- Before abandoning a shot hole, all portions of charges should be blown from the hole and container wrappings and other wastes from blasting should be removed for proper disposal. The ground surrounding an abandoned shot hole should be restored to the original condition.

- All litter, shot wire, paper, bags, styrofoam cups, etc. should be bagged and removed from the seismic line and disposed of at an approved municipal landfill. Daily patrols should be made to remove materials that may be potentially harmful to wildlife.

- Activities should be completed and access should be closed off to limit disturbance across large areas of caribou range during late winter.

3.7.5   Fisheries

- Fording of streams should be avoided and temporary stream crossings should be used.

- If fording is necessary it should be limited to streams with gravel or sand beds that are no wider than 15 m.

- Hazards to and from other ships and fishing boats should be minimized; for inshore waters, surveys are to be only conducted in daylight hours. In addition, where feasible, main shipping routes are to be avoided.

- A fisheries liaison officer should be appointed before and throughout the seismic operation.

- Most oil spills and leaks during seismic operations occur during fuel loading. The risks can be managed by agreeing in advance – emergency shutdown procedures; continuous monitoring of transfer; loading in daylight hours only; providing means to clean up spills on deck; and establishing written procedure for dealing with and reporting of oil spills.

- When considering the possible impacts of seismic sounds on the marine ecosystem, it makes sense to embed these considerations within the larger framework of the impact of all anthropogenic noise on the ecosystem.

- Any physical damage by anchoring should be avoided.

- When necessary, diving centers and dive boat operators should be informed of timing and locations of surveys. They should be advised of minimum diving distance limitations from the survey boat. In these circumstances, consideration should be given to using at least two chase boats.

- All vessels must comply with the MARPOL regulations on waste disposal, oily water discharge, and sewage controls.
- Access to stream crossings, particularly fords, should be designed so that it is not on the outside of any bend in the watercourse.
- Lines, trails, and rights-of-way should be located and constructed parallel to streams a minimum of 30 m from any stream except at crossings.
- The number of stream crossings should be minimized. Ideally they should be avoided except at existing crossing sites.
- Approaches to stream crossings should be as level as possible and the stream crossing itself has to be at a 90-degree angle to the stream.
- Winter crossings should not impede water flow at any time of the year and be v-notched prior to Spring break-up.
- Seismic lines should be set back from major rivers by 500–1000 m, unless they are crossing.
- Explosives should not be used to conduct seismic exploration in areas of water bodies that are not frozen to the bottom.
- No materials should be stored on any ice surface of a water body or within 30 m of such a water body.
- Fuel containers should be located at least 100 m from the high-water mark of any water body to prevent spills and leaks from entering the water.
- Water intake from water bodies should utilize screens on intake hoses to prevent disturbance to stream or lake bottoms and to prevent the entrainment of fish.
- Operators should protect fisheries habitat by limiting activities that may result in bank erosion and streambed disturbance at watercourses.

3.7.6 Corporate/management guidelines

Rojas (1998) described guidelines for the corporate management with respect to petroleum operations. Some of the guidelines related to seismic operations are listed below:

- A corporate policy on safety, health, environmental, and social issues should be formally adopted. The corporate policy expresses the company's intentions and principles in relation to overall safety, health, and environmental performance; and commitment to implementing strategies to address social concerns. Senior management must provide the direction and resources (people and money) to implement the policies.
- A Safety, Health, and Environmental Management System (SHEMS) should be developed and implemented. The SHEMS is the part of the overall business management system that includes the organization, planning, responsibilities, processes, and procedures for achieving the commitments made in the SHE policy. The SHEMS provides confidence to stakeholders that there is management commitment to environmental performance, safe operations, and maintaining good community relations. It provides the framework, guidance, and requirements necessary to ensure that management of non-technical issues is integrated into business operations.
- A consistent and comprehensive risk assessment process should be implemented. During the risk assessment process, social and environmental impacts need to be thoroughly assessed through the environmental impact assessment (EIA) and social impact assessment (SIA).
- Relationships with the government, communities, non-government organizations (NGOs), and other stakeholders to implement social programs in the project area should be forged. Partnerships bring stakeholders together and build trust between the company and these external organizations.

3.7.7   Risk factors

- The risks associated with seismic operations can be minimized dramatically through simple and controlled pre-planning and risk analysis. A formal risk assessment that addresses applicable risk factors must be completed by the helicopter operator and mitigation controls presented to the company representative for review prior to the start of operations.

- Risk factors will vary with each activity depending on items such as terrain type, density altitude, local weather conditions, restriction to visibility, type of external loads carried, search and rescue resources, and the performance characteristics of the type of helicopter used. This should be ascertained.

3.7.8   Research and technological

The dearth of scientific information regarding field experiments on fish, invertebrates, and the larger marine mammals, makes it extremely difficult to evaluate the impact of a particular type of seismic sound, or more generally noise, on particular species. Thompson (1998) reported the missing scientific knowledge about seismic explorations. Some of the suggested studies are described below:

- More work is required to determine the sound characteristics and environmental conditions under which seismic effects on behavior, physiology, and physical well-being of all types of marine species might occur.

- The available information on the effectiveness of mitigation measures needs to be more fully evaluated, as a basis for both interim advice on appropriate operational requirements in the short term and additional research needs to increase knowledge in the longer term.

- In addition to targeted research, there is great value in linking a program of structured collection of information to the conduct of seismic surveys, to facilitate learning-by-doing. However, such information collection programs must be well coordinated, and accompanied by the resources to analyze, interpret, and apply the new information, as it is collected and submitted to scientific authorities.

- A few representative studies on distance-effect relationships for all taxa, but particularly eggs and larvae, would greatly aid understanding of potential risks posed by seismic sound. The potential for effects stemming from sound exposure level (cumulative over a survey), as well as peak received sound pressure level, should be considered, including under conditions of 3D surveys.

- Specific research is needed on the level of received sound experienced by sessile invertebrates, and the effects of seismic sounds on such organisms. The physics of the sound levels to which benthic organisms are exposed is complicated due to shear effects interacting with pressure effects, and the proximity to the bottom substrates. Hence results of generic sound propagation models are likely to be misleading with regard to exposure levels of sessile benthic species. However, the errors could be in any direction, and in sites of complex bathymetry there could be patchy distributions of areas with higher intensities of exposure than predicted by sound propagation models and other areas with lower intensities.

- There is a specific absence of information on the effects of seismic sounds on molting of invertebrates with hard exoskeletons.

- There is a need to further clarify the best sound propagation models for the areas likely to host seismic exploration, and how the habitat characters should influence model selection. Generic models also need to be evaluated in relation to the sensitivity and precision of their predictions relative to requirements for evaluating potential impacts, although site-specific implementations of generic models will continue to be desirable.

- Better data input is needed during modeling of the expected pattern of spread of seismic sounds during surveys. Near- and far-field sound measurements should be encouraged as part of planned seismic operations for an area that has not been surveyed previously, or if previous models have been shown to be inaccurate. Further research on potential impacts of seismic sound on marine mammals is urgently needed.

Following the above guidelines, most environmental and ecological problems can be avoided or minimized, leading to sustainable explorations in petroleum operations. To achieve sustainability, other operational phases are also needed. After exploration operations, drilling and production operation are the next two phases. The sustainable management of drilling and production operations is discussed in Chapter 4.

Chapter 4

Drilling and Production Operations

4.1 Introduction

Drilling and production are the two major steps in petroleum operations, which are considered to be highly costly and risk involving. Drilling is the most expensive operation in the long journey of hydrocarbon production. Tremendous difficulties, involving high cost, mission delays and, in some cases and injuries casualties, are inevitable with the current method of drilling. For example, many drilling problems occur before running the casing and thus completing the cement job. The unsealed borehole can allow formation fluids to flow to the wellbore or to allow drilling fluid to escape into the formation. Gas kicks, lost circulation, formation damage, borehole swelling, stuck pipes, formation fracture, and borehole collapse are among other associated problems.

The best available drilling and production operations end up producing numerous gaseous, liquid, and solid wastes and pollutants (Khan and Islam 2003b; Holdway 2002; Veil 2002; EPA 2000), none of which has been completely remedied. Therefore, it is believed that drilling operations have negative impacts on habitat, wildlife, fisheries, and biodiversity (Wenger et al. 2004; Holdway 2002; Khan and Islam 2003a; Currie and Isaacs 2004; Schroeder and Love 2004). The success of these high-risk drilling and production operations depends on the use of sustainable technology (Khan et al. 2005a; Khan and Islam 2005b). As in all industrial sectors, the present drilling and hydrocarbon production development processes are developed for reasons of economic success. Environmental well-being does not receive proper attention in the conventional technology development process.

Current drilling techniques were developed at the beginning of the last century. Many problems persist with these methods, including downtime due to dull bits, the lack of precise vertical or horizontal wells, and formation fluid leakage during drilling due to the lack of a seal around the hole. The need for a new method of drilling oil and gas wells is immense. As alternatives, some non-conventional techniques, such as laser and water jet drilling, are now receiving attention.

Laser drilling is a new technology that has been proposed as a method to eliminate the current problems with drilling and to provide a less expensive alternative to conventional methods (Bjorndalen et al. 2003). Although lasers have found widespread use in many industries, it is only recently that research in this area has been directed to the oil and gas industry (Agha et al. 2003). Laser drilling can increase the penetration rate by more than 100 times over conventional rotary drilling methods and the problems and associated downtime linked to dull drill bits can be eliminated, as well as wastes created from drilling mud.

Water jet drilling is also reported as an alternative. This has many advantages that will be discussed further in Section 4.6.2. The combination of laser and water jet drillings can open up great opportunities in the future. This process can be best characterized as guided water drilling, as the laser beam principally creates a very narrow opening, followed up by a water jet. This combined process removes some of the shortcomings of using each technique alone, by creating a complementary process that is very efficient.

This chapter examines the current drilling and production practices. The technological limitations and environmental consequences of the current practices are discussed. The chapter also provides a brief overview of sustainability in drilling and production operations. The sustainability of current technologies is evaluated. Other alternatives that remove limitations of current drilling techniques are also discussed. Scotiz Shelf offshore activities are used as an example for various analyses. With this, different guidelines are developed to achieve sustainability in the drilling and production sectors. Finally, a series of research ideas for rendering production operations sustainable are given.

4.2   Current Practices of Drilling and Production

Seismic surveys provide a general idea of the presence of hydrocarbons based on the structure and properties of rocks. It is the exploratory drilling that confirms the presence of hydrocarbons. Production drilling takes place if the project site is determined to be economically feasible, based on the quality and quantity of the hydrocarbon reservoir. There are basic differences between exploratory and production drillings. In exploratory drilling, the diameter of the drill hole is smaller than in that of production drilling, resulting in different drilling-cuts being generated (Groot 1996).

For a drilling operation, suitable rigs are selected based upon environmental and geotechnical conditions. There are submersible and semi-submersible rigs, jackup rigs, as well as drill ships. For example, submersible rigs are used in shallow water (up to 25 m) and jack-up rigs operate in deeper water, such as 20–125 m. Rigs consist of self-contained legs, which are lowered to make contact with seabed. A mat-supported rig is used on soft seafloor sediments. Generally, submersibles and drill ships are used in deeper water. Security and risk related to these structures have become important issues after an accident involving the world's largest, 40-storey-tall offshore oil platform, in Brazil on March 20, 2001. Quantitative risk assessment is now applied to offshore installations (Ramsay 1994).

Regardless of the type of rig used, drilling operations are similar. The main task of a drill rig is performed by the hosting, circulating, and rotation system, backed-up by the pressure-control equipment. A common drilling operation is depicted in Figure 4-1, clearly showing the drill bits, drill mud input, and screening of drill cuttings. A drill bit is attached to the end portion of a drill pipe. Motorized equipment rotates the drill pipe to enable it to cut into the rocks. During drilling, numerous pumps and prime movers circulate drilling fluids from tanks through a standpipe into the drill pipe and drill collar to the bit. The muds flow out of the annulus above the blowout preventer over the shale shaker (a screen to remove formation cutting) and back into the mud tanks (Figure 4-1).

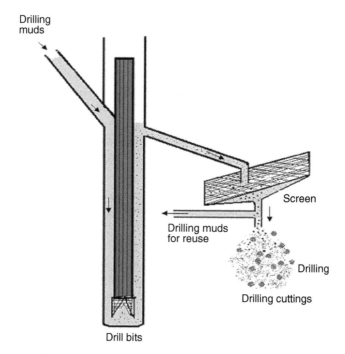

Figure 4-1   A representation of drilling that includes drill bits, shaker-screen, and input of drilling muds.

In a drill hole, a casing is needed as the drill pipe goes deeper. Generally, for the first 1000 m, casing is not required. For greater depths, the drill string is removed and a pipe, the "well casing", is inserted into the well. As drilling progresses, the well is lined with additional casing, both to prevent rock from crumbling into the hole and to contain any high-pressure gases and liquids. On top of the casing, the blowout preventer is added to close the well in the event of uncontrollable pressures.

Mainly water-based, oil-based, and synthetic-based drilling muds are used in exploration and production drilling (Wenger et al. 2004; Khan and Islam 2003a). The composition ranges from a simple clay-water mixture to a complex blend of minerals chemically suspended in water and oil. The composition of drilling muds is shown in Table 4-1 (data source: Zwicker et al. (1983); Patin (1999); Wenger et al. 2004). Water-based mud is composed of water and bentonite, with heavy minerals added for weight. Chemical additives are mixed in order to stabilize the drilling fluids during use, and to reduce corrosion and bacterial activity (Table 4-1). Figure 4-2 shows the composition of oil-based and water-based drilling muds in percentage weight, excluding density control.

Table 4-1   Composition of drilling muds used in oil and gas exploration

Product	Composition	Concentration
Base fluid	Water, pentonite clay (sodium montmorillonite), caustic soda (sodium hydroxide)	As needed
Additives	Lignosulfonate, phosphates (sodium acid pyrophosphate and tetrasodiumpyrophosphate), plant remains (predominant usage of quebracho), lignite	1–2.7 kg/bbl
Density control	Barite (natural barium sulfate ore), ferrophosphate ore, calcite, siderite, hematite, and heavy metals.	0–317.5 kg/bbl
Fluid-loss control	Starch (corn and potato), polyanionic cellulose polymer, xanthum gum, sodium carboxymethyle-cellulose, lignite.	<0.45–4.5 kg/bbl
Lost Circulation	Ground nut shells, micas, ground cellophane, diatomacheous earth, cottonseed hulls, ground or shredded paper	0.9–13.6 kg/bbl
Corrosion and scale control	Sodium sulfite, zinc chromate, tall oil, amines, sodium hydroxide, phosphates, bactericides	0.11–2.7 kg/bbl
Solvents	Isoprophanol, glycerol, isobutanol, ester alcohols, diesel oil	–
Lubricant	Asphalts, diesel oil, fatty soaps, gilsonite, glass beads, rosin soap, enhanced mineral oil, synthetic oil (no aromatic content)	0.09–2.7 kg/bbl

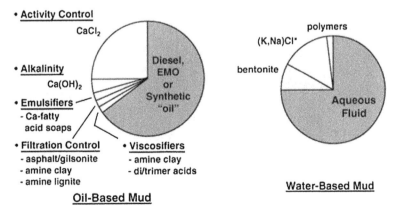

Figure 4-2   Composition of oil-based and water-based drilling muds (in weight %, excluding weighting agents) used in offshore oil and gas operations. Source: Wenger et al. (2004).

The oil-based muds are composed of mineral oils, barite, and chemical additives (Table 4-1). Oil-based muds are used for deeper well sections and where the well is drilled at an angle. In the case of synthetic-based mud, mineral oil is replaced by oil-like substances. Such replacement is Uaimed to make the mud more environmentally acceptable. Requirements as to appropriate drilling mud depend on well depth and the geological conditions of the drilling area.

With the advances in drilling technology, drilling can now be carried out in much deeper formations. Figure 4-3 shows the total number of wells drilled, and their depths, in a study area of the Scotian Shelf. As noted in Chapter 3, drilling activities on the Scotian Shelf were initiated in 1967. By 2004, 199 wells had been drilled, out of which 56 were drilled for production, 136 for exploration, and the few remaining for delineation purposes. This is in contrast to what happened in the North Sea and Gulf of Mexico (Managia et al. 2005). According to Groot (1996), in the North Sea, 1530 exploration, 997 appraisal, and 2256 production wells were drilled. On the Scotian Shelf, most well drilling periods were relatively short, i.e., less then three months. At present, the industry is actively drilling in the deeper slope waters. The peak drilling activities took place in the years 1970–1979 (CNSOPB 2005).

Drilling an offshore well releases drilling muds, cuttings, and produced waters into the marine environment. This remains an issue of concern (Khan and Islam 2003a; Cranford et al. 2003; Rezende et al. 2002). In Figure 4-4, the systematic drilling activities are shown. Small arrows with numerical numbers represent the different types of waste released into environment.

Drilling depths vary with location. A better way to understand the physical impact might be in terms of distance drilled (Groot 1996). In Figure 4-3, depth drilled by each well is presented. Until 2004, a total 796,283 m distance was drilled. The average well depth on the Scotian Shelf is about 4000 m below the sea floor (Figure 4-3). A well about this depth typically generates 1000–5000 m^3 of wastes (Groot 1996; Patin 1999). A production platform generally consists of 10–15 wells, which generates (12 × 5000 m^3) 60,000 m^3 of wastes.

All rock removed from a well is deposited on the seabed. For an average well, this is equivalent to a column 5000 m deep, with a diameter of 90 cm at the surface, to about 20 cm at the bottom. The volume of rock can range from 300 to 1200 m^3, and the volume of mud and cuttings combined can reach 3200 m^3 from each exploratory well (estimates based on Groot (1996), Putin (1999)). However, it is unclear how long it takes for the cutting piles to become totally eroded by wave action and tidal currents (Groot 1996; Kingston 2002).

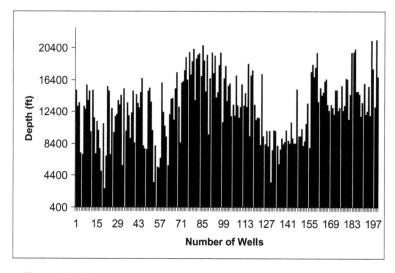

Figure 4-3   Number of wells drilled on the Scotian Shelf, 1967 to 2004.

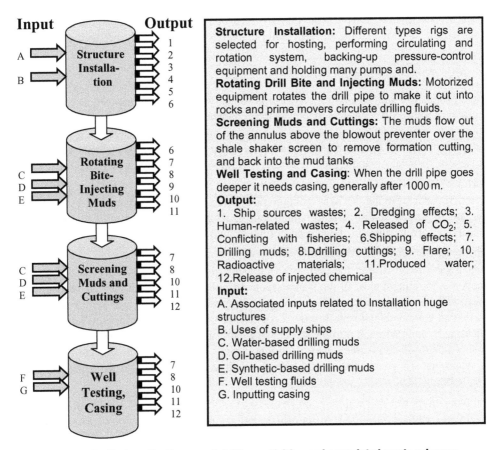

Figure 4-4   Systematic diagram of drilling activities and associated waste releases.

4.3   Production

After an oil or gas discovery has been made, there is an enormous amount of evaluation and planning to be done before an energy company can decide on the best way to produce the hydrocarbons and develop the field. To provide a general idea of oil and gas development one of the largest production facilities is discussed. The Sable Island Offshore Energy Project (SIOEP) on the Scotian Shelf is the largest offshore gas project of its kind in North America today. The technological details described below focus on this project in particular. However, deviations are noted, given the size and scope of other offshore oil and gas projects that we are studying. This will provide an outline of what to expect in terms of the technological requirements of offshore oil- and gas-extracting and processing units.

SIOEP, once it is finished, will consist of six platforms (one at each field). The Central Thebaud Platform Complex is the only manned platform, and the rest are unmanned (Figure 4-8). The unmanned platforms are called satellite platforms. Production platforms are the combination of jackets and associated equipment (top side facilities), which manage production. Many types of production platforms are found offshore, reflecting the specific needs of the field and hydrocarbon source. The "jacket" includes the pipes or casings, which surround and protect the well on the seabed. They remain fixed to the seabed for the life of the project. Topside facilities include the equipment needed to capture liquid or gaseous hydrocarbons, as well as crew quarters, generators, cranes, etc.

In order to extract gas out of the ground, the natural pressure in a reservoir usually forces oil and gas to the surface. However, as reservoirs of liquid hydrocarbons, such as crude oil, become depleted, drillers inject water into the well below the level at which the hydrocarbons are found. Because oil is lighter than water, the oil is pushed up and flows to the surface. Natural gas reservoirs normally produce their own pressure.

4.3.1 Central Thebaud Platform "topside facilities"

The Thebaud production and processing platform is the only permanently manned structure for the entire offshore portion of SIOEP. It consists of two platforms connected by a walkway. One of the platforms houses the Thebaud wells, while the other contains processing and production equipment for gas from Thebaud, as well as gas brought in from the satellite platforms, and living quarters (sleeps 40), and a helicopter landing pad. Steel jackets, their legs reaching the ocean floor, support both of these platforms.

The production complex's main function is the dehydration and separation of the produced gas, that is, to remove as much water from the gas as is technologically possible. The gas that is extracted is cooled by an inlet cooler, and separated in a three-phase inlet group separator into water, condensate (oil), and gas. Two of these separators are located on the central complex, with one equipped to handle low pressure in the future, when gas pressure in the wells decreases.

The combined gas stream is then fed into two triethylene glycol (TEG) gas dehydration contractor trains whose function is to remove most of the water vapour from the gas. The TEG is then recovered by boiling the water off, and sending it back to the contractor. The condensate, which was removed back at the inlet separators, is then combined and sent through a condensate coalescer and stripper to remove water. Dewatered condensate is pumped back into the gas downstream of the triethylene glycol contractors. This final product of dewatered gas and condensate (oil) is sent through a pipeline to shore, and on to the gas plant at Goldboro. Water vapor is vented into the air and any liquid water is sent to a water separation and treatment system, finally being discharged into the sea.

The central platform power and water systems consist of multiple generators capable of running on natural gas or diesel fuel (only used for the start-up of the platform or as emergency power) and filtered seawater. Its fire protection and safety systems consist of a fixed extinguishment system, inert gas fire suppression, foam firefighting capabilities, seawater deluge, survival suits, lifeboats, temperature and pressure monitoring and control, and ventilation and pressurization detection systems.

4.3.2 Satellite platform facilities

The satellite platforms are designed to be unmanned and to contain as little processing equipment as possible. This enables most of the machinery to be centered on the Thebaud platform. Gas produced at the satellite platforms is separated in a three-phase group separator (water, gas, condensate). Produced water is treated in a hydrocyclone separator, which separates out more condensate by using centrifugal force, and then discharged overboard. The gas and condensate are then recombined and sent to the Thebaud complex (via subsea interfield pipelines) for further processing. Monoethylene glycol and corrosion inhibitors are injected into the flow to prevent the formation of hydrates in the gas and corrosion of the pipelines.

4.3.3 Scale of production operation

Going from exploration to production is a major step in the development of an offshore reserve. The difference is mostly one of size: an offshore development has many of the same activities as a single exploratory well (Figure 4-6) (Section 3.1) but on a larger scale and for a longer period of time. Offshore "production" involves all the activities needed to develop an oil or gas reserve. It is more expensive and takes longer than exploration. Unlike exploration, it requires the installation of permanent platforms at the wellhead. Given that the SIOEP is the largest project of its kind in North America today, the figures for scale will be higher than the average offshore oil and gas project in Nova Scotia. Figure 4-5 shows the production volume of the SIOE project (CNSOPB 2005), depicted as the monthly average production of gas and associated water from January 2004 to September 2005.

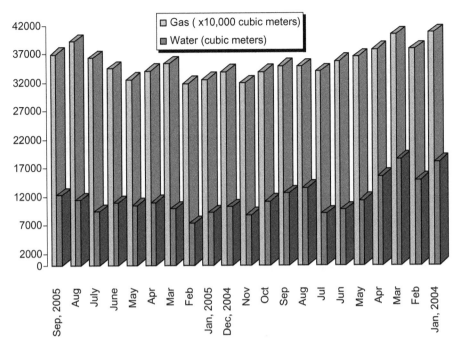

Figure 4-5 Average Monthly Production of raw gas (ten thousand cubic meters) and produced water (cubic meters) in the Sable Energy Project from January 2004 to September 2005. Data source: CNSOPB (2005).

The SIOEP is located 200 km off Nova Scotia, 10–40 km north of the Scotian Shelf. Drilling below the seafloor can go as deep as 5 km. The platforms required for offshore oil and gas are separated by many kilometers. For example, in SIOEP, the distance between the central Thebaud platform and the farthest satellite platform, Alma, is 50 km, and the closest two platforms (Venture and South Venture) are separated by more than 5 km. The amount of materials required to construct the production facilities is immense. For example, the SIOEP project used more than 900,000 m of pipe, 109 million kg of steel, 53,000 m of chrome tubing, and 160 million kg of concrete. This project is expected to yield over 100 billion m^3 of natural gas over the next 20–25 years.

4.4 Problems with Current Practices

Drilling and production activities can have adverse effects on the environment in various ways. For example, blow-out and flaring of produced gas waste energy, carbon dioxide emissions into the atmosphere, and careless disposal of drilling muds and other oily materials, can have a toxic effect on terrestrial and marine life. To achieve sustainability, the drilling and production activities should not cause any damage to the environment across all of its activities.

Before drilling and production operations are allowed to go ahead, the Valued Ecosystem Component (VEC) level impact assessment should be done to establish the ecological and environmental conditions of the area proposed for development and assess the risks to the environment from the development. The VEC assessment provides what is called an ecological baseline. During the development and subsequent operations, ongoing environmental studies can detect whether the environment has suffered in any way.

For drilling and production operations VEC assessment, the first thing is the selection of VECs having the potential to be affected by operation activities. Selection of VECs depends on the operation site. For example, if the operation site is an offshore field then fisheries is the important area. If an operation site is within a

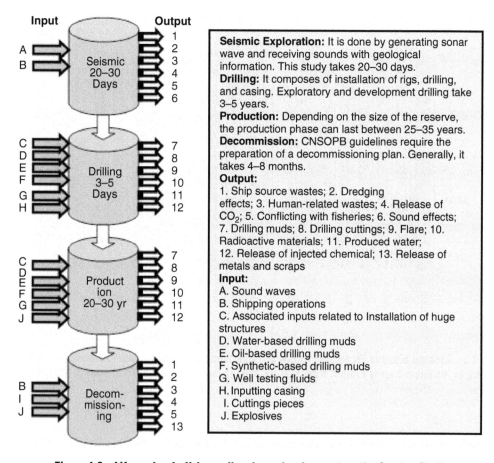

Figure 4-6   Life cycle of offshore oil and gas development on the Scotian Shelf.

forested area, wildlife is an important component that should be analyzed. For analyzing the status and effects levels on each VEC, the distributions, policy commitment, applicable monitoring protocol, and typical pre-mitigation are discussed. Figure 4-7 presents the different factors used to analyse for each VEC.

To help understand the VEC assessment, a case study conducted on the Scotian Shelf is shown below. The VEC is done on a specific site, using a standard method. As a result, this procedure can be applied in any other part of the world. The VECs selected for this study are the species at risk, ecological and economical vulnerability as related to fisheries, carbon, and a special habitat (Sable Island).

4.4.1   *Status and distribution*

4.4.1.1   **Species at risk**

Endangered species are already at risk due to the ecological changes caused by different human activities. For studying species at risk, three important species are selected, the leatherback turtle, Atlantic salmon, and piping plover. The leatherback turtle, *Dermochelys coriacea*, lives in marine waters off Nova Scotia. The diadromous species Atlantic salmon, *Salmo salar*, uses marine, coastal, and estuarine habitats. The coastal-nesting, intertidal-feeding piping plover, *Charadrius melodus*, is the subspecies melodus. Table 4-2 presents an overview of pertinent aspects of the life histories of the selected endangered species. Each species

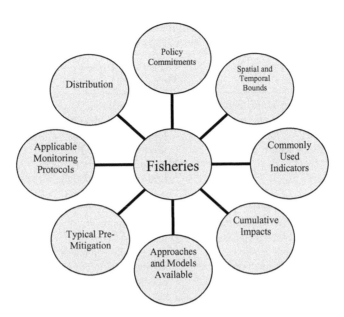

Figure 4-7   Multiple factors of valued ecosystem components.

is absent from Nova Scotia for the winter months. The proportion of habitat used by these species while in or near Nova Scotia is small; not because of habitat preference, but because of limited populations. The habitat that can be used by the selected species consists of a large proportion of the area suitable for petroleum development. In the case of the plover and salmon, valuable habitats exist in sections along much of the coast, and all marine waters considered in this report have had turtle sightings.

Fisheries: Human Resources Development Canada (HRDC) estimated the number of jobs, employment prospects, and wage levels for most Canadian occupations and industries, including those related to fisheries. In Nova Scotia overall, the fishing sector contributed 1.7% of the province's overall employment in 1998 (HRDC 2001). However, for many coastal communities, a greater percentage of jobs are derived from fisheries. For example, in Cape Breton 15% of jobs are related to the fisheries sector (Collins et al. 2001a). The total fish landings in Nova Scotia were worth US$ 542.513 million in 1999 (DFO 2002). In the last four years, the value of Nova Scotian fish landings increased in some parts. This suggests a significant increase in the money generated from the fisheries sector. It also suggests that fisheries are an emerging economic sector, which will generate more money according to the increasing trend of landing values. The increase of landing might cause other fisheries collapse, but according to fisheries biologists, the increase of landing is due to increase in stock as well as catch of non-conventional species (Biron 2000).

Groundfish habitat includes coastal inlets, estuaries, and deeper waters of the continental shelf. Ground species include cod, flatfish (e.g., flounder, plaice, halibut), redfish, haddock, pollock, and hake (DFO 2002). Pelagic species include Atlantic herring, tuna, and Atlantic mackerel (NHNS 2002). Invertebrates include snow crab, northern shrimp, and short fin squid.

The Nova Scotian fisheries zone surrounds the province of Nova Scotia, and extends to the shelf edge, more than 200 nautical miles from the coast at some points. To the north, the Laurentian Channel separates the Nova Scotian fisheries zone from the Newfoundland Labrador Shelf. To the south, it extends to the Fundian Channel. According to the North American Fisheries Organization, the Nova Scotian fisheries zone lies on the 4VN, 4W, 4X, 4T, and 5Y (NAFO 2002). The Scotian Shelf has a complex topography consisting of numerous offshore shallow banks and deep mid-shelf basins (Zwanenburg et al. 2001).

Table 4-2   Summary of current status and relevant life history features of selected endangered species of Nova Scotia

	Atlantic Salmon *DFO, 2001	Leatherback Turtle **Cook, 1984	Piping Plover ***Environment Canada, 1989
Estimated population	River-specific. Fundy and Atlantic coast are in strong decline. Gulf of St. Lawrence in slight decline or stable*.	Declining numbers, currently 20–30,000 females; $1/3$ of 1980s level (NOAA 2001). Most common marine turtle in Nova Scotian waters (Gilhen 1984)	Considerable recovery since 1917 *Migratory Birds Convention Act*;*** subsequent decline; 470 in Atlantic region, 4,070 in North America (Plissner and Haig 2000)
Date of listing	2001	1970	1985
Occurrence	Pelagic; western Cape Breton Island stable, Scotian Shelf and Fundy populations extirpated or endangered.*	Pelagic; throughout marine and coastal waters. Most common marine turtle in Nova Scotian waters.**	Shorebird, feeds in intertidal zone, nests near beach dunes. Northumberland Strait, and South Shore.***
Seasonal distribution (some variation exists among regions of province)	Aquatic for first year, then over-winter off Greenland from October to May. Return and summer in coastal and estuarine waters, enter rivers and spawn in fall.*	June to October in local waters, peak numbers in August. Most northerly marine turtle. Several decades to mature, nest and lay eggs on beaches south of South Carolina.**	Arrive in April, May, nest on beach, hatch in June, winter feeding grounds along Atlantic coast and Gulf of Mexico, shore bird with most southerly nesting distribution.***
Feeding	Aquatic insects in early years, chiefly marine invertebrates as adults.	Predominately jellyfish (Caurant et al. 1999).	Aquatic and marine invertebrates along shore.
Lifespan	Maximum 9 years	50–100 yrs	Average 5 yr, up to 14 yr
Known local causes of decline	Water quality of river habitat resulting from land use, pollution, physical disruption, acid precipitation, and buffering capacity of water (Kessler-Taylor 1986).	Entanglement in fishing equipment (Gilhen 1984), propeller injury (NOAA 2001).	Habitat loss, vulnerability of nesting habits, human disturbance, predation (Melvin et al. 1991). Potential threat to nesting habitat by sea level rise.
Causes of decline external to Nova Scotian waters	Commercial fishery off Greenland (Kessler-Taylor 1986).	Meat source, beach habitat disturbance (Peterson 1993), tangling in fishing equipment.**	Loss of habitat, human disturbance, and predation.***

4.4.2   *Policy commitments*

4.4.2.1   Species at risk

In Canada, protection of wildlife falls within the bounds of the *Canada Wildlife Act*, and the *Wild Animal and Plant Protection and Regulation of International and Interprovincial Trade Act* (Environment Canada 1999). Canada has no federal endangered species act. However, federal, provincial, and territorial governments do agree in principle to compliment their work through the accord for the Protection of Species at Risk (Environment Canada 1999). In 1996, Nova Scotia introduced a *Species at Risk Act*. The Department of Fisheries and Oceans (DFO) manages the salmon population through the regulation of recreational catch, access to fisheries by First Nations fishery, while also conducting management in accordance with the international North Atlantic Salmon Conservation Organization agreement in 1989. The piping plover is further protected by *the Migratory Birds Convention Act 1994*.

4.4.2.2 Fisheries

Fisheries on the Scotian Shelf have been regulated under the International Commission for the Northwest Atlantic Fisheries (ICNAF) since 1945. On January 1, 1977, following a series of lengthy international negotiations at UN Law of the Sea conferences, Canada's jurisdiction over coastal waters was extended to 200 nautical miles. The federal Department of Fisheries and Oceans (DFO) has regulated the fisheries of Nova Scotia since the extension of fisheries jurisdiction (SCF 1993). A variety of fishing plans were introduced by the Ministry of Fisheries, such as Community-Based Management, ITQ Fleets, and the popular TAQ (DFO 1999). This management plan provides some commitment to community.

The 1982, the Law of the Sea Convention required that catch possibilities in addition to domestic needs be made available to foreign parties. This gives the commitment, not only to Canadians, but also to other nations to access the fisheries. For example, in recent years, foreign (Cuban and Russian) participation in the Scotian Shelf silver hake fishery has been allowed, as required under this provision (DFO 2002). According to treaty rights and constitutional commitments, an Aboriginal Fisheries is also active.

The Canada-Nova Scotia Offshore Petroleum Board (CNSOPB) is responsible for protection of the environment during all phases of offshore petroleum activities, from initial exploration to abandonment (CNSOPB 2002). Under the *Canadian Environmental Assessment Act*, the CNSOPB is now officially a federal authority. It is also involved in initiatives led by the DFO related to marine protected areas and integrated management planning under the *Oceans Act* (CNSOPB 2002). While some other federal departments deal with fisheries matters, the DFO takes chief responsibility for water and its resources.

4.4.2.3 Special habitat: Sable Island

Figure 4-8 shows the location of Sable Island. The island currently receives legal protection from the *Canada Shipping Act*, Sable Island Regulations and the Migratory Bird Sanctuary (MBS) Regulations under the *Migratory Birds Convention Act*. These provide a relatively high degree of protection for the island. The conservation value of the MBS designation is limited primarily to migratory birds and their nests, therefore, the MBS regulations are only effective as a conservation tool when the migratory birds are actually nesting, and have little effect at other times of the year (Beson 1998).

The Sable Island Regulations set out a list of activities that are prohibited on the island and control access. The federal Fisheries Act administered by the DFO also extends protection and management jurisdiction

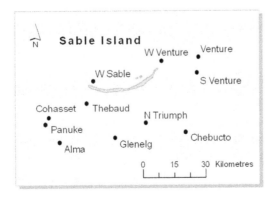

Figure 4-8   Sable Island – oil and gas operations.

over marine mammals. Particularly relevant to Sable Island is the protection the Fisheries Act extends to seal populations (Beson 1998). There is also the Sable Island Preservation Trust, which is a non-profit, charitable organization incorporated under the *Societies Act of Nova Scotia*. Its principal governing documents are its bylaws, its agreement with the federal Departments of Environment and Fisheries and Oceans signed on March 29, 2000, and the Conservation Strategy for Sable Island. Some of the roles of the trust include maintaining a human presence on Sable Island and managing the provision of transportation, communications, food, and other life support requirements; managing island infrastructure; carrying out meteorological programs; and developing and implementing programs, policies, and procedures to promote protection of ecosystems (Sable Island Preservation Trust 2001).

The CNSOPB has regulated a one nautical mile "no-drilling" zone around the island (CNSOPB 2001). Other petroleum activities within that zone or on the island require the operator to prepare a code of conduct, which prohibits non-essential activities and limits others to specific areas or rights-of-way.

Carbon: The Government of Canada, by signing the Kyoto Protocol, has committed itself to limit and reduce emissions of six greenhouse gases (including carbon dioxide) during the period 2008–12. Canada has agreed to limit its average annual emissions during the same period to 94% of its baseline emissions, a reduction of 6% from the baseline and 20–30% from the projected emissions in 2010. In October 2000, the province of Nova Scotia along with the other provinces and the federal government, accepted a national implementation strategy on climate change (Government of Nova Scotia 2001).

According to the Canadian Association of Petroleum Producers, the industry is only responsible for 20% of the total greenhouse gas emissions in Canada during the production and distribution of oil and gas, with the other 80% attributed to end-use consumption (CAPP 2002).

In 1998, the government of Nova Scotia estimated that 20,100 kilotonnes (kt) of CO_2 was emitted by the province, as well as projecting an increase to 21,100 kt by 2010. This model suggests that if Nova Scotia plans to achieve Kyoto's target the province will have to reduce emissions by 16% (Government of Nova Scotia 2001).

The CNSOPB is responsible for regulating CO_2 emissions, which result from oil and gas activities in offshore Nova Scotia. The new draft of CNSOPB regulations, Section 2(2.2) states that companies pursuing any type of activity in the offshore should "provide an estimate of annual quantities of greenhouse gas (GHG) that will be emitted from its offshore installation(s) and a description of plans to control and reduce these emissions" (CNSOPB 2002).

It should also be noted that the Canadian Association of Petroleum Producers (CAPP) has made a voluntary commitment toward reducing GHG emissions from operational facilities. Since 1995, most oil and gas companies in Canada have voluntarily participated in Canada's Voluntary Challenge and Registry (VCR) program, where companies are encouraged to limit their net GHG emissions and report to CAPP annually on their progress toward achieving such a goal (NCCS 1998; CAPP 2002).

4.4.3 *Spatial and temporal bounds for impact assessment*

4.4.3.1 **Species at risk**

Choosing boundaries carefully is critical to accurate EIA (Peterson et al. 2001). Because of vulnerability of endangered species, the importance of this choice is heightened and it is recommended that the broadest practical scope be selected in the interest of applying the concepts of the precautionary principle and sustainability. Migratory species pose an added challenge in addressing the boundaries of assessment, as distant activities may have a significant relevance for local activities. All three selected endangered species will require this consideration, although with different temporal considerations because of differing life spans.

Marine and coastal species should be considered at the individual, population, and community levels, with a focus on stability of communities (persistence, resistance, and resilience) (Dayton 1986). Addressing species at risk will require a shift in focus to the individual level because of fragmentation of populations and communities; this would encompass consideration of territoriality and distribution (Dayton 1986). Because the protection of an endangered species' habitat is essential for their preservation (Barla et al. 2000), EIA must integrate this consideration, in particular for habitats of vulnerable life stages such as breeding and migration.

Appropriate spatial scope of EIA in coastal seawater habitats is determined by how biological components interact with patterns of water movement such as boundary currents, gyres, eddies, and upwelling (Dayton 1986). In EIA, prevailing winds and weather patterns must also be accounted for because of their influence in dispersion of population, nutrients, and impacts (Dayton 1986). The concept of "natural oceanographic units" serves as a concept valuable in determining spatial scope while integrating the above factors. Based on movement of matter and the distribution of strong interrelationships among components in Nova Scotian marine waters, three units can be identified; the Scotian Shelf, Bay of Fundy, and Southern Gulf of St. Lawrence. Specific habitats of species at risk (Table 4-3), particularly at sensitive life stages, may be considered within these larger units. However, migration patterns of these species link the Scotian Shelf directly with the other two water bodies.

Much relevant data have been collected to date in conditions for which the features of land proximity, water depth, and water circulation are dissimilar to those that exist locally. This may be significant when considering the Southern Gulf of St. Lawrence and the Bay of Fundy because of the impact of these conditions on wave action and circulation, which will greatly effect the dispersion of toxic materials within the oceanographic units (Dayton 1986). Of the three species selected for study, considerations specific to Atlantic salmon use the river populations as the appropriate unit of analysis (Cronin and Bickham 1998) and the importance of aquatic sediment quality to breeding requirements (Murphy et al. 1999).

Diurnal variation is important to EIA in terms of sampling. However, accounting for seasonal variation in species vulnerability through cycles of migration, nesting, and feeding is critical for accurate results (Dayton 1986). Seasonal variation in environmental effects on production impacts and mitigate actions in offshore petroleum activity includes the effects of extreme weather, ice movement and cover, as well as localized wind events such as the Suêtes in Cape Breton. Annual changes in currents are significant (Peterson 1993). Episodic events may have long-term, and perhaps unobserved impacts upon a population (Dayton 1986). Based on a historic cluster of earthquake epicenters at the northern limit of Nova Scotian waters, including a 1929 event measuring 7.2 on the Richter scale (SIEAP 1983), geological seismic activity is another significant episodic potential. Longevity of organisms must be considered with respect to each of these temporal considerations.

Table 4-3   Summary of status and relevant life history features of selected endangered species of Nova Scotia

	Atlantic Salmon* (FOC 2001)
Estimated population	River-specific. Fundy and Atlantic coast are in strong decline. Gulf of St. Lawrence in slight decline or stable.*
Date of listing	2001
Occurrence	Pelagic; western Cape Breton Island stable, Scotian Shelf and Fundy populations extirpated or endangered.*
Seasonal distribution (some variation exists among regions of province)	Aquatic for first year, then over-winter off Greenland from October to May. Return and summer in coastal and estuarine waters, enter rivers, and spawn in fall.*
Feeding	Aquatic insects in early years, chiefly marine invertebrates as adults.
Lifespan	Maximum 9 years
Known local causes of decline	Water quality of river habitat resulting from land use, pollution, physical disruption, acid precipitation, and buffering capacity of water (Kessler-Taylor 1986).
Causes of decline external to Nova Scotian waters	Commercial fishery off Greenland (Kessler-Taylor 1986).

4.4.3.2 Fisheries

Generally, offshore oil and gas projects using fixed platform drilling rigs, are located within the shallow coast of Nova Scotia but the impacts on the fisheries might be felt much farther away than the Nova Scotian coastal areas. The pelagic fish, because of its highly migratory nature, spends some of its life cycle off the Scotian Shelf. The effects of hydrocarbon accidents and pollution may extend further than the legal boundaries. If there is a release of effluents or oil spills, phytoplankton and zooplankton production will be destroyed, which in turn will severely affect the plankton feeding fish and fish larvae. The effects on the productive coastal and estuarine productive system will be harmful for the overall marine ecosystem.

The interaction of oil and gas development projects with fisheries has crucial effects. Effects might extend for the short term during exploration or construction, but activity during breeding and migration periods of some fish species will have greater effects. Herring, tuna, and mackerel are the most migratory fish. Mackerel approach the Atlantic coast in late May in large shoals and leave again in the fall for overwintering (NHNS 2002). Bluefin tuna spend winters in warm southerly waters and move to Nova Scotia as the season progresses. If the seismic operation or other exploratory operations are taking place in this period then the effects will be greater. The overall effects on fisheries will be continuous throughout the life cycle of an offshore oil and gas project.

4.4.4   *Commonly used indicators in impact assessment*

4.4.4.1 Species at risk

Population size, mortality, and reproduction rate of the fish species in question can be used as indicators, along with impacts upon habitat:

*   The Department of Fisheries and Oceans uses surveys of egg concentration and fishing returns on individual rivers as a basis for predicting salmon population (FOC 2001).

*   Concentrations of toxins that are potentially released by industry into muscle tissue are measured (Zhou et al. 1997), and enzyme responses in liver and gills of salmon are used as indicators of toxin presence (Gagnon and Hardway 1999).

*   Impact upon salmon habitat is examined by monitoring oil residue and PAH levels in sediments of spawning areas (Murphy et al. 1999; Stagg et al. 2000).

*   Indicators for impact on the leatherback turtle have typically been through extensive and ongoing record-keeping of stranding events (Peterson 1993), while water quality and concentrations of toxins in turtle tissue and their food source can also provide information on impacts (Caurant et al. 1999). The most reliable population estimates are nest counts on beaches in their southern breeding grounds (Peterson 1993).

*   The International Piping Plover Census in 1991 and 1996 have been used to quantify levels of piping plovers in breeding grounds and wintering grounds. These also serve to identify habitats important to plover survival (Plissner and Haig 2000).

4.4.4.2 Fisheries

In nearshore environments, petroleum hydrocarbons resulting from oil spills or operational discharges are more likely to reach the seabed and be incorporated into bottom sediments. Toxic hydrocarbons, particularly medium and high molecular weight aromatics and heterocyclics, may persist for a long period of time.

Through measuring the chemical tracers of drilling discharge in sediments, (e.g., barium, chromium, aromatic hydrocarbons), we can estimate toxicity levels, which will reflect the impacts of the hydrocarbons (Spies 1987). Mussels are sessile, bottom dwelling organisms and so serve as primary indicators of hydrocarbon contamination because they become tainted when affected by oil or hydrocarbon effluents.

There is no evidence that an oil spill has direct effects on a stock as reflected by mortality of eggs and larvae (Boesch et al. 1987). However, there is concern that a large spill occurring during a critical recruitment period could seriously diminish recruitment to the stock for the following year. This could particularly apply to the eggs and larva concentrated in the surface water. For example, haddock have only one year-class out of every five to ten that contributes substantially to the fishery. The loss of a good year-class could be disastrous to population numbers. Carefully studying stock size, fish landing data, and species composition will help us to detect if there are any major changes due to hydrocarbon pollution.

4.4.5   Approaches and models available for impact prediction

4.4.5.1   Species at risk

Approaching EIA from habitat considerations within natural oceanographic units allows the application of Ecosystem Stability Models. Relevant modeling work is founded chiefly on the behavior of toxins in the environment. However, only limited modeling of the effects of impacts upon populations exists. The assessment of modeling efforts has proved to be difficult for endangered species because they are mobile and widely distributed, and are therefore difficult to monitor (Plissner and Haig 2000).

A simple but useful approach has been to correlate mortality with exposures to activity, based on previous observations, and applying this work in the prediction of impacts. Existing records, such as leatherback turtle strandings, have been used in assessing the impacts of fishing activities, particularly when viewing changes in regulations as in "experiments" in fisheries' impacts on turtle population (Peterson 1993). There is also considerable data from accidental spills, particularly on salmon on North America's west coast (Murphy et al. 1999) and on Scottish salmon farms (Ackman et al. 1996).

4.4.5.2   Relations to fisheries

It is necessary to conduct inventories and detailed chemical analysis of a variety of produced waters to determine potential toxicity and to obtain environmental tracers to aid in fate prediction. This can be determined by measuring biological effects at individual, population, and community levels. Individual induction of enzyme systems and biochemical and physiological stress, indicates hydrocarbon effects on marine fish.

There are some common methods for studying the population age and size structure, reproduction, and recruitment that can help us predict the impact of oil and gas on fisheries. Three popular models are the Catch per Unit Effort (CPUE) in the Inshore Fishery, Electronic Length Frequency Analysis (ELEFAN), and the Environmental Model.

The CPUE is mainly used in the inshore fisheries (Zimmerman 1999). Using this model can help us to estimate the status of a stock by observing changes in the availability of fish to fishermen.

The ELEFAN computer software takes length frequency data of fish as input, and gives the status of a stock. It is one of most convenient methods to study the status of fisheries. By carefully collecting fisheries data, prediction of the impact on the fisheries may be aided.

The Environmental Model is used to predict relative direction of the annual harvest with respect to the historical average. The model uses physical parameters such as air temperature and rainfall along with the fisheries data as parameter input (Zimmerman 1999). This model gives a more accurate picture of stock status by considering environmental parameters. It might be useful in predicting the impact of oil and gas on the fisheries.

4.4.6   Typical pre-mitigation impacts

4.4.6.1   Species at risk

Caution must be exercised in applying current knowledge to the local context. Much information on the impacts of activities associated with petroleum development has been gathered in other conditions dissimilar to the study area. The degree of applicability to nearshore, shallow, and variably confined water bodies present in Nova Scotia should be considered in each case.

The impacts of installation will be based on the amount of traffic and changes to water quality as a result of emissions and pollution from boats. Marine installation activities will occur in areas that are critical to migration of the two marine species under consideration. However, habitat degradation is a factor to consider for all three:

- The selection and handling of drill muds are significant in determining the toxicity of materials introduced into the environment. Oil-based muds are to be avoided unless necessary. They can be re-used; and if used, mineral oils are preferable because of their lower toxicity (WCOEEAP 1986). Mud release causes smothering of bivalves in benthic habitats and would only impact the endangered species under consideration through overall ecosystem degradation (WCOEEAP 1986).

- The protocol for disposal of drill cuttings is also important for water quality.

- Related land-based activities such as storage, traffic, helicopter pad, and dock facilities are an essential feature of the activity (HEAP 1986) and could have an impact upon piping plovers and the aquatic breeding habitat for salmon. Helicopter noise causes fleeing of birds from nests (WCOEEAP 1986), causing the effects to be seasonally significant.

Drilling and production wastes are a major environmental issue and must be considered for their potentially significant effects (SIEAP 1983):

- Routine discharge includes grey and black water, ballast/preload water, bilge water, deck drainage, machinery discharge, garbage, and cooling water (Collins et al. 2001b).

- The normal range of pollutants released by nearshore petroleum activities will exist at higher concentrations at sites closest the operation (Collins et al. 2001b).

- Nutrient enrichment and warm water plumes of discharge will attract fish (SIEAP 1983) and can attract turtles and salmon, potentially lengthening their exposure to higher concentrations of toxins. They will also disrupt fish distribution patterns, affecting overall environmental conditions for turtle and salmon.

- All products transported and used by the activity have potential for hazard through accidental spillage.

- Migrating birds are attracted to marine lights and structures and may suffer injury as a result (Collins et al. 2001b).

- Because of significant boat traffic noise already present, petroleum development is not described as a significant contributor to noise levels (Collins et al. 2001b).

- Flaring dramatically decreases air quality (Gabos 2000). This deterioration of the environment does not threaten marine species immediately, but is a potential threat to the plover.

- The contents of petroleum reserves cannot be identified definitively before actual drilling (Collins et al. 2001b), thus the impacts of both oil and gas release into the marine environment need to be considered. Their potential impacts include direct harm to individual organisms and impact on populations due to degraded habitat quality. The risk of blowouts is estimated at 0.56%, and those that may involve the release of oil into the marine habitat are estimated at 0.02% (S.L. Ross Environmental Research Ltd. 2001). The bulk of gas released to the marine environment escapes to the atmosphere within 24 hours, thus drastically reducing its potential to harm marine species (S.L. Ross Environmental Research Ltd. 2001). There is a need to locate specific data on actual residue remaining in the water column, and its potential impact hazardous effects on species. The focus of EIA for spills is at the water surface and coastal habitats.

There are considerable data available as a result of oil spills on fish species:

- The potential damage of marine oil lies in the variety of oil products, which differentially disperse in tissue, substrate, and food. Ackman et al. (1996) provide data on the effects of petroleum spills in salmon farms.

- Lethal effects are expected during blowouts in nearshore environments (SIEAP 1983). Long-term sublethal effects (HEAP 1986) and reproductive dysfunction (Peterson 1993) are cited as potential impacts.
- A decline in efficiency of feeding through impaired food conversion is also documented (Vignier et al. 1992).
- The persistence of weathered oil, which releases mutagenic PAHs, is a direct threat to the environment (Cronin and Bickham 1998). It is particularly crucial in coastal and estuarine feeding habitats if there is landfall of spills.
- Marine turtles develop lesions and suffer damage to eye, nose, and lung tissue in response to oil exposure (Collins et al. 2001b). Lung and digestive capacity are also noted to suffer (Collins et al. 2001b). Impact on food quality is important as dietary preferences are thought to be the main influence in variation in toxicity levels found in tissues among turtle species (McKenzie et al. 1999).
- The impact of landfall of oil slicks during breeding and rearing periods of the plover would be critical in fouling nesting areas and food.

4.4.6.2 Fisheries

All groundfish are carnivores, feeding on benthic invertebrates such as worms, molluscs, and crustaceans. Hydrocarbon pollution directly affects the benthic invertebrates, which are prime food of groundfish, so could potentially deplete the groundfish stock.

The pelagic fisheries are complex in nature. Most of the pelagic fish species are migratory moving along the entire Atlantic coast of Nova Scotia. If there is any oil pollution in any part of the Nova Scotian coast, it could be devastating for the local fisheries as well as the Newfoundland or US fisheries.

Pelagic species generally feed on zooplankton and smaller fish species. Herring and mackerel feed on the planktonic crustaceans and fish eggs and larvae, and may also filter with their gillrakers when food is suitable (NHNS 2002). Each of these components will potentially be contaminated in the event of an oil spill.

4.4.7 *Typical mitigation measures and residual impacts*

4.4.7.1 Species at risk

Avoidance of sensitive life stages, such as breeding and migration, and the critical habitats, such as breeding and migration, would be the cornerstone of mitigative measures. Seasonal variation is the key factor to consider in reducing the negative impacts of any phase of activity.

For installation:

- Land-based activities should be placed in existing industrial zones (HEAP 1986);
- Air traffic should avoid plover nesting areas; aircraft exclusion zones should be established;
- Careful management of effluent;
- Choice of lower-toxicity drill muds;
- Recycling and onshore disposal of muds;
- Onshore disposal of drill cuttings; and
- 500 m exclusion zone around rig recommended (Collins et al. 2001b).

For production:

- Routine discharge – There are a number of processes by which routine discharge can be reduced; "Zero discharge" can be applied to some products. However, in the case of the Sable Gas Project, dilution upon entry to the marine environment was considered to be an effective mitigative measure (SIEAP 1983).

- Noise and Risk Mitigation – Efforts to reduce incidence of accidental spills and noise can be heightened; however, there may be little done to mitigate the effect of structures.

- Flaring – The collection and use of gases otherwise burned off can be executed (Aycaguer et al. 2001); while burning at higher temperature also reduces toxicity (Gabos 2000).

- Blowouts – Chemical dispersion is considered as a mitigative measure in marine spills to speed biological degradation of the spilled material, and to lessen the impact of landfall. Reducing the size of oil droplets decreases the physical impact of oil upon marine species. However, the associated increase of surface area increases toxicity of oil in the environment (Environment Canada 1973). Dispersion is more effective for lighter-weight products, thus not effective in reducing the toxic release of PAHs from weathered oil (Environment Canada 1973).

The requirement to surface for air places the turtles at risk in the vicinity of spills, and it would seem that spotting and removing turtles would be the most effective method, if we wish to reduce the effect of the spill.

In the case of spills, all efforts to avoid landfall are critical to the piping plover. Landfall impacts are mitigated with on-site burning, dispersion, or hot-water or cold-water cleaning of the sediment (Environment Canada 1973). However, the effects of traffic, toxic nature of dispersants, and damage resulting from hot and cold water cleaning of substrate are thought to be as destructive as the impact of the oil (Environment Canada 1973). Clean-up requires attention to the life cycle stage in order to reduce impact upon the piping plover.

4.4.7.2 Fisheries

In most EIA studies for oil and gas development it is generally suggested that a contingency plan be drafted in the case of accidents, blowouts, or oil spills (Collins et al. 2001a). In this report, some new approaches are offered to mitigate the impacts of fixed rig drilling to the fisheries resources.

Many fish species, especially groundfish species, have declined in number in Nova Scotia over the last ten years (SCF 1993). There is potential for further reductions in fish stocks due to the future effects of oil and gas development. This will augment the already existing problem of depleting fisheries in some communities. The aim of artificial recruitment is to compensate the depleted fish stock. After identifying which species are affected, an initiative might be taken to produce the fish larvae in a hatchery. After rearing the larvae up to stocking size, they can be released into the open seawater. How much larvae will be needed for restocking can be estimated by considering the mortality rate of larvae in the open water. This compensation technique may help recover the fish species stocks.

One approach to site remediation is to decommission offshore oil and gas platforms by dismantling them completely and disposing of all component parts onshore (Pulsipher and Daniel 2000). The alternative approach of the traditional decommissioning might be to develop artificial reefs by using the abundant rigs. The artificial rigs might be used for fishing or diving, or constitute platforms for offshore wind-energy projects.

Marine scientists have argued that trawling is detrimental to the sea bed. Les and Norse (1998) reported that bottom trawling is comparable to clear-cutting of forest lands. However, clear-cutting, the authors note, annually affects an area the size of the state of Indiana, while trawling takes place over an area twice the size of the contiguous United States. It is also reported that fishing gear is damaged due to the presence of drilling structures. Thus platform positioning practices, which limit the area of the sea bed open to trawling, may well benefit the overall marine environment. Presently operating or abandoned oil and gas development zones might be designated as marine protected areas.

Fish breeding grounds should not be open to oil and gas exploration. Destruction of breeding not only displaces fish, but also reduces the juvenile recruitment to the stock. Oil effluents have had great impacts on larval fish. It is reported that larval metamorphosis is hampered due to the effects of oil effluents. Extensive fish kills have been associated with only a few spills. In one spill in Florida large numbers of scup and tomcod were washed ashore (Spies 1987).

4.4.8 Applicable monitoring protocols

4.4.8.1 Species at risk

Maintenance of current habitat protection measures must continue, along with meticulous record-keeping. Data should be collected on status and behavior that may indicate reactions to activities. These may prove helpful in further developing an understanding of the impact of nearshore petroleum development on species at risk. Surveys must be non-intrusive because of the vulnerability of the species. The migratory component of each of the species' life cycle calls for strengthening of the international initiatives, which are already active, and the need to coordinate knowledge sharing at an international level (Peterson 1993).

4.4.8.2 Fisheries

The food chain is the most important component in the ocean's biological process (NERC 2001). The major player in the food chain is the primary producer, which generates carbohydrates through photosynthesis. All other organisms in an oceanic environment depend on this primary producer (Maine Shore Stewards 2001). Thus the productivity of the ocean, for example, fertility and nutrients for fish and other organisms of the ocean, is determined by the primary producer of the ocean, chlorophyll-pigmented microscopic phytoplankton (Cobscook Bay resource Centre 1993). Microbes and plankton are sensitive to any kind of environmental change. By studying the concentration of plankton we can estimate the status of fish stocks or productivity. The plankton/microbe populations will fall quickly in response to medium- or long-term exposure to oil. Using satellite data, it is now possible to directly monitor or trace any effluent release, especially oil spills from rigs or transporters.

In addition to the satellite's estimation of ocean productivity, the biomass of the study area, especially close to oil rigs, should be monitored by using the light and dark bottle method (chemical methods) (CNSOPB 2002). This will aid in understanding the effects of oil and gas development on the food chain, and indirectly on fisheries.

Fish tissue should be taken and analyzed to monitor the hydrocarbon accumulation in fish bodies. Besides tissue sampling, it might be necessary to study the bioavailability of petroleum hydrocarbons and their degradation to various trophic groups, especially the benthos, which is an important food for groundfish.

4.4.9 Other projects/activities and cumulative impacts

Petroleum operations cannot be separated from other human activities. Khan and Islam (2007) outlined the impact of petroleum activities on other projects/activities, as well as the cumulative impact of such activities on other species. Because any degradation of the ecosystem will ultimately affect human beings, it is important to evaluate overall impact of petroleum activities. Table 4-3 summarizes the impact of petroleum activities on selected endangered species that are subject to risks associated with many habitats during their migration journey.

Potential seismic impacts on munitions dumps within licensed areas have been studied by MacNeil (2002). Similarly, cumulative chemical effects of petroleum activities have been documented as early as the 1990s (e.g., "Pigmented Salmon Syndrome" reported by Croce et al. 1997). Also more recent investigations indicate that cumulative noise effects on behavior, hearing, and health can be expected. It is noted that changing patterns of fishing due to displacement by the petroleum industry may alter their effects on fish and turtle populations. It is well known that beach traffic may be influenced by the petroleum industry, impacting piping plover. Pursuit of activities, which threaten endangered species, may adversely affect the attraction of naturalists, anglers, and eco-tourists to the region, impacting the tourist and service industries. Increased boat traffic will increase risk of boat collision and potential for spills of various substances (WCOEEAP 1986).

4.4.9.1   Fisheries

Cumulative effects assessment is a vital tool to ensure progress towards sustainability (HAL and GMA 2000). It is important to study cumulative impacts in the fisheries sector. This will help in identifying which other sectors are also responsible for contributing to impacts felt by Nova Scotia fisheries. We identified marine transportation (traffic, gear damage), land and air transportation, fisheries, and recreation and tourism as the main sectors which are also responsible for the cumulative impacts on fisheries. Seabed mining is also a sector with potential for development along the Nova Scotian coast, which may contribute to cumulative effects.

4.5   Operations in the Ecologically Sensitive Areas

Previous sections discussed the VECs-based assessment for drilling and production operations. Recently, oil and gas operations have been extending to special areas that are ecologically sensitive. Conventionally, petroleum operations are not allowed in these areas because of their ecological value. However, by developing special measures, drilling and production can be carried out. In this section, an approach is discussed to operate drilling and production in such ecologically sensitive areas. A case study is presented in the Sundarban mangroves, which is a world-famous ecologically sensitive ecosystem. It shows that by undertaking proper protection and mitigation measures, oil and gas operations can be developed, while still preserving ecological quality and protecting wildlife. Since there is limited research that focuses on environmentally sensitive areas, this work can serve as a basis for understanding the potential effects and required remediation of oil and gas in such an area. The findings of this case study are applicable in any environmentally sensitive area.

4.5.1   Sundarbans sensitive ecosystem

Bangladesh has the world's largest mangrove forest. The Sundarbans is one of the most productive and biologically diverse wetlands in the world. Figure 4-9 shows the position of the Sundarbans. This unique coastal tropical forest is among the most threatened habitats on Earth. Its importance lies in its floristic composition, resource and economic value, and as a precious wildlife reserve. It is native to some 40,000 wildlife species, including endangered royal Bengal tigers, rare freshwater dolphins, and crocodiles (Table 4-4). It is a nursing and breeding ground of over 400 fish species, over 300 bird species, and many others.

Mangrove ecosystems are one of the most productive and biodiverse wetlands on Earth. In contrast, these unique coastal tropical forests are among the most threatened habitats (Fall 2005). Due to commercial exploration they may be disappearing more quickly than the inland tropical rainforest, and so far, with little public notice (Dewalt et al. 1996; Fall 2005).

The Sundarbans is the largest single-tract mangrove ecosystem in the world, with an area of about 10,000 km^2 (FAO, 1994). It contains a large variety of genera and species of plants, wildlife, and an astonishing biodiversity (Saha et al. 2006). It stretches across Bangladesh and India, over the northern part of the Bay of Bengal (Figure 4-9). Due to its great significance, the Sundarbans was declared a World Heritage Site and Biosphere Reserve. However, the future of the Sundarbans remains uncertain due to logging, human settlement, and other numerous exploitation activities (Saha et al. 2006).

In order to improve a struggling economy, the Bangladesh government has been looking into using the oil and gas resources in the country. After the recent discovery of huge amount of proven gas reserves, Bangladesh received attention from many multinational oil companies. It is considered that its reserves are placed at 15.3 trillion cubic feet (TCF), while the US Geological Survey estimates that it contains an additional 32.1 TCF "undiscovered reserves" (Kumaraswamy and Datta 2006). The most terrestrial and offshore areas of Bangladesh have been divided into 23 blocks and are gradually being leased to multinational oil and gas companies.

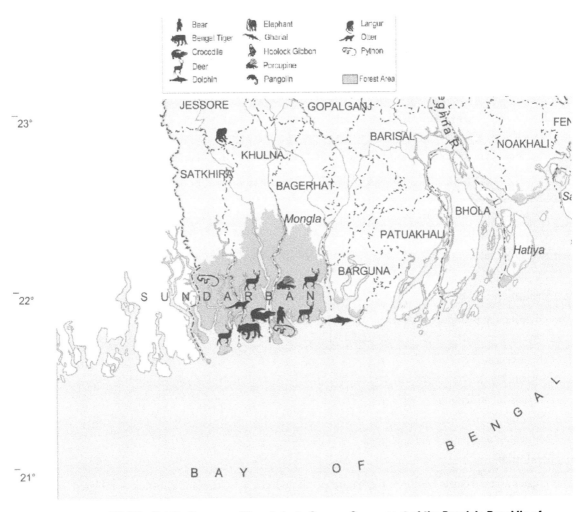

Figure 4-9 **Wildlife distribution map of Bangladesh. Source: Government of the People's Republic of Bangladesh (2006).**

Table 4-4 **The population of the large terrestrials mammals of Sundarbans**

Name	Population
Spotted Deer	80,000
Wild Boar	20,000
Tiger	350
Rhesus Macaque	40,000
Otter	20,000

Source: ESCAP (1987)

The Bangladesh government recently completed the initial signing of a production sharing contract with two multinational oil giants to begin oil and gas exploration in block number five within the Sundarbans. The companies plan to conduct seismic and aerial surveys. In Bangladesh the gas and oil exploration business is in its infancy and therefore environmental impacts have not yet been felt. However, the impact of hydrocarbon pollution has been clearly reported in many other parts of the world (Stejskal 2000; Teresa et al. 2002; Cheevaporn and Menasveta 2003; Guzzella et al. 2005). Oil and gas exploration might pose a new and, for the most part, unknown threat to the Sundarbans ecosystem.

4.5.2   *Environmental engineering planning and impact analysis*

The environmental engineering, when planning an impact analysis for oil and gas activities in a certain area, should consider its physical, social, economic, and environmental impacts as a whole in order to foster a healthy and efficient society in the investigated region.

Environmental engineering planning is a continuous process. It is composed of the following stages (Figure 4-10):

1. Goals statement
2. Data collection and analysis
3. Program establishment
4. Resources allocation
5. Actions implementation
6. Evaluation.

The environmental impact assessment of an oil and gas activity comprises the following steps (Figure 4-11):

1. Site location
2. Activity types
3. Impacts identification
4. Proposal
5. Mitigation and impact management
6. Evaluation.

The impact assessment is based on the nature of oil and gas activities and substances released while performing such activities.

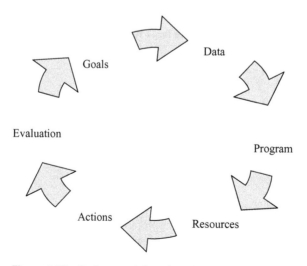

Figure 4-10   Environmental engineering planning stages.

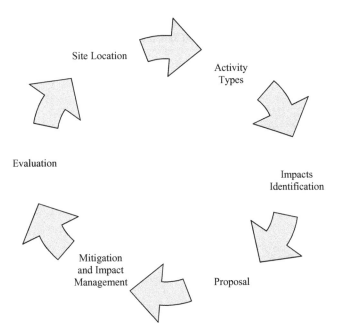

Figure 4-11 Environmental impact analysis steps.

4.5.3 Drilling and production activities

To evaluate the drilling and production operations in the Sundarban mangrove sensitive area, all activities are analyzed (Table 4.5). Oil and gas field production begins if a commercially feasible amount of oil and gas is discovered during drilling. It starts by constructing many production and transportation facilities and replacing temporary facilities. Based on the size of the oil reservoir, immense amounts of materials are necessary to erect production facilities. For example, the Sable Island offshore energy project in Canada used more than 900 km of pipe, 109,000 tonnes of steel, 53 km of chrome tubing, and 160,000 tonnes of concrete (Khan et al. 2006c). The potential development and production activities for the Sundarbans mangroves are shown in Table 4-5.

4.5.4 Potential impacts on ecologically sensitive sites

Many studies report that mangrove ecosystems are especially unique and among the most productive ecosystems. To identify the potential impact of oil and gas development, the valued ecosystem components (VEC) of Sundarbans are identified. Considering the values and services provided by Sundarbans, five VECs are discussed in this section.

4.5.4.1 Wildlife

As discussed above, the Sundarbans is a unique habitat for wildlife and supports a rich and diverse fauna (Saha et al. 2006). There are 315 species of birds, 400 species of fish, 53 species of reptiles, and 8 species of amphibians (Hendrichs 1975; Mukherjee 1975; Sarker and Sarker, 1988; Islam 2001). Species diversity and wildlife population are shown in Tables 4-6 and 4-7.

Since current oil and gas production operations cause pollution, it is expected that such activities will adversely impact the wildlife of the Sundarbans (Bolze and Lee 1989). For example, the presence of development-associated structures may interfere with movement (Table 4-6). If it is severe or prolonged, it may

Table 4-5   Potential environmental disruptions resulting from oil and gas and development activities in the Sundarbans

Activities	Potential Development						
	Noise	Aircraft	Human intrusion	Traffic and access	Structure	Habitat alteration	Harmful material
Seismic wave production			+				
Clearing vegetation	++	+	+			+	
Road construction	+++		++	+		+	
Mobilization of truck/equipment	++		+	+			
Site development	+	+	++	+		++	+
Drill pad construction	+	+	+		+++	+	
Excavation of storage/mud pits	++						
Drilling and related activities	+		+				
Water supply	+				+		+
Borrow pit excavation	+		+		+		
Wellhead/pump until installation	+						+
Construction of process plant and storage facilities	+		+++	+		++	+
Installation of flow lines			+		++	+	
Erection of power lines	+		+				
Communication system					+		
Drainage facilities	+		+		+	+	
Operation of process facilities					++	+	
Pipe string							
Trenching and pipe installation	+		+				
Pipe burial and backfill	+		+		+		
Maintenance and inspection	+		+				
Accidents	+					++	+
Secondary recovery	+		+			+	
Air trip/helipads	+	+					
Worker accommodations		+	+				
Increase in local population			+++			+++	
Development of ancillary	+				+++		
Industry	+		+	++		+++	
Well plugging			+		+++		
Site restoration/renegotiation	+		+				

result in reduced reproductive success or loss of available habitat (Bolze and Lee 1989). Increased human/ wildlife encounters may cause animals to avoid an area. It is also reported that effects on courtship behavior reduce existing stock and permanent loss of habitat use (Bolze and Lee 1989). The noise of helicopters and other crafts will be a serious threat to the huge annual winter migration of birds. A detailed impact on wildlife is shown in the Table 4-7.

4.5.4.2   Species at risk

There is an amazing amount of wildlife throughout the forest, but a large number of valuable animals have already become extinct (Sarker and Sarker 1988). For example, out of the 49 mammal species, no less than 4 major species, the Javan rhinoceros (*rhinoceros sonadacius*), wild buffalo (*Buhalus huyabails*), swamp deer (*Cervus duvauecli*), and hog deer (*Axis porcinus*) have become extinct since the beginning of the last century (Sarker and Sarker 1988).

Table 4-6 Species diversity in the Sundarbans

Group of Species	No. of Species
Mangroves	27
Mangrove associate plants	69
Phytoplankton and algae	37
Ichthyoplankton	200
Fish and shrimp	400
Shark	3
Shrimp	24
Skates and rays	5
Crab	7
Lobster	3
Molluscs	26
Mammals	49
Reptiles	53
Amphibians	8
Birds	310

Sources: Hendrichs 1975; Mukherjee 1975; Ahamed 1984; Islam 2001

Table 4-7 Potential impacts of oil and gas development activities on the Sundarbans' VECs

Activities	Potential Impacts Level on VECs						
	Wild-life	Species at Risk	Especial habitat	Forest	Fish-eries	Aqua-culture	Solar salt
Clearing vegetation	5	7	7	6	6	3	0
Road construction	4	5	6	3	3	1	1
Mobilization of truck/equipment	3	4	3	1	1	1	1
Site development	5	5	5	3	3	0	1
Drill pad construction	4	4	5	3	2	1	0
Excavation of storage/mud pits	4	3	4	3	1	1	1
Drilling and related activities	7	7	6	2	8	6	7
Water supply	0	2	4	5	1	3	2
Borrow pit excavation	2	2	2	1	1	1	2
Wellhead/pump until installation	2	2	2	1	1	0	0
Construction of process plant and storage facilities	3	3	3	2	4	0	0
Installation of flow lines	4	3	3	2	2	1	0
Erection of power lines	2	3	4	3	0	0	0
Communication system	1	2	1	2	0	0	0
Drainage facilities	5	2	5	3	3	2	7
Operation of process facilities	6	3	4	2	5	3	7
Pipe string	4	3	3	2	2	2	1
Trenching and pipe installation	2	4	3	6	1	3	2
Pipe burial and backfill	2	3	2	3	1	2	2
Maintenance and inspection	2	4	1	1	2	1	2
Accidents	7	10	10	7	8	9	10
Secondary recovery	5	5	7	5	3	3	4
Airstrip/helipads	3	3	4	2	1	2	0
Worker Accommodations	4	4	4	1	1	1	0
Increase in local population	5	5	7	3	3	1	0
Development of ancillary industry	5	6	5	3	2	1	1
Well plugging	2	3	2	1	2	1	2
Site restoration/renegotiation	2	4	2	2	2	2	3

The tiger is one of the most endangered species in the world. Today, there are only about 7000 tigers left in the wild, down from more than 100,000 a century ago (UN 2002). Almost 550–600 tigers reside in the Sundarbans. The following Sundarbans species are those considered to be globally threatened as of late 1993 according to the IUCN status categories:

- White-bellied Heron (*Ardea insignis*);
- Lesser Florican (*Eupodotis indica*);
- *Panthera tigris* (Tiger);
- Hispid Hare (*Caprolagus hispidus*);
- Hoolock Gibbon (*Hylobates hoolock*);
- Green Turtle (*Chelonia mydas*);
- Hawksbill Turtle (*Eretmochelys imbricata*);
- Olive Ridley (*Lepidochelys olivacea*);
- Batagur (*Batagur baska*).

Oil and gas developments can further endanger the species already at risk. Several studies from different parts of the world show that oil and gas development activities have negative impacts on species at risk (Stejskal 2000). Table 4-7 shows the detailed impacts of oil and gas operations on the Sundarban wildlife. Similar effects are also discussed below.

4.5.4.3 Special habitats

The Sundarbans provides a critical habitat for diverse marine and terrestrial flora and fauna (Table 4-6). It provides ecological services and works as an essential habitat, nutrient producer, water purifier, nutrient and sediment trap, storm barrier, shore stabilizer, aesthetic attraction, and energy storage unit (Hossain 2001; Islam 2003). Mangrove sediments appear to be efficient in retention and accumulation of phosphorus (Silva and Mozeto 1977).

Mangroves have a significant importance for fisheries. Many species use the mangrove ecosystem in various ways (Hossain 2001; Islam 2003). Some fish, obligatory species, spend their entire life cycle in this environment. Migratory species spend a crucial part of their life (e.g., peneaid prawns) using it as a shelter or as a source of food. Others, the facultative species, survive and reproduce even in the absence of mangroves but show a preference for the habitat and the nutrients provided therein (Hong 1996).

Healthy mangrove forests are the key to healthy marine ecology. Deterioration of the Sundarbans mangroves due to oil and gas exploration might directly and indirectly cause declines of fisheries, degradation of clean water supplies, salinization of coastal soils, erosion, and land subsidence (Hossain 2001; Islam 2003). This coastal ecosystem is so specialized that any minor variation in its hydrological quality causes noticeable mortality (Blasco et al. 1996). Potential impacts of oil and gas drilling and production are presented in Table 4-7.

4.5.4.4 Mangrove forests

The Sundarbans form 45% of the productive forest of the country. It contributes half of the forest-related revenue and is an important source of wood and no-wood resources (Katebi and Habib 1989). A wide range of forest products such as timber, fuelwood, pulpwood, matchwood, and thatching materials are obtained from the Sundarbans. In 1992–1993, US$ 7.0 million revenue was earned from the area (FAO, 1994).

The natural vegetation of the Sundarbans is composed of halophytic tree species, commonly termed as mangrove. There are 56 species of mangroves and mangrove associates in the Sundarbans (Table 4-6) (Chaffey and Sandom 1985; Khatun and Alam 1987; Saenger and Siddiqi 1993). The major predominant tree species are;

Sundaries (*Heritiera fomes*), and
gewa (*Exocoecaria agallecha*)

followed by minor proportions of:

passur (*Caraoa obovata*)
keora (*Sonneratia apetala*)
baen (*Avicennia officinalis*)
kankar (*Pruguiera gymorhiza*)

and a few other species such as:

goran (*Ceriaps rexburghias*)
hantal (*Pheoenix paludosa*)
shingra (*Cynometra remiflora*)
khalisi (*Ameiceras maius*), and
bhola (*Hibiscus tiliaceus*).

Oil can have a significant negative impact on the mangroves. Potential impacts on mangrove forests are shown in the Table 4-7. If an oil slick were to enter the mangrove forest when the tide is high, it would be deposited on the aerial roots and sediment surfaces as the tide proceeded. Oil that covers the trees' breathing pores can asphyxiate the surface roots and so lead to the death of the mangrove (Katebi and Habib 1989). The trees can also be killed by the toxicity of the hydrocarbons. It is reported that a lighter fraction of the oil, considered to be the most toxic, generally evaporates or degrades rapidly and thus the heavier fraction is the cause of most of the chronic impacts (Teal and Howarth 1984). It is known what portion of these imparts are due to the use of toxic catalysts during the refining process (Eakhal et al. 2007)

Chronic exposure to hydrocarbons results in damage to aerial roots, 25–64.5% defoliation depending on the species, 100% mortality in seedlings, reduction in propagate density accompanied by atrophy and malformations, and finally a serious reduction in basal area (Lamparelli et al. 1997). Recovery from the effects of oil spills begins ten years after an incident. It is reported that an oil spill in 1994 near the Sundarbans spread about 15 km downstream from the ship and caused the instant mortality of seedlings of *Heritiera*, *Exoecaria*, and many species of grasses (Karim 1994).

4.5.4.5 Coastal fisheries

In the Sundarbans mangroves, an average of 10,000 tons of fish and shellfish were caught annually in the late 1970s and early 1980s (KU 1995). With 169,908 km^2 of waterways, approximately 53 kg/ha of fish are being caught within the mangrove area itself (Penn 1983; Mahmood et al. 1994; KU 1995). To this figure should be added the portion of the offshore fisheries of mangrove development species. Whereas over 120 species of fish and shellfish are known to be caught of the mangrove-dependent species in the Sundarbans area (Table 4-6), the main portion consists of shrimps and hilsha (*Tenolosa* and *Hilsha* sp.).

A recent estimate of the total yield from the inshore-estuaries and offshore fishery of the Sundarbans reserved forest records 11,700 tons in 1993 (KU 1995). Penn (1983) estimated the potential annual production of demersal finfish in the offshore waters of Bangladesh to be 10,000 to 20,000 tons, and that of the shrimps to be 200 to 4000 tons. He mentioned that available yields per unit length of the coastline were relatively high by world standards. One of the main reasons for this high yield is the influence of the Sundarbans.

Some of the potential negative effects of oil and gas development on coastal fisheries are listed below. For a general list of such effects, see Table 4-7:

- Altered reproduction of fish species;
- Damage to fish eggs and larvae;
- Reduced fish stock;

- Change in the quality of fish and shrimp due to tenting;
- Effects on the fish migration;

4.5.4.6 Coastal aquaculture

Coastal shrimp farming is one of the major economic activities in Bangladesh and shrimp ponds are generally developed in the mangrove areas. Chakaria Sundarbans, the southeast mangrove areas of Bangladesh, are almost (90%) destroyed through shrimp farming (Mahmood et al. 1994). Shrimp farming is not permitted in the southwest Sundarbans reserve forests but recently several aquaculture farms have been established at the border of the forest (FAO 1984a). Frozen food export is the second largest sector in Bangladesh. A major portion of the earnings come from farmed shrimp export. Every year around 30,000 tons of tiger shrimp is exported to Japan, North America, and Europe.

Exploration and development of oil and gas may cause environmental problems, such as tainted product quality, increased disease rate due to the oil pollution, closure of aquaculture production if pond soil is affected, and transfer of hydrocarbons to the human body. Details of impacts are shown in Table 4-7. In the table, a predefined schematic scale is followed. The degree of importance is within the range of 1 to 10, where 10 is the highest score and 1 is the lowest.

4.5.4.7 Salt production

Solar salt production is a traditional and important industry in Bangladesh, and producing table salt. Traditionally, salt is produced in the muddy coastal land (where water retention power is high) by evaporating the seawater (salinity 35 ppt). In 1997, from December 12 to the end of March, about 500,000 tonnes of refined salt were already produced in the coastal areas of Bangladesh (AMITECH 1997). A total of 50,000 acres has been brought under salt production in these coastal areas (AMITECH 1997).

Oil and gas operations in the Sundarbans as well as surrounding coastal areas can affect salt production, causing quality deterioration in edible salt, hydrocarbon transfer to the human body, salt bed contamination, closing of production, and pressure on the foreign currency (US dollar) to import salt. More detailed impacts are shown in Table 4-7.

In addition to analyzing, assessing effects of drilling and production activities, and developing mitigation plans, sustainable technology can play a positive role in minimizing the impacts on the ecosystem. The following section introduces emerging technologies in production and operations.

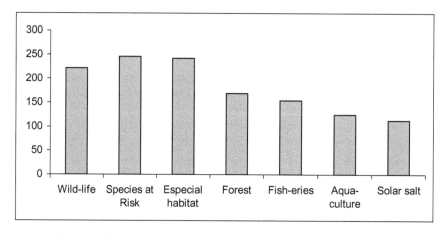

Figure 4-12   Impact levels of oil and gas development on the VECs.

4.6 Emerging Technologies and Future Potential

4.6.1 *Potentials of laser drilling*

Recently, lasers have been mentioned as a viable option to improve drilling technologies (Islam and Wellington 2001). Mustafiz et al. (2002) studied the benefits of laser drilling, current laser types, and the experimental work that has been conducted. By eliminating the use of a drill bit with laser drilling, the downtime created by dull bits will be drastically reduced and the waste created due to drilling mud will be eliminated. This is especially important in offshore drilling where space limitations hamper operations. In fact, laser drilling can increase penetration rates by greater than 100 times over the conventional rotary drilling methods (Graves and O'Brien 1999a). This can be of tremendous benefit. Lasers can penetrate many different types of rocks including sandstone, limestone, and shale (O'Brien et al. 1999a), thus reducing the problems that occur when conventional horizontal drilling is unable to achieve a perfectly straight borehole.

Lasers may have the ability to melt the rock in a way that creates a ceramic sheath in the wellbore (USDOE 2002), which would eliminate the problems encountered with formation fluid seepage. Besides, a significant reduction in drilling costs can be expected with laser application that is capable of mono-diameter drilling deep into the hole from the surface. Additionally, laser drilling has the potential to be used for perforation drilling (Graves and O'Brien 1998a). A study conducted by Graves and O'Brien (1998b) for the Gas Technology Institute determined that laser drilling in the oil industry, especially with infrared lasers, is promising due to its excellent heat transfer characteristics. O'Brien et al. (1999a) pointed out that a combination of laser and rotary drilling might be advantageous due to the resulting increase in bit life. The development of downhole laser drilling machines, laser-assisted drill bits, laser-perforation tools, and sidetrack and directional laser drilling devices are all possible with the advancement of laser technology (Gaddy et al. 1998).

4.6.1.1 Background

The word Laser is an acronym for Light Amplification by Stimulated Emission of Radiation. Basically, lasers convert different forms of energy such as chemical, electrical, heat, etc. into energy of light (intense photon). It is a source of electromagnetic coherent radiation where coherency stands for the temporal and spatial phase correlation of the propagation method. Petroleum applications of lasers include temperature measurement inside the wellbore (Woodrow and Drummond 2001), mud cake thickness measurements (Amanullah and Tan 2001), and asphaltene remediation in carbonate formations (Zekri et al. 2001). Islam (2001) outlined the application of lasers for realtime monitoring and eventual dynamic reservoir management. There are two categories of lasers, Continuous Wave (CW) and Pulse or Repetitive Pulse (RP). Table 4-8 lists the various types of high-power lasers that are viable for the oil and gas industry (O'Brien et al. 1999b, Graves and O'Brien 1998a).

The processes of destroying rocks using laser involves the transfer of radiant energy into solids. This transfer is determined by three basic phenomena: reflection, scattering, and absorption (Graves and O'Brien 1998a). Absorption is the process by which heat is collected by the rock, eventually resulting in the destruction of that rock. Scattering is a function of the wavelength and roughness of the surface while the composition of the solid and the planar surface determines the reflection (Gahan et al. 2001). Both scattering and reflection represent the energy that is not absorbed by the rock or the losses in the system. Therefore, the following criteria were outlined by Graves and O'Brien (1998a) for the feasibility of laser drilling:

- Low reflectivity of rocks, resulting in good coupling of laser radiation with rocks;
- Deep penetration of laser energy into rock, resulting in volumetric absorption of laser energy;
- Low thermal conductivity of rocks, resulting in effective heating.

Physical or chemical changes that may occur in the rock during lasing are determined by rock composition, crystalline lattice structure, porosity, water saturation, hardness, strength, elasticity, laser power density, wavelength, band of radiation absorption, and laser pulse length (Batarseh 2001).

Table 4-8   Laser types and wavelengths

Laser Type	Wavelength, μm	Comments
HF(DF) – Hydrogen Fluoride and Deuterium Fluoride	2.6–4.2	A continuous wave (CW) laser. Mid-Infrared Advanced Chemical Laser (MIRACL) is this type.
COIL – Chemical Oxygen Iodine Laser	1.315	A continuous wave (CW) laser developed in 1977. It is considered to have the potential to remove drilling and re-completing petroleum problems such as sidetracking, directional drilling, or well-control.
CO_2 Laser	10.6	It can be either continuous wave (CW) or repetitive pulsed (RP). The average power is 1 MW. It is durable and reliable, although its large wavelength makes it inefficient.
CO Laser	5–6	It can be either continuous wave (CW) or repetitive pulsed (RP). The average power is 200 kW.
FEL Laser – Free-Electron Laser	Any wavelength	It is a continuous wave (CW) laser. Since it can attain any wavelength, the laser can be adjusted to minimize reflection, scattering and plasma screening.
Nd : YAG – Neodymium : Yttrium Aluminum Garnet	1.06	Operate at a power of 4 kW.
KrF (Excimer) Laser – Krypton Fluoride Excimer	0.248	It is a repetitive pulsed (RP) laser. The maximum average power is 10 kW.

There are many factors in conventional drilling that determine the rate at which the borehole will be created. These factors include the weight-on-bit (WOB), mud circulation rate, rotary speed, hydraulic horsepower bit design, and hole size. With laser drilling, this rate may only depend on the delivered power and the hole size (O'Brien et al. 1999a), thus eliminating the complexity of current drilling systems.

4.6.1.2   Experimental review

Experimental work pertaining to laser drilling dates back to the late 1960s and early 1970s (Moavenzadeh et al. 1968; Carstens and Brown 1971). Moavenzadeh et al. (1968) used a laser on marble and granite to test its ability to weaken the rock. The laser operated at a power output of 750 watts and the rocks were irradiated for 1, 3, 5, 10, and 30 s. Two diameters of beam were applied, 0.3 cm and 3 cm. It was found that the length of lasing time was correlated with the extent of damage. It was discovered that the damage to the rock was not only surface damage but the laser also had an effect on the interior of the sample. Also, the 0.3 cm-diameter beam resulted in more visual damage. Failure tests were conducted and it was determined that granite that has been exposed to lasers for 3 to 5 s showed a reduction in strength by 85% and for marble by 100%. Carstens and Brown (1971) used a CO_2 laser to determine its ability to cut rocks. Two different types of tests were run: kerfing and penetration. It was found that all rock types examined, granite, limestone, basalt, dolomite, and concrete, were cut by the laser although the amount of success in granite was hindered by the development of a viscous glassy melt. In 1990, Mauer et al. (1990) suggested that laser drilling was not feasible, but that was based on the old information provided by the early low-powered lasers.

Due to advances in laser technology, more experimental work on this topic has recently been undertaken. Graves and O'Brien (1998a) conducted an experiment to determine the feasibility of the US Army's MIRACL (Mid-Infrared Advanced Chemical Laser) for drilling gas wells. MIRACL is a continuous wave laser with a wavelength of 3.8 μm and a power output of 600 to 1200 kW. Two 9-Darcy dry sandstone samples of $12 \times 12 \times 3$ were used to drill 6-inch and 2-inch diameter holes. The 6-inch diameter sample was created to determine the amount of rock removed by a full laser beam. The laser was directly pointed at the slab and a 2.5-inch deep hole was created after only 4.5 s of exposure. This resulted in the removal of 5.5 lbs of rock at an equivalent rate of penetration (ROP) of 166 ft/hr. The purpose of drilling the 2-inch diameter drill was to test the laser's ability to drill horizontally for perforation purposes. The output power for this test was 500 kW. The rock was exposed to the laser for 2 s and a 6-inch deep hole was produced during this time. The equivalent ROP for this run was 450 ft/hr. Also, after the completion of the experiment, the initial permeability was compared to the final permeability around the hole and no change was found.

O'Brien et al (1999a) tested various types of cores under various conditions, with the US Air Force's Chemical Oxygen-Iodine Laser (COIL). This laser was designed to track and destroy missiles at a range of 31 miles. Since gas wells are drilled to approximately 3 miles, this laser was chosen as a welcome candidate. COIL is a continuous wave laser with power levels between 5 and 10 kW and a wavelength of 1.315 μm. For each experimental run, a $1/4$-in diameter hole was made by 8 s of irradiation. Sandstone, limestone, shale, salt, and granite were lased and the specific energy (SE) of each run was determined. It was found that the SE ranged from 10 to 40 kJ/cm^3 for 3 different power levels (100%, 50%, 35%), for all types of rock. The salt sample used the least SE for the 100% power level, whereas limestone used the least SE for the other power ranges. Also, shale consistently used the greatest amount of SE. Although there is a correlation between the shale samples and SE, many of the samples were taken from different formations so it is difficult to determine a correlation between the SE needed, laser power, and rock type.

Batarseh (2001) compared the SE to ROP and found that consistently the lower the SE the greater was the ROP for seven rock types including sandstone, shale, and granite. Gahan et al. (2001) determined that limestone SE values were greater than that of shale. They concluded that this may be due to the color of the rock and hence its reflective properties. Since limestone is lighter in color, it is less able to absorb the laser energy. To understand the effect of rock materials on SE, a short drilling time can be useful. The ablation of rock depends on the composition of the rock materials (Mauer et al. 1990). Since there are many factors, especially secondary effects that are involved with laser drilling activities, minimization of lasing time can be a better indication of the rock's effect. A longer period usually causes more penetration (depending of the rock type) and after attaining a certain depth, the purging system can no longer clean the by-products.

Sandstone, limestone, shale, salt, and granite were also lased to determine the effect of beam periodicity, which is continuous wave or chopped wave (O'Brien et al., 1999a). These test results were inconclusive, as some chopping resulted in higher SE while others resulted in lower. The authors gave no explanation as to why this might be.

The impact of fluid saturation was studied by O'Brien et al. (1999a) and Batarseh (2001) with Berea Yellow Sandstone. Cores were saturated with fresh water, brine, oil, and gas. The effect of the saturation test showed that a greater amount of SE was needed to penetrate a saturated core than a non-saturated core but the difference was not that considerable.

The hole penetration limitations test was performed to determine if the hole depth and vapor contamination in the wellbore would have an effect on the ROP (O'Brien et al. 1999a). The samples were lased for 3, 6, 9, 12, and 15 s and it was found that the greater the exposure time, the greater the SE required. Gahan et al. (2001) also came to this conclusion when they irradiated shale, sandstone, and limestone for 0.5, 1.0, and 1.5 s with an Nd:YAG pulsed laser. O'Brien et al. (1999a) and Batarseh (2001) cited that plasma screening may be a factor. Plasma screening basically means a change in the atmospheric conditions inside the rock or the formation of liquid or gas inside the hole. The gas that is formed due to this phenomenon has toxic potential (Batarseh 2001). Gahan et al. (2001) concluded that an increase in exposure time allows development of melted rock to act as a barrier, which inhibits the laser from interacting with the virgin face of the rock and consuming laser energy. The greater the percentage of quartz in the rock, the greater SE required due to this melting factor (Batarseh 2001). Batarseh (2001) observed that different-sized holes were created for different types of stones and that glass was found in some rock types such as feldspar and quartz after lasing.

Table 4-9 summarizes the hole depth and volume obtained for different types of rocks after 40 s of lasing time. Also, the diameter of the hole, beam shape, and geometry, also known as fluence, may effect the SE

Table 4-9   Effects of a CO_2 laser on hole depth and volume after 40 s (Batarseh 2001)

Hole Depth, cm	Hole Volume, cm^3	Rock
2.5–5.0	1.5–3.0	Limestone, dolomite, marl, travertine, gypsum, anhydrite
2.0–2.5	1.0–1.5	Limonite, granite, aplite, liparite, pegmatite, rhyolith, obsidian, tuff, tuffite, agglomerate
1.0–2.0	0.5–1.0	Argillite, clay, kaolinite, mica schist, gneiss sandstone, graywacks, alevrolite
0.5–1.0	0.1–0.5	Quartzite, jasper, siliceous shale, anthracite, boghead
0.1–0.5	<0.1	Graphite shale, eclogite

required. Gas purging directed at the rock can also affect the SE requirements (O'Brien et al. 1999a). Gahan et al. (2001) stated that gas purging minimizes the melted material and reduces gas condensation and debris (secondary effects) in the wellbore, thus a more accurate SE can be found. The experimental work by Gahan et al. (2001) created a shallow hole whereas the experimental work by Batarseh (2001) involved melted deposits in a deeper hole, thus eliminating some of the beneficial effects of gas purging.

Hallada et al. (2000) discussed the importance of gas purging and described the function of the purging system. The functions include providing a transparent medium for the laser to pass through, the ability to clean the hole of debris, and the ability to move molten rock into fractures to seal them as well as sealing of the wellbore wall. In the sandstone tests performed by Gahan et al. (2001), when the exposure time was increased from 0.5 s to 1.5 s, the 1.5 s sample exhibited a greater SE for pulse repetition rate (Wp) of 400 pulses/s. However, for a Wp of 200 pulses/sec, an opposite phenomenon was observed. This can be explained by Hallada et al.'s (2000) observations. They compared their test results to those of O'Brien et al.'s (1999b) and found that the point where SE began to increase the hole depths were similar but the exposure time was not. Thus it was concluded that the main factor in determining efficiency was not exposure time but the depth of the hole. This conclusion was re-iterated when a comparison of depth-drilling speeds and volume-drilling speeds were conducted. Therefore, even though the exposure time increased from 0.5 to 1.5 s, the hole depth was not sufficient enough to have an effect on SE.

O'Brien et al. (1999a) also studied the hole penetration limitations associated with the effect of confining pressure with Berea Yellow Sandstone and the effect of gas atmosphere with Dry Berea Gray Sandstone. The effect of confining rock stress tests involved stresses applied in the horizontal, vertical, and both horizontal or vertical simultaneously. The results showed that the unstressed sample had the lowest SE. Interestingly, the sample with both horizontal and vertical stresses showed lower SE than the samples with just horizontal and vertical. The authors offered no explanation for this and the amount of stress applied was not reported. During the effect of gas atmosphere tests the atmosphere was composed of nitrogen and argon gases. The results showed the nitrogen atmosphere needed slightly less energy compared to the argon gas atmosphere.

A comparison between vertical and horizontal shots was made by O'Brien et al. (1999a) in both continuous and chopped waveforms to determine which orientation required the greatest SE. For the continuous wave, shale was the rock type used and for the chopped wave tests shale, sandstone, and limestone were used. In all tests, the vertical shot required more energy than the horizontal shots. The authors gave no explanation for this but it may be because of gravitational effects. The vertical holes can become more easily contaminated with vapors from the laser and rocks whereas with horizontal shooting, the vapor is allowed to escape because gravity is not working against the base of the wellbore.

Gahan et al. (2001) conducted two sets of experiments with an Nd:YAG laser. These are initial linear track tests and fixed laser parameter tests. In the linear track test, rectangular samples of Berea Grey Sandstone, Frontier Shale, and Ratcliff Limestone were selected. The continuous movement of the slab was followed under a fiber-optic beam with varying beam focal positions with respect to the slab surface. This test exhibited different rock-laser reaction zones (areas of identical physical reaction) from the state of melting to scorching. Also, the associated parameters for each zone were determined in achieving the threshold parameters (corresponds to minimum SE). This experiment was also marked with losses during the fiber-optical delivery. The actual delivered power was in the range of 43%–73% of the calculated average power.

The effect of pulse width (Wp) and pulse repetition rate (R) are reported to affect SE (Gahan et al. 2001). An increase in pulse width causes an increase in energy absorption per unit area at a particular intensity. This reduces the SE value. Also, as the pulse width becomes larger there is less time for the sample to cool down, thus raising the heat of the sample. Eventually, thermal stresses reduced and SE decreased. An increase in pulse repetition rate also decreased the SE value. Since low R indicates a higher cooling time between successive pulses, the rock will cool down with fissure formation and SE will increase. The authors also stated that the dominant feature on SE is the effect of the Wp over R (in a finite range of R, SE drops more for higher Wp). However, this is not a convincing conclusion since one comparison was made with varying Wp and peak intensity (I).

Batarseh (2001) discussed the effect of lasing on permeability, porosity, and elastic moduli. Changes were determined with samples of sandstone, limestone, and shale with COIL, as well as sandstone with MIRACL. All samples were 5 cm long. It was shown that all samples had an increase in permeability throughout the sample length with the exception of one point. This result differed from that of Graves and O'Brien (1998a) who found no change in permeability. Due to the thermal conductivity of the rock, it was revealed that there was a permeability change in the stone even where not in contact with the laser. The size of the hole created was not discussed and therefore the distance between the laser and the permeability change cannot be determined from this study. The results of the change in porosity tests showed that the most significant increase was 2 cm from the top of the sample. This was related to an increase in fractures. The size of the hole created was also not stated for this test. The elastic moduli test showed that Young's moduli, shear moduli, and bulk moduli were reduced for all rock types tested.

4.6.1.3 Advantages of laser drilling

Drilling operations are currently very active. In the United States alone, over 20,000 wells (Graves and O'Brien 1998) have been drilled in one year, which approximates to 9,000 miles penetration. It was estimated that only 50% of the rig time was spent on deepening the well, the other half being spent on rigging-up and down, pulling and lowering the drill string in the hole and casing, cementing, mud conditioning, and other operations. The average cost is still at US$ 388/ft and the average rate of penetration (ROP) is at 22.6 ft/hr. Drilling activities are expected to increase with the ever-growing world demand on energy.

The foreseen advantages of introducing laser drilling over more traditional methods, based in the light of the experience with rotary drilling and limited experimental and numerical research studies on laser drilling, include:

- Dramatic increase in the rate of penetration (up to 100 times);
- Significantly reduced rig time;
- Elimination of casing, cementing, and bits;
- No blowout and gas kick hazards;
- Cost effective drilling;
- Increased safety and environmental sensitivity.

4.6.1.4 Analysis

In order to develop the mathematical formulation of the laser drilling process appropriate for the laser rock interaction, the following assumptions were made (Agha et al. 2004):

(a) The laser beam is radially symmetric and is focused on the surface. The energy flux arriving at a particular location is independent of the height of the surface on which it impinges.

(b) The convective heat losses from the surface to the environment are treated via an interfacial heat transfer coefficient.

(c) The radiation heat losses from the surface are treated by considering the Stefan Boltzman law.

(d) The sandstone is assumed to be infinite in all in-plane directions. A null heat transfer condition is set up at the lateral boundaries to the modeled domain.

(e) Material removal is assumed to occur only by volatilization, i.e., melt ejections are neglected.

(f) All the thermo-physical properties are assumed to be independent of temperature.

(g) It is assumed that the pulse on time is much shorter than the pulse off time and, therefore, all the plasma generated will be extinguished between pulses.

(h) The velocity of the liquid metal inside the hole is neglected and the temperature distribution within the liquid is assumed to be linear.

(i) The incident laser energy is instantly converted into heat at the target material (sandstone) and the laser beam does not penetrate into the sandstone, i.e., the sandstone is assumed to be opaque.

4.6.1.5   Mathematical formulation

Figure 4-13 shows a schematic diagram of the laser drilling (LD) process. A laser beam is incident upon the top surface of the target (rock), which absorbs a fraction of the incident light energy, causing melting of material followed by vaporization. The physical process taking place during the material removal is a combination of heat transfer, thermodynamics, and fluid flow with a free surface boundary at the liquid–vapor interface (S_{LV}) and a moving boundary condition at the liquid–solid interface (S_{LS}). This kind of problem, along with its moving boundary conditions, is termed the Stefan problem.

The laser-rock interaction can be divided into three main stages. The first stage is called the heating-up period during which the temperature is below the melting temperature and no melting or vaporization will occur. The solid absorbs the incident laser energy and, as a consequence, its temperature increases with time.

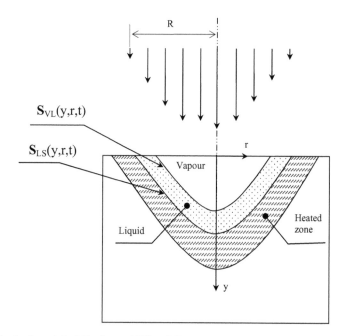

Figure 4-13   Physical model of the laser drilling (LD) process. Redrawn from Bjorndalen et al. (2003).

The second stage represents the melting stage, which starts as soon as the highest temperature (at the center of the laser beam, i.e., (0,0)) reaches the melting point. All the laser energy absorbed during this stage will result in melting more solid as well as increasing the liquid temperature. When the highest temperature reaches the vaporization temperature, evaporation will occur at the liquid surface and the third stage starts.

Since the process of material removal by laser drilling consists of three main stages, an energy balance equation was used for each stage.

First Stage – Preheating: This stage can be classified as a pure conduction stage with a convective–radiative boundary condition at the surface. The energy equation used during this stage is the simple heat conduction equation:

$$\frac{\partial T}{\partial t} = \frac{k\left[\dfrac{\partial^2 T}{\partial r^2} + \dfrac{1}{r}\dfrac{\partial T}{\partial r} + \dfrac{\partial^2 T}{\partial y^2}\right] + q_{Laser}}{\rho C_P} \tag{4.1}$$

And the thermal penetration depth is evaluated by

$$\delta = \sqrt{\alpha \cdot t}$$

Second Stage – Melting: During this stage, the liquid surface temperature is below the saturation temperature and, therefore, the liquid–vapor interface velocity is zero. An energy balance at the solid–liquid interface during this stage is expressed as

$$\frac{\partial S_{SL}}{\partial t} = \frac{\left[K_S \dfrac{\partial T_S}{\partial y} - K_L \dfrac{\partial T_L}{\partial y}\right]\cdot\left[1 + \left(\dfrac{\partial S_{SL}}{\partial r}\right)^2\right]}{\rho\, h_{SL}} \tag{4.2}$$

Third Stage – Evaporation: Evaporation starts when the liquid–surface temperature reaches the saturation temperature. The mathematical equations involved in this stage are the more complicated because vaporization, melting, and conduction equations have to be solved simultaneously. The energy balance at the liquid–vapor interface can be expressed as

$$\frac{\partial S_{LV}}{\partial t} = \frac{\alpha_{abs} q_{Laser,i} - \dfrac{K_L}{H_{Liq.}}(T_{Sat} - T_m)\cdot\left[1 + \left(\dfrac{\partial S_{SL}}{\partial r}\right)^2\right]}{\rho[h_{LV} + C_{PL}(T_{Sat} - T_m)]} \tag{4.3}$$

Knowing that the thickness of the liquid layer is given by

$$\mathbf{H}_{Liq} = \mathbf{S}_{SL} - \mathbf{S}_{LV}$$

4.6.1.6 Laser energy transfer

There are three main processes by which lasers might transfer energy into a rock target, absorption, reflection, and scattering. Here, it is assumed that the laser source is operated in pulse mode only and so the laser power is both a function of space and of time. Thus:

$$q_{Laser}(r,t) = I(r) \cdot T(t)$$

where $I(r)$ and $T(t)$ are arbitrary functions of space and time, respectively. The laser operating characteristics determine these shape functions. The spatial intensity profile was assumed to take the following Gaussian profile (Bauerle 2000):

$$I(r) = \frac{2q_0}{\pi R^2} \cdot e^{-2\left(\frac{r}{R}\right)^2}$$

In addition, the following two phenomena affect the laser energy transfer:

1. *Blackbody radiation* – As its temperature rises, the sandstone turns into an intense source of radiation. At high rock temperatures over an extended area of heating, a substantial fraction of the incident energy will be emitted back by the surface of the rock as blackbody radiation.

2. *Plasma screening* – High power laser radiation can cause the formation of plasma (ionized gas) over the surface irradiated. The laser plasma reflects, scatters, and absorbs the incident laser radiation, preventing energy from reaching the rock face.

4.6.1.7 Numerical modeling

Bjorndalen et al. (2003) and Agha et al. (2004) solved the govening equations using the Crank–Nicolson method, in which old and new time temperature values are employed using the iterative implicit method.

A FORTRAN computer code was developed to solve the above finite difference equations. Figure 4-14 gives a simplified flow chart for the solution algorithm employed. The solution procedure starts by calculating the

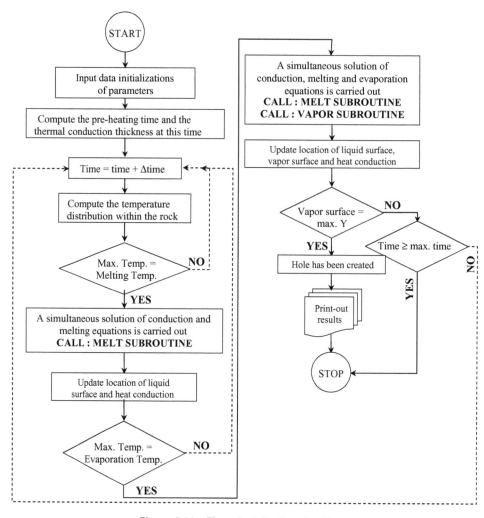

Figure 4-14   Flow chart for the algorithm.

temperature within the solid rock using energy Equation (4.1) at all points inside the material. When the highest temperature T(0,0,t) exceeds the melting temperature, the second stage starts and simultaneous solution of the heat conduction Equation 4.1 along with Equation 4.2 were carried out to obtain the temperature field as well as the recession velocity of the surface $\left(\dfrac{\partial S_{SL}}{\partial t} \right)$.

The solution procedure continues as more solid is converted into liquid, and as soon as the maximum temperature in the liquid layer reaches and exceeds the evaporation temperature, evaporation starts, consequently forming the required keyhole.

4.6.1.8 Detailed investigations about laser drilling

The intensity of laser power incident on the rock surface is the dominant factor in determining the drilling rate of penetration (ROP) and the final depth of the drilled hole. Figure 4-15 shows the effect of laser power on ROP for both sandstone and limestone. It can be seen that limestones have higher ROP and increasing the laser power increases ROP. This is attributed to the lower melting temperature as well as lower evaporation temperature of limestone.

Studying the effect of lasing time can further manifest the effect of laser power on the drilling speed. Figure 4-16 shows the effects of two different laser power levels on limestones and sandstones. It can be seen that there is a rapid decrease in drilling speed as the lasing time increases up to 25 s. Increasing the lasing time beyond this point does not seem to have a noticeable effect on ROP. This is attributed to the fact that more lasing time means more laser interaction with the rock, resulting in a more penetration depth and higher energy losses.

However, Figure 4-17 shows the total depth of the drilled hole as a function of lasing time. It should be mentioned that the effect of plasma formation and gases in the lased hole, which is expected to increase the energy losses, was not considered while constructing Figure 4-17.

For the purpose of obtaining the energy requirements and the optimum drilling speed for different materials, the specific energy requirements were calculated for both materials. The specific energy was calculated as the ratio between the incident flux of laser power and the drilling speed. The results for the specific energy requirements are given in Figure 4-18.

Figure 4-18 shows a comparison between specific energy required to penetrate a sandstone rock sample and that required to penetrate limestone when laser power intensity and lasing time are kept constant. Predictions of the numerical model have been compared to documented experimental data (O'Brien et al. 1999; Graves and O'Brien 1999), Figure 4-19 demonstrates such comparison in terms of specific energy consumption to penetrate sandstone and limestone samples as predicted numerically and tested experimentally. Figure 4-20 compares numerical and experimental results of depth of hole drilled in sandstone and limestone rocks. It

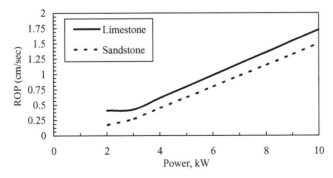

Figure 4-15   Effect of Laser Power on Drilling Rate of Penetration, ROP(m/Sec) for 8 seconds lasing time. (Redrown from Bjorndalen et al. 2004.)

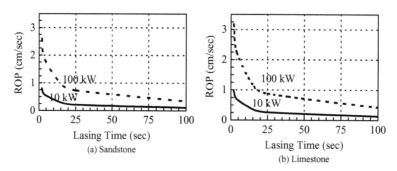

Figure 4-16a   The effect of lasing time on drilling rate of penetration.

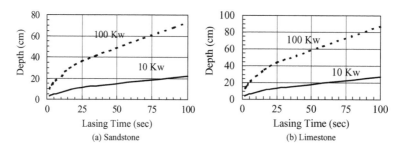

Figure 4-16b   The effect of lasing time on the final depth of drilled hole.

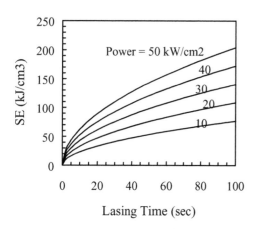

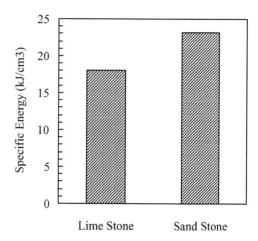

Figure 4-17   The variation of specific energy with lasing time under different incident laser power intensity – for sandstone.

Figure 4-18   Specific energy requirement, for a laser power intensity of 10 kW/cm2 and 10 s lasing time.

clearly shows that the numerical model prediction of energy consumption per unit volume of rock removed is higher than suggested by the experimental results.

Moreover, the same energy consumption numerical model predicted less depth drilled of key hole, which is attributed to the assumption of neglecting the melt expulsion process imposed in the numerical model. The authors believe that this assumption is valid, knowing that as the hole gets deeper the chance to remove

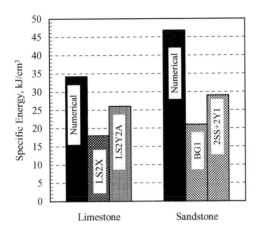

Figure 4-19 Comparison between the predicted specific energy consumption and that obtained experimentally (O'Brien et al. 1999).

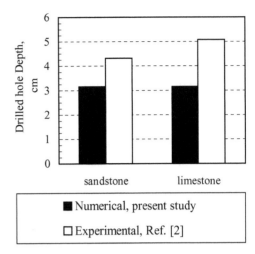

Figure 4-20 Comparison between numerical and experimental results of final depth of hole drilled (Agha et al. 2004); experimental data from O'Brien et al. 1999).

Table 4-10 Thermodynamic properties of rocks used in the numerical model (Graves and O'Brien 1998)

Property	Sandstone		Limestone
Melting Temperature, °C	1540		1260
Vaporization Temperature, °C	2200		2000
Latent Heat of Melting, kJ/cm^3		4.5	
Latent Heat of Vaporization, kJ/cm^3		30	

melt rocks as a melt is not foreseen, but rather any removal is expected to be by the vaporization process. It is not practical to assume other means of rock melt removal, as evaporation seems to be the only effective process for material removal, and this will result in higher energy consumption (compare 20–40 kJ/cm^3 to 3–6 kJ/cm^3). Table 4-10 shows the ratio of latent heat of the melting process to that of vaporization used in the numerical model.

A numerical simulation model describing the transient thermal behavior of limestone and sandstone under the influence of laser radiation has been developed. Considering the results of this model for each set of circumstances, the following conclusions were drawn:

- Using laser technology in drilling oil and gas wells is feasible considering the efficiency of the process and its benefits on the economics, drilling time, and environmental aspects supported by both prediction of numerical modeling and experimental data.

- Although in this last study only sandstone and limestone were considered, other types of rock are expected to behave in the same manner.

- Rate of penetration can be increased dramatically by increasing laser power for both sandstone and limestone. The rate of increase is about the same while the sandstone consumes little more power than limestone. However, with longer lasing times, both sandstone and limestone rate of penetration increases similarly.

- Numerical prediction results show more energy consumptions and less rate of penetrations when compared to published experimental data. The reason for this discrepancy is the assumption that the vaporization process will remove the melted rock rather than the melting process, since it is more likely to take place in reality.

- Effect of rock fluid saturation on rate of penetration (ROP) and energy consumption has not been addressed. Future investigation is required to cover this aspect.

4.6.2   Water jet drilling

Drilling is the most important and one of the oldest technologies on Earth. A parameter of principal importance in any drilling process is the "weight on bit". This is the axial force acting on the bit during the cutting process. Normally this force is relatively large and may be generated by proper anchoring of the drill machine to the drilled surface. Alternatively, weight on bit may be provided by the self-weight of the drill unit structure. Water jet drilling does not require any weight on bit. The water jet systems have little loss of pressure or power throughout its drill pipes. The bit power is essentially equal to the power available at the surface. The energy loss is nominal which is remarkable.

The prime advantages of this are that high pressure water jet drilling (HPWD) does not require any torque or thrust during jet erosion. For this reason, HPWD supplies an exclusive capability for drilling a constant radius directional hole without any steering corrections. Moreover, pure water jet drilling is less sensitive to formation changes than rotary drilling because cutting is controlled by the bit orientation. HPWD present high rates of penetration because the power available at the bit is extremely high. Rotary drilling provides slightly higher drilling rates. However, this approach generates torque loads that can cause the trajectory of the hole to spiral. HPWD is capable of rapidly drilling small-diameter holes in a wide range of erosion-resistant rock types. Finally, a HPWD system can be made very lightweight because the thrust and torque requirements are nominal. Currently, normal water jet drilling is used as an accessorial activity in many industries, including the oil and gas industry. The technique is normally used to remove cuttings, rock chips, mud cake, and to clean the formation of the reservoir as well as the surface.

Existing rotary drilling systems are capable of drilling shallow directional holes. The equipment is heavy, drilling rates are low, and costs are high. In addition, toxic chemical additives are used in a rotary drilling system. For any drilling system, the factors such as type of formation, depth of drilling, and depth of desired screen setting should be considered when selecting an appropriate drilling method. At present, drilling technology has modernized well profiles and directions. HPWD technique is now used for horizontal drilling. Horizontal wells are being drilled across the reservoir by exposing a relatively large reservoir area, which is used as drainage or injection. Most of the new wells are completed without cementing or liner. Due to the long opening, acidization is used to remove mud cake. Since this process is costly, the HPWD process is being used more frequently. HPWD is a cost effective and simple to handle process in drilling technology.

Normally, high strength, high permeability rock types such as Berea Sandstone have a low specific energy and threshold pressure. Medium strength, low permeability rocks such as limestones and sandstones have intermediate specific energy. Finally, high-strength, low permeability rocks such as granite, quartzite, and basalt have high specific energy and threshold pressures. Because of its benefits over other drilling systems and its versatile usage in the industry, HPWD has generated great interest among researchers.

4.6.2.1   Background review

Maurer and Heilhecker (1969) conducted a series of experiments by water jet on different rock samples. They concluded that water jets can successfully drill sedimentary rocks. They used a high pressure pump up to 13,500 psi, which can give 200 to 300 ft/hr penetration rates. They have also concluded that water jet is economical for drilling oil wells. They found that the hydraulic jet drilling rate is influenced by nozzle size, nozzle pressure, and rock strength. Fenn (1989) investigated the use of water jet for use in conjunction with free-rolling cutters. He conducted a series of laboratory tests with disc and button cutters to determine the effect of variations in the jet and cutting parameters on the cutter performance. His results indicate that no additional improvement in cutting performance is gained by an increase in jet pressure above 40 MPa. Ho-Cheng (1990) studied water jet drilling to model an optimal water jet pressure, which is a function of hole depth and material parameters. He found reasonable agreement with data obtained from water jet drilling of graphite epoxy laminate. He concluded that the predicted optimal water jet pressure can be applied in a control scheme for maximizing the productivity of water jet drilling of composite laminates. Hood et al. (1990) studied high-pressure water jet for developing a better understanding of the erosion mechanisms to cut the rock materials. They developed an empirical model to describe the different parameters involved with the system. This model is defined as the rock erosion by a high-pressure water jet.

Yasuda and Hoshina (1993) studied the fundamentals of the application of the ultra high pressure water jet for rock drilling and have developed an ultra high water jet boring system using ultrahigh pressure water jet. Aslam and Alsalat (2000) have discussed the theoretical aspects of the HPWJ technology, case histories, and well performance data. They have also pointed out that HPWJ can be used for steel cutting or to make holes by using abrasive materials such as sand and beads. It has been reported in the literature that the efficiency of the process depends on four factors:

1. stand-off distance;

2. fluid velocity;

3. jet stream profile; and

4. rotation.

Buset et al. (2001) described the penetration effects on formation zone caused by water jet technology. Lia et al. (2001) conducted experiments on water jet and polycrystalline diamond compact (PDC) for rate of penetration in hard rocks. They pointed out that the combined cutting effect of water jet and PDC is effective in hard rocks. Arangath et al. (2002) discussed the high hydraulic horsepower jetting tool used for scale removal. They have also investigated water jetting in horizontal well drilling. Dunn-Norman et al. (2002) discussed processes for sustainable recovery of heavy oil from ultra-shallow reservoirs, using low cost, innovative horizontal drilling and completion methods. They have argued that the use of water jet drilling over 15,000 ft (5,000 m) was more favorable compared to conventional rotary drilling system. They also concluded that water jet drilling methods appear most preferable for drilling horizontal wells in ultra-shallow reservoirs.

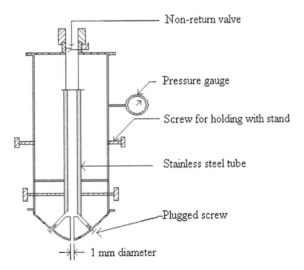

Figure 4-21   Stainless steel drill bit. Redrawn from Hossain et al. (2006a).

Recently, Hossain et al. (2006a) have described the contributions of different researchers on water jet drilling and the importance of this technology. They have developed some empirical relationships using water jet drilling on rate of penetration, depth of penetration (Hossain et al. 2006a, 2006b). They have studied the variation of depth, ROP, temperature, flow rate, and pressure with time. The variations of depth and ROP with temperature, side effects, and thermal exposure time have also been studied. They also studied the effects of gap variation between drill bit tip and sample top surface (Hossain et al. 2006b).

4.6.3   Smart drilling systems

"Smart" drilling is a self guided drilling system capable of sensing and adapting to conditions throughout the drilling operation and around and ahead of the drill bit (DOE 2005). These systems operate in realtime, i.e., they analyse and react to data collected during drilling and are designed to modify the trajectory, speed, and operation of the bit when parameters measured by the sensors so dictate.

According to DOE (2005), smart drilling systems must be able to:

- sense the properties of the rock at and just ahead of the bit;
- sense the condition of the bit and other critical drilling system components;
- collect the data in a form that can be easily and quickly transmitted to the surface;
- use these data to adjust drilling rate and/or trajectory to changing conditions; and
- perform these functions under extreme conditions.

The smart drilling system is able to determine the properties of the rock through which the drill bit is cutting. Sensing the rock properties helps to take necessary steps in adjusting ROP and replacing worn out drill bits. The conventional drilling systems have a limited ability to analyze the rock or the conditions of the bit downhole.

During drilling, the objectives are to drill the hole and to collect the geological information in different depths. This is a problem in conventional drilling, especially in deep formations (Wenger et al. 2004). Used drilling muds alters the geological formation to be sampled. However, a smart system accurately measures parameters such as formation porosity, pore pressure, temperature, mineralogy, rock strength, and stress state. As a result, smart drilling helps estimate the formation's elastic properties (brittleness, ductility, hardness) and energy dissipation rate (wave attenuation). In addition, smart drilling adjusts drilling parameters (e.g., weight on the drill bit and rotary speed, torque, pump rate) to reduce stress on the bit and prevent problems (e.g., blowout, loss of circulation, drilling off-course). Most drilling systems today have a limited ability to analyse the rock or condition of the bit downhole as rock properties change with depth (DOE 2005).

Figure 4-22 represents smart drilling accessories, such as a drilling rig and drill bit. In addition, there is a downhole sensor attached to the drill bit that collects data and transfers it to a computer system containing advisory software. Telemetry methods are used to transfer this information with high speed. The data are then analyzed with robust and flexible software that can accurately assess conditions and properly adjust the drilling parameters. The subsurface sensor and transferring system are capable of operation under extreme conditions, such as temperatures exceeding 350°F (175°C) and pressures greater than 10,000 psi (DOE 2005).

Using this integrated technique allows the smart drilling system to anticipate problems and either make adjustments or take preventative action to avert mechanical problems.

4.6.4   Deep gas drilling

Recently, high temperature electronic components have been developed that can be used in instrumentation in deep gas drilling systems (DOE 2005). A high-temperature (approaching 400°F), high-pressure, measure-

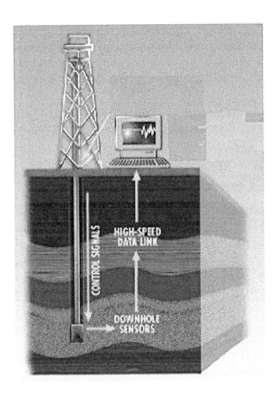

Figure 4-22   Smart drilling. Courtesy: DOE (2005).

ment-while-drilling tool, will be able to provide direction, inclination, toolface, and gamma ray measurements continuously in realtime. The tool will be fully retrievable while the drillstring is downhole, eliminating the need to remove the entire drillstring assembly to retrieve the measurement-while-drilling (MWD) equipment. This new system improves the economics of deep well drilling by reducing down time and boosting the overall rate of penetration in deep hostile environments (DOE 2005). The attached wireless, electromagnetic (EM) telemetry communications system transmits data to and from downhole equipment in realtime, enabling both surface processing of downhole sensor data and direct surface control of downhole tools.

4.6.5   Innovative materials for deep drilling

More drilling operations are moving toward offshore and wells are getting deeper. However, most of the construction and drilling materials still remain conventional. For example, the drilling rigs can weigh thousands of tons, the drilling muds can contain numerous toxic additives, and gas flaring still accounts for some 25% pollution by the petroleum industry. Recently, better construction materials have become available. Innovations in metallurgy and plastics technology and in the development of composite materials show great potential for being adapted in drilling construction materials. For example, the recently developed composite carbon fiber-epoxy resins have been manufactured into lightweight drill pipes (DOE 2005).

NETL (2004) studies show that lighter weight drillstring components will extend the depth rating of conventional drilling rigs, currently limited by the weight of the suspended drillstring that they are designed handle. This could reduce the need to build new large rigs to meet increased depth requirements. In addition, lightweight components can help reduce torque and drag, factors that reduce bit life under difficult drilling conditions.

During the drilling of deep wells, high pressures and temperatures, corrosive formation fluids, and abrasive cuttings in the drilling fluids can cause conventionally manufactured welded pipe to collapse, erode, or leak (DOE 2005). Similar problems can occur in tubing or casing after a deep well is completed. The development of new ways to manufacture seamless coiled tubing and drillpipes can improve the performance of wellbore tubulars.

Finally, with the development of ultra-light material, in the future it may be possible to carry the whole drilling equipment in a helicopter. This would be practical when reaching ecologically sensitive areas for petroleum development.

4.6.5.1   Tubulars for deep gas well drilling

According to DOE (2005), a continuous microwave process for the manufacture of seamless coiled tubing and drill pipe is being developed. The process will employ microwaves to sinter continuously formed/extruded steel powder into seamless metal tubes, both efficiently and economically. Tubular drillstring components made by this process are expected to demonstrate superior quality and performance relative to conventionally manufactured tubular products, resulting in improved life cycle, reduced erosion rates, and lower failure rates. These improvements will result in a better rate of penetration.

4.6.5.2   New drillbit technology

One of the main problems with conventional drilling is the slower rate of penetration (ROP), especially in deeper wells. At greater depth the ROP is around 3 to 5 ft/hr. Initially, the tricone bits with hardened inserts used for drilling hard formations at shallower depths were applied as wells went deeper (DOE 2005). Tricone bits failure (bearing failure) is common in the deeper well due to application of greater weight on the bit. The failure rate is high and its causes are difficult to diagnose, resulting in lost cores, more frequent trips, higher costs, and lower overall penetration rates.

According to DOE (2005), fixed cutter bits with polycrystalline diamond compact (PDC) cutters constitute a solution to the problems inherent with tricone bit moving parts. The PDC cutting surface employs synthetic polycrystalline diamonds bonded to a tungsten-carbide stud or blade. This type of bit now holds the record for single-run footage in a well of 22,000 feet. It drills several times faster than tricone bits, particularly in softer formations, and PDC bit life has increased dramatically over the past 20 years. But PDC bits have their own set of problems in hard formations.

To overcome these problems, after much testing and computer modeling of stresses, high temperatures, hydraulics, and wear mechanisms, a reliable thermally-stable polycrystalline (TSP) diamond bit is being developed (DOE 2005) (Figure 4-24). A new drill bit should include not only improvements in conventional

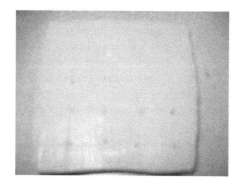

Figure 4-23   Composite drill pipe. Photo courtesy of DOE (2005).

Figure 4-24   New drill bit technology. Photo courtesy of DOE (2005).

drill bit efficiency, durability, and longevity, but also advancement in the development of less conventional drilling tools. New features, such as heat, pressure, chemicals, or electro-hydraulic discharges are used to break and remove rock from the wellbore, as opposed to the shearing and grinding mechanisms of conventional bits.

4.7   Sonic While Drilling

Sonic-while-drilling is vital in oil and gas operations. It should be followed and investigated thoroughly in order to accurately assess the petrophysical and thermodynamic conditions ahead of the bit, characterize the oil and gas reservoir correctly, and then control well production efficiently and economically. Therefore, the data obtained following this procedure should be processed immediately and the uncertainties surrounding them should be dealt with in a commonsense manner to guarantee a sound decision-making process.

Sonic logging functions on the principle that sound waves travel through denser, and less porous, rock more quickly than through lighter, and more porous, rock. This technique uses a transmitter and receivers in one downhole tool to measure, in microseconds, the time differences required for sound pulses to pass through formation beds. A sonic log displays travel time of P-waves versus depth. Sonic logs are recorded by pulling a tool on a wireline up the wellbore while drilling. The tool emits a sound wave that travels from the source to the formation and back to a receiver.

Sonic log data are used to characterize oil and gas reservoirs, estimate porosity and shaliness of reservoir rocks, enhance interpretation of seismic records, locate zones with abnormally high pressures, and assess the mechanical integrity of surrounding reservoir rocks, among others. This section proposes a stochastic evaluation of rock properties such as porosity based on the sonic-while-drilling data.

4.7.1   Benefits of sonic-while-drilling

Information derived from drill bit seismic can help predict upcoming drilling hazards, eliminate casing strings, and ensure that vital horizons are cored. When employed correctly, the method can lead to substantial cost savings and enhance the safety of drilling operations. More importantly, data acquired can be used to characterize the reservoir in a realtime fashion. This latest aspect has received little attention in the past.

The technique requires only surface sensors and does not interfere with the drilling process itself. This is one of the reasons why there appears to be few reported applications of the seismic-while-drilling technique for offshore applications. It is indeed difficult to maintain a receiver on the ocean bed, especially for semi-submersible or floating drilling rigs. This is not the case for onshore drilling.

Once fully developed, the proposed technique will lead to the assessment of the petrophysical and thermodynamic conditions ahead of a drill bit. For offshore applications, such knowledge is worth millions of dollars per hour of drilling time. The benefit could be further increased if a technique were developed that would help perform realtime reservoir characterization. Finally, if more modern techniques such as the excitation and frequency analysis technique are introduced, the accuracy of the technique will improve greatly.

Mustafiz et al. (2006a) recommended the following steps in order to develop a comprehensive monitoring and data processing tool:

1. Coupling of the 3D imaging with a drilling control system, for realtime control of drilling activities. This is a refinement of the diagnostic while drilling (DWD), recently developed by Sandia National Laboratories in the United States. Three areas of improvement should be sought. They are:
    i.   increase the speed of the data inversion for volume visualization;
    ii. increase the capacity of data mining through the use of latest advances in large volume data processing (borrowed from medical and human genome researchers); and
    iii. develop strategies for tapping into computers faster than the one currently available.
2. Develop a dynamic characterization tool, based on seismic-while-drilling data. This will use inversion to determine permeability data. At present, cuttings need to be collected before preparing petrophysical logs. The numerical inversion requires the solution of a set of nonlinear partial differential equations. Conventional numerical methods require these equations to be linearized prior to solution, in which process many of the routes to final solutions are suppressed (note that a set of nonlinear equations should lead to the emergence of multiple solutions). Recently, Mustafiz et al. (2006b, 2006c, 2006d) proposed the use of a relatively new technique (Adomian domain decomposition) for solving nonlinear partial differential equations. This method has gained popularity in tracking the various routes of multiple solutions of a set of nonlinear equations.
3. Couple the sonic method with the excitation and frequency analysis method. Recently, the ADNR Explorer of Switzerland has marketed a new process that uses excitation and frequency analysis in a subsonic frequency range between 0.1 and 20 Hz. With this method, they have been able to predict subsurface fluid composition with high accuracy. In practice the necessary energy at the required frequencies is generated with a seismic vibrator coupled to the surface soil. The response from the structure is detected with a group of sensors with 3D characteristics (Sensor Array). Digital post processing and filtering separate the signal from the noise. If proven successful, this method can revolutionize realtime data acquisition and subsequent analysis and control of both drilling and reservoir management.
4. Investigate the possibility of using 3D sonogram for volume visualization of the rock ahead of the drill bit. In order to improve resolution and accuracy of prediction ahead of the drill bit, the 3D sonogram technique can be extremely beneficial. The latest in ultrasound technology gives the ability to generate images in 4D (time being the 4th dimension).
5. Couple 3D imaging with a comprehensive reservoir model. This will allow the use of drilling data to develop input data for the simulator with high resolution.

4.7.2   A novel stochastic method for evaluating sonic-while-drilling data

Ketata et al. (2005) presented a novel method for stochastic evaluation of rock properties, such as porosity, by processing sonic-while-drilling data. Stochastic feature describes the randomness of such data in time and space.

4.76.2.1 Rock porosity

Rock porosity is the ratio of a volume of void spaces within a rock to the total bulk volume, V_T, of that rock (Figure 4-25). It is commonly expressed as a percentage. All the void space is referred to as pore volume, V_P. Percent porosity is calculated as:

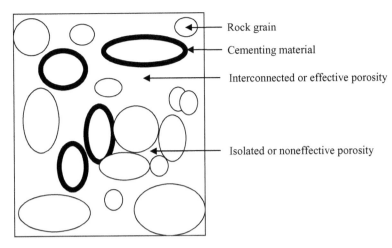

Figure 4-25   Rock porosity.

$$\phi = 100V_P/V_T \tag{4.4}$$

where the total pore volume is the sum of the interconnected or effective pore volume, V_{EP}, and the isolated or non-effective pore volume, V_{NP} (Figure 4-25):

$$V_P = V_{EP} + V_{NP} \tag{4.5}$$

Porosity is estimated from a sonic log as a function of the sonic travel time or interval transit time, Δt. The latter is the reciprocal of the velocity of a sonic wave, and calculated as follows:

$$\Delta t = 10^6/v \tag{4.6}$$

where Δt is measured as microseconds per foot, and v as feet per second.

The total rock bulk volume is the sum of the solid volume and interstitial fluid volume:

$$V_T = V_S + V_F \tag{4.7}$$

where

$$V_S = (1 - \phi)V_T \tag{4.8}$$

$$V_F = \phi V_T \tag{4.9}$$

The total sonic travel time:

$$V\Delta t = V_S\Delta t_S + V_F\Delta t_F \tag{4.10}$$

then

$$\Delta t = (V_S\Delta t_S + V_F\Delta t_F)/V \tag{4.11}$$

which equals

$$\Delta t = (1 - \phi)\Delta t_S + \phi\Delta t_F \tag{4.12}$$

Thus, the rock porosity equals

$$\phi = \frac{\Delta t - \Delta t_S}{\Delta t_F - \Delta t_S}$$

(4.13)

As a result:

$$\frac{1}{v} = (1 - \phi)\frac{1}{v_S} + \phi\frac{1}{v_F}$$

(4.14)

where

$$\phi = \frac{v_F(v_S - v)}{v(v_S - v_F)}$$

(4.15)

Therefore, the average velocity of sonic wave comes to

$$v = \frac{v_S v_F}{(1 - \phi)v_F + \phi v_S}$$

(4.16)

4.7.2.2 Stochastic processes

A stochastic process is a process showing random behavior. For example, the sonic wave velocities v, v_S, and v_F are stochastic. As a result, the porosity ϕ is stochastic. A stochastic process can be represented by an autoregressive integrated moving average (ARIMA) model (Ketata et al. 2005a).

An ARIMA model is defined as

$$\phi_p(B)(1 - B)^d z_l = \theta_q(B)a_l$$

(4.17)

where l is the length or the depth of the wellbore, B the backshift operator,

$$\phi_p(B) = 1 - \phi_1 B - \ldots - \phi B^p$$

(4.18)

$$\theta_q(B) = 1 - \theta_1 B - \ldots - \theta_q B^q$$

(4.19)

and a_l a white noise, $WN(0, \sigma^2_a)$.

The sonic wave velocities, the interval transit times, and the rock porosity follow ARIMA models (Figures 4-26–4-28).

4.8   Knowledge-based Optimization of a Rotary Drilling System for the Oil and Gas Industry

Drilling a well is necessary to reach an oil and gas reservoir. Rotary drilling is the main technique used to achieve this task. This section proposes a novel expert system that optimizes the rotary drilling system. It is named the Rotary Drilling System Optimizer (RDSO). It is a knowledge-based optimization system using M.4, a knowledge engineering tool. It comprises knowledge entries that are facts and rules. Based on the knowledge base and the data given by the user, the expert system inspects the correctness of the information supplied, and determines the rock penetration rate and the drilling time, among other variables. It is tested successfully by verifying its components and validating its performance and functionality and is proved to be correct, reliable, and useful.

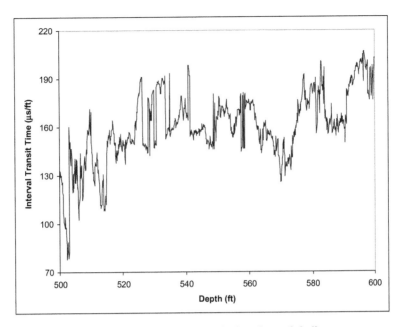

Figure 4-26A   Interval transit time for rock bulk.

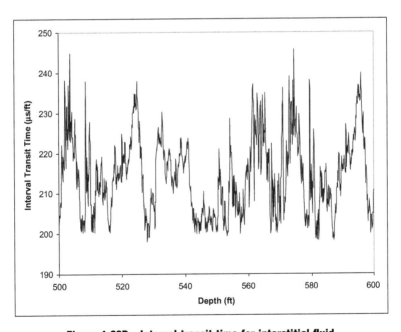

Figure 4-26B   Interval transit time for interstitial fluid.

4.8.1   Computer view of a drilling operation

In rotary drilling, the drill string is supported by a derrick. The string is rotated by being coupled to the rotating table on the derrick floor. The drill bit is located at the end of the string. It is generally designed with three cone-shaped wheels tipped with hardened teeth. More drill pipes are added to the drill string as the bit penetrates further into the Earth's crust. The force required to cut into the Earth comes from the

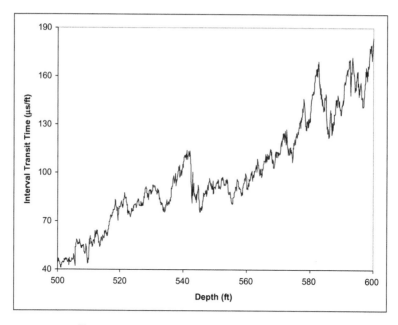

Figure 4-27   Interval transit time for rock matrix.

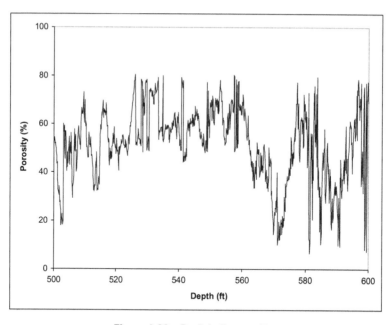

Figure 4-28   Rock bulk porosity.

weight of the drill string itself. To remove the cuttings, drilling mud is circulated down through the drill pipe, out through nozzles in the drill bit, and then up to the surface through the space between the drill pipe and the wellbore.

The rotary drilling system is composed of the well parameters, rock properties, rotary drill parameters, and bit-rock interaction parameters. The well parameters are its direction, diameter, and depth or length. The rock properties are its type, composition, bulk density, porosity, angle of internal friction, compressive strength, and abrasiveness, among others. The rotary drill parameters are the weight on bit, rotary speed, and bit diameter. The bit-rock interaction parameters are the bit energy transfer ratio, or mechanical efficiency, and specific energy.

Optimizing the rotary drilling system is better achieved using rotary drilling expertise available in the oil and gas industry (Ketata et al. 2005a, 2005b, 2005c). An expertise-based system, also called expert system, is a branch of artificial intelligence. The ultimate objective is a device programmed to act as an intelligent agent, which is enabled to emulate the human thought and reasoning process. This section introduces an expert system developed to optimize the rotary drilling system using M.4 (Teknowledge Corporation 1993; 1995; 1996).

4.8.1.1   Well parameters

The well parameters taken into account are direction, diameter, and depth or length. The drilling direction can be vertical, slanted, or horizontal (Figure 4-29).

4.8.1.2   Rock properties

The rock properties considered in this study are type, composition, bulk density, porosity, angle of internal friction, compressive strength, and abrasiveness. Table 4-11 illustrates an example of rock types and properties included in the expert system.

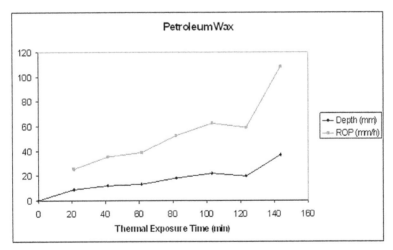

Figure 4-29   Vertical, slanted, and horizontal drilling directions (left to right).

Table 4-11   Rock properties

Rock	Quartz percentage (%)	Bulk density (g/cc)	Porosity (%)	Permeability (md)
Mesaverde Sandstone	75	2.64	20	115
Berea Yellow Sandstone	90	2.03	23	6000
Berea Gray Sandstone	85	2.08	21	480
Limestone	0	2.43	0.6	0.02
Shale	35	2.49	3	0.2

4.8.1.3 Rotary drill parameters

The rotary drill parameters are the weight on bit, rotary speed, and bit diameter. The weight on bit is the amount of downward force placed on a bit by the weight of the drill stem.

4.8.1.4 Bit-rock interaction parameters

The bit-rock interaction parameters are the bit energy transfer ratio, or mechanical efficiency, and specific energy.

The mechanical energy transfer ratio equals

$$\eta_E = E_e/E_r \tag{4.20}$$

where η_E is the mechanical energy efficiency, E_e the effective rotary energy, and E_r the actual rotary energy.

The specific energy (*SE*) is the amount of rotary energy required to drill a unit volume of rock. It is calculated using the formula:

$$SE = \eta_E WN/DR \tag{4.21}$$

where W is the weight on bit, N the rotary speed, D the well diameter, and R the rate of rock penetration.

The drilling time required to accomplish this task is calculated as

$$t_d = L_w/R \tag{4.22}$$

where L_w is the well depth or length.

4.8.1.5 Expert system components

The major components of an expert system are the knowledge base, the inference engine, and the user interface (Figure 4-30).

4.8.2 Expert system development

The development stages of an expert system are identification, conceptualization, formalization, implementation, validation, and maintenance. The Rotary Drilling System Optimizer (RDSO) has a knowledge base composed of knowledge entries that are facts and rules. A fact is a knowledge base entry of the form:

$$expression = value\ of\ integer \tag{4.23}$$

that provides a value for the expression. The fact is qualified by a certainty factor that indicates the degree of certainty with which the fact is believed. If no certainty factor is specified, the fact is assumed to be definite, with a certainty factor of 100. If no value is specified, then the value is assumed to be yes.

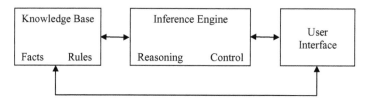

Figure 4-30   Expert system components.

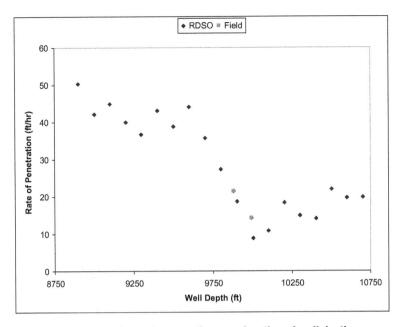

Figure 4-31   Rate of penetration as a function of well depth.

A rule is a knowledge base entry of the form:

$$if\ premise$$
$$then\ conclusion. \tag{4.24}$$

where premise is generally a proposition of the form:

$$expression = value \tag{4.25}$$

or a combination of such propositions joined by the Boolean connectives *and*, *or*, or *not*, *conclusion* is made up of propositions of the form 4.

The RDSO was tested successfully. It was verified and validated by comparing its performance in optimizing the rotary drilling system to the experts' cases (Pessier and Fear 1992; Graves et al. 2002). One hundred cases were used to test this expert system by inspecting its correctness and results based on the experts' knowledge, literature reviews, lab experiments, and field data. For example, Figure 4-31 shows the concordance between the RDSO and field data. It compares the penetration rate between the RDSO and the field information for shale. It proves that the RDSO answers match those of the experts and the field cases.

4.9   Future Research in Production Operations

In a series of articles on future needs in petroleum engineering research, including those involved in petroleum production operations (Islam 2000; Islam and Wellington 2001; Taheri et al. 2005), production operations were identified as prime candidates for research as current state-of-the-art is considered to be inadequate for "greening" petroleum operations (Bjorndalen et al. 2004). Production operations research should include wellbore as well as corrosion problems in addition to classical production engineering problems. Note that the transition between laboratory and field is the shortest for this research due to the urgency of the problem as well as the ease of implementation of a production project. The following projects are considered to be innovative ones that will give the highest return for funds spent.

4.9.1   Use of an expert system for remediating wellbore problems

Wellbore problems are numerous. A remediation technique that is beneficial to one can be detrimental to others. An expert system will optimize a remediation technique with greatest reliability to-date. At present, design criteria do not include other processes, often leading to implementation of unsustainable processes.

4.9.2   Truly intelligent systems for controlling stochastic processes

Controlling a wellbore offers some of the most formidable challenges in a control system. Artificial intelligence fails to control petroleum applications, because computers do not operate the same way as human brains do (Ketata et al. 2006a). There is a great need to include true (natural) intelligence in devising decision making tools (Ketata et al. 2006b).

4.9.3   Novel methods to mitigate asphaltene precipitation, sulfur and wax deposition, and sour gas corrosion methods

The deposition of these materials is one of the most difficult problems in the petroleum industry. Even the theory of deposition and propagation of asphaltene, sulfur and others has not reached maturity. This research should develop theories, validated with experimental evidence, in order to propose techniques capable of mitigating the problems related to solid deposition. Recently advances have been made by our research group in using natural materials for remediating production problems (Al-Darbi et al. 2002; Al-Maghrabi et al. 1999). Now a more aggressive research thrust in this direction is warranted.

4.9.4   Novel cementing technology to remediate downhole corrosion.

The extent of downhole corrosion has been related to cementing deficiencies only recently (Talabani and Islam 2000). New methods should be investigated in order to develop natural cementing agents that can minimize downhole corrosions (Al-Darbi et al. 2006).

4.9.5   Novel pumping and gas-lift methods for improving oil production from a well

These "well established" techniques have a number of inherent problems associated with them. These difficulties arise from the surge in sand content in heavy oil formations and water content in light oil formations. Research in this topic will involve new methods that deal with the use of surface active agents for improving performance during pumping or gas-lift operations, for heavy oils and real time monitoring for light oil (in order to allow timely intervention).

4.9.6   Bioremediation of wellbore problems

This environmentally appealing process should be investigated in order to increase the probability of success during a field operation. Thermophilic bacteria, along with others, should be tested in order to produce from non-thermal as well as well as thermally active wells. Advances in using natural energy and mass sources have been made recently by our research group (Al-Maghrabi et al. 1999; Chaalal et al. 2005).

4.9.7   Sand control and sand production

Sanding problems are far from being resolved. In heavy oil reservoirs, it is considered that productions should be under uncontrolled sand production. However, recent restrictions on produced sand disposal is causing concern to the petroleum industry. Even in light oil reservoirs, sanding problems continue to hamper production (Nouri et al. 2006a; 2006b). There is a need to develop techniques that can minimize sand production and maximize produced sand utilization. In controlling sand production, attention should be given to environment-friendly techniques that avoid the use of toxic chemicals. Bio-restoration offers an attractive avenue of research and should be investigated (Zhong and Islam 2005a, 2005b).

4.9.8 High water-oil-ratio (WOR) problems

Various WOR controlling techniques should be investigated, including the chemical methods. However, it is important to note that chemicals that are of synthetic origin should be avoided. These chemicals are not sustainable in the long-run and are expensive. Instead, focus should be given to natural chemicals that are derived from waste materials (Rahman et al. 2004; Rahman et al. 2006). Also, to be considered is the use of greenhouse gases that have attractive surface active properties (Newman-Bennett 2007).

4.9.9 Risk analysis of production strategies

Current risk analysis tools only consider short-term implications. Because sustainable development requires the implementation of time extending to infinity, the risk analysis procedure must be revised, to include intangibles. Recently, our research group has made progress in this regard (Hossain et al. 2005, 2007; Mehedi et al. 2007a, 2007b), but more research must be performed in order to develop truly integrated analysis that will ensure sustainability.

4.9.10 Use of new materials for minimizing environmental impact of produced oil and gas

Both gas–gas and liquid–liquid separation can be carried out with natural materials. This separation has a number of benefits, such as that it is environmentally benign (hence sustainable), inexpensive (particularly if naturally occurring materials are found locally), and efficient (especially if global efficiency is considered). The conventional practice is to develop membranes that are made of petroleum derivatives. Such membranes are expensive and are toxic in the long run. There is a need to explore separation materials from commonly found natural materials.

4.9.11 Novel techniques for drilling/production waste management

Conventionally, waste management is not considered to be a production operation concern. Consequently, production engineers often underestimate the fate of various contaminants. There is a need for the involvement of production engineers in managing drilling/production wastes, including liquid wastes, sludges, and emulsions. New chemical methods, which are economically attractive and environmentally appealing, should be investigated. Recent trends dictate that petroleum waste be converted into useful by-products. Of particular importance is the remediation of heavy metals. Despite advances made recently in developing techniques for remediating heavy metals, a truly sustainable technique has eluded researchers (Khan and Islam 2007).

4.9.12 Manufacturing and custom design of novel tools

Only recently, the research team of Islam and Taheri (Dalhousie University, Canada) have launched an ambitious research project in order to develop an array of novel tools to be applied during the production operations. If this research is successful, a series of novel tools for petroleum well completion will be developed. Taheri and Hassan (2003), clearly demonstrated that major savings can be achieved by using such a design strategy. This research involves improvements in the areas of sand control, composite materials for well completion, built-in stiffener, novel perforation tools, and others.

4.10 Benefit of Sustainable Drilling and Productions

The success of high-risk drilling and production operations depends on the use of appropriate technologies. In this chapter, many emerging technologies have been proposed for drilling and production operations. The main objectives of these proposed technologies are to make development of oil and natural gas cost effective, more efficient, and more protective of the environment. Uses of appropriate technologies will also help find new reserves, improve drilling efficiency, reduce costs, and increase production.

The proposed, emerging technologies have had positive environmental benefits in reducing negative impacts on lands, surface waters and aquifers, wildlife, and air quality. Innovations in drilling technology will significantly reduce the environmental impact. This can be achieved, for example, by using smaller drilling pads, smart wells, and measurement while drilling technologies. Better drilling technology can produce more oil and gas from fewer wells. Fewer wells means less land disturbed by drilling operations and the associated surface infrastructure and transportation systems.

It is reported that microhole drilling opens up a domestic resource of 218 billion barrels of oil found at depths of less than 5000 ft (NETL 2004). The microhole drilling can be conducted using helicopter transport to field sites, which will further reduce the footprint and reduce environmental disturbance. The small rig size also contributes to the reduced visual impact of drilling that has caused public objections in some areas. Helicopter transport of equipment, materials, and personnel to sensitive areas will benefit the wildlife and ecosystems by eliminating contact. This is particularly important to migrating species and species that require specific areas for calving grounds or nesting.

Use of sustainable technology to produce oil and to meet environmental regulations has developed new, improved techniques and strategies that accomplish both goals. Sustainable drilling and production operations also make good business sense and help protect the environment. Many oil companies have learned that going that one step farther to protect sensitive environments and avoid pollution pays back in increased benefits and improved public relations. In the next chapter (Chapter 5), sustainability status of these technologies is elaborated.

Sustainable Waste Management

5.1　Introduction

The basic drilling technology currently used by the petroleum industry was developed 100 years ago when rotary drilling surpassed cable tool drilling as the standard method for reaching gas and oil formations. Major problems persisting with this method are technological insufficiency, cost-ineffective, and wastes (Mustafiz et al. 2004; NETL 2004; more limitations are identified in Chapter 4). O'Brien et al. (1999a) reported many technological problems, such as differential pressure sticking, lost circulation, swelling, and collapse. NETL (2004) reported the limitations of conventional methods in deep drilling. Also, the cost of drilling deep wells (20,000 ft) almost doubled between the years 1979 and 1994, with the cost of drilling offshore between 1974 and 1994 having tripled (Graves and O'Brien 1998a).

In addition to their high cost and technological limitations, conventional drilling and production methods generate an enormous amount of wastes (Veil 1998; EPA, 2000; Khan and Islam 2003, 2006c). To manage these wastes there are many regulations (OWTG 2002), guidelines (OWTG 2002), and techniques (Veil 1998) that have been developed in different parts of the world. Existing management practices are mainly focused to achieve sectoral success and are not coordinated with other operations surrounding the development site. For example, there is still no suitable management strategy for integration of offshore waste management with coastal zone management (Khan and Islam 2003b). Offshore oil and gas waste management is better off being a part of the national integrated coastal zone management program.

Khan and Islam (2005b) recently introduced sustainability in the energy sector, especially in the oil and gas development. Introducing sustainable technology might play an enormous role in waste reduction. It is reported that reducing drilling wastes and produced water discharge from offshore oil and gas operation are technically achievable and, in most cases, economically advantageous (Patin 1999; Veil 2002). It is not particularly difficult to reduce discharges to zero by using modern technologies, such as re-injection, recycling and closed-loop drilling, waste treatment, and disposal systems.

This chapter elaborates on different aspects of waste generation and management of petroleum operations, especially during drilling and production phases. It starts by identifying types of wastes generated, followed by details of the methods of waste estimation. Then the current practices of waste management are discussed. Best available technologies for waste management are evaluated using a variety of practical and extensive methods. The chapter ends with the analysis of the process.

5.2　Drilling and Production Wastes

Table 5-1 shows many types of wastes that are generated during the drilling and production phases. Some of them are generated in huge quantities and are discussed below.

5.2.1　Drilling muds

Drilling muds are condensed liquids that may be oil- or synthetic-based wastes, and contain a variety of chemical additives and heavy minerals that are circulated through the drilling pipe to perform a number of functions. These functions include cleaning and conditioning the hole, maintaining hydrostatic pressure in

Table 5-1   Types of wastes generated in drilling and production phases of oil and gas development

Drilling and installation	Production
– Drilling muds	– Produced water
– Drillings cuttings	– Treatment, and completion fluids
– Produce sands	– Deck drainage
– Storage displacement water	– Produced sand
– Bilge and ballast water	– Storage displacement water
– Deck drainage	– Bilge and ballast water
– Well treatment fluids	– Deck drainage
– Naturally occurring radioactive materials	– Well treatment fluids
– Cooling water, desalination brine	– Naturally occurring radioactive materials
– Water for testing fire control	– Cooling water, desalination brine
– Other assorted wastes,	– Water for testing fire control
– Accidental discharges: air emission: oil spills, chemical spills, blowout	– Accidental discharges: air emission: oil spills, chemical spills, blowout
– Human wastes: sanitary wastes, kitchen wastes, and sink and shower drainage, laundry wastes, and trash	– Other assorted wastes,
– Other industrial wastes: cardboard, empty container, scrap metal, wood pallets, used chemicals and paint, sandblasting grit and paint, and cooling water.	– Human generates wastes: sanitary wastes, kitchen wastes, and sink and shower drainage, laundry wastes, and trash
	– Other industrial wastes.

Table 5-2   Composition of drilling muds which are used in oil and gas exploration

Product	Composition	Concentration (kg/bbl)
Base fluid	Water, bentonite clay (sodium montmorillonite), caustic soda (sodium hydroxide)	As needed
Additives	Lignosulfonate, phosphates (sodium acid pyrophosphate and tetrasodiumpyrophosphate), plant remains (predominant usage of quebracho), lignite	1–2.7 kg/bbl
Density control	Barite (natural barium sulfate ore), ferrophosphate ore, calcite, siderite	0–317.5 kg/bbl
Fluid-loss control	Starch (corn and potato), polyanionic cellulose polymer, xanthum gum, sodium carboxymethyle-cellulose, lignite.	<0.45–4.5 kg/bbl
Lost circulation	Ground nut shells, micas, ground cellophane, diatomacheous earth, cottonseed hulls, ground or shredded paper	0.9–13.6 kg/bbl
Corrosion and scale control	Sodium sulfite, zinc chromate, tall oil, amines, sodium hydroxide, phosphates, bacteriocides	0.11–2.7 kg/bbl
Solvents	Isoprophanol, glycerol, isobutanol, ester alcohols, diesel oil	
Lubricant	Asphalts, diesel oil, Fatty soaps, gilsonite, glass beads, rosin soap	0.09–2.7 kg/bbl

Source: Modified from Zwicker et al. (1983); Patin (1999)

the well, lubrication of the drill bit and counterbalance formation pressure, removal of the drill cuttings, and stabilization the wall of the drilling hole. Differences in the composition of the water-based, oil-based, and synthetic-based muds are shown in the Table 5-3. Additionally, Table 5-2 outlines the various chemical additives and heavy metals that may be added to the drilling muds to stabilize the drilling fluids during use, and to reduce corrosion and bacterial activity.

Water-based muds (WBMs) are a complex blend of water and bentonite. Oil-based muds (OBMs) are composed of mineral oils, barite, mineral oil, and chemical additives (Table 5-2). Synthetic-based muds (SBMs) are characterized by the replacement of mineral oil with an oil-like substance, and are free of inherent con-

Table 5-3 Composition of water-based, oil-based, and synthetic-based drilling muds

Water-based	Oil-based	Synthetic-based
– Bentonite (0 to 50) – Barite (0 to 500) – Caustic soda (0 to 5) – Soda ash (0 to 3) – Sodium bicarbonate (0 to 3) – Seawater (any portion) – Freshwater (any portion) – Drill solids (0 to 100)	– Barite (% 60.8) – Base oil (31.3%) – CaCl (3.3%) – Emulsifier (2.2%) – Filtrate control/wetting agent (1.8%) – Lime (0.2%) – Viscosifier (0.2%)	A drilling fluids whose continuous phase is composed of one or more fluids produced by the reaction of specific purified chemical feedstock, rather than through physical separations such as cracking and hydro processing

Source: GESAMP 1993
Source: Patin 1999
Source: OWTG 2002

taminants such as radioactive components and toxic heavy metals. SBMs are considered benign environmentally, have the potential to biodegrade under aerobic conditions, and therefore their use is preferred over WBMs and OBMs (OWTG, 2002). To minimize the quantity of oil discharged into the marine environment, use of water-based or synthetic-based mud is encouraged (OWTG 2002). Oil-based muds may be required in exceptional circumstances for deeper well sections and where the well needs to be drilled at a vertical angle.

Typically a single well may lead to 1000–6000 m^3 of cuttings and muds depending on the nature of cuttings, well depths, and rock types (CEF 1998). A production platform generally consists of 12 wells, which generate (62 × 5000 m^3) 60,000 m^3 of wastes (Patin 1999; CEF 1998).

Spent synthetic- or oil-based muds remaining from a drilling mud changeover or after drilling program completion are recovered and recycled, re-injected downhole, or transferred to shore, based on the composition and presence of oils in the muds. Onshore disposal requires approval by local regulatory authorities. According to the Canadian Offshore Waste Treatment Guidelines, under no circumstances is approval granted for the disposal of whole oil-based muds directly into the sea (OWTG 2002). Generally, water-based drilling muds are discharged onsite from offshore installations without treatment. The recommended level of polycyclic aromatic hydrocarbon content is less than 10 mg/kg in muds (OWTG 2002) for discharge directly in to the sea to be permitted; in excess of this the muds must undergo treatment prior to discharge.

5.2.2 Drill cuttings

Drill cuttings, transported to the platform via drilling muds, are tiny rock particles, about the size of sand and gravel, generated through drilling activities. In some parts of the world, problems occurring from drill cuttings are due to unplanned management. For example, in the Sakhalin Shelf the drill cuttings form huge piles as a consequence of long-time drilling operation (Patins 1999). According to CEF (1998), the volume of rock can range from 300 to 1200 m^3, and the volume of mud and cuttings combined can reach 3200 m^3 from each exploratory well.

Under current Canadian regulations, drill cuttings associated with the use of water-based drilling muds are usually discharged at sea (OWTG 2002). Due to the high concentrations of oil, cuttings associated with oil-based muds are not approved for such disposal. With respect to cuttings associated with synthetic or enhanced mineral oil-based muds, disposal may be via re-injection or direct disposal into the marine environment. Prior to disposal, cuttings must be treated to ensure that the concentration of oil on solid is below 6.9 g/100 g wet solid. The decision to practice re-injection is dependent on economic feasibility. If it is not economically feasible, cuttings are discharged at sea (OWTG 2002).

5.2.3   Produced water

During the oil and gas production, produced water is brought up from the oil and gas reservoir. It includes formation water, injection water brine, and any chemicals circulated downhole or added during the processes of oil–water separation. The amount of produced water is shown in Figure 5-1. This increases with the production life, initially generating small amounts, which increase with time.

Different countries tend to have different guidelines for the discharging of produced water. These include subsurface re-injection, subsea separation, and downhole separation. The Canadian (OWTG 2002) guidelines ensure that the 30-day weighted average of oil in discharged produced water does not exceed 40 mg/L and that the 24-hour arithmetic average of oil in produced water does not exceed 60 mg/L.

5.2.4   Produced sand

Produced sand originates from within geological formations and consists of the accumulated formation sands that are coated with reserve hydrocarbons. These sands are generated during production and the slurried particles used in hydraulic fracturing. Based on the concentration of oil in produced sand, it may be allowed to discharge, or may need to be treated to reduce oil concentration. Approval is dependent on both the concentration and aromatic content of oil associated with the sands (OWTG 2002). The general recommended level of oil concentration in produced sand is 10 mg/kg.

5.2.5   Storage displacement water

Storage displacement water is that pumped into and out of oil storage chambers on certain types of production installations during oil production and off-loading operations (OWTG 2002). Generally, in this type of water, oil concentration is higher. If the oil concentration is 15 mg/L or less, then this water can be discharged into the sea.

5.2.6   Bilge and ballast water

In the offshore operations huge amount of seawater may seep or flow into the structure through various points in the offshore installation, during the long operation periods. This water is known as bilge, whereas

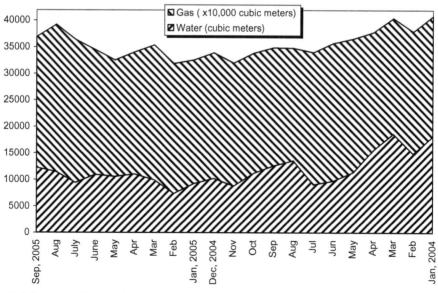

Figure 5-1   Total amount of gas and water produced in the Sable Island offshore energy project on the Scotian Shelf from December 2001 to September 2005.

ballast water is the water used to maintain the stability of an offshore facility. Bilge and ballast water contain high concentrations of hydrocarbon. According to Canadian regulations, if the oil concentrations of either bilge or ballast water exceeds 15 mg/L, disposal at sea is not permitted (OWTG 2002).

5.2.7 Deck drainage

Water reaches the platform deck through precipitation, wave-action, and sea spray, or from routine operations such as wash-down and fire drills, referred to as deck drainage. If the concentration of oil is higher than 15 mg/L, under the OWTG, it is required that the water be treated prior to disposal.

5.2.8 Well treatment fluids

For the well treatment operations such as well stimulation, well workover, well completion, and formation fracturing, different types of fluids are used, known as well treatment of fluids. These fluids mix with produced water in the well and are brought to the surface during production. Prior to disposal the well treatment fluids are treated as produced water, and as a result cannot be discharged until the concentration of oil is decreased to the acceptable level of 30 mg/L oil (OWTG 2002). Well treatment fluids containing diesel or other highly aromatic oils are not approved for use unless they can be recovered and recycled at the site, or shipped to shore for disposal. In addition, fluids must be treated to reduce salinity to below pH 5.0, before they can be lawfully discharged into the marine environment (OWTG 2002).

5.2.9 Naturally occurring radioactive materials

It is reported (OWTG 2002) that radioactive elements such as uranium, potassium, thorium, radium, and radon may be present in oil and gas products, posing potential risks to human health. These materials occur naturally in the deeper layers of the Earth's substratum. These naturally occurring radioactive materials can be precipitated during the separation process of oil, gas, and water.

5.2.10 Cooling water

In cooling water, biocide or chlorine occurrences must be treated to reduce the chlorine level. Generally, the use of biocide in cooling water is discouraged (OWTG 2002).

5.2.11 Desalination brine

Brines are produced during the desalination process in the offshore, to obtain potable water. The brine is directly discharged into the sea without treatment.

5.2.12 Water for testing fire control

Routine exercises are performed in readiness to combat accidental fire in offshore structures. Certain amounts of this water may be contaminated. Such water can be discharged without treatment.

5.2.13 Other assorted wastes

There are numerous other wastes that may be generated on offshore installations, including sludge from oil–water separation systems, hydrostatic test water, completion fluids, mono-ethylene glycol, spent lubricants and plastic materials, along with excess or damaged supplies. Depending on the material, these wastes may be reused, recycled, or discharged directly to the sea.

Table 5-4   Major oil accidents in the marine environment

Date	Name, Location (consistency with names of spills, and place)	Quantity (tons)
December 7, 2004	2004 Selendang Auy, Unalaska, Alaska	424,000 gallons intermediate fuel oil, 18,000 gallons marine diesel.
Nov, 2002	Tanker Prestige, Spain	70,000
1992	La Coruna, Spain	74,000
1991	Heaven, Genoa, Italy	144,000
Jan 25, 1991	Sea Island, Kuwait	1.5 million (tons)
Mar 24, 1989	Port Valdez, Alaska	41,000
Feb 1983	Persian Gulf	600,000
June 3, 1979	Gulf of Mexico	600,000
Feb 1983	Persian Gulf	600,000
1942	US East Coast	590,000
July 1, 1979	Trinidad and Tobago	300,000

Source: GESAMP (1993); Bregman (2000)

5.2.14 Human-derived wastes

Like other domestic places, wastes are generated from daily human activities on offshore facilities. These include sanitary wastes, kitchen wastes, laundry wastes, sink and shower drainage, and trash.

5.2.15 Other industrial wastes

As in other industries, common industrial wastes are generated from offshore facilities. Cardboard, empty containers, scrap metal, wood pallets, used chemicals, sandblasting grit, and paint are some examples.

5.2.16 Accidental waste releases

Oil tanker accidents, although not considered part of direct activities of oil and gas operations, are one of the major contributors to reported oil spills in the marine environment (Table 5-4). Other spills may occur, especially during transfers to and from supply vessels. Spills from oil and gas operations may consist of a variety of materials or supplies, other than oil or gas. These include diesel during connection or disconnection of fuel lines, or the contents of containers being lifted from vessel to vessel.

The chances of a blowout are much higher in an exploration well than in a production well, though still rare. Approximately 1% of exploratory wells worldwide have had blowouts, with small releases. In over 22,000 of all kinds of wells drilled in US coastal waters from 1971 to 1993, only 5 blowouts were reported, resulting in a total discharge of 170 m^3 (CEF 1998).

5.3   Waste Estimation

For the purposes of the waste estimation, information about the life cycle of petroleum operations and its different activities was collected through review of government documents and reports, and published papers, such as CNSOPB activities reports (CNSOPB 2005), Khan and Islam (2003a, 2006d), and EPA (2000). Severe limitations were encountered in finding information about quantitative data, such as amount of waste generation, amount of emissions, and use of toxic compounds. To overcome this restriction, different types of waste generation were estimated following the methodology of EPA (2000).

Total amounts of drilling wastes, such as drill cuttings, drilling fluids, and releases of oils were estimated for the four different types of wells used for exploratory and development drilling in shallow and deep waters. Drilling wastes have consequently been estimated using Equations 5.1 to 5.8 (EPA 2000).

The dry drill cuttings volume is estimated based on Equation 5.1. In this estimation, the dry drill cuttings are equivalent to gauge hole volume plus washout.

$$\text{Drilling hole volume (ft}^3) = \{\text{length (ft)} \times \pi \, [\text{diameter (ft)}/2]2\} \times (1 + \text{washout fraction of } 0.075) \quad (5.1)$$

$$\text{drill cuttings (bbls)} = \text{hole volume (ft3)} \times 0.1781 \text{ bbls/ft}^3 \quad (5.2)$$

$$\text{drill cuttings (lbs)} = \text{drill cuttings (bbls)} \times 910 \text{ lbs/bbl} \quad (5.3)$$

Waste Components are estimated following Equations 5.4 and 5.5. The units are estimated in lbs. The algebraic calculation of lbs of waste components in the given drilled:

$$\text{Total Wastes (TW)} = (\text{base fluid}) + (\text{water}) + (\text{barite}) + (\text{drill cuttings}) \quad (5.4)$$

$$\text{TW} = (\text{RF} \times \text{TW}) + \{[\text{RF} \times (\text{WF/SF})] \times \text{TW}\} + \{[\text{RF} \times (\text{BF/SF})] \times \text{TW}\} + (\text{DF} \times \text{TW}) \quad (5.5)$$

where

TW = total waste (whole drilling fluid + dry cuttings), in lbs
RF = retort weight fraction of synthetic base fluid
WF = water weight fraction from drilling fluid formulation
SF = synthetic base fluid weight fraction from drilling fluid formulation
BF = barite weight fraction from drilling fluid formulation
DF = drill cuttings weight fraction, calculated as follows

$$\text{DF} = 1 - \{\text{RF} \times [1 + (\text{WF/SF}) + (\text{BF/SF})]\} \quad (5.6)$$

In order to calculate TW, Equations 5.4 and 5.5 are first used to calculate DF (Equation 5.6). Then TW is calculated following Equation 5.7.

$$\text{TW} = \text{drill cuttings (lbs)} / \text{DF} \quad (5.7)$$

Input data to estimate the emissions of petroleum operations are shown in Table 5-5. These data have been gathered from different sources, such as EPA (2002) and Wenger et al. (2004). In this estimation, 10.2% (wt/wt) standard (baseline) solids control has been taken into account (EPA 2000).

The whole drilling fluid volume are estimated following Equation 5.8.

$$\text{whole SBF volume (bbls)} = \text{synthetic base fluid (bbls)} + \text{water (bbls)} + \text{barite (bbls)} \quad (5.8)$$

The formation oil in whole mud discharged is 0.2% (vol.), calculated based on Equation 5.8.

$$\text{Formation oil (bbls)} = 0.002 \times \text{whole SBF volume (bbls)} \quad (5.9)$$

This waste estimation process has been used during offshore oil and gas development, in the Scotian Shelf of Canada. The left arrows represent the input of chemicals or activities. The right arrows show the qualitative emissions or impacts on the environment. Major emissions and related activities have been identified (Figure 5-2). From the drilling and production phases, the major wastes are ship source wastes, human related wastes, release of CO_2, conflict with fisheries, dredging effects, sound effects, drilling mud, drill cuttings, flare, radioactive materials, produced water, release of injected chemicals, and release of metals and scraps. Similar types of wastes have been reported by Khan and Islam (2003a). It is also reported that generation of wastes and other pollutant emissions is the main contentious issue in oil and gas development (Khan and Islam (2003a, 2003b, 2004, 2005a, 2005b, 2005c). These emissions have negative effects on fisheries, biodiversity, special habitat, wildlife, and benthic habitat, as substantiated by Holdway (2002), Schroeder and Love (2004), Khan and Islam (2004), and Wells et al. (1995).

The phases of drilling and production generate the 7th and 8th types of wastes (Figure 5-2). Of all the phases, the drilling phase generates the highest volumes of wastes and toxicity levels. Therefore this section focuses on quantitative waste generation from drilling activities. Wastes from a typical drilling well consist of drilling mud, cuttings, barite, formation fluid, organic compounds (including oil compounds), and heavy metals concentrations. Estimated levels of wastes are shown in Tables 5.6–5-8.

Table 5-5   Characteristics of SBF drilling muds (EPA 2000)

Waste Characteristics	Value	Waste Characteristics	Value
SBF formulation	(by weight)	SBF density	280 lbs/bbl
Synthetic base fluid	47%	SBF drilling fluid density	9.65 lbs/ga
Barite	33%	Barite density	1,506 lbs/bbl
Water	20%	Water	350 lbs/bbl
Percent (vol.) formation oil	0.2%	Drill cuttings	910 lbs/bbl
Priority pollutant organics	lbs/bbl of SBF	Non-Conventional Organics	lbs/bbl of SBF
Naphthalene	0.0010024	Alkylated benzenes	0.0056429
Fluorene	0.0005468	Alkylated naphthalenes	0.0530502
Phenanthrene	0.0012968	Alkylated fluorenes	0.0063859
Phenol	0.000003528	Alkylated phenanthrenes	0.0080683
		Alkylated phenols	0.0000311
		Total biphenyls	0.0104867
		Total dibenzothiophenes	0.0004469
Priority pollutant metals	mg/kg Barite	**Non-Conventional**	**mg/kg**
Cadmium	1.1	**Metals**	**Barite**
Mercury	0.1	Aluminium	9,069.9
Antimony	5.7	Barium	588,000
Arsenic	7.1	Iron	15,344.3
Beryllium	0.7	Tin	14.6
Chromium	240.0	Titanium	87.5
Copper	18.7	**Conventional**	**lbs/bbl of**
Lead	35.1		**SBF**
Nickel	13.5	Total oil as SBF	190.5
Selenium	1.1	Total oil as formation oil	0.588
Silver	0.7	TSS as barite	133.7
Thallium	1.2		
Zinc	200.5		

Generally, water-based, oil-based, and synthetic-based mud/fluids are used in drilling. The detailed compositions of different types of drilling muds are shown in Figure 5-4. Due to considerations of toxicity and environmental degradation, synthetic-based mud types are recommended in Europe, the United States, and Canada (EPA 2000; CNSOPB 2000, Wenger et al. 2004). Due to its high toxicity level, OBM is not recommended for use. In the case of waste estimation, SBM is favored because it has been widely used (EPA 2000; Khan and Islam 2006d). The oil retention level of SBM was considered to be only 10.2%, which corresponds to conventional uses (EPA 2000).

On the Scotian Shelf, exploratory and development wells have been drilled to different depths (Figure 5-3). Drilling data of the Scotian Shelf from 1964–2004 has been analyzed. For waste estimation purposes, the wells are divided into shallow water wells (<1000 ft) and deep water wells (>1000 ft). Exploratory and development drilling has been conducted in both shallow and deepwater. Four categories of wells have been separately estimated and presented in Tables 5-6–5-8.

Table 5-6 shows that the major waste components are drill cuttings, SBS, formation oil, water, and barite. Figure 5-5 shows comparative waste generation of deep and shallow water during exploratory and development drilling. For all types of wells, exploratory drilling in deepwater generates the maximum amount of wastes. The solid wastes associated with drilling are dry-cuttings (includes 7.5% washout), cuttings and adherent drilling fluid generated from SBF/OBF interval (wet-cuttings), and barite. The solid wastes, including dry-cuttings, wet-cuttings, and barite that are generated from deepwater exploratory drilling are 71,772, 784,674, and 1,002,165 kg, respectively. The estimated amount for the other three types of wells is pres-

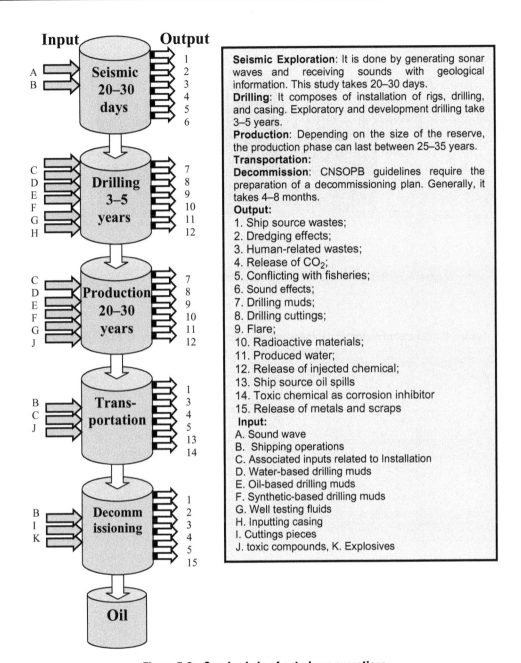

Figure 5-2 Supply chain of petroleum operations.

ented in Table 5.6. Barite is another solid waste that is generated in high volumes from the same well (Figure 5-5).

The total amount of SBF used in shallow-exploratory, shallow-development, deep-exploratory, and deep-development drilling are 63,666, 30,381, 102,221, and 45,975 kgs per well, respectively. SBF is the main cause of concern because it contains purified oil and other substances. Its detailed composition is shown in Figure 5-4 and Table 5.7. The detailed result of SBF wastes is shown in Figures 5-5 and 5-6.

Barite is generated as a solid waste in drilling. It is used as a weight agent in all kinds of drilling fluids, mainly composed of priority pollutant and non-pollutant metals. Barite is one of the most harmful pollutants (Wenger et al. 2004; EPA 2000). The composition of a variety of metals is shown in Table 5-8. The priority

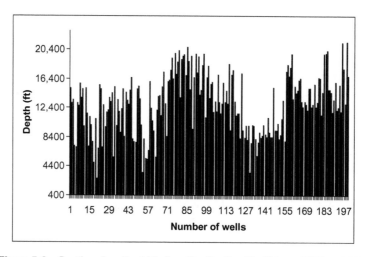

Figure 5-3   Depths of wells drilled on the Scotian Shelf from 1967 to 2003.

Table 5-6   Estimated major waste components of shallow and deep water development and exploratory drilling wells

Waste Component	Shallow Water (kgs)		Deep Water (kgs)	
	Development	Exploratory	Development	Exploratory
SBF	30,381	63,666	45,975	102,221
Water	12,928	27,092	19,564	43,498
Barite	21,332	44,702	32,280	71,772
D-cuttings	233,215	488,719	352,918	784,674
A-cuttings	297,856	624,178	450,737	1,002,165
W-SBF	64,641	135,460	97,819	217,491
F-fluid	94	196	142	316

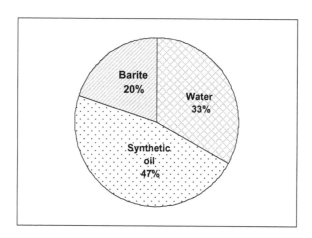

Figure 5-4   Different composition of drilling mud.

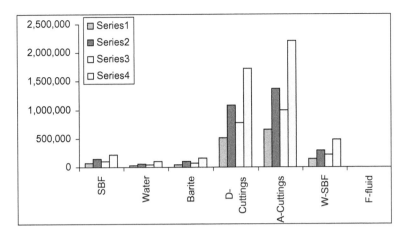

Figure 5-5   Waste components of shallow and deep water during exploratory and development drilling.

Table 5-7   Estimated priority pollutants from shallow water and deep water drilling wells

Priority Pollutant Organics	Shallow Water (kgs)		Deep Water (kgs)	
	Development	Exploratory	Development	Exploratory
Cadmium	2.34599178	4.91621664	3.5501159	7.8933036
Mercury	0.21327198	0.44692879	0.3227378	0.7175731
Antimony	12.1565029	25.4749407	18.396055	40.901664
Arsenic	15.1423106	31.7319437	22.914385	50.947687
Beryllium	1.49290386	3.1285015	2.2591647	5.0230114
Chromium	511.852752	1072.62908	774.57074	1722.1753
Copper	39.8818603	83.5756828	60.35197	134.18616
Lead	74.858465	156.872004	113.28097	251.86814
Nickel	28.7917173	60.335386	43.569604	96.872362
Selenium	2.34599178	4.91621664	3.5501159	7.8933036
Silver	1.49290386	3.1285015	2.2591647	5.0230114
Thallium	2.55926376	5.36314542	3.8728537	8.6108766
Zinc	427.61032	896.092214	647.08931	1438.734
	Non-Conventional Metals (kgs)			
Aluminum	19343.5553	40535.9939	29271.997	65083.158
Barium	1254039.24	2627941.26	1897698.3	4219329.5
Iron	32725.0924	68578.0936	49521.858	110106.56
Tin	31.1377091	65.2516026	47.11972	104.76567
Titanium	186.612983	391.062687	282.39558	627.87642

pollutant metals contained in drilling mud are cadmium, mercury, antimony, arsenic, beryllium, chromium, copper, lead, nickel, selenium, silver, thallium, and zinc. The estimated composition of these heavy metals for exploratory drilling in deep water is shown in Figure 5-7.

Khan and Islam (2004) found that many organic compounds (oil compounds) are emitted with other wastes. The detailed result of organic compounds released from four drilling wells is shown in Table 5-7. According to EPA (2002), emitted organic compounds are divided into two types, priority pollutants and non-conventional organic compounds (Table 5-8). Priority pollutants are naphthalene, fluorene, phenanthrene, and phenol. Figure 5-8 shows the comparative amount of priority pollutant in the four different wells. Naphthalene was found in the highest amount in all four wells, while phenol had the lowest levels for

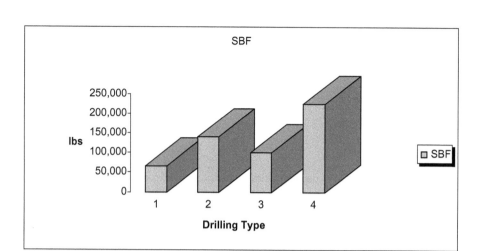

Figure 5-6   Total amounts of synthetic-base fluid (SBF) used in the shallow and deep water wells.

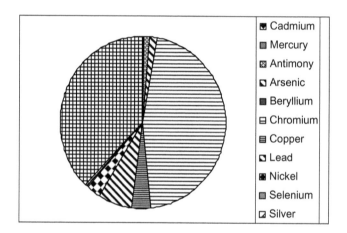

Figure 5-7   Priority pollutant metals in a deep water development well.

all wells. Non-conventional organic compounds found were alkylated benzenes, alkylated naphtalenes, alkylated fluorenes, alkylated phenanthrenes, phenols, total biphenys, and total dibenzothiophenes. The estimated composition of priority and non-conventional organic compounds in drilling wastes is presented in Table 5-8.

During the production period, two major emissions occurred. One was from flaring and the other from produced water. Table 5.9 shows the pollutant types, emission factor, and total amounts emitted from flaring in a single well on the Scotian Shelf. CO_2 constituted the highest amount of pollutants – 1100 tons. After flaring, the next highest emissions resulted from the drilling platform and associated equipments. CNSOPB (2005) estimated the release of CO_2 from a drill ship and drilling project to be at 22,443 and 16,161 tons per well respectively for a 90-day operation.

Produced water emits the highest amount of pollutants released from petroleum operations (Khan and Islam 2005a). Figure 5-9 shows monthly production of produced water, along with gas produced from the Sable Island Offshore Energy Project (SIOEP) and the Scotian Shelf from 2001 to 2005. Produced water contains heavy metals, oil, and brine solutions (Khan and Islam 2005a).

Table 5-8 Estimated organic compounds in shallow and deep water development and exploratory wells

Priority Pollutant Organics	Shallow Water (lbs)		Deep Water (lbs)	
	Development	Exploratory	Development	Exploratory
Naphthalene	0.5924184	1.2409712	0.8961456	1.9937736
Fluorene	0.3231588	0.6769384	0.4888392	1.0875852
Phenanthrene	0.7664088	1.6054384	1.1593392	2.5793352
Phenol	0.00208505	0.00436766	0.003154	0.0070172
	Non-Conventional Organics lbs			
Alkylated benzenes	3.3349539	6.9859102	5.0447526	11.223728
Alkylated naphthalenes	31.3526682	65.6761476	47.426879	105.51685
Alkylated fluorenes	3.7740669	7.9057442	5.7089946	12.701555
Alkylated phenanthrenes	4.7683653	9.9885554	7.2130602	16.047849
Alkylated phenols	0.0183801	0.0385018	0.0278034	0.0618579
Total biphenyls	6.1976397	12.9825346	9.3751098	20.858046
Total dibenzothiophenes	0.2641179	0.5532622	0.3995286	0.8888841

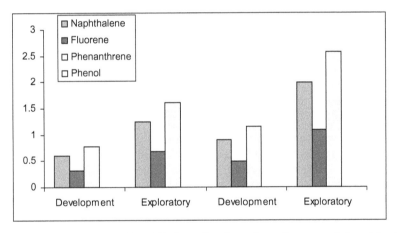

Figure 5-8 Comparative amounts of naphthalene, fluorine, phenanthrene, and phenol in shallow and deep water development and shallow wells.

In addition to the above mentioned wastes, huge amounts of waste are produced when decommissioning a project. These wastes are produced by the large amount of materials used in the construction of production facilities. For example, the SIOEP project used more than 900,000 m of pipe, 109 million kg of steel, 53,000 m of chrome tubing, and 160 million kg of concrete. According to present existing guidelines, the structure has to be removed based on the depth of the water (Khan and Islam 2007). However, even after following the accurate guidelines, a lot of materials are left at the production site.

5.4   Environmental Fate of Petroleum Wastes

From above, it is revealed that during the oil and gas production large amounts of water are brought up from the reservoir, known as produced water. It includes formation water, injection water, brine, and any chemicals circulated downhole or added during the processes of oil–water separation or oil recovery enhanced (Khan and Islam 2003a). Produced water is the largest amount of waste released and its amount significantly increases with the production age of the reservoir (Figure 5-9). Generally, different countries

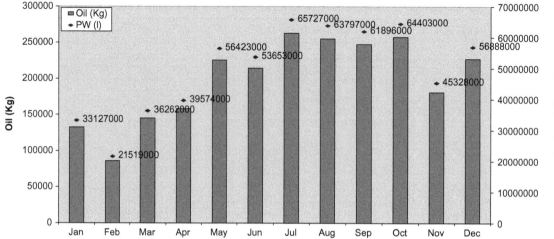

Figure 5-9   Average monthly volumes of produced water and oil disposed in 2001 at Hibernia.
Data source: CNOPB (2001).

have different guidelines for the discharge of produced water. These are subsurface re-injection, subsea separation, and downhole separation. The Canadian waste management guidelines ensure that the 30-day weighted average of oil in discharged produced water does not exceed 40 mg/L and that the 24-hour arithmetic average of oil in produced water does not exceed 60 mg/L (OWTG 2002). Even though these guidelines are followed, a larger amount of oil has been released into the marine environment (Figure 5-9). Considering the larger amount and long time of production, it is necessary to understand the consequences of produced water.

It is reported that produced water and its oil compounds can cause direct physical damage to wildlife, fisheries, aquaculture, and important habitats, i.e., fisheries habitats, mangroves, shores, and coral reefs. It also has toxicological effects on exposed organisms (Holdway 2002; Kingston 2002; Wells et al. 1995; GESAMP 1993). The effect of produced water and oil in the marine ecosystem has been extensively researched to assess its toxicity, weathering, degradation, and bioremediation (Berry and Wells 2004; Kingston 2002; Patin 1999; Holdway 2002; Khan and Islam 2003b). These types of studies have been conducted through direct experimental measurements, as well as using environmental models. Considering the dynamic nature of the aquatic/marine ecosystem, environmental modeling is a suitable tool for studying the fate and effects of toxic compounds.

5.4.1   Produced water discharges

The impacts of produced water can vary depending not only on its quantity and chemical composition, but also on the ability of the environment to absorb, use, or render inactive the constituents of produced water. The principal environmental concerns, as summarized from discussions of Stephenson (1992) and Continental Shelf Associates Inc. (1997), may be categorized as follows:

5.4.1.1   Oil content

The impact of dispersed oil into the aquatic environment depends on the location of the discharge point relative to the water column. If the discharge point is close to the bottom, then local sediment contamination will occur, with associated effects on the benthic community. Field measurements show that produced water dilutes rapidly. However, its rate of dilution is highly variable. It is influenced by the energy of the water, expressed as waves, currents, vertical mixing, and general flow of water masses.

Dispersed oil droplets that rise and eventually reach the surface lose their more toxic compounds through volatilization. Generally, the environmental impact of dispersed oil is not so detrimental, since most of it is removed with treating equipment, designed to remove discrete droplets of oil from the water phase. Its primary impact is the increase in biological oxygen demand (BOD), especially near the mixing zone.

5.4.1.2 Dissolved hydrocarbons and non-hydrocarbon organic compounds

Apart from locally increasing BOD, each of the components of soluble oil has a different fate in the environment. The light, one-ring aromatic hydrocarbons (benzene, toluene, xylenes, naphthalene) volatilize and dilute rapidly after the discharge of produced water. However, dilution is less rapid when a large volume of highly saline water is discharged into estuarine, low-saline, poorly mixed water.

Generally, the light aromatic hydrocarbons do not accumulate in sediments. However, higher molecular weight aromatic and aliphatic hydrocarbons may be found in high concentrations in sediments near produced water discharges. Phenols are diluted rapidly. They are degraded by bacteria and sunlight, which convert them to simpler hydrocarbons, and can be metabolized further to carbon dioxide and water. These factors, in combination with relative concentrations of each of the constituents in produced water, have each played some role in ensuring that, until now, little or no adverse long-term impacts on the environment has been observed from soluble oil.

5.4.1.3 Heavy metals

Barium from produced water precipitates as barite (barium sulfate) as it mixes with seawater, which is rich in sulfate. Soluble forms of iron and manganese present in produced water at high concentrations, tend to precipitate as insoluble oxyhydroxides when anoxic produced water mixes with oxygenated natural water. Many of the other metals present in produced water tend to co-precipitate with barium sulfate, iron, and manganese oxyhydroxides and are thereby removed rapidly from the water column. In high energy or deep-water environments, it is unlikely that metals from produced water will accumulate at concentrations higher than the natural background. Instead, they are probably diluted in the water column.

5.4.1.4 Radioisotopes

Concentrations of radioisotopes in produced water depend on the geological formation of the reservoir rock. Studies have indicated no detectable harmful impact of these components on fish, mollusks, or crustaceans.

5.4.1.5 Treating chemicals

The effects of treating chemical in the aquatic environment depend on the type of chemical and the way it is used (quantities, type of treatment, and method of discharge). For example, production-treating chemicals such as scale inhibitors, corrosion inhibitors, and biocides, can be used either in continuous treatment mode or in distinct large amount mode. In the first case, the concentrations produced are generally nontoxic, while in the second case the concentrations can exceed toxic levels. The adverse impact on the environment also depends on the way the operator handles the discharge. Sometimes the discharge is treated separately and at other times a continuous treatment is followed.

5.4.1.6 Salinity

When produced water of lower salinity is discharged into the sea the impact would be the same as with rain. Even when water of higher salinity is discharged into the sea, rapid dilution occurs so that no harmful impact is added to the aquatic environment.

5.4.1.7 Dissolved oxygen

Produced water is initially de-oxidized to prevent operational problems resulting from the presence of oxygen. In high-energy environments, the low oxygen content of produced water does not seem to be a problem. However, in low-energy environments, the problem of low oxygen content can be controlled if the discharge occurs above the surface of the receiving water, so that some oxygenation occurs before entering the aquatic environment. Thus, while some of the constituents of produced water have lesser impact than others, some can have serious toxic impacts depending upon their concentration and quantity. Chemicals that reach toxic levels in the aquatic environment have to be carefully assessed and soundly treated by the operator.

5.4.2   Studying fate and behavior of produced water compounds

Fugacity modeling is one of the better ways to simulate the fate and behavior of chemicals (Mackay 2005). Fugacity is a measure of the tendency of a toxic chemical to escape from one phase to another. For example, spilled oil is transferred from water to the atmosphere or sediments or into the tissues of aquatic organisms. This model is a valuable tool for understanding how contaminants migrate into different phases. The Level III Fugacity-Based Environmental Model (CEMC 2003) was used in this research to predict the fate and transport of compounds of crude oil by incorporating not only the single medium fate and transport but the inter-medium transfers as well. Khan and Islam (2005a) presented a robust model, which was developed based on Mackay and Paterson (1983); Mackay (2001), and Webster et al. (2003).

The model was run using an environmental condition similar to Atlantic Canada. The physical size (primary compartment) of the spill areas was hypothetically given, but all other environmental data/properties required for this study, such as organic carbon fraction, lipid fraction, volume fraction, and transport velocities are similar to the data of the Atlantic maritime region of Canada. These data are collected from the CEMC (2003) database. Table 5-9 provides the environmental properties of the studied area.

Table 5-9   Environmental properties of the study area

Total Surface Area		377,000 m^2	Organic Carbons	g/g
Air height		2 km	Particles in Water	0.2
Average water depth		30 m	Soil solids	0.02
Average soil depth		10 cm	Sediment solids	0.04
Average sediment depth		30 cm	Fish lipid	0.048

Bulk/Sub Compartments	Volume m^3	Densitykg/m^3	Transport Velocities	m/year
Air	2.52E+08	1.185413	Air side air–water MTC	43,800
Water	3,750,000	1,000.007	Water side air–water MTC	438
Soil	100	1,500.237	Rain rate	8.76
Sediment	37,500	1,280	Aerosol dry deposition velocity	87,600
Air vapor	2.52E+08		Soil air phase diffusion MTC	175.2
Air particles	0.00504	1.185413	Soil water phase diffusion MTC	0.0876
Exclusive water	3,749,978	2,400	Soil air boundary layer MTC	43,800
Suspended particles	18.75	1,000.007	Sediment–water MTC	0.876
Fish	3.75	1,000	Sediment deposition velocity	0.00438
Soil air	20	2,400	Sediment resuspension velocity	0.001752
Soil water	30	1.185	Soil water runoff rate	0.438
Soil solids	50	1,000	Soil solids runoff rate	8.76E-05
Pore water	30,000	2,400	Scavenging ratio (unitless)	200,000
Sediment solids	7,500	1,000		

Hypothetically, an area of 377,000 m² was considered as the study site where oil containing produced water had been continuously released year round. The area was characterized as an oceanic environment with an average water depth is 30 m. The total water body of this site is 125,000 km² with an average 30 cm sediments contents. The characteristics of the coastline were kept similar to the average Atlantic Canada coastal environment. The bulk compartments of this studied area were divided into water, soil, sediment, groundwater, coastal water, and biota (flora and fauna). There were 10 sub-compartments, namely, air vapor, air particles, water, water particles, biota, soil air, soil water, soil solids, pore water, and sediment solids. The total volume and densities of these compartments are shown on the Table 5-9.

The model uses other environmental data, such as water surface area and total volume depth of active layer of sediment, particle concentrations in inflow water, aerosols in air, and volume fraction of solids in surface sediment, particle densities (suspended sediment, sediment solids, aerosols), organic carbon fraction, water inflow and outflow data, sediment deposition, re-suspension and burial rates, atmospheric deposition parameters (aerosol dry deposition velocity, scavenging ratio, rain rate), and mass transfer coefficients (volatilization – airside and waterside sediment-water diffusion).

Transport velocities used in the model were taken as the value of this region. In this study, they are mass transfer coefficients (MTC), flows per unit area, or ratios of diffusivity to path length, and have the dimensions of velocity (m/h). As a first approximation, transport velocities are independent of the chemical, although they do contain diffusion rates that are somewhat chemical dependent. However, the chemical-to-chemical variability of diffusivity is usually negligible when compared with the variability in Z values. Where Z is the fugacity capacity of pollutant in each compartment, each unit is mol/m³-Pa. In this simulation, deposition was treated as three parallel processes: dissolution in rain water; wet deposition of particles by rain with a default scavenging ratio of 0.2×106 or snow with a scavenging ratio of 106; and dry deposition of particles with a default deposition velocity of 10.8 m/h (0.3 cm/s) (CMEC 2003).

5.4.3 Simulation tests of produced water

Existing waste management guidelines (Khan and Islam 2003a) mainly take into account the composition of the oil in produced water. The rates of water production were examined and monthly production data was gathered from CNOPB (2001). Based on the oil composition in produced water and their toxicity level, two major compounds of crude oil, benzene and naphthalene were selected for this simulation study running through the Fugacity-Based Environmental Model Level III (Mackay 2001; Webster et al. 2003). These chemicals are highly water-soluble compounds in the BTEX and PAH group and considered to have higher risk quotient value (Berry and Wells 2004). The primary reason of toxic concern of these chemicals is their aqueous-solubility, low molecular-weight, and higher concentration in the produced water (Berry and Wells 2004; NRC2003).

In this simulation study, the chemical properties of these selected chemicals, such as molecular mass, LogKow, LogKoa, Kaw, water solubility, vapor pressure, melting point, Hennery's Law constant, reaction half-lives (air, water, soil, and sediment), and partition coefficient for all the sub-compartments were required as model input. The chemical properties of benzene and naphthalene were collected from Mackay et al. (2000) (Table 5-10). The half-life value of benzene for air, water, soil, sediment, suspended particles, fish, and aerosol are 17, 170, 530, 1700, 170, 170, and 17 hrs, respectively. The half-life value of naphthalene for air, water, soil, sediment, suspended particles, fish, and aerosol are 210, 1650, 1300, 1800, 700, 340, and 210 hrs, respectively. The detailed chemical data of these two oil compounds are given in Table 5-11.

Several simulation tests were run for benzene and naphthalene in the hypothetical environment. The physical size of the study area was imaginary. The study area was designed as a coastal region by providing marine characteristics. In modeling, the amount and flow rate of benzene and naphthalene were used in a similar fashion to real oil and gas operations (Figure 5-10).

For the simulation, it was estimated that total 72,238 kg benzene was released in a 377,000 m² study area along with 598.59 million liters of produced water. It was assumed that this oil compound was directly

Table 5-10   Chemical properties of benzene and naphthalene (Mackay et al. 2000; CEMC,2003)

	Benzene		Naphthalene	
Molar mass	78 g/mol		128 g/mol	
Data temperature	25 C		25 C	
LogKow	2.13		3.37	
Vapour pressure (VP)	12,700 Pa		12.26 Pa	
Melting point	5.35 C		128 C	
Fugacity ratio	1		0.009506	
Sub-cooled liquid VP	12,700 Pa		148.88 Pa	
Water solubility	1,780 g/m³		31.5 g/m³	
Henry's Law Constant	556.51 Pa.m³/mol		0.057 Pa.m³/mol	
Partition Coefficients	Dimensionless	L/kg	Dimensionless	L/kg
Octanol–water (Kow)	134.8963	–	2,344.228	–
Organic carbon–water (Koc)	–	55.30749	–	961.1336
Air–water (Kaw)	0.224509	–	2.34E-05	–
Soil solids–water	2.65476	1.10615	46.13441	19.22267
Sediment solids–water	5.309519	2.2123	92.26882	38.44534
Suspended particles-water	26.5476	11.0615	461.3441	192.2267
Fish-water	6.475023	6.475023	112.523	112.523
Aerosol-air	472.4409	–	40,300.25	–

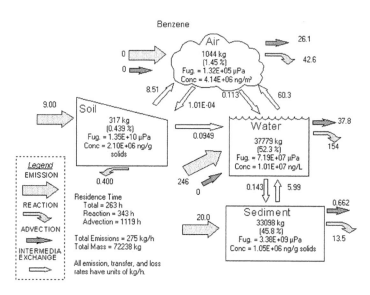

Figure 5-10   Diagram of fugacity model output for benzene.

released from the oil and gas platform along with produced water released into the open sea. This oil compound came mixed with water, therefore the release of hydrocarbons in the air was considered negligible. The emission rates of this compound in water, soil, and sediment were decided as 246, 9 and 20 kg/hr, respectively. The inflow rates in water and sediment were 246 and 20 kg/hr. The inflow concentrations were 2 ng/m³ and 3 ng/L in the air and water. The result of the benzene simulation is shown in Figure 5-10. In the case of naphthalene simulation, it was decided that 239,585 kg was emitted in the study area every year with an emission rate of 246, 9, and 20 kg, in water, soil, and sediment, respectively. Considering the pro-

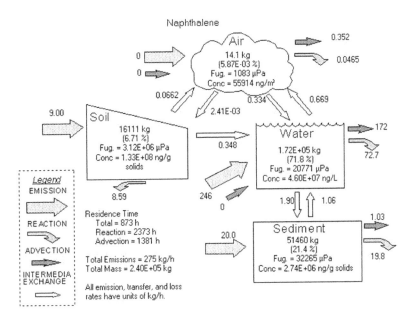

Figure 5-11 **Diagram of fugacity model output for naphthalene.**

duced water, it was estimated that major amount of these chemicals were released into the water area. The emission result of naphthalene is shown in the Figure 5-11.

There are three rectangular and one cloud-shaped box in both Figures 5-10 and 5-11, representing the main compartments of the study area. These are soil, water, sediments, and air. Different types of arrow-signs in these figures represent the emission, reaction, advection, and intermediate exchanges of the chemicals among the compartments. Inside the boxes, the value of residing amounts, the percentage fugacity values, and concentrations are also shown. These pictorial presentations give us the comparative scenario of all compartments. This may be helpful in deciding which compartment will require the most attention in the case of an accident. Other existing models (non-fugacity models) merely assess the effect level within a particular compartment rather than considering the whole picture (Mackay 1985). This is the most positive side of the model, which gives the whole picture of the total environment. For example, some models use only the emission status of water and then need another model to study soil, sediment, or air (Mackay and Paterson 1991). This environmental modeling is referred to as the concept of "unit world" consisting of at least air, water, soil, and sediment compartments (Mackay 2004).

Based on the emission rate, inflow concentration, and other physiochemical parameters, the total amounts of benzene found in air, water, soil, and sediment were 1044, 37,779, 317, and 33,098 kg respectively. Initially, it was reported that benzene did not emit in the air, because this compound came in soluble conditions within produced water. However, in the air compartment, 1044 kg benzene was recorded, believed to have come from intermediate exchange, mainly from water and soil. Water and soil media released benzene at rates of 60.3 and 8.51 kg/hr to the air. The total overall residence time of benzene in the system is 263 hours, whereas its reaction time is 343 hours and advection time is 1119 hours. Benzene is a highly water soluble compound and as a result it can transmit or accumulate in organisms faster than any other chemical. Therefore, it is considered as a highly toxic compound (NRC 2003; Mackay 1987; Neff 2002). The higher residence time of this compound will cause more damage and risk to inhabited flora and fauna in the released area.

A grand total of 239,585 kg of naphthalene was released to the study area. The highest amount of this compound was found in the water component (172,000 kg) followed by sediment, soil, and air compartments. With the same process of intermediate exchange, identified in benzene release mechanisms, naphthalene is found to make its way into the air, although not through direct release. However, naphthalene was found in much smaller quantities than benzene, although both compounds had similar flowrates of 246 kg/h in water, 20 kg/h in sediment, and 9 kg/h in soil. Their total masses were different for each compartment, indicating more naphthalene transmission to the system (Figure 5-12). This difference is attributed to the different chemical properties of these compounds.

Detailed investigations of benzene and naphthalene are shown in Table 5-12, which shows the fugacity, amount of chemicals, and concentration for all compartments and sub-compartments. Interestingly, in the case of both studied compounds, the highest amount of chemicals was found in water, followed by sediments and soil. In the air the amount of chemicals are low for both cases. The amount of the chemicals is also determined for each sub-compartment. For example, in the case of naphthalene simulation, the bulk water compartment has net 172,000 kg of the chemical. This water compartment also has 3 sub-compartments, which are suspended particles, fish, and exclusively water. The naphthalene concentrations in these 3 compartments are 3,100.73, 151.2551, and 1,344,207 kg, respectively (Table 5-12). The unique observation in the water compartment is that most of the chemical (63%) is accumulated in suspended particles (sub-compartment). The suspended particles of water are mainly phytoplankton and zooplankton, or silt contents.

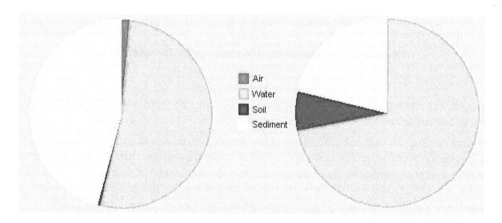

Figure 5-12   Concentration of benzene (left) and naphthalene (right) in air, water, soil, and sediment compartments.

Table 5-11   Phase properties of the benzene and naphthalene

		Benzene			Naphthalene		
		Z mol/m³.pa	Amount kg	Conc.g/m³	Z mol/m³.pa	Amount kg	Conc.g/m³
Air:	Bulk	4.03E-04	13385.7	4.14E-03	4.03E-04	110.0803	5.59E-05
	Vapor	4.03E-04	13385.7	4.14E-03	4.03E-04	110.0802	5.59E-05
	Aerosol	0.190591	1.26E-04	1.957413	16.25784	8.87E-05	2.253339
Water:	Bulk	1.80E-03	484341.2	10.0743	17.29927	1347459	45.99328
	Water	1.80E-03	484273.8	10.07296	17.25763	1344207	45.88255
	Particles	4.77E-02	64.28191	267.4128	7961.705	3100.73	21167.65
	Fish	1.16E-02	3.135703	65.22263	1941.879	151.2551	5162.841
Soil:	Bulk	3.00E-03	4068.313	3173.284	403.2626	125865.2	161107.5
	Air	4.03E-04	109.2367	426.023	4.03E-04	2.52E-02	0.16117
	Water	1.80E-03	729.8375	1897.577	17.25763	1615.921	6894.595
	Solid	4.77E-03	3229.238	5037.612	796.1704	124249.3	318078.1
Sediment:	Bulk	3.35E-03	424332.6	882.6117	332.2743	402029.3	1372.26
	Water	1.80E-03	182322	474.0372	17.25763	16704.45	71.2723
	Solid	9.54E-03	242010.5	2516.91	1592.341	385324.9	6576.211

In the region of study, the silt concentration is comparatively lower because of insignificant discharge of the mass from the river network. Therefore, the suspended particles are mainly composed of the plankton communities in the water.

Based on these findings, we can conclude that in case of real hydrocarbon releases from the produced water, severe damage of the plankton community may occur. Planktons are the primary food source in the marine ecosystem, occupying the top position in the food pyramid as the primary producer. If the food pyramid is severely affected, the complete food cycle cannot be sustained. As a result, the overall productivity of the ecosystem will be significantly reduced and finally lead to the breakdown of the food chain in the accident region. This is an important issue that has been observed in this study, which requires more detailed studies to evaluate its affect on the food chain. In this simulation study, the effects on different species (especially the effects on different components of the food chain, i.e., producers, consumers) could not be determined. Even though the concentration of benzene in fish is only an approximation, the size of the fish population, their migration and movement, and other biological data were not considered in this study. These are the major drawbacks of this model. However, other existing models commonly used in oil spill estimation, such as the OIL IN ICE model, PC Oil Slick simulation, SIMAP, and NRDAM/CME (McCay and Payne 2001; French 1998, 2001; McCay 2003), do not consider the food chain based assessment at all.

Fugacity (Pa) values of benzene and naphthalene, for all environmental compartments, are shown in Table 5-12. It was found that the fugacity value of benzene was much higher in the sediment and air than in any other compartments. The fugacity of naphthalene was also estimated through the model for all compartments and sub-compartments. The highest concentration was found in sediments and the second highest in soil compartments. The fugacity value in the air is negligible.

The transport rates for spilled compounds are calculated (Table 5-12). The highest transport rate for benzene is between air and water, and for naphthalene between the water and sediment, though both chemicals were released in a similar fashion. This transport pattern of oil compounds in between different compartment and sub-compartments helps us understand which one will be adversely affected in case of a spill and will help in the development of a management plan accordingly. If we know in which compartment the spill has occurred, it will be easier to judge the effect level by observing the transport rates. For example, a spill onto the soil is less dangerous than a spill into water, because the transport rate between oil to other media is less than that of the water to media. Managers can take precautionary measures and determine vulnerability of marine zones by taking such transport rates into account.

So far, the used model only assesses the oil compounds of produced water in the major compartments as of air, water, soil, and sediments. It also assesses fish in water as a major biotic component. However, this model completely ignores the overall impacts on species level. Considering this drawback, an integrated model is proposed (Figure 5-13), where the fugacity model is integrated with a food chain model. This integrated model initially screens the effects of oil compounds using a fate model and then food chain model to specifically identify their impacts on species level. The food chain component is a multiphase model

Table 5-12 Calculated intermediate transport and individual process D values

Intermediate Transport	Benzene (kg/h)	Naphthalene (kg/h)
Air to water	12.09228	2.742805
Air to soil	108.6537	19.79159
Water to air	8.04E-02	16.42926
Water to sediments	118.8296	176.8371
Soil to air	0.022074	0.143905
Soil to water	8.734469	2.614213

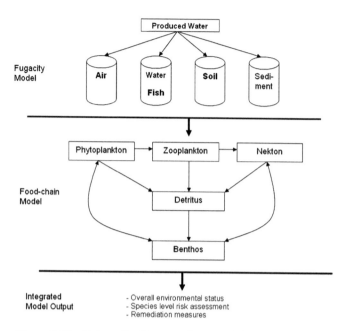

Figure 5-13 Schematic diagram of proposed integrated model.

capable of assessing the probable impacts of oil on the food chain components, such as primary producers (phytoplankton), primary consumers (zooplankton), secondary consumers (fish, squid), and higher tropical consumers (higher carnivores) (Figure 5-13).

5.5   Current Practices of Waste Management

5.5.1   Direct discharge

Some of the wastes generated from offshore development are approved for direct discharge into the sea, following proper treatment and acceptable concentrations of contaminants. Generally, Canadian regulations permit the discharge of biodegradable wastes. There are standard guidelines to follow terms of accepted levels of polycyclic hydrocarbon contaminations within the waste. If the contamination level is higher than this level, then the waste needs to be treated in order to bring down the contamination below the acceptable level.

Table 5-13 shows the quantity of different wastes generated during drilling and production and which ones are allowed to be directly discharged into the sea.

5.5.2   Re-injection of drilling wastes

Recent strict regulations do not allow dumping of contaminated wastes directly into the sea. As a consequence, re-injection of exploratory and drilling wastes, especially the drill cuttings and produced water, is becoming a more common practice when possible. For re-injection, drill cuttings are slurried by milling and adding seawater. The slurry is then disposed of by pressurized pumping into the disposal well. The concept of disposal of waste drill cuttings, by means of re-injection downhole, has already been applied successfully by several operators worldwide. The technique has proved to be viable in many different areas and formations around the world, with the most activity in the North Sea, Alaska, Gulf of Mexico, and Venezuela (Patin 1999; Veil 2002; Islam 2003; Schmidt et al. 1999). The process is recognized as highly environmentally friendly and has proven to be more economical than the disposal of drill cuttings onshore.

Table 5-13 Amount of major waste produced and discharge in the sea

Waste Type	Amount of Waste release (tones)	Discharge
Drilling muds	15–30 (periodically/single well) 150–400 (bulk at end/single well) 45,000 (production site/multi-wells)	Mostly water based muds
Drill cuttings	200–1,000 (production, multi-wells)	Water-based muds associated with cuttings
Produced waters	1,500/day (production/multi-wells), increase with age.	– If oil concentration is higher then treated before discharge
Accidental discharges	In over 22,000 wells of all kinds drilled in US coastal waters from 1971 to 1993, only 5 blowouts occurred and the total discharge was 170 m^3 – is the average/year	– Oil spills, gas blowouts, chemical spills
Other wastes	– Data not available	Storage displacement water, Bilge and ballast water, Deck drainage, Well treatment fluids, Naturally occurring radioactive materials, Cooling water, desalination brine, Water for testing fire control, Other assorted wastes
Platforms	Data not available	More than 4,000 tonnes does not carry to shore for disposal.

Source: Patin (1999); CEF (1998)

Several different approaches are used for injecting drilling wastes into underground formations for permanent disposal. Slurry injection technology involves grinding or processing solids into small particles, mixing them with water or some other liquid to make a slurry, and injecting the slurry into an underground formation at pressures high enough to fracture the rock. There are two common forms of slurry injection. Figure 5-14 shows both types of slurry injection. One is annular injection and the other is injection into a disposal well.

Annular injection introduces the waste slurry through the space between two casing strings (Figure 5-14). At the lower end of the outermost casing string, the slurry enters the formation. The disposal well alternative involves injecting to either a section of the drilled hole that is below all casing strings, or to a section of the casing that has been perforated with a series of holes at the depth of an injection formation.

When the slurry is ready for injection, the underground formation is prepared to receive it. First, clear water is rapidly injected to pressurize the system and initiate fracturing of the formation. When the water starts flowing freely at the fracture pressure, the slurry is introduced into the well. Slurry injection continues until an entire batch of slurried material has been injected.

At the end of this batch, additional water is injected to flush solids from the well bore, and then pumping is discontinued. The pressure in the formation will gradually decline as the liquid portion of the slurry bleeds off over the next few hours, and the solids are trapped in place in the formation. Slurry injection can be conducted as a single continuous process or as a series of smaller-volume intermittent cycles.

Different types of rocks have different permeability characteristics. Slurry injection relies on fracturing, and the permeability of the formation receiving the injected slurry is a key parameter in determining how readily the rock fractures, as well as the size and configuration of the fracture. When the slurry is no longer able to move through the pore spaces, and the injection pressure continues to be applied, the rocks will crack or fracture. Continuous injection typically creates a large fracture consisting of a vertical plane that moves outward and upward from the point of injection. Most annular injection jobs inject into shale or other low-permeability formations, and most dedicated injection wells inject into high-permeability sand layers.

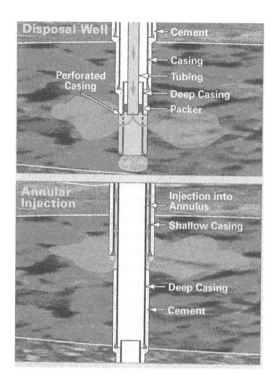

Figure 5-14   Two types of slurry injection. Photo courtesy of DOE (2005).

5.5.3   Onshore disposal

There are established onshore techniques for cleaning drill cuttings. Generally, onshore facilities treat solids using techniques such as grinding, direct thermal desorption, and indirect thermal desorption. Emerging solids treatment techniques, such as micro-emulsion, supercritical extraction using liquid natural gas or liquid carbon dioxide, are considered as alternatives (Veil 1998, 2002). At present, offshore wastes, such as cuttings, only go for onshore disposal when solid and water treatment rates and the potential for reuse of recovered oil are economically feasible.

5.5.4   Separation of mud from cuttings

Table 5-6 shows that huge amounts of drilling muds are used in drilling operations. The drilling mud, in addition to carrying out numerous drilling functions, also carries out rock cuttings to the surface. To reuse the mud and in order to environmentally manage the solids, the cuttings need to be separated. The first step in separating the cuttings from the mud involves circulating the mixture of mud and cuttings over vibrating screens called shale shakers (NETL 2005) (Figure 5-15). After the liquid mud has passed through the screens, it is used for recirculation through the drill pipe, the separated rocks remaining on top of the shale shaker screen. These cuttings are directed down the screen using a vibration action.

For better performance, or to completely separate cuttings from mud, several shale shakers are used. For example, first, second, and third shale shakers are known as primary, secondary, and tertiary shale shakers. The primary shakers use coarse screens to remove the larger cuttings. The secondary shakers use fine mesh screens to remove much smaller particles, and tertiary shakers are used to separate very fine particles. Generally, primary and secondary shale shakers are used. Using more shakers might create problems. For example, strong vibration of shakers can break up the cuttings into comparatively much smaller particles, which can in turn become very difficult to remove from mud. According to NETL (2004), in general, the

Figure 5-15   Shale shaker. Photo courtesy of NETL (2005).

Figure 5-16   Mud tank. Photo courtesy of NETL (2005).

separated drill cuttings are coated with a large quantity of drilling mud roughly equal in volume to the cuttings. Figure 5-16 shows the mud tank containing separated mud from drilling method.

Drilling mud is an expensive material, which requires costly and tedious disposal and remediation mechanisms. As a result, reuse of mud is one of the best options in waste management. For reuse, mud cuttings must be separated. For further separation of the drill cuttings, additional mechanical processing is often used in the mud pit system. According to NETL (2005), three types of mechanical equipment are used:

1. hydrocyclone-type desilters and desanders;

2. mud cleaners (hydrocyclone discharging on a fine screened shaker); and

3. rotary bowl decanting centrifuges.

The present waste management guidelines restrict the disposal of drill cuttings coated with mud. In order to comply with these guidelines, the solids are further treated with drying shakers using high gravitational separation, vertical or horizontal rotary cuttings dryers, screw-type squeeze presses, or centrifuges (NETL 2005). The cuttings dryers recover additional mud and produce dry, powdery cuttings. Different drying methods are used for drying the cuttings. Figure 5-17 shows a drying system used for drying cuttings. Figure 5-18 shows dried cuttings piled up next to the dryer.

Figure 5-17   Vertical cuttings dryer. Photo courtesy of NETL (2005).

Figure 5-18   Dried cuttings. Courtesy of NETL (2006).

5.5.5   Solidification and stabilization

The separation of cuttings from mud does not solve all problems. Further treatment is required to remove mud from separated cuttings in order to make them suitable for reuse. Sometimes cuttings separated from the mud by the shale shakers are difficult to transport due to leaching. In current practice, these cuttings are solidified using various chemicals. There are also mechanical processes that are used in solidification. The main objective of solidification is to reduce the surface areas and to restrict contamination. The stabilization process makes the wastes into less soluble and mobile form, but this process does not necessarily change their chemical nature. Different materials, such as cement, fly ash, lime, and calcium oxide are used for solidification and stabilization.

The solidification and stabilization of waste come with their limitations. For example, by adding cementing materials into drilling waste, the overall volume of waste further increases. According to NETL (2004), these methods of waste management are also not cost effective. These methods are rather more suitable for land based waste management. The waste management methods require a lot of space, which is not appropriate in the offshore petroleum operations. In addition to the need for long-term management, there is risk of leaching pollutants from solid/stabilize form.

5.6 Evaluation of Current Waste Management Technologies

5.6.1 Produced water management techniques

Worldwide, produced water is the largest volume waste generated in association with oil and gas production. In Norway, a series of technologies, i.e., re-injection, treatment, water shut-off, down-hole separation, removal, or replacement of process chemicals are considered in order to meet the "zero impact" challenge (Johnsen et al. 2000; CAPP 2001).

Reduction of the amount of dispersed oil in produced water aims to ultimately decrease the marine environmental impacts. The ever-increasing volumes of produced water, which now constitutes the largest single fluids stream in exploration and production operations, warrant a structural and integral approach toward its management. Prior to the produced water disposal, the quality should be upgraded through treatment, to the locally required standards. Produced water treatment starts with primary three-phase separation (oil, gas, and water) where water is removed from the bulk produced fluid. Primary technologies available to deal with produced water generally fall into one of three categories, any combination of which may be employed in a given field:

- Conformance control measures;
- Conventional disposal methods;
- Downhole oil/water or gas/water separation and disposal methods.

Most operators rely on a combination of technologies. A number of new options have been proposed for reducing the costs or increasing the efficiency of surface water treatment and disposal processes. Mostly, operators continue to rely on conventional disposal solutions and focus on finding ways to reduce the cost of those operations as much as possible. Over the past several years the third category, downhole separation and disposal, has seen an increased number of installations and attention. Produced water requires treatment of a number of constituents depending on the intended use. Because the produced water is delivered at around 160°F, pathogenic microorganisms are of no concern (Funston et al. 2001; Lang, 2000).

Biological treatment can be used for the actual removal of organic material. The major limitation of any biological treatment system is that long retention periods are required for the degradation process. The only potential alternative to discharging produced water entails underground injection. While injection is a well-established practice onshore, it is problematic offshore. There are many factors to consider, including the availability of space and/or load capacity for added equipment required for injection, additional tankage, water treatment equipment, and injection pumps (CAPP 2001).

5.6.2 Existing regulations

Produced water discharges are regulated differently in different parts of the world – in some countries, arbitrarily and unpredictably, social-political conflicts hinder consistent implementation. The Oslo-Paris Commission (OSPAR) and the US Environmental Protection Agency (EPA) play the main role in produced water regulation worldwide. The existing Canadian Offshore Waste Treatment Guidelines (OWTG), derived from OSPAR Recommendation 92/6, are currently under review. The guidelines with regards to

Figure 5-19   Poster illustrating produced water re-injection at a North Sea drilling site (STATOIL 2000).

Selection of Chemicals Intended to be used in Conjunction with Offshore Drilling and Production Activities (Offshore Chemical Screening Guideline, OCSG) also draws heavily upon the Harmonized Mandatory Control System (HMCS) developed by OSPAR (Decision 96/3). Canada depends heavily on the OSPAR guidelines for the North Sea, following the OSPAR regulatory regime in the east coast offshore: "currently, the applicable limit is a 30 day average dispersed oil and grease content of 40 mg/L. Additionally, if dispersed oil and oil and grease content averages over 80 mg/L for any 48-hour period, it must be reported. Produced water must be monitored every 12 hours and the 30 day average must be calculated daily" (CAPP 2001).

In Canada, the use of treatment chemicals in production operations is regulated through application of the OCSG. The OCSG provide a "clearing house" approach to regulation. Once a chemical is determined as acceptable, only tracking of usage is required. To be approved for discharge, 13 screening criteria must be passed. These include specific identification of the chemical, proposed use, and domestic and international classifications. Additional information on toxicity and environmental information may be required. The OWTG also require approval of the location of the produced water discharge point.

Other requirements include the execution of an Environmental Effects Monitoring (EEM) program that is intended to determine the adequacy of ongoing disposal procedures and the effectiveness of the waste treatment technologies in place. To date, little substantive EEM data are available – a situation ascribed to the early stage of development in Canada's offshore. Under such a constraint, experience from other areas of the world must be relied upon to make preliminary decisions relative to produced water discharges as the OWTG are reviewed. Different oil and gas producing countries have different regulatory procedures dealing with the release of produced water into the marine environment. Some of these procedures are more developed than others. For example, the basis for national laws governing the discharge of offshore drilling wastes in oil-producing coastal states of Western Europe is the OSPAR Harmonized Offshore Chemical Notification Format (HOCNF). Non-parties also follow the HOCNF, which include those that do not have legislation requiring environmental testing, such as Angola, or those countries that have legislation, such as Australia (Baker Hughes Inc. 2001). The United States, Mexico, and China follow separate regulations on testing requirements.

Table 5-14   Technology to clean up produced water (Wills 2000)

Technology	Process	Advantages	Disadvantages	Costs (US $)
Carbon adsorption	Modular granular activated carbon systems	Removes hydrocarbons and acid, base and problem; produces neutral compounds; low energy requirements; higher throughput than other treatments (except biological; treats a broad range of contaminants; very efficient at removing high Mwt. Organics	Fouling of carbon waste granules is a stream of carbon and backwash; requires some pre-treatment of produced water stream.	"Middle range" of costs
Air stripping	Packed tower with air bubbling through the produced water stream	Can remove 95% of volatile compounds as well as benzene, toluene, naphthalene, phenanthrene, anthracene, pyrene, and phenols; H_2S pyrene, and can be stripped but pH must be adjusted; higher\temp. improve removal of semivolatiles; small size, low weight and low energy requirements; simple to operate; well-known technology.	Can be fouled by oil; risk of iron and calcium scales forming; generates an off-gas waste stream that may require treatment; requires some pre-treatment of produced water stream.	Low capital operating costs; overall treatment cost $0.02 to $0.10/1000 gallon, plus $0.50 to $1.50/1000 gallon if off-gas control by activated carbon is required.
Filtration	Very fine membranes.	Effective removal of particles and dispersed and emulsified oil; small size, low weight and low energy requirements; high throughput rates.	Does not remove volatiles or dissolved compounds. Does not affect salinity; oil, sulfides or bacteria may foul membrane, which requires daily cleaning; waste streams may contain radioactive materials; requires some pre-treatment of produced water stream.	Low capital and operating costs (similar to air stripping).
Ultra-violet light	Irradiation by UV lamps	Destroys dissolved organics and both volatile and non-volatile organic compounds, including organic biocides; does not generate additional waste stream; handles upset or high loading conditions.	Will not treat ammonia, dispersed oil droplets, heavy metals or salinity; relatively high energy requirements; UV lamps may become fouled; residues may be toxic if peroxide used; requires some pre-treatment of produced water stream.	Similar capital costs to chemical oxidation with ozone but operating costs lower because no waste stream.

Table 5-14   Continued

Technology	Process	Advantages	Disadvantages	Costs (US $)
Chemical oxidation	Ozone and/or hydrogen peroxide oxidation	Removes H_2S and particulates; treats hydrocarbons, and acid, base and neutral organics, volatiles and non-volatiles; low energy requirements if peroxide system used; straightforward to operate.	High energy inputs for ozone system; oil may foul catalyst; may produce sludge and toxic residues; requires some pretreatment of produced water stream.	"Middle range" of costs.
Biological treatment	Aerobic system with fixed film biotower or suspended growth (e.g., deep shaft)	Treats biodegradable hydrocarbons and organic compounds, H_2S, some metals and, in some conditions, ammonia;"fairly low" energy requirements; handles variable loadings, if acclimated.	Large, heavy plant required for long residence times; build-up of oil and/or iron may hinder biological activity; aeration may cause calcium scale to form; may produce gas and sludge requiring treatment; requires some pretreatment of produced water stream.	"No costs estimated"

Table 5-15   Produced water treatment standards comparison (CAPP 2001)

Country	"BAT"	Effluent Limits	Monitoring Requirements	Exception Thresholds	Routine Reporting
United States	Gas flotation	29 mg/L monthly avg. 42 mg/L daily max	Total O&G Gravimetric	Any exception	Annual
Canada	Not stated	40 ppm 30 day avg. 80 ppm 2 day avg.	Dispersed O&G 2x/day	Any exception	Monthly
Norway	Gas flotation hydrocyclone	40 ppm monthly avg.	Dispersed O&G 1/day composite O&G 1/yr comprehensive	>40 ppm monthly avg	Quarterly O&G Annual comprehensive
United Kingdom	Gas flotation hydrocyclone	40 ppm monthly avg. 30 ppm annual avg	Dispersed O&G 1/day composite O&G 1/yr comprehensive	>100 ppm	Monthly O&G Annual comprehensive

5.6.3   Evaluation method

Multicriteria Decision Making, a powerful tool based on mathematical programming under multiple objectives to help in the process of searching for decisions that meet and fulfill a multitude of conflicting objectives (Gal et al., 2004), was applied in this evaluation. Decision analysis looks at the paradigm in which an individual decision maker (or decision group) contemplates a choice of actions in a certain environment. The theory of decision analysis is designed to help the individual make a choice among a set of pre-specified alternatives. According to Keeney (1992), the first step in a Multicriteria Decision Making process is to

Table 5-16   Typical quantities of wastes discharged during offshore oil and gas exploration and production activities

Source	Approximate Amounts (Tonnes)
Exploration sites (ranges for a single well):	
Drilling mud – periodically	15–30
– bulk at end	150–400
Cuttings (dry mass)	200–1,000
Base oil on cuttings	30–120 (a)
Production site (multiple wells):	
Drilling mud	45,000 (b)
Cuttings	50,000 (b)
Production water	1,500/day (c)

(a) Actual loss to environment may be higher (Chenard et al., 1989);

(b) Estimate based on 50 wells drilled from a single offshore production platform, drilled over 4 to 20 years (Neff et al., 1987)

(c) From a single platform (Menzie 1982)

Table 5-17   Testing requirements that pertain to China regarding the toxicity of drilling fluids

Test	Organism	Requirements
Assessment of the toxicity to the marine alga, *Chaetoceros mulleri* (48 and 96 hours)	*Chaetoceros mulleri* (marine algae)	6,400 mg/l
Assessment of the toxicity to the marine shellfish, *Pinctada martensii* (48 and 96 hours)	*Pinctada martensii* (pearl oyster)	10,000 mg/l
Assessment of the toxicity to the marine fish, *Ctenogobius gymnachen* (48 and 96 hours)	*Ctenogobius gymnachen* (marine fish)	10,000 mg/l

identify the evaluation criteria following a value-focused thinking approach. The decision-making process relies largely on information about the alternatives.

5.6.3.1 Identification/selection of criteria

The criteria are scored based on the management issues related to water produced. The selection of criteria is an important step in managing the produced water, as it will help to success this process. Environmental safety and its sustainable management will be the main decision-making criteria.

While selecting the appropriate technologies for the management of produced water, their sustainable management, along with environmental safety, was used as threshold criteria. The options that were not suitable for the better management of produced water were rejected. To compare and assess the different options along these lines the decision-making criteria are used. The decision-making criteria were again divided into three major categories, costs, technical performance, and regulations and environmental monitoring. An evaluation measure under each option was done through different measuring devices. Qualitative and quantitative values were used during the rating process.

5.6.3.2 Identifying alternatives

It is essential for each technology option to pass the above-mentioned criterion. Accordingly, the technologies using various types of treatment and disposal mechanisms including biological treatment, filtration, re-injection, were chosen. The strategies including downhole separation, ecological risk assessment, re-injection, gas flotation, and hydrocyclone were selected.

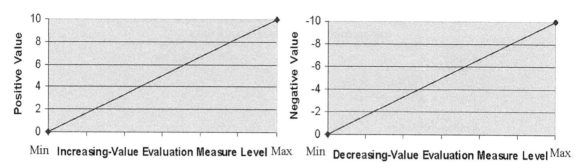

Figure 5-20   Single-value functions used in normalizing quantitative scores.

5.6.3.3 Data collection

To perform a comparative evaluation of existing produced water management technologies, the relevant data were collected from different sources, i.e., journals, proceedings, websites, as well as from personal communications with representatives of different government agencies and NGOs.

5.6.3.4 Assigning weight

A direct weighting method developed by Parnell et al. (1999) was followed to assign weights to the criteria. In this method, weights that represent the relative importance of the criteria are directly assigned to the criteria. The assigned weights are shown as the number in brackets out of a total weight of 100 (Figure 5-20). In this assessment, equal importance was given to the criteria such as costs and regulations and environmental monitoring as the weight was given 35 in each category. In order to achieve the goal of environmental safety and its sustainable management, they were divided into three criteria, such as costs, technical performance, and regulations and environmental monitoring. Again these three criteria were grouped into several segments so as to assign weights among themselves.

5.6.3.5 Scoring

Qualitative and quantitative were the types of data used in the scoring process. Capacity of management systems, treated here as a quantitative data, were normalized to interval scaled values before being used in the evaluation. The normalization was conducted using linear value functions. The quantitative data may have values which are either increasing (a higher value indicates a higher ranking) or decreasing (a lower value indicates a higher ranking). Therefore, a negative sign was assigned to the scores with decreasing values. This results in two different ranges of normalized scores, 0 to 10 for increasing values and –10 to 0 for decreasing values. Figure 5-20 shows the linear single value functions used in normalizing quantitative scores (Thanyamanta et al. 2003).

General equation:

$$Score\ (in\ 0-10\ range) = \frac{\pm (Quantitative\ data\ value - Minimum\ Value) \times 10}{Maximum\ Value - Minimum\ Value} \tag{5.10}$$

Subjective rankings were implied to assess the option with regard to qualitative data. In order to compare options, the subjective rankings were converted into numbers within the range of 0 to 10 by using conversion charts. The conversion charts (Figure 5-20) were divided and marked with the numbers from 0 to 10, where the higher value represents a preferred characteristic.

5.6.3.6 Overall value model

In order to model the overall value, the widely-used additive value model was selected because of simplicity and robustness (Hobbs and Meier 2000) (Figure 5-21). This model provides a single index that can be used to compare alternatives.

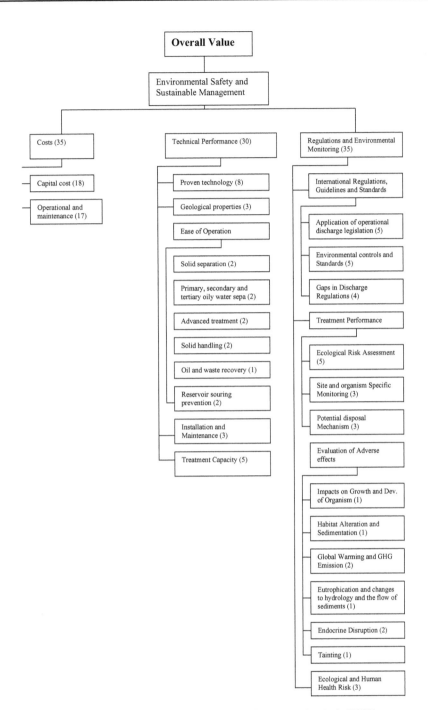

Figure 5-21   Value model. Modified from Thanyamanta et al. (2003).

The overall value for each cuttings management option can be calculated by using the following equation (Thanyamanta et al. 2003):

$$Overall\ Value, V(X) = \sum_{i=1}^{n} W_i X_i \tag{5.11}$$

where i is the criterion identification number, X_i represents the score under the i-th criterion, W_i is the corresponding weighting factor, and N is the total number of criteria.

With the weights assigned and the ranges of the scores, the overall values may range from −600 to 450. The minimum overall value of −600 is obtained by the worst option with the lowest scores (0 or −10) under all of the criteria, while the maximum overall value (450) is the overall value of the best option obtaining the highest scores (0 or 10) for all criteria.

The options related to the management of produced water stream were ranked and compared using overall values. According to the calculated overall values of each management option, the three best alternatives were the re-injection, downhole separation, and Ecological Risk Assessment with overall values of 295, 279, and 230, respectively. Therefore, based on the overall values alone, the three options were considered the optimum alternatives for the management of produced water.

In the cost category, the re-injection scored the highest and showed significant offshore applicability. Re-injection is treated as a cost-effective option to comply with environmental regulations relating to disposal. Table 5-19 shows a comparison of the total scores of each major category of the criteria. Downhole separation received the highest score in the technical performance category. In the regulations and environmental monitoring category, Ecological Risk assessment got the highest score.

The relationship between offshore applicability and cost was considered, to compare the cost-effectiveness of each option. The offshore applicability was defined as the applicability of a treatment option to be used offshore with respect to technical performance and regulations and environmental monitoring. The sum of the two category scores, under technical performance and regulations and environmental monitoring, were used to represent the applicability score. The values were then plotted against cost category scores of the evaluated options to outline the performance of each management option compared with their costs (Figure 5-23).

Re-injection and downhole separation provided high values of offshore applicability with the lowest costs (Figure 5-23). The environmental risks associated with produced water discharges will increase with increasing water production, unless improved water treatment (including removal of dissolved organics) or produced water re-injection (PWRI) is implemented. ERA, despite its slightly higher costs, showed significantly better offshore applicability. Therefore, ERA was the third most cost-effective option for this comparison.

Table 5-18   Proven technology and availability criterion

Proven Technology	Rating Value (Score)
Used waste/similar types of waste	3.5
Toxicological studies in lab	4.8
Field study	5.0
Currently in operation	7.0
Commercially available	8.5
Offshore application	9.0

Table 5-19   Category scores of each technology option

Technologies	Overall Values
Re-injection	295
Carbon adsorption	200
Downhole separation	279
Coconut shell filtration	168
Ecological risk assessment (ERA)	230
Water shut off	211
Biological treatment	224
Warm softening	185
Hydrocyclone	228
Gas floatation	210

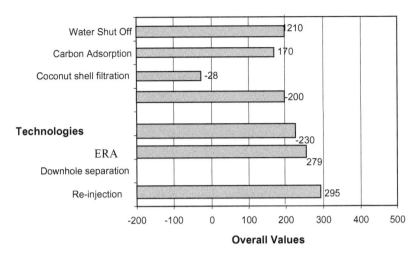

Figure 5-22 Overall values of the alternatives.

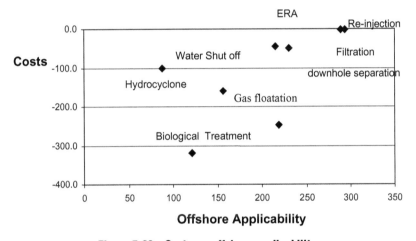

Figure 5-23 Costs *vs* offshore applicability.

5.6.4   *Key management issues*

There are several questions that need to be addressed in order to manage produced water effectively. The environmental manager is faced with this first question: Is it possible to determine the quantities of produced water that are going to be released into the marine environment during the lifetime of the drilling operation? Many times, the exact amount is not easy to predict, but an approximation of the order of magnitude of the release can help the manager set an effective, strategic environmental protection plan. The quantities involved will not only determine the long-term impacts on the marine environment but will also determine the technical options of the manager. The economic and technical feasibility of the different methods in dealing with produced water is highly influenced by the quantities of produced water involved.

A second question, for which the environmental manager wants an answer, is: Is it correct to reach conclusions about the toxicity of produced water from data dealing with the toxicity of its constituents? Many times, when conducting a risk assessment, the impacts of the constituents of produced water are examined. It is assumed that the impacts of the mixture can be generated by summing up the impacts of the individual constituents of produced water. The environmental manager needs to know how the mixture will behave since produced water is released as a mixture into the marine environment.

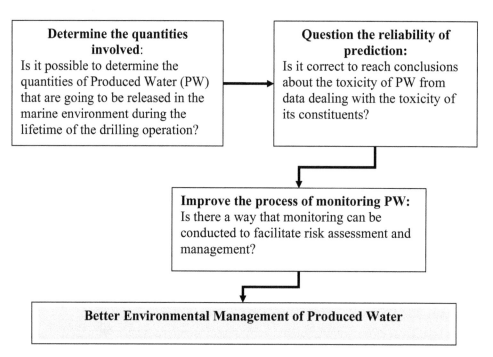

Figure 5-24   Key environmental management questions for produced water (Katsimpiri 2001).

Oil and gas extraction generates large volumes of produced water. This water naturally exists in the reservoir and being denser, lies under the hydrocarbons. One factor that determines the quantity of produced water is the quantity of formation water that naturally exists in the reservoir. Its quantity and composition is influenced by the geological formation and geological history of the reservoir. The different geological phases and processes that the reservoir has gone through determine the quantity of formation water that remains trapped. Generally, larger reservoirs have larger quantities of produced water. Furthermore, oil reservoirs contain large volumes of formation water, while gas reservoirs tend to produce small quantities (UKOOA 1999).

Data on toxicity is important in the risk assessment process and is used by the risk assessor to generate conclusions. However, this data and the conclusions derived do not always directly address the management needs. The manager needs to know how the organisms in the area of the drilling operation are affected. The third question that the manager needs to generate an answer to is: Is there a way that monitoring can be conducted to facilitate risk assessment and management?

These three questions are set from the management point of view and aim to use the available scientific knowledge to better manage produced water (Figure 5-24). The answers to these questions will ultimately describe different dimensions of produced water.

The quantities of produced water are also influenced by the characteristics of the rock and the stage of production. Most new commercial oil and gas reservoirs are initially free flowing because of the underground pressure. The rate of flow depends on a number of factors, the combination of which is unique for that reservoir, including permeability of the reservoir rock, underground pressure, viscosity of the hydrocarbon, and oil/gas ratio. However, these factors are not constant throughout the lifetime of the reservoir and when the flow is not adequate, additional pressure is required. This external source of pressure may be provided by injecting water into the reservoir rock. The quantity of water injected into the reservoir depends on the stage of production. The more depleted the reservoir, the more injected water is required to maintain its pressure. Therefore, throughout the lifetime of the reservoir the quantities of produced water vary, and the ratio of formation/injected water changes. Aging fields and the rising rate of hydrocarbon extraction, in general, generate a dramatic increase in produced water.

With quantities of this magnitude it becomes apparent that the discharge of produced water is an important operating and environmental issue. It is an operating issue when the manager has to determine the best way to handle produced water and is an environmental issue when the manager is trying to determine and possibly reduce the effects of produced water in the area of concern.

At the problem formulation stage of the risk assessment process, the amounts of produced water from a well should be established. The manager has to be aware of the expected quantities of produced water, the spatial distribution, and temporal distribution of formation water/sea water during the lifetime of exploration of the reservoir. Taking into account the parameters that influence the quantities of produced water, four important phases are identified as important in determining the quantities of produced water. The order has been set according to increasing detail and the number of studies required. These are:

- Determination of the type of reservoir (i.e., oil or gas);
- Determination of the type of rock;
- Determination of the size of the reservoir;
- Determination of the stage of production.

These phases give the manager an idea as to the quantities of produced water with which he will be dealing. This will enable the manager to make the best choice as far as alternative technologies are concerned. Handling large quantities of produced water from a small oil reservoir over a short period of time requires the use of a different technology compared to handling small quantities of produced water discharged over a long period of time and generated from a large gas reservoir. The quantities of produced water involved not only determine the technologically feasible but also the most economical option. The construction of installations for treating produced water is a significant expenditure that has to be justified.

Furthermore, the quantities forecasted along the temporal and spatial scales will determine the extent of the associated environmental effects. Generally, guidelines describe the permissible concentrations of certain constituents of produced water. When these concentrations are not exceeded the discharge of produced water is allowed in the marine environment from offshore installations. These guideline values rely heavily on the fact that produced water is rapidly diluted. However, for the manager who is familiar with the area of concern (i.e., open or enclosed water masses, existing or non-existing mechanisms) it is important to know the quantities that are expected to be diluted in that area. This information will help the manager to have a site-specific perspective on environmental concerns.

5.6.5 Toxicity and risk assessment of produced water

A six-step approach that considers the variability of the type and concentrations of constituents of produced water is suggested, to better assess and manage the toxicity of produced water (Figure 5-25).

The first step is to directly measure the toxicity of produced water. If the measurement demonstrates elevated toxicity, the second step should be conducted. Otherwise, no particular measurement is required by the risk manager. In the second step, the mixture should be fractionated and the individual components investigated. In step three, the components that are found in toxic concentrations should be removed and the toxicity of the whole mixture re-measured. If the toxicity of produced water is still elevated then the assessor proceeds to step four. In this step, the types of interactions among the constituents are investigated. The mixture is fractionated and steadily re-fractionated. The assessor also investigates the type of interactions that exist and the level of threshold concentrations that influence these interactions. Once these are identified, the assessor is in a position to determine the major components that contribute to the toxicity of produced water. The manager can then set priorities for clean up and source-monitoring. By this process the toxicity of produced water is examined in a manner that follows a specific order. This process is more or less dictated by the unique characteristics of produced water.

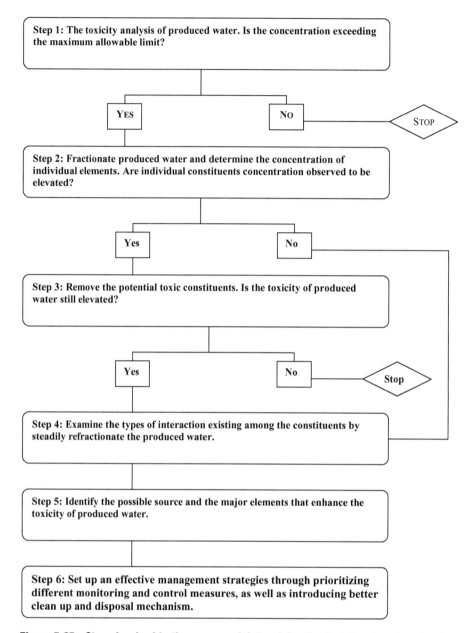

Figure 5-25   Steps involved in the process of determining the toxicity of produced water.

One of the most important components of an environmental risk assessment is the data that it is based on. The process of data collection constitutes a monitoring program. The way monitoring is designed and conducted can affect the results of an ecological risk assessment.

The most effective monitoring to determine the effects of produced water on aquatic organisms during offshore drilling would be an organism- and site-specific monitoring program that takes into account the site characteristics of the area and the organisms that live there. Even though a monitoring program might seem a straightforward process, its actual design is complex. The considerations that the risk assessor and the risk manager have to incorporate when designing a site- and organism-specific monitoring program are summarized as:

- spatio-temporal design;
- total area sampled;
- boundaries of the area of environmental concern;
- treatment of habitat heterogeneity within sites;
- selection of appropriate organisms based on scientific criteria;
- philosophical support for choosing certain organisms (i.e., importance for the area);
- scope of the assessment for the communities and habitats,

For the monitoring program to be useful in environmental assessment, it has to be carefully designed by the risk assessor and the risk manager. The points of sampling and the spatio-temporal frame have to be established first to make the monitoring effective. The total area sampled has to be decided and its boundaries determined. The risk manager is constricted by political boundaries that seldom match the ecological boundaries. Both political and ecological boundaries have to be considered in the process of designing a site-specific monitoring program. The sampling and the monitoring in general has to be designed in such a way that will incorporate the habitat heterogeneity within sites. The above reasons make the selection of the sampling site and the selection of appropriate organisms a thoughtful process, rather than a coincidental one based on the data that already exists.

The selection of appropriate organisms should be supported by theory. The basic consideration of a risk manager's choice is the importance of certain organisms, while that of the risk assessor is ensuring that organisms are appropriate ecosystem indicators and cover different trophic levels in the marine environment. The matching of these choices can result in an effective organism-specific monitoring mechanism. It is important to try to incorporate the scope of the assessment for the human and nonhuman communities of the area when designing a site- and organism-specific monitoring program.

5.7   Alternative Waste Management

5.7.1   *Waste minimization*

In the previous section it was shown that deep well drilling generates huge amount of wastes, consisting of mud and cuttings. The generation of cuttings is positively related to the size of drilling hole.

Figure 5-26 shows the oil and gas drilling mechanism and multi-layer piping. The figure shows the larger diameter on the top, gradually becoming smaller toward the bottom. The hole diameter can be 20 inches or larger for the uppermost sections of the well, followed by different combinations of progressively smaller diameters. Some of the common hole diameters are 17.5, 14.75, 12.25, 8.5, 7.875, and 5.5 inches (DWMIS 2005). Generally, above the aquifer (drinking water zone) the large hole is cemented in between drill well and casing. Then, a smaller diameter hole is drilled and repetition of cementing is continued (Figure 5-26). Different waste minimizing technologies are discussed in the following section.

5.7.1.1   Directional drilling

Instead of conventional drilling, directional drilling opened up many new possibilities for improving production and minimizing wastes by reaching target reservoirs. There are three types of directional drilling, extended-reach drilling, horizontal drilling, and multiple laterals off a single main well bore. Figure 5.27 shows different types of advanced drilling technology.

Extended-reach drilling: In order to minimize the drilling of multiple wells, extended-reach drilling is an alternative. For example, using a single platform or drilling pad, multiple extended-reach wells can be drilled instead of many conventional drilling wells. Using extended-reach drilling allows many wells to be completed from a single location and avoids the environmental impacts of multiple surface structures (DOE 2005).

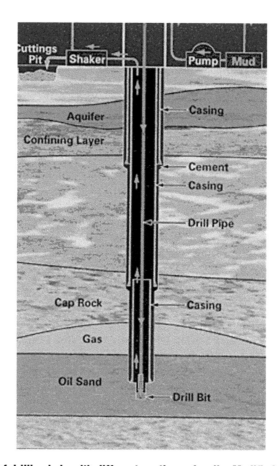

Figure 5-26   Size of drilling hole with different sections of wells. Modified from DWMIS (2005).

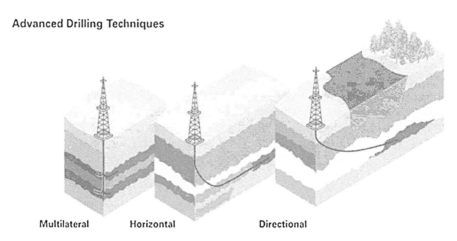

Figure 5-27   Pictorial view of advanced drilling technologies. Source: DOE (2005).

Horizontal drilling: When hydrocarbon reserves are not thick but extend over a large lateral area, then horizontal drilling is appropriate. In conventional drilling, resources are considered economically non-feasible, because many vertical drilling are required to develop this site. However, a single horizontal drilling can serve this purpose and a horizontal well generates less waste than several vertical wells (DOE 2005).

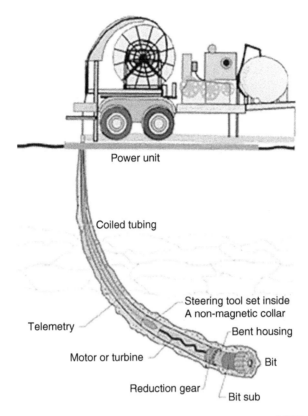

Power unit

Coiled tubing

Steering tool set inside
A non-magnetic collar

Telemetry

Bent housing

Motor or turbine

Bit

Reduction gear

Bit sub

Figure 5-28   Coiled tubing drilling system. Photo courtesy of DOE (2005).

Multiple laterals: Multiple laterals technology helps to reduce drilling wastes by decreasing the number of drilling wells. Fewer wells mean less wastes. In convention vertical drilling, multiple wells are required in the case of multiple, small, oil-bearing zones or zones at several different depths. Using multiple laterals techniques, the number of wells required can be reduced drastically. The total volume of drilling waste is lower than would be generated if several full wells were drilled (EPA 1999).

5.7.1.2 Drilling smaller-diameter holes

To minimize the waste generation, several technologies can be applied to drill with small holes. Some of them are closer spacing of successive casing strings, slimhole drilling, and coiled tubing drilling.

Closer spacing of successive casing strings: Earlier, only one standard size of drill bit was used, being comparatively larger. Nowadays, there are many different sizes available. In addition, various casing sizes also avail to fit with different drill bits. This technology significantly reduces the generation of drill cuttings and uses of muds.

Slimhole Drilling: According to DOE (1999), slimhole wells are defined as wells in which at least 90% of the hole has been drilled with a bit six inches or less in diameter. Although slimhole technology has been available since the 1950s, it was not commonly used because the small-diameter well bore restricted stimulation, production, and other downhole manipulations. Development of modern technology has overcome these disadvantages. In addition to generating less drilling waste, slimhole rigs have a smaller surface footprint on a drilling pad (DOE 1999).

Coiled tubing drilling: Figure 5-28 presents a coiled tubing drilling system. In this drilling system a continuous length of tubing is fed off a reel and sent down the hole (DOE 1999), as opposed to the conventional drilling, where individual sections of drill pipe are assembled together. The coiled tubing drill is comparatively smaller than the traditional drill pipe. As a result, a smaller volume of cuttings is generated. In addition to reducing waste volumes, the surface footprint is smaller, the noise level is lower, and air emissions are reduced (DOE 2005).

5.7.1.3   Drilling with less fluid

The discussion on waste estimation revealed that drilling mud is one of the major wastes created from drilling operations (Tables 5-7 and 5-8). At the end of the drilling job, the disposing of this toxic mud is of major concern. Using advanced technology, drilling can proceed with minimal or no drilling mud. For example, in pneumatic drilling methods, wells can be drilled using air or other gases as the fluid that circulates through the drilling system (DOE 2005). According to DOE (1999), there are four different types of pneumatic drilling, air dust drilling, air mist drilling, foam drilling, and aerated mud drilling. All of these techniques rely on gas or blends of gas and mud to lift cuttings to the surface.

5.7.1.4   Other waste minimization techniques

Drilling and production involve many activities and many people. Properly coordinated management techniques can also significantly reduce wastes. There are many relatively simple processes that can be used on drilling rigs to reduce the amount of mud that is discarded or spilled (DOE 2005). Use of pipe wipers, mud buckets, and vacuuming of spills on the rig floor can avoid and reduce significant amount of wastes. In addition to controlling the muds, these devices allow clean mud to be returned to the mud system and thus save additional cost. Uses of simple tools, such as drums and sacks in the manufacturing, storing, and transporting of muds to a drilling location can also reduce waste generation. These simple techniques can benefit waste minimization efforts.

5.7.2   Environmentally friendly muds and additives

As mentioned before, different types of drilling mud are used, for example, synthetic-based muds (SBMs), water-based muds (WBMs), and oil-based muds (OBMs). Selection of suitable drilling mud can reduce environmental hazards significantly, and also affect the overall volume of used muds and cuttings that are generated (DOE 2005). Figure 5-29 shows the mud systems in a petroleum drilling operation.

5.7.2.1   Alternative muds

SBMs use nonaqueous fluids instead of oils as their base. These base fluids are olefins, esters, linear alpha-olefins, poly alpha-olefins, and linear paraffin. These components are environmentally more acceptable (biodegradable, low toxicity) than previously used oil compounds. In addition, SBMs drill a cleaner hole than WBMs, with less sloughing, and generate a lower volume of drill cuttings. SBMs are recycled to the greatest extent possible, while used WBMs are generally discharged into the sea at offshore locations. Overall, OBMs are the most toxic to the environment (Khan and Islam 2007).

5.7.2.2   New drilling fluid systems

Even though SBMs are more effective technically, WBMs are comparatively more environmental friendly than SBMs and OBMs. The main drawback of WBMs is that they require five times the volume of the drilled hole. However, SBMs and OBMs do not require as much. According to DOE (2005), a new generation of WBMs has been developed and used effectively in the past few years. They have demonstrated improved drilling performance and significantly reduced dilution rates.

Other alternative fluids can be developed with suitable drilling properties that contain fewer components or additives that would inhibit subsequent breakdown by earthworms or microbes. As a result, environmentally

Figure 5-29   Mud additives system in the drilling operation. Photo courtesy of DOE (2005).

sustainable drilling fluids can be used as soil supplements. For example, conventionally cow dung has been using as a drilling mud for drinking water wells in Asia.

In 2005, DOE suggested that drilling fluids based on format brines are more environmentally friendly than traditional fluids. Common examples of these brines are cesium format ($HCOO^-Cs^+$), potassium format ($HCOO^-K^+$), and sodium format ($HCOO^-Na^+$). In the conventional drilling, barite is the most commonly used weighting agent where many heavy metals are added. Hematite (Fe_2O_3), ilmenite ($FeTiO_3$), and calcium carbonate ($CaCO_3$) can be used instead of barite.

5.7.3   *Beneficial reuse of drilling wastes*

5.7.3.1   Recycling of muds

Table 5-20 shows the amount of WBMs used in drilling, most of them discharged after finishing the drilling job. In contrast, many OBMs and SBMs are recycled whenever possible. By adding required compounds the muds can be used as new muds. Using simple processes that can be used on drilling rigs, clean mud can be captured and returned for reuse. Examples include pipe wipers, mud buckets, and vacuuming of spills on the rig floor. Recovery of mud during tank cleaning may also allow the mud to be reused. Solids control equipment, such as centrifuges, can be used to remove solids from the recirculating mud stream (DOE 2005).

5.7.3.2   Reuse of cuttings

Table 5-20 shows the enormous amount of cuttings generated from drilling operations. The present guidelines do not allow the release of heavily oil coated cuttings, but it is necessary to dispose of these wastes.

Table 5-20   Amount of major waste produced and discharged in the sea

Waste Type	Amount of Waste release (tones)	Discharge
Drilling muds	15 to 30 (periodically/single well) 150 to 400 (bulk at end/single well) 45,000 (production site/multi-wells)	Mostly water based muds
Drill cuttings	200 to 1,000 (Production, multi-wells)	Water-based muds associated with cuttings
Produced waters	1,500/day (Production/multi-wells), increase with age.	If oil concentration is higher then treated before discharge
Accidental discharges	In over 22,000 wells of all kinds drilled in US coastal waters from 1971 to 1993, only five blowouts occurred and the total discharge was 170 m^3 – is the average/year available?	Oil spills, gas blowouts, chemical spills
Other wastes	Data not available	Storage displacement water, Bilge and ballast water, Deck drainage, Well treatment fluids, Naturally occurring radioactive materials, Cooling water, desalination brine, Water for testing fire control, Other assorted wastes
Platforms/rigs	Data not available	More than 4000 tonnes does not carry to shore for disposal.

Source: Patin (1999); CEF (1998)

Reuse is one of the better options. Before reusing the cuttings, it should be ensured that the hydrocarbon content, moisture content, salinity, and clay content of the cuttings are suitable for the intended use of the material. Some important reuse options are given discussed.

Road surfacing: Oily cuttings serve the same function as the traditional tar-and-chip road surfacing. Due to their oily contents, the cuttings can be used as anti-corrosion agents.

Construction material: Treated cuttings can be used for different construction materials, such as fill material, daily cover material at land fills, and aggregate or filler in concrete, brick, or block manufacturing. Other possible construction applications include use in road pavements, bitumen, and asphalt or use in cement manufacture. Cuttings can also be used for plugging of abandoned wells (DOE 2005).

Use of cuttings for fuel: It is reported that oily cuttings can be used as a fuel in a power plant (DOE 2005). The cuttings that are blended in at a low rate with coal, may serve as primary fuel source. This use decreases the cost of oil remediation.

5.7.4   Waste disposal

5.7.4.1 Wastes burial

Burial is one of the popular disposal methods, pits and landfills being the two major waste disposal sites. This is also the most common onshore disposal technique used for disposing of mud and cuttings. The burning of drilling wastes is advantageous because it offers simple, low-cost technology for uncontaminated solid wastes. It requires limited surface area requirements. There are also some disadvantages, such as the potential for groundwater contamination if burial is not done correctly or if contaminated wastes are buried, and the resulting liability costs.

Pits: Pit burial is a low-cost, low-tech method, not requiring wastes to be transported away from the well site. The use of earthen or lined pits is integral to drilling waste management. Generally, the solids are buried in the same pit (the reserve pit) used for collection and temporary storage of the waste mud and cuttings after the liquid is allowed to evaporate (Figure 5.30).

Figure 5-30   Oil field waste pits. Source: US Fish and Wildlife Service (DOE 2005).

Figure 5-31   Commercial oil field waste landfill. Photo courtesy of DOE (2005).

Different factors should be considered when it comes to pit burial. It is not suitable if wastes contain high concentrations of oil, salt, biologically available metals, industrial chemicals, and other materials with harmful components that could migrate from the pit and contaminate water resources. Areas with shallow groundwater are not appropriate; a pit location of at least five feet above any groundwater is recommended to prevent migration to the groundwater. The top of the burial cell should be below the rooting zone of any plants that are likely to grow in that area in the future (normally about three feet). Low-permeability soils such as clay are preferable to high-permeability soils such as sand.

Landfills: Landfills are also used for disposal of drilling wastes and other oil field wastes. At the landfill, a bottom liner overlaid by a geological barrier is developed to prevent contamination of the soil. A top liner, which is drawn over the wastes during non-active periods, will be installed permanently after the landfill is closed. Two collection pits collect rainwater and subsequent leachate (Morillon et al. 2002) (Figure 5-31).

In some oil field areas, large landfills are operated to dispose of oil field wastes from multiple wells. Burial usually results in anaerobic conditions, which limits any further degradation when compared with wastes that are land-farmed or land-spread, where aerobic conditions predominate.

Runoff and leaching should be prevented from wastes in landfill sites, with appropriate types and degree of controls in place. Natural barriers or manufactured liners placed between the waste material and the ground-water help control leaching.

5.7.4.2 Land application

Land application of drilling wastes is a form of bioremediation allowing the in-sita microbial population of the soil to metabolize, transform, and assimilate waste constituents in place. While using the land application techniques, the amount of waste application should be within the assimilate capacity of soil and should not destroy soil integrity, create subsurface soil contamination problems, or cause other adverse environmental impacts (DOE 2005). There are two types of land applications that are popularly used, known as land farming and land treatment, both similar in principle.

Land farming is a biological process where oily petroleum wastes are treated. The complete land farming process may take several years to remediate. It involves the controlled and repeated application of wastes to the soil surface, using microorganisms in the soil to naturally biodegrade hydrocarbon constituents, dilute and attenuate metals, and transform and assimilate waste constituents (DOE 2005).

Land farming is a relatively low-cost drilling waste management approach. Some studies indicate that it does not adversely affect soils and may even benefit certain sandy soils by increasing their water-retaining capacity and reducing fertilizer losses (EPA 1999). Inorganic compounds and metals are diluted in the soil, and may also be incorporated into the matrix (through chelation, exchange reactions, covalent bonding, or other processes) or may become less soluble through oxidation, precipitation, and pH effects. The attenuation of heavy metals (or the taking up of metals by plants) can depend on clay content and cation-exchange capacity.

The addition of water, nutrients, and other amendments (e.g., manure, straw) can increase biological activity and aeration of the soil, thereby preventing the development of conditions that might promote leaching and mobilization of inorganic contaminants. During periods of extended dry conditions, moisture control may also be needed to minimize dust.

Periodic tillage of the mixture (to increase aeration) and nutrient additions to the waste-soil mixture can enhance aerobic biodegradation of hydrocarbons. After applying the wastes, hydrocarbon concentrations are monitored to measure progress and determine the need for enhancing the biodegradation processes. Application rates should be controlled to minimize the potential for runoff.

Pretreating the wastes by composting and activating aerobic biodegradation by regular turning (windrows), or by forced ventilation (biopiles) can reduce the amount of acreage required for land farming (Morillon et al. 2002).

The land treatment processes are similar to those in land farming, also known as land spreading. In land treatment, a one-time application of waste is made to a parcel of land. The objective is to dispose of the waste in a manner that preserves the subsoil's chemical, biological, and physical properties, by limiting the accumulation of contaminants and protecting the quality of surface and groundwater (DOE 2005).

5.7.5   *Thermal treatment technologies*

Using high temperatures, thermal technologies reclaim or destroy hydrocarbon-contaminated materials. This technology is considered as the most efficient treatment for destroying organics (in petroleum operations oil), and also reduces the volume and mobility of inorganics such as metals and salts (Bansal and Sugiarto 1999). However, metals and salts cannot be removed by this technology and need to be treated further based on the condition of the wastes.

Potential application of thermal technology in the drilling and production wastes is high, because these waste streams are high in hydrocarbons, with typical values ranging from 10 to 40% (DOE 2005). For example, oil-based muds are good candidates for thermal treatment technology. DOE (2005) also suggested that the thermal treatment can be an interim process to reduce toxicity and volume and prepare a waste stream for further treatment or disposal (e.g., landfill, land farming, land spreading), or it can be a final treatment process resulting in inert solids, water, and recovered base fluids. Thermal treatment is considered to be cost effective, ranging from $75 to $150/ton, with labor being a large component (Bansal and Sugiarto 1999).

Thermal treatment technologies can be grouped into two categories, which are incineration and thermal desorption. Incineration is not commonly used for drilling wastes but has greater applicability for materials such as medical waste. Incineration technologies oxidize (combust) wastes at high temperatures (typically 1200–1500°C) and convert them into less bulky materials that are nonhazardous or less hazardous than they were prior to incineration (Morillon et al. 2002). Incineration is typically used to destroy organic wastes that are highly toxic, highly flammable, resistant to biological breakdown, or that pose high levels of risk to human health and the environment.

Incineration is commonly done by two different methods. One is rotary kilns and the other cement kilns. Most incineration of drilling wastes occurs in rotary kilns, a mature and commercially available technology, which is durable and able to incinerate almost any waste, regardless of size or composition (DOE 1999). A rotary kiln tumbles the waste to enhance contact with hot burner gases. Figure 5-32 shows the rotary kilns thermal treatment method of drilling waste.

A cement kiln can be an attractive, less expensive alternative to a rotary kiln. In cement kilns, drilling wastes with oily components can be used in a fuel-blending program to substitute for fuel that would otherwise be needed to fire the kiln. The cement kiln temperature range is 1400–1500°C. Cement kilns may also have pollution control devices to minimize emissions. The ash resulting from waste combustion becomes incorporated into the cement matrix, providing aluminum, silica, clay, and other minerals typically added into the cement raw material feed stream.

The second group of thermal treatment is known as thermal desorption, in which heat is applied directly or indirectly to the wastes, to vaporize volatile and semivolatile components without incinerating the soil (Figure 5-33). Because thermal desorption depends on volatilization, treatment efficiency is related to the volatility of the contaminant. Thus, thermal desorption easily removes light hydrocarbons, aromatics, and other volatile organics, but heavier compounds such as polycyclic aromatic hydrocarbons are less easily

Figure 5-32   Rotary kiln. Source: DOE (1999).

Figure 5-33   Thermal desorption. Photo courtesy of USEPA (1999).

removed. In some thermal desorption technologies, the off-gases are combusted, and in others, such as in thermal phase separation, the gases are condensed and separated to recover heavier hydrocarbons. Thermal desorption technologies include indirect rotary kilns, hot oil processors, thermal phase separation, thermal distillation, thermal plasma volatilization, and modular thermal processors.

5.7.5.1   Indirect rotary kilns

These kilns use hot exhaust gases from fuel combustion to heat the drilling wastes. The technology consists of a rotating drum placed inside a jacket. Heat is supplied through the wall of the drum from the hot exhaust gas that flows between the jacket and the drum. The drilling wastes are agitated and transported through the processor inside the rotating drum. Treated solids are recirculated to prevent the formation of an isolating layer of dried clay in the inside of the drum. Because the overall heat transfer from the exhaust to the material is low, relatively large heating surfaces are required, and the process units are correspondingly large. The units typically heat the wastes to about 500°C, which provides for the efficient removal of oil from the wastes, but which can also lead to thermal degradation and decomposition of residuals in the recovered solids. The process typically retains the wastes for about 30 to 150 minutes.

5.7.5.2   Hot oil processors

In these processors, heat is transported to the drilling wastes by circulating hot oil inside hollow rotors. The rotors also agitate and create the required axial transport in the bed. Conventional fuels provide the primary heat source for the hot oil. Some units augment the heat from the hot oils with electric heating on part of the heat surface to reach the temperature needed for complete removal of the oil in the waste. Retention times for complete removal of oils are about 30 to 150 minutes (DOE 2005).

5.7.5.3   Thermal phase separation

TPS consists of five subsystems. In the first, the drilling wastes are screened to remove foreign matter prior to delivery to the desorption chamber. Next, the shell of the chamber is heated externally with a series of burners fueled by propane, natural gas, diesel, or recovered drilling fluid. The drilling wastes are heated indirectly to raise the temperature of the drilling waste to the boiling point of the hydrocarbons (usually about 220°C, but sometimes up to 500°C), where they are volatilized and separated from the host matrix under a vacuum. Screw augers, which slowly draw the wastes through the inner heating shell, ensure suitable agitation and thorough heating of the solids matrix. The water vapor and gaseous hydrocarbons extracted in the desorption chamber are rapidly cooled by direct contact with water sprays fed with recirculated process water. The condensed liquids and recirculated quench water are then sent to an oil-water separator, where the recovered fluid is collected, analyzed, and recycled.

Advantages of TPS over rotary kilns or directly fired desorption systems are more sophisticated air emissions control, the ability to treat materials with up to 60% undiluted oil (because there is no potential for combustion), and the opportunity of visual inspection during operations. Treated solids are contained and tested prior to use as an onsite fill material. TPS processing removes 99% of hydrocarbons from the feedstock (Zupan and Kapila 2000).

5.7.5.4 Thermal distillation

Because constituents of liquid mixtures evaporate at different temperatures, thermal distillation allows for the separation of solids, liquids, and the different constituents of liquids. In high-temperature thermomechanical conversion and cracking, drill cuttings are distilled and cracked to boil off water and oil.

5.7.5.5 Thermal plasma volatilization

Thermal plasma results when a common gas is heated to extremely high temperatures (up to 15,000°C). It is also used to treat oil-contaminated soils that include substances such as chlorides, which are unsuitable for a combustion process because of their potential to generate dioxins and furan compounds as byproducts. The process uses a plasma reactor, containing a plasma torch operating in an inert atmosphere. The waste material is fed into the reactor. In the reactor, the torch, with a jet temperature of about 15,000°C, is used to heat the waste to up to 900°C without combustion, causing any hydrocarbons to volatilize.

5.7.5.6 Modular thermal processors

The process is designed to flash-evaporate the fluid phase from drilling wastes. It uses a combination of electrical and mechanical energy (through a hammer mill) to evaporate the fluid phases. The process is optimized at temperatures in the range of 250–260°C, but this range can vary depending on the type of wastes and the boiling point of associated hydrocarbons (DOE 1999). The evaporated fluids are retained and then condensed, allowing selective recovery of the individual fluids (typically hydrocarbons and water). The remaining solids are discharged as an inert powder, and the recovered fluids can be reused or recycled. The process generates no atmospheric emissions.

5.8 Sustainability of Waste Management

Sustainable drilling and production operations requires a sustainable supply of clean and affordable energy resources that do not cause negative environmental, economic, and social consequences. The concept of sustainability was brought into the core of social, economic, and environmental debates after the well-known definition of sustainable development devised by the World Commission on Environment and Development (WCED 1987). However, sustainable development has remained an obscure term due to the absence of explicit guidelines for sustainability. Recently, Khan and Islam (2005b) and Khan et al. (2005c) developed an innovative criterion for achieving true sustainability in technological development. This reliable concept can be applied equally effectively to offshore technology development. This kind of new technology should have the potential to be efficient and functional far into the future, in order to ensure true sustainability. Sustainable development is seen as having four elements, economic, social, environmental, and technological. Delivery is the overarching imperative that drives both implementation and further strategy development (Khan and Islam, 2005b).

5.8.1 Application of the green supply chain

To achieve sustainability in the petroleum wastes management, the "Olympic" green supply chain is applied. The major components of the proposed supply chain are based on the conventional Olympic logo. It consists of five zeros (zero-wastes), which are:

(i)   zero emissions

(ii) zero resource wastes

(iii) zero wastes in activities

(iv) zero use of toxics, and

(v)   zero waste in product life cycle.

The basic concept of the "Olympic" green supply chain model is shown in the Figure 5-34. The inputs are in the form of energy, humans, and materials, which are green. The output of the mode is petroleum products and by-products, which are also green. The rectangular box illustrated as a processing unit shows that toxic compounds cannot be used in the chain. Production and administrative activities can be wasted with regards to inefficiency in processing. The three arrows on top of the rectangular box indicate that the chain emits no gaseous, solid, and liquid emissions or discharges.

This model considers complete product cycle. For example, this mode considers beyond the product's end of the project life. Overall, the system shows green products that generate no product life cycle, in the form of transportation, use, and end-of-use of the product. Thus, this model will help achieve sustainability in waste management and over all petroleum operations. This is because such an approach is always the norm in Nature, in which no system releases toxic wastes. Moreover, output of one system is the input for another, and as a result there is no waste generated. Finally, we can tell that Nature is zero-waste and the proposed "Olympic" green model is based on this same principle. Detailed description of "Olympic" green supply chain is discussed in Chapter 2. Following sections discuss how to achieve the five zero waste objectives.

5.8.2   *Zero emission of air, water, solid wastes, soil, toxic wastes, and hazardous wastes*

It is evident from earlier discussions (Section 5.3) that drilling and production operations generate huge amounts of solid, liquid, and gaseous wastes. Figures 5-6 to 5-9 and Tables 5-6 and 5-7 clearly illustrate how every activity emits wastes. In the proposed "Olympic" green supply chain model (Figure 5-34), it has been demonstrated that no wastes are allowed to be released into air, water, and soil. Major emissions and

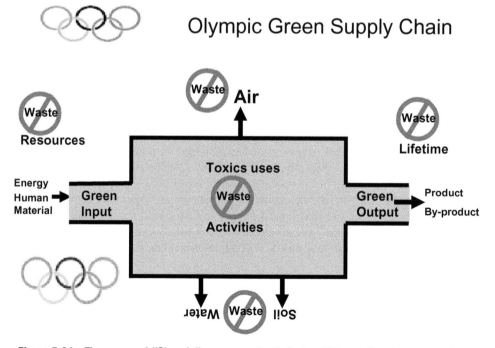

Figure 5-34   The proposed "Olympic" green supply chain for offshore oil and gas operations.

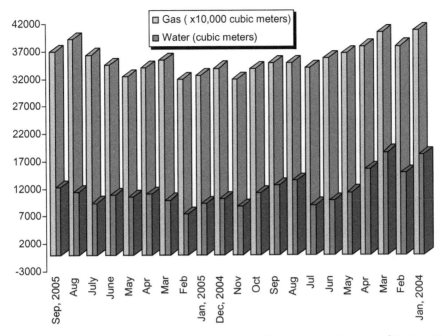

Figure 5-35   **Average monthly production of raw gas (10,000 m³) and produced water (m³) in the Sable Energy Project, from January 2004 to September 2005.**

wastes discharged from drilling and production are detailed in Section 5-3. The main wastes of petroleum operations that have an impact on the environment are drilling-waste fluids or muds, drilling-waste solids, produced water, and volatile organic compounds. Moreover, produced water constitutes the highest amount of wastes released in the whole of petroleum operations (Figure 5-2) (Khan and Islam 2005a, 2007). Figure 5-35 shows the monthly average production of produced water along with gas production.

The objective of the "Olympic" green supply chain model has not been achieved in current petroleum wastes management. These waste production and discharge practices are considered to have negative impacts on the environment (Holdway 2002; Khan and Islam 2003a, 2004, 2005a, 2005b, 2005c; Colella et al. 2005; Currie and Isaacs 2005).

According to the "Olympic" model (Figure 5-34), to achieve sustainability, these waste discharges and emissions should be stopped. Current wastes management practices have been analyzed and evaluated to examine the technological efficiencies of stopping emissions and wastes management. At the end of Sections 5-5 and 5-7, it is revealed that no technology is currently available that can completely eliminate emissions and achieve the target of the "Olympic" green supply chain. Khan and Islam (2007) also found similar results while evaluating offshore seismic technologies. It is acknowledged that there are some technologies capable of reducing the wastes to some extent. For example, re-injection of slurries, thermal treatment, and land application are considered effective technologies. However, while these technologies solve problems at some levels, the risk remains of creating problems at other levels. Therefore, the need to develop sustainable technology is extremely important (Khan et al. 2005b).

5.8.3   *Zero waste of resources (energy, materials, and human)*

The second objective of the "Olympic" green supply chain is achieving zero waste of resources (Figure 5-34). It requires zero waste generation from input materials, involvement of humans, and use of energy. Current drilling and production management are identified as high energy consuming methods (Section 5.7). However, there is still room to improve energy use, for example, by reducing the drilling and production

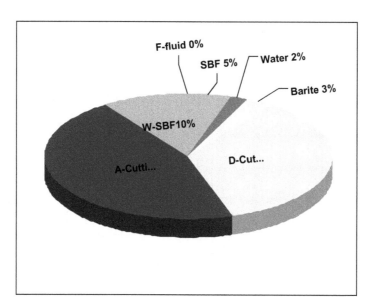

Figure 5-36   Total drilling wastes from a deep water development well.

Table 5-21   Emissions associated with flaring in one well (CEF 2005)

Pollutant	Emission Factor (kg/1000 m^3)	Total Emissions (tonnes)
NO$_x$	1.1	0.616
SO$_2$	0.27 *(H$_2$S in pppmv/100	0.00756
PM 2.5	0.61	0.342
Benzene	0.0025	0.0014
Total PAHs	0.000048	0.000027
CO$_2$ (CO$_2$E)	1913	1100
CH$_4$ (CO$_2$E^2)	0.04*23	0.515
N$_2$O (CO$_2$E^2)	0.04*296	663

periods. Drilling and production periods can take a long time, often ranging up to 20 years. By using efficient technology, this period should be reduced.

In the drilling phase huge amounts of drilling fluids are used (Table 5.6). By using proper technology, drilling fluids can be recycled, reused, and regenerated. As a result, an enormous amount of resources can be saved in the current operations practices. It is reported that large amounts of fluids are washed out with drill cuttings. A single development well in deepwater loses 217,491 kg of SBF and 71,772 kg of barite (Table 5-6) (Section 5.3). In conventional waste management practices to reduce drill cutting wastes, more shale screen and centrifuger may be used. In addition, some alternative technologies have also been suggested where the use of drilling fluid can be avoided altogether.

Flaring is a management technique commonly used in the production phases. No-flare design can be achieved by the separation of solids, liquids, and gases. To achieve this, new innovative techniques needed to be developed, which will be both environmentally appealing and economically attractive. Bjorndalen et al. (2003) recently proposed new techniques for solid–liquid separation that include the EVTN system as well as surfactants from waste materials. They also suggested liquid–liquid separation, in which human hair waste can be used as a demulsifier. Their study concluded that successful no-flare design is possible, which will save a lot of resources from burning.

Loss of life and injuries are often reported, often from drilling and production operations. These are frequently associated either with lifting incidents on decks or with drilling activity. According to Powell (2004), although no fatalities were reported between April 2002 and March 2003, three deaths have occurred since. Recently, Shell was fined £900,000 following the death of two workers on the Brent Bravo platform in the North Sea in September 2003 (BBC 2005).

5.8.4 Zero wastes in activities (administration, production)

Drilling and production activities are extensive works that are carried out in the field. For example, when drilling, a field engineer has to take instant decisions based on the drilling situation. Incidentally, it has been reported that drilling activities had to be stopped in an Ecuador oil field to await a simple decision from the company head office in China (Lojan 2005). This long process of decision making leads to significant loss in revenue. This is an example of wasteful administrative activities that involve high costs. To achieve the objectives of "Olympic" green supply chain, the company management should change to avoid wasting any activities.

In production and drilling phases numerous accidents also take place, such as blowout and oil spills. By taking precautions in advance and involving local communities in hazards and risks management, many resources can be saved. Khan et al. (2006c) suggested community-based energy system management as an alternative solution to the existing bottom-up approach in petroleum operations. According to them, the community-based approach is effective in securing administrative, production, and other environmental benefits.

This partnership can be a community-based model using a multi-stakeholder approach and consensus decision making. It will suggest appropriate management measures to combat the waste problem. The main goal of the stakeholder is to protect and promote environmental quality in their ecosystem. In the meantime, economic development is not ignored.

According Khan and Islam (2003a) most of the offshore oil and gas operators are large multinational corporations. It is important to develop partnerships between industry operators and the community to combat pollution issues in the marine environment. This kind of partnership offers opportunities in joint investment in the development and financing of sustainable waste management programs, facilities, and services as well as in the evaluation of management options. Bridging the gap between operators and other marine stakeholders can further improve the economic and environmental benefits. In the hydrocarbon operating regions, stakeholders or community members may take a "middle ground" approach, balancing between environmental, social, and economic considerations.

It is important to realize that oil producers can play a beneficial role in managing their environment practices, by exceeding governmental guidelines. For example, according to environmental guidelines, if oil content in the drill cuttings of 10 mg/L is allowable, then the company should maintain it at less than 5% or better. In this general point of view, it seems as if companies are spending more money to maintain higher environmental standards. However, in the actual calculations, their production costs will decrease through avoiding potential environmental compensation (Veil 1997; Ofiara 2002), saving management costs, and avoiding environmental effects monitoring (EEM) programs, etc. It will also significantly decrease the processing time by the regulatory agencies. The truth is that if a company maintains better environmental standards than the mandatory guidelines requires, this will minimize the costs and regulatory processing time. It also can gain an intangible value, by improving its reputation in the eye of the general public through good will.

5.8.5 Zero use of toxins

The "Olympic" green supply chain model demands avoiding toxic compounds in drilling and production operation activities. Many toxic compounds are used in drilling fluids, well testing chemicals, and corrosion inhibitors in these operations. Our present society increasingly depends on chemical approaches in every arena of life, using more toxic chemicals to neutralize less toxic chemicals (Veil 1997). There are better ways of solving problems through other means. For example, vitamin C as tablet and vitamin C from natural sources acturally such differently. The same rule applies to chemical antibiotics and natural antiboitics (e.g. olive oil). Only recently has the fact that natural solutions are the only solutions come to light (Chbetri et al. 2007). Similar problems exist in the oil sector as well. To treat oil spills, many other different toxic chemicals as well as hot steam are being used. It is reported that the use of other chemicals and hot steam are more harmful than the original oil spill would have been (Wells et al. 1995). Also, to prevent oil pipeline corrosion, billions of dollars are spent on treatment by chemicals, which are more harmful to the environment. This problem might be solved through using natural, organic compounds (Veil 1997).

Tables 5-7 and 5-8 show the types and amounts of organic pollutants used in the drilling process. Instead of using toxic compounds, natural non-toxic compounds can be used. Al-Darbi et al. (2002) found that natural vegetable components are highly efficient in controlling pipeline corrosion. They suggested replacing conventional toxic corrosion inhibitors with these natural components. Figure 5-4 shows the composition of drilling fluids which have many toxic compounds, including carcinogens. Recently, many of these compounds were replaced by vegetable-based compounds (EPA 2000).

5.8.6   Zero waste in product life cycle

It is revealed from the alternative waste management discussion (Section 5.7) that uncontrolled and unjustified wastes management approaches can bring environmental hazards that will persist over long periods of time. For example, untreated wastes in landfills can leach out toxic compounds for the whole lifetime of products. The "Olympic" green supply chain demands a stop to wastes in the life cycle, including transportation, usages, and end-of-life of a product (Figure 5-34). Appropriate and better technologies are needed to avoid the wastes, not only during the operation time but to stop life-cycle emissions also.

By implementing the "Olympic" green supply chain for drilling and productions operations, the major goals of sustainability are supported through improved technology, economic well-being, environmental protection, and social benefits. From the above analysis, it is revealed that current technologies and wastes management techniques are not suitable to completely achieve the zero wastes objectives, and that there is a need to develop new technologies to meet this goal. These new technologies must ensure sustainability following the recently developed criteria of sustainability in drilling and productions operations and wastes management.

Chapter 6

Reservoir Engineering and Secondary Recovery

6.1   Introduction

Reservoir engineering explores the engineering operations for developing and production of oil and gas reservoirs. It involves analyzing the production behavior of oil and gas wells, including well performance engineering, reservoir aspects of well performance, restricted flow into the wellbore, rate decline analysis, and fundamentals of artificial lift. Secondary recovery often refers to either water or gas injection in order to arrest the pressure decline in the reservoir and/or to increase displacement efficiency. Historically, three distinct phases in petroleum reservoir engineering are identified as primary, secondary, and tertiary or enhanced recovery.

Primary recovery involves recovering oil by natural driving mechanisms of the reservoir. Such natural drives may be through solution gas expansion, water drive, gas-cap drive, or simply gravity drainage. During primary recovery, the natural pressure of the reservoir or gravity drive oil into the wellbore, combines with artificial lift techniques, such as pumps that bring the oil to the surface. Commonly, only about 10% of the reservoir's original oil in place is typically produced during primary recovery (DOE 2004).

Secondary recovery is the oil recovery technique in which gas or water is injected in order to maintain the reservoir pressure. It is generally accomplished by injecting water or gas to displace oil and drive it to a production wellbore. General estimates show that the secondary recovery is 20 to 40% of the original oil of a reservoir.

This chapter presents an overview of the reservoir engineering operations. Current practices of reservoir engineering are examined and special emphases are given to well testing, well logging, and core analysis. Limitations of current practices are also discussed. Finally, a guideline is provided in order to improve the predictive ability of various testing techniques.

6.2   Well Test Analysis

The first complete attempt at well testing and analysis was published by Matthews and Russell in 1967, and this has become a standard reference for many petroleum engineers. Many additional technical papers have been published extending the scope of well test analysis, which deal with many new problems, providing solutions of previously unsolved problems, and changing the approach of some phases of well test analysis. Therefore, it is necessary to update the advances in well testing and discover more accurate well test analysis technology for the near future. The basic test method is to create a pressure drawdown in the wellbore that causes formation fluid to enter the well. If measured, the flow rate and pressure during production or the pressure during the shut-in period will provide enough information to characterize the tested well. In well testing, mathematical models are used to relate pressure response with flow rate history of a reservoir and the design and interpretation of a well test is dependent on its objectives.

6.2.1   Objectives of well testing

Reliable information about reservoir conditions is important in many phases. As a result, well testing is an important branch in petroleum engineering. The objectives of a well test can be classified into three major categories, reservoir evaluation, reservoir management, and reservoir description (Horne 1997). Well testing is necessary to determine the ability of a formation to produce reservoir fluids. It is also important to determine the fundamental reason for well productivity. A properly designed, executed, and analyzed well test will provide information about formation permeability, extent of wellbore damage or stimulation, reservoir pressure, and heterogeneities. The production engineer must know the provisions of production and injection of the well to entice the best possible performance from the reservoir. Much of that information can be obtained from well testing. The information obtained from this testing includes wellbore volume, damage and improvement, reservoir pressure, permeability, porosity, reserves, reservoir and fluid discontinuities, and other related data (Earlougher 1977).

6.2.2   Types of well tests

There are several methods used in well test analysis:

- pressure drawdown tests
- pressure build-up tests
- type curve analysis
- interference and pulse tests
- drillstem and wireline formation tests
- gas well tests.

6.2.2.1   Pressure drawdown tests

Pressure drawdown tests can be defined as a series of bottom-hole pressure measurements completed during a period of flow at a constant producing rate. Many traditional analysis techniques are derived using the drawdown test as a basis. Generally, the well is closed in earlier to the flow test period of time because it is necessary to allow the pressure to become equal throughout the formation. Moreover, the well is shut-in until it reaches a constant reservoir pressure before testing. In a drawdown test, a well, now static, stable, and shut-in, is open to flow. It is completed by producing the well at a constant flow rate while continuously recording bottom-hole pressure. When a constant flow rate is attained, the pressure measuring equipment is lowered into the well. Figure 6-1 shows the production and pressure history during a drawdown test. It may take a few hours to several days, depending on the objectives of the test. Drawdown tests are normally recommended for new wells. If a well has been closed for some reason, a drawdown test may also be done. It is also recommended for a well where there are uncertainties in the pressure build-up interpretations. The main advantage of drawdown testing is the possibility for estimating reservoir volume. The shortcomings of this method are:

1. It is difficult to build the well flow at a constant rate, even after it has stabilized.
2. The well condition may not initially be static or stable, especially if it was recently drilled or had been flowing previously.
3. A single permeability value is obtained for the entire well.

6.2.2.2   Pressure build-up tests

A pressure build-up test is conducted by shutting in a well when it is in production at a constant rate for some time. The procedure of this test is to shut-in the well at the surface, allow the pressure to build up in the wellbore, and record the downhole pressure as a function of time. Figure 6.2 shows the rate and pressure behavior for an ideal pressure build-up test, with t_P as the production time and Δt *as* running the shut-in time. The pressure is measured immediately before shut-in and recorded as a function of time during the shut-in period. The resulting pressure build-up curve is analyzed for reservoir properties and wellbore condi-

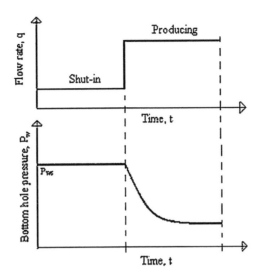

Figure 6-1 Idealized rate schedule and pressure response for drawdown testing. Redrawn from Earlougher (1977).

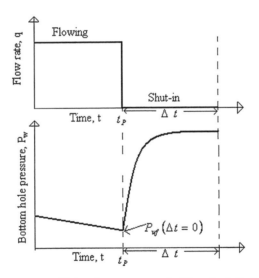

Figure 6-2 Idealized rate and pressure history for a pressure build-up test. Redrawn from Earlougher (1977).

tions. Knowledge of surface and subsurface mechanical conditions is important in this test. Therefore, tubing and casing sizes, well depth, packer locations, etc. need to be known before the start of the test. It is important to stabilize well conditions, otherwise standard data analysis techniques may give erroneous information about the formation.

The main advantage of build-up testing is that the constant flow rate condition is more easily achieved because the flow rate is zero during a build-up test. This test has a number of shortcomings:

1. It is difficult to achieve the constant rate production prior to the shut-in. It may be necessary to close the well briefly to run the pressure tool into the hole

2. Production is lost while the well is shut-in.

3. A single permeability value is obtained for the entire well

6.2.2.3   Type curve analysis

Type curves are normally used to determine formation permeability and characterize damage and simulation of the tested well. Some are used to determine the beginning of the middle time region (MTR). There are several types of curves, such as Ramey et al.'s type curves, McKinley's type curves, Gringarten et al.'s type curves (Lee 1982). Most of these curves are constructed by simulating constant-rate pressure drawdown tests. Basically, a type curve is a preplotted family of pressure drawdown curves. It (Ramey's) is a plot of dimensionless pressure change, p_D vs dimensionless time, t_D. The curve has two parameters that differentiate the curves from one another. These two are skin factor, s and a dimensionless wellbore storage constant, C_{SD}. Type curves are generated by solving the diffusivity equation with a specific set of initial and boundary conditions. Some are analytical and others are based on reservoir simulators. The field data plotting matches with any type curve by tracing the plot and finding out the desired parameter value from the curve. Type curve matching may be used for drawdown, build-up, interference, and constant pressure testing (Earlougher 1977).

6.2.2.4   Interference and pulse tests

When one well is closed in and its pressure is measured while others in the reservoir are being produced, the test is termed an interference test (Matthews and Russell 1967). This test monitors pressure changes at the closed well in the reservoir, which is at some distance from the original producing well. Interference tests are based on the principle of super position. These tests are designed to determine reservoir properties from the observed response in one well to production from one or more other wells in a reservoir (Lee 1982). This test is useful to characterize reservoir properties over a greater length scale than that of a single well test. This type of test can give information on reservoir properties, which cannot be obtained from ordinary pressure build-up or drawdown tests. This test also gives information about the direction of flow. A variation of this test is called pulse testing (Johnson et al. 1966). In this method, a production well near the observation well is alternately produced and then closed in to give a series of pressure pulses. The pulses are detected at the observation well by an accurate pressure gauge. These tests have two shortcomings:

1. They require sensitive pressure recorders and may take a long time to carry out due to low pressure differential between the wells;

2. A single permeability value is obtained for the entire well.

6.2.2.5   Drillstem and wireline formation tests

A drillstem test (DST) is a way of estimating formation and fluid properties before completion of a well (Lee 1981). The DST tool comprises an arrangement of packers and valves placed at the end of the drillpipe. This arrangement can be used to isolate a formation zone, allowing it to produce into the drillpipe and drillstem. A fluid sample is normally collected during the test and thus can generate the information about the types of fluids. A pressure recorder on the DST device can record pressures during the shut-in periods. Figure 6.3 shows the rate and pressure behavior for a typical drillstem test. At point A, the tool is lowered into the hole. Between points A and B, the tool is at the bottom. When the packers are set, the mud column is compressed and a still higher pressure is recorded at point C. The tool is opened for an initial flow period, and the pressure drops to point D as shown. As fluid accumulates in the drillstem above the pressure gauge, the pressure rises. Finally, at point E, the well is shut-in for an initial pressure build-up test. After a suitable shut-in period, the well is reopened for a second final flow period, from point G to point H. This final flow period is followed by a final shut-in period (from point H to point I). The packers are then released, and the hydraulic pressure of the mud column is again imposed on the pressure gauge. This testing device is then removed from the hole (point J to point K).

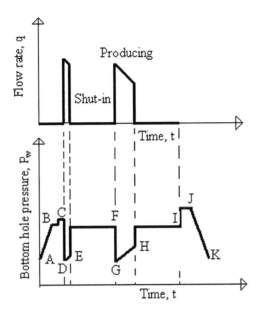

Figure 6-3 Idealized rate and pressure history for a two-cycle drillstem test. Redrawn from Earlougher (1977); Lee (1982).

In many areas, hole conditions prohibit the use of DSTs as temporary wellbore completions. In these areas and in others where the costs of the required number of DSTs for complete evaluation is prohibitive, wireline formation tests are often used in formation evaluation.

6.2.3 Future of well testing

All well testing analyses use several layers of linearization. Because well test analyses were initially performed with analytical models, such linearization was necessary. However, such linearization often continues to take place despite the availability of numerical tools that can solve nonlinear problems (Mustafiz et al. 2006). Even though in certain cases, such linearization is justified because strictly nonlinear solutions give similar results to the linear models, there are situations for which such linearization is not acceptable. It is important to understand the assumptions behind various levels of linearization.

6.2.3.1 Darcy's law

Because practically all well testing models (including single-well reservoir simulators) include the use of Darcy's law, it is important to understand the assumptions behind this momentum balance equation. The following assumptions are inherent to Darcy's law and its extension (Darcy, 1856):

1. The fluid is homogenous, single-phase, and Newtonian;
2. No chemical reaction takes place between the fluid and the porous medium;
3. Laminar flow condition prevails;
4. Permeability is a property of the porous medium, which is independent of pressure, temperature, and the flowing fluid;
5. There is no slippage effect; e.g., Klinkenberg phenomenon;
6. There is no electro-kinetic effect.

Rigorously speaking, none of the above approximations is valid. Therefore, it is important to note this assumption and make necessary corrections whenever data are available. Typically, scientists have recognized these shortcomings and numerous technical papers have been written in order to eliminate one assumption at a time. However, to date, there is no study available demonstrating the contributions of each of the assumptions (Mustafiz and Islam 2005).

The applicability of Darcy's law is further restricted in fractured formations. The fundamental question to be answered in modeling fracture flow is the validity of the governing equations used. The conventional approach involves the use of dual-porosity, dual-permeability models for simulating flow through fractures. Choi et al. (1997) demonstrated that the conventional use of Darcy's law in both the fracture and matrix of the fractured system is inadequate. Instead, they proposed the use of the Forchheimer model in the fracture while maintaining Darcy's law in the matrix. However, their work was limited to single-phase flow. In future, the present status of this work can be extended to a multiphase system. It is anticipated that gas reservoirs will be suitable candidates for using the Forchheimer extension of the momentum balance equation, rather than conventional Darcy's law. Similar to what was done for the liquid system (Cheema and Islam 1995), opportunities exist in conducting experiments with gas as well as multiphase fluids, in order to validate the numerical models. It may be noted that in recent years several dual-porosity, dual-permeability models have been proposed based on experimental observations (Tidwell and Robert 1995; Saghir et al. 2001).

6.2.3.2   Mathematical method

Material balance: Material balance equations are known to be the classical mathematical representation of the reservoir. According to this principle, the amount of material remaining in the reservoir after a production time interval is equal to the amount of material originally present in the reservoir minus the amount of material, due to production plus the amount of material added to the reservoir due to injection. This equation describes the fundamental physics of the production scheme of the reservoir. There are several assumptions in the material balance equation:

1. Rock and fluid properties do not change in space;
2. Hydrodynamics of the fluid flow in the porous media is adequately described by Darcy's law;
3. Fluid segregation is spontaneous and complete;
4. Geometrical configuration of the reservoir is known and exact;
5. PVT data obtained in the laboratory with the same gas-liberation process (flash *vs* differential) are valid in the field;
6. It is sensitive to inaccuracies in measured reservoir pressure. The model breaks down when no appreciable decline occurs in reservoir pressure, as in pressure maintenance operations.

Decline curve: The rate of oil production decline tends to follow one of the mathematical forms, exponential, hyperbolic, and harmonic. Two assumptions apply to the decline curve analysis:

1. The past processes continue to occur in the future;
2. Operation practices are assumed to remain same.

Statistical method: In this method, the past performance of numerous reservoirs is statistically accounted for to derive the empirical correlations, which are used for future predictions. Ertekin et al. (2001) described it as a "formal extension of the analogical method". The statistical methods have the following assumptions:

*   Reservoir properties are within the limit of the database;
*   Reservoir symmetry exists;
*   Ultimate recovery is independent of the rate of production.

In addition, Zatzman and Islam (2007) pointed out a more subtle, yet far more important shortcoming of the statistical methods. Practically all statistical methods assume that two or more objects based on a limited number of tangible expressions makes it legitimate to comment on the underlying science. It is equivalent to stating if effects show a reasonable correlation, the causes can also be correlated. As Zatzman and Islam (2007) pointed out, this poses a serious problem as, in absence of time space correlation (pathway rather than end result), anything can be correlated with anything, making the whole process of scientific investigation spurious. They make their point by showing the correlation between global warming (increase) with a decrease in the number of pirates. The absurdity of the statistical process becomes evident by drawing this analogy. Shapiro et al. (2006) pointed out another severe limitation of the statistical method. Even though they commented on the polling techniques used in various surveys, their comments are equally applicable in any statistical modeling. They wrote:

> Frequently, opinion polls generalize their results to a US population of 300 million or a Canadian population of 32 million, on the basis of what 1000 or 1500 "randomly selected" people are recorded to have said or answered. In the absence of any further information to the contrary, the underlying theory of mathematical statistics and random variability assumes that the individual selected "perfectly" randomly is no more or less likely to have any one opinion over any other. How perfect the randomness may be determined from the "confidence" level attached to a survey, expressed in the phrase that describes the margin of error of the poll sample laying plus or minus some low single-digit percentage "nineteen times out of twenty", i.e., a confidence level of 0.95. Clearly, however, assuming in the absence of any knowledge otherwise, a certain state of affairs to be the case, *viz.*, that the sample is random and no one opinion is more likely than any other, seems more useful for projecting horoscopes than scientifically assessing public opinion.

Analytical method: Analytical methods can only apply to linear equations. Most approaches simplify boundary conditions and apply to single-phase flow in order to solve the governing equations.

Differential calculus: The history of differential calculus dates back to the time of Leibnitz and Newton. In this concept, the derivative of a continuous function to the function itself is related. At the core of differential calculus is Newton's formula that has the approximation attached to it that the magnitude and direction change independently of one another. There is no problem in having separate derivatives for each component of the vector or in superimposing their effects separately and regardless of order. That is what mathematicians mean when they describe or discuss Newton's derivative being used as a "linear operator". Newton's difference-quotient formula states that when the value of a function is inadequate to solve a problem, the rate at which the function changes sometimes becomes useful. Therefore, the derivatives are also important in reservoir simulation. In Newton's difference-quotient formula, the derivative of a continuous function is obtained. This method relies implicitly on the notion of approximating instantaneous moments of curvature, or infinitely small segments, by means of straight lines. It becomes obvious that this derivative is a linear operator precisely because it examines change over time (or distance) within an already established function (Islam, 2006). This function is applicable to an infinitely small domain, making it nonexistent. However, when integration is performed, this nonexistent domain is assumed to be extended to the finite and realistic domain, making the entire process questionable (Abou-Kassem et al. 2007).

6.2.3.3 Analytical models with nonlinear terms

The governing equation in any well test problem is the diffusivity equation (Ertekin et al. 2001). For well testing applications, the diffusivity equation is a radial flow equation. Over the years, numerous methods, empirical equations, and theories have built up analytical and design techniques in well testing. Although the connection between reservoir simulation and well testing is understood, rarely is a bridge between the two detected. Only recently, Mustafiz et al. (2005) worked on these two areas on the common ground of the diffusivity equation and wrote the complex but more realistic nonlinear partial differential equation (PDE). This section introduces their work and presents the solutions of the nonlinear PDEs. Such solutions were possible because they used a semi-analytical tool, the Adomian Domain Decomposition Method (ADM) to solve the nonlinear equations in their original form, without linearization.

Formulation: The diffusivity equations for a two-phase (oil-water) system in their 3D forms are expressed by the following equations:

Oil phase

$$\frac{\partial p_0}{\partial t} = \frac{k_0}{\phi \mu_0 S_0 c_0} \left[\frac{\partial^2 p_0}{\partial r^2} + \frac{1}{r} \frac{\partial p_0}{\partial r} + \frac{1}{r^2} \frac{\partial^2 p_0}{\partial \theta^2} + \frac{\partial^2 p_0}{\partial z^2} \right] - \frac{1}{S_0 c_0} \frac{\partial S_0}{\partial t} \tag{6.1}$$

Water phase

$$\frac{\partial p_W}{\partial t} = \frac{k_W}{\phi \mu_W S_W c_W} \left[\frac{\partial^2 p_W}{\partial r^2} + \frac{1}{r} \frac{\partial p_W}{\partial r} + \frac{1}{r^2} \frac{\partial^2 p_W}{\partial \theta^2} + \frac{\partial^2 p_W}{\partial z^2} \right] - \frac{1}{S_W c_W} \frac{\partial S_W}{\partial t} \tag{6.2}$$

Equations 6.1 and 6.2 assume that permeability and viscosity are constants and the medium is isotropic. Furthermore, in non-dimensional expressions, when the pressure profile is homogeneous and independent of angle, θ (Craft and Hawkings 1991; Panawalage et al. 2004), Equations 6.1 and 6.2 turn into Equations 6.3 and 6.4, respectively:

$$\frac{\partial p_{oD}}{\partial t_D} = \frac{k_{ro} c_t}{[1 - S_{wi} - S_{WD}(1 - S_{wi} - S_{or})]c_0} \left[\frac{\partial^2 p_{oD}}{\partial r_D^2} + \frac{1}{r_D} \frac{\partial p_{oD}}{\partial r_D} + \left(\frac{r_w}{h} \right)^2 \frac{\partial^2 p_{oD}}{\partial z_D^2} \right]$$
$$- \frac{2\pi k h}{q \mu_o c_o} \frac{(1 - S_{wi} - S_{or})}{[1 - S_{wi} - S_{WD}(1 - S_{wi} - S_{or})]} \frac{\partial S_{wD}}{\partial t_D} \tag{6.3}$$

$$\frac{\partial p_{wD}}{\partial t_D} = \frac{\mu_o k_{rw} c_t}{\mu_w S_w c_w} \left[\frac{\partial^2 p_{wD}}{\partial r_D^2} + \frac{1}{r_D} \frac{\partial p_{wD}}{\partial r_D} + \left(\frac{r_w}{h} \right)^2 \frac{\partial^2 p_{wD}}{\partial z_D^2} \right] + \frac{2\pi k h}{q \mu_w c_w} \frac{(1 - S_{wi} - S_{or})}{[S_{wi} + S_{WD}(1 - S_{wi} - S_{or})]} \frac{\partial S_{wD}}{\partial t_D} \tag{6.4}$$

Solutions of the nonlinear equations: In recent years, the ADM has been successfully applied to solve wave equations (Biazar and Islam, 2004), and a system of ordinary differential equations (Biazar et al. (2004), etc. Equations 6.3 and 6.4 can be approached using the same technique where any governing equation is converted into recursive relations. In this section, Equation 6.5, which is similar to Equations 6.3 and 6.4, is applied to explain how decomposition method works:

$$\frac{\partial p}{\partial t} = E \left[\frac{\partial^2 p}{\partial r^2} + \frac{1}{r} \frac{\partial p}{\partial r} + F \frac{\partial^2 p}{\partial z^2} \right] + G \frac{\partial S}{\partial t} \tag{6.5}$$

In canonical form, Equation 6.5 is written as

$$Lp = E \left[\frac{\partial^2 p}{\partial r^2} + \frac{1}{r} \frac{\partial p}{\partial r} + F \frac{\partial^2 p}{\partial z^2} \right] + G \frac{\partial S}{\partial t} \tag{6.6}$$

where, $L = \dfrac{\partial}{\partial t}$ with the inverse $L^{-1} = \int\limits_0^t (.)dt$, which means $L^{-1}L = I$.

Applying the inverse operator to Equation (6.6), we derive the canonical form of the equation:

$$P(r, z, t) = P(r, z, 0) + E \int\limits_0^t \left(\frac{\partial^2 p}{\partial r^2} + \frac{1}{r} \frac{\partial p}{\partial r} + F \frac{\partial^2 p}{\partial z^2} \right) dt + G(S_t - S_0) \tag{6.7}$$

Similarly, Equation 6.5 can be rearranged to apply the inverse operators $L^{-1} = \int\limits_{R_2}^r \int\limits_{R_2}^r (.)drdr$, where $L = \dfrac{\partial^2}{\partial r}$

and $L^{-1} = \int\limits_{Z_2}^z \int\limits_{Z_2}^z (.)dzdz$ where, $L = \dfrac{\partial^2}{\partial t}$, which result in two more pressure expressions:

$$P(r, z, t) = P(R_2, z, 0) + \frac{1}{E} \int\limits_{R_2}^r \int\limits_{R_2}^r \left(\frac{\partial p}{\partial t} - C \frac{\partial S}{\partial t} \right) - \frac{1}{r} \int\limits_{R_2}^r \int\limits_{R_2}^r \frac{\partial p}{\partial t} dr - F \int\limits_{R_2}^r \int\limits_{R_2}^r \frac{\partial^2 p}{\partial z^2} dz \tag{6.8}$$

$$P(r, z, t) = P(r, z_2, t) + \frac{1}{EF} \int_{z_2}^{z} \int_{z_2}^{z} \left(\frac{\partial p}{\partial t} - C \frac{\partial S}{\partial t} \right) dz - \frac{1}{Fr} \int_{z_2}^{z} \int_{z_2}^{z} \frac{\partial p}{\partial r} dz - \frac{1}{F} \int_{z_2}^{z} \int_{z_2}^{z} \frac{\partial^2 p}{\partial r^2} dz \qquad (6.9)$$

Now, Equations 6.7 to 6.9 can be averaged to obtain the expression for pressure. The ADM considers the solution as the sum of the following series:

$$p(r, z, t) = \sum_{n=0}^{\infty} p_n \qquad (6.10)$$

where p_0 is defined to be the summation of the terms on the right-hand side of Equation 6.8, derived from initial conditions and those terms that do not depend on the unknown function p.

$$p_0 = \frac{1}{3} [P(r, z, 0) + P(R_2, z, 0) + P(r, z_2, t) + G(S_t - S_0)] \qquad (6.11)$$

The other terms of the series are defined as:

$$P_{n+1} = A_n(p_0, \ldots, p_n) \; n = 0, 1, 2 \ldots \qquad (6.12)$$

where $A_n \, (p_0, p_1, p_2, p_n)$ or the Adomian polynomials are computed using an alternate algorithm:

$$\begin{aligned}
A_n(p_0, \ldots, p_n) = \frac{1}{3} \Bigg[& E\int_0^t \left(\frac{\partial^2 p_n}{\partial r^2} + \frac{1}{r} \frac{\partial p_n}{\partial r} + F \frac{\partial^2 p_n}{\partial z^2} \right) dt + \frac{1}{E} \int_{R_2}^{r} \int_{R_2}^{r} \left(\frac{\partial p_n}{\partial t} - C \frac{\partial S}{\partial t} \right) drdr \\
& - \frac{1}{r} \int_{R_2}^{r} \int_{R_2}^{r} \frac{\partial p_n}{\partial r} drdr - F \int_{R_2}^{r} \int_{R_2}^{r} \frac{\partial^2 p_n}{\partial z^2} dzdz + \frac{1}{EF} \int_{z_2}^{z} \int_{z_2}^{z} \left(\frac{\partial p_n}{\partial t} - C \frac{\partial S}{\partial t} \right) dzdz \\
& \frac{1}{Fr} \int_{z_2}^{z} \int_{z_2}^{z} \frac{\partial p_n}{\partial r} dzdz - \frac{1}{F} \int_{z_2}^{z} \int_{z_2}^{z} \frac{\partial^2 p_n}{\partial r^2} dzdz \Bigg]
\end{aligned} \qquad (6.13)$$

To derive the solution we have to evaluate the parameters E, F, and G. This requires information on fluid and matrix properties, which can also be available. Relationships about phase-saturation constraint and capillary pressures as functions of phase saturations for an oil-water system will be necessary to solve Equations 6.3 and 6.4.

$$S_o + S_w = 1 \qquad (6.14)$$

$$P_{cow} = P_o - P_{ow} = f(S_w) \qquad (6.15)$$

Example: In this section, a solution to the simplest form of diffusivity equation is given. The solution of this classic equation is important as Equations 6.1 and 6.2 originate from it. In radial form, the dimensionless diffusivity equation is given by

$$\frac{\partial p_D}{\partial t_D} = \frac{\partial^2 p_D}{\partial r_D^2} + \frac{1}{r_D} \frac{\partial p_D}{\partial r_D} \qquad (6.16)$$

The initial condition is assumed as

$$P_D(r_D, 0) = A.r_D \qquad (6.17)$$

Let, $L = \dfrac{\partial}{\partial t_D}$ with the inverse $L^{-1} = \displaystyle\int_0^{t_D} (.) dt_D$. So, we have

$$P_D(r_D, t_D) = P_D(r_D, 0) + \int_0^{t_D} \left(\frac{\partial^2 p_D}{\partial r_D^2} + \frac{1}{r_D} \frac{\partial p_D}{\partial r_D} \right) dt_D \tag{6.18}$$

Using the procedure described previously, we have

$$A_n(p_{0D}, p_{1D}, \ldots, p_{nD}) = \frac{\partial^2 p_{nD}}{\partial r_D^2} + \frac{1}{r_D} \frac{\partial p_{nD}}{\partial r_D} \tag{6.19}$$

The solution by ADM consists of the following scheme:

$$P_{0D} = A r_D \tag{6.20}$$

$$P_{(n+1)D} = \int_0^{t_D} \left(\frac{\partial^2 p_{nD}}{\partial r_D^2} + \frac{1}{r_D} \frac{\partial p_{nD}}{\partial r_D} \right) dt_D \tag{6.21}$$

where, $n = 0, 1, 2, \ldots$

$$A_0(P_{0D}) = \frac{\partial^2 p_{0D}}{\partial r_D^2} + \frac{1}{r_D} \frac{\partial p_{D0}}{\partial r_D} = \frac{A}{r_D} \tag{6.22}$$

$$P_{1D} = \frac{At_D}{r_D} \tag{6.23}$$

$$A_1(P_{0D}, P_{1D}) = \frac{At_D}{r_D^3} \tag{6.24}$$

$$P_{2D} = \frac{At_D^2}{2r_D^3} \tag{6.25}$$

$$A_2(P_{0D}, P_{1D}, P_{2D}) = \frac{9At_D^2}{2r_D^5} \tag{6.26}$$

$$P_{3D} = \frac{3At_D^3}{2r_D^5} \tag{6.27}$$

In general, the expression can be written as

$$A_{n-1}[P_{0D}, P_{1D}, P_{2D}, \ldots P_{(n-1)D}] = \frac{1^2 * 3^2 * \ldots * (2n-3)^2}{(n-1)! r_D^{2n-1}} At_D^{n-1} \tag{6.28}$$

and

$$p_{nD} = \frac{1^2 * 3^2 * \ldots * (2n-3)^2}{n! * r_D^{2n-1}} At_D^n \tag{6.29}$$

Therefore,

$$p_D(r_D, t_D) = A \left[r_D + \sum_{n=1}^{\infty} \frac{1^2 * 3^2 * \ldots * (2n-3)^2}{n! * r_D^{2n-1}} t_D^n \right] \tag{6.30}$$

Equation 6.30 is the solution of the governing equation, Equation 6.16. From the solution, the dimensionless pressure *vs* time graph can be plotted (Figure 6-4).

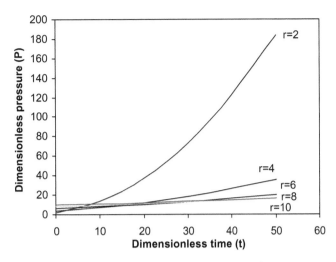

Figure 6-4 Dimensionless pressure *vs* time.

Table 6-1 Compressibility values used in the Exponential Integral method

Type	Compressibility, Kpa^{-1}		
	PS technique	PD technique	Casting technique
Sample A	0.000015	0.000027	0.000029
Sample B	0.00003	0.00006	0.00006
Sample C	0.0008	0.0008	0.0015

Table 6-2 Values used in the Exponential Integral method

Parameter(s)	Values
Sample radius, R, (cm)	2.86
Perforation radius, r, (cm)	0.68
Sample height, h, (cm)	10
Flow rate, q, (mm^3/min)	30,000
Fluid viscosity, μ, (mPa.s)	1.519
Formation volume factor, B_0	1

This solution to the classical diffusivity equation with few assumptions, therefore the ADM can be immensely useful to solve equations of higher dimensions, i.e., multiphase, multidimensional problems.

The same solution technique has recently been used by Rahman et al. (2006) in the context of well testing. They solved the single-phase equation for various cases of pressure build-up tests and compared their numerical results with experimental ones. The following figures show that the comparison between experimental and mathematical results is reasonable. The discrepancy between the theoretical and experimental data can also be attributed to only a 1 D approximation, which is used in the simulation. Circumferential and vertical flow is not considered during modeling and the second-order nonlinear term is also neglected. The samples are assumed to be isotropic, fully saturated, and under constant temperature, whereas the sensitivity analysis suggests that permeability is the most important parameter. All relevant data are reported in Tables 6-1 and 6-2.

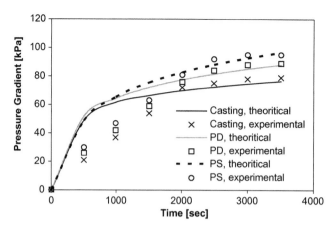

Figure 6-5   (A) Comparison of experimental and exponential integral based observations for transient differential pressure of sample A.

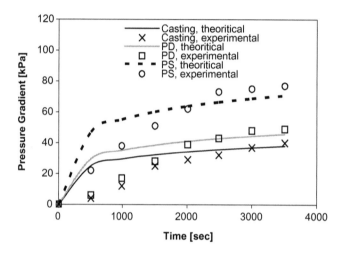

Figure 6-5   (B) Comparison of experimental and exponential integral based observations for transient differential pressure of sample B.

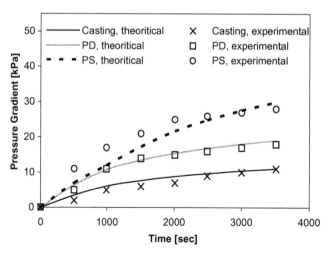

Figure 6.5   (C) Comparison of experimental and exponential integral based observations for transient differential pressure of sample C.

6.2.3.4 Nonlinearity in numerical models

Because of the severe limitations of the analytical models, numerical models are often used to interpret well test data. This is a typical procedure using a single-well reservoir simulator. Even though a reservoir simulator has the ability to model multiphase flow, often only the single phase option is activated to keep the procedure simple.

Claims are made that nonlinear equations are solved with a numerical simulator, but the solution techniques used in reservoir simulation are appropriate only for systems of linear equations. Therefore, the use of these techniques requires linearization of the finite difference equations. Such techniques include the explicit treatment of the transmissibility terms, simple iteration of the transmissibility terms, extrapolation of the transmissibility terms, and fully implicit treatment of the transmissibility terms. Transmissibility, T, is a combination of viscosity, μ, formation volume factor, B, and geometric factor, G, where μ and B are functions of pressure, P, and temperature, T. That is, while modeling single-phase flow, these techniques are used to linearize the pressure-dependent properties. The explicit method does not offer any improvement in the value of the nonlinearities as iterations progress.

The simple iteration method presents improvement in the value of nonlinearities in a stepwise manner. In the fully implicit method, the improved value of nonlinearities falls on the tangent of the nonlinearities at the previous iteration as the iteration continues. The same methods are equally applicable to multiphase flow problems. Other linearization methods, such as the linearized-implicit method (MacDonald and Coats 1970) and the semi-implicit method of Nolen and Berry (1972) are not applicable to single-phase problems. These two methods are used in multiphase problems to deal with nonlinearities due to fluid saturation only (Abou-Kassem et al. 2006).

A number of algorithms can be used to solve a given system of linear equations. The finite difference equations, if written for n simulation grids, produce an n x n coefficient matrix. The direct and iterative are the two general methods used in reservoir simulation. Theoretically, the direct methods are capable of obtaining an exact solution after a fixed number of computations have been carried out. However, it is assumed that the computer is able to process an infinite number of digits. In reality, no such computer, including the super ones, has the ability to round-off errors, no matter how insignificant they are. Numerous articles have been reported to improve the direct methods (i.e., Gaussian elimination, Gauss – Jordan reduction) including the Crout reduction, Thomas' algorithm, etc.

Moreover, features such as sparse-matrix technique, pivoting, multiple known vectors, and iterative improvements have also appeared to improve the features of direct procedures (Ertekin et al. 2001). The iterative method has emerged as a solution scheme, in which each iteration shows a gradual reduction of error if the scheme is converging. However, to reduce the slow convergence rate, and in turn the monotony of the iterative procedure, significant advancements have been made. Factors such as convergence requirements, speed, and storage needs have been addressed through the early attempts of Jacobi iteration, Gauss–Seidel iteration, successive over-relaxation (LSOR), block-successive over-relaxation (BSOR), alternating-direction-implicit-procedure (ADIP), etc. However, the development of more powerful, yet linear solution techniques, such as conjugate-gradient-like (CGL), replaced the previously popular techniques in the 1960s and 1970s (Ertekin et al. 2001).

Mustafiz et al. (2006) investigated the effects of the variation of fluid and formation properties, the value of the time interval Δt, and the simplification in formulation, on the pressure of a single-phase (compressible fluid) flow problem. They showed that only for single-phase flow was the linearization of flow equations justified. They recommended that a new approach be used for determining the full spectrum of solutions of nonlinear equations that should naturally give rise to multiple solutions. Further details of this procedure are available in Islam et al. (2007).

6.3   Current Practice in Well Logging

Electrical well logging has been used in the oil industry for over 75 years. Many additional and improved logging devices have been developed and used throughout these decades. Hundreds of technical papers have been written to present the various logging methods, their application, and their interpretation. At present, the focus is on the techniques involving digitization of data, with some research being done on replacing radioactive elements from well testing procedures.

A well log may be defined as a tabular or graphical representation of any drilling conditions or subsurface features encountered, which relate to either the progress or evaluation of an individual well (Gatlin 1960). Well logs record the types of formations encountered, any pertinent fluid flows or oil and gas shows observed, and other related operational remarks.

6.3.1   Types of well logging

There are several methods used in well logging:

- Resistivity logs
- Spontaneous potential logs
- Gamma ray logs
- Gamma ray absorption logs
- Neutron logs
- Sonic porosity logs.

6.3.1.1   Resistivity logs

The resistivity of a formation is a key parameter in determining hydrocarbon saturation. As formations contain conductive water, electricity can pass through easily. Formation resistivities are usually from 0.2 to 1000 ohm-m. Resistivities higher than 1000 ohm-m are uncommon in permeable formations but are observed in impervious, very low porosity formations. Formation resistivities are measured by sending a current into the formation and measuring the ease of the electrical flow or by inducing a current into the formation and measuring how large it is. The resistivity of a formation depends on resistivity of the formation water, amount of water present, and pore structure geometry (Schlumberger 1991). Resistivity logging methods are used to measure the resistivity of the flushed zone and the true resistivity of the uninvaded virgin zone. Resistivity devices display apparent resistivity values. The apparent resistivity is affected by the resistivity and geometry of borehole, adjacent beds, and invaded and un-invaded zones of the bed of interest (Bassiouni 1994). The apparent resistivity value should be corrected for borehole and adjacent-bed effects.

6.3.1.2   Devices and principle of resistively logs

Figure 6-8 shows a conventional resistivity log device, for which a constant intensity current is passed between two electrodes, A and B. The resultant potential difference is measured between two other electrodes, M and N. Electrodes A and M are on a sonde. B and N are, theoretically, located at an infinite distance away. In practice, B is the cable armor and N is an electrode on the bridle (the insulation-covered lower end of the cable) far removed from A and M. The distance A to M is called the spacing (16-in spacing for the short normal, 64-in spacing for the long normal), and the point of inscription for the measurement is at O, midway between A and M. There are different devices used to measure the resistivity of the formation. These are lateral devices, focusing electrode tools, dual laterolog-R_{ox} tool, spherically focused log (SFL) device, induction logging tool, phasor induction SFL tool, and microresistivity devices (Schlumberger 1991; Bassiouni 1994).

The principle of a resistivity log is that currents are passed through the formation by means of current electrodes and voltages are measured between measure electrodes. These measured voltages provide the

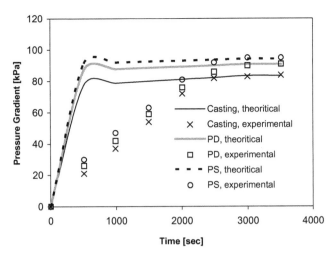

Figure 6-6   (A) Comparison of experimental and Adomian decomposed observations for transient differential pressure of sample A.

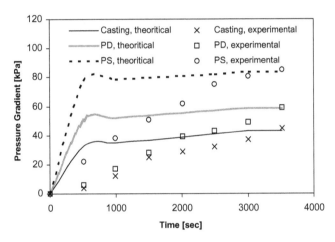

Figure 6-6   (B) Comparison of experimental and Adomian decomposed observations for transient differential pressure of sample B.

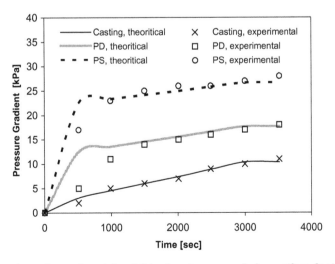

Figure 6-6   (C) Comparison of experimental and Adomian decomposed observations for transient differential pressure of sample C.

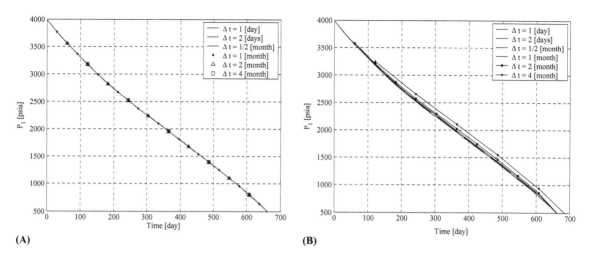

(A)   **(B)**

Figure 6-7   The effect of the time step in computation: A) Cubic spline interpolation using original formulation; B) Cubic spline interpolation using linear formulation (from Mustafiz et al. 2006).

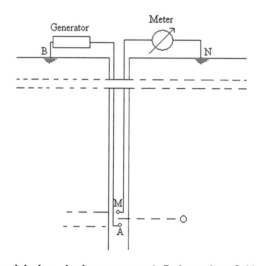

Figure 6-8   Normal device – basic arrangement. Redrawn from Schlumberger (1991).

resistivity recorded in each device. In a homogeneous, isotropic formations of infinite extent, the equipotential surfaces surrounding a single current-emitting electrode (A) are spherical. The voltage between an electrode (M) situated on one of these spheres and the one at infinity is proportional to the resistivity of the homogeneous formation, and the measured voltage can be scaled in resistivity units.

6.3.1.3   Spontaneous potential logs

The spontaneous potential (SP) logs record naturally occurring physical phenomena within rocks. Naturally occurring electrical potentials are observed at the Earth's surface and in its subsurface. They are associated with weathering of mineral bodies, variation of rock properties at geological contacts, bioelectric activity of organic material, thermal and pressure gradients in underground fluids, and other phenomena. The SP curve records the electrical potential (voltage) produced by the interaction of formation connate water,

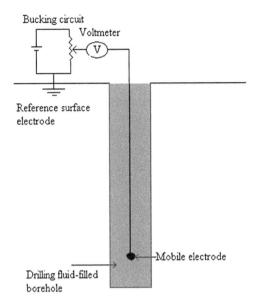

Figure 6-9   Schematic of the circuit used to measure the SP logs. Redrawn from Bassiouni (1994).

conductive drilling fluids, and certain ion-selective rocks (shale). Figure 6-9 shows the circuit used to measure the SP. The main circuit components are a downhole mobile electrode, a fixed surface electrode, a voltmeter, and a bucking circuit composed of batteries and a variable resistor. The absolute voltage measured between the two electrodes could be several hundred millivolts. The SP measurement emphasizes the variation of potential rather than its absolute value because bucking circuit is used. The sensitivity of the linear scale is chosen so that the variations remaining on the track can be read easily.

The SP log can be used to distinguish between impermeable shales and permeable and porous sands. The boundaries of each permeable bed can be defined and their thickness calculated. In thick beds, the deflection reaches a maximum and stabilizes as a plateau. The response in thin beds falls short of the plateau and exhibits a rounded shape. Maximum SP deflection is also affected by shaliness. The SP observed in boreholes is of electrokinetic and electrochemical origin. Electrochemical potential, believed to be the major contributor to the SP, consists of two components, the diffusion potential and the membrane potential.

The movement of ions, which causes the SP phenomenon, is possible only in formations having a certain minimum permeability. There is no direct relationship between the value of permeability and the magnitude of SP deflection, nor does SP deflection have any direct relation to the porosity. The slope of the SP curve at any level is proportional to the intensity of SP currents in the borehole mud at that level. The shape of the SP curve and the amplitude of the deflection opposite to a permeable bed depend on several factors. Some factors affect the distribution of SP current lines and the potential drops taking place in each of the media through which SP current flows. These factors are the thickness and true resistivity of the permeable bed, resistivity and diameter of the zone contained by mud filtrate invasion, resistivity of the adjacent shale formation, and resistivity of the mud and diameter of the borehole (Schlumberger 1991).

Sometimes, a small-amplitude sine-wave signal is superimposed on the SP, such as when some mobile part of the winch becomes accidentally magnetized. An intermittent contact between casing and cable armor may also cause false spikes on the SP curve. In these situations, the SP curve should be read in such a way that the sine wave amplitude or skipe is not added to or subtracted from the authentic SP deflection. It is sometimes difficult to record a good SP on offshore or barge locations, as passing ships, cathodic protection devices, and leaky power sources may all contribute to a noisy SP. On land, proximity to power lines and pumping wells may have a similar effect on the SP curve. Many of these disturbances can be solved by a careful choice of the ground-electrode location.

6.3.1.4   Gamma ray logs

Certain elements show nuclear disintegration by emitting energy in the form of alpha, beta, and gamma particles. Alfa particles are helium nuclei and beta particles are electrons. Both of these have relatively low penetrating power and are stopped by a small thickness of solid materials. Gamma rays are similar to X-rays (electromagnetic waves) and are able to penetrate several inches of rock or steel. The gamma ray (GR) log is a measurement of the natural radioactivity of the formations. The principal radioactive families of elements are the uranium-radium series, the thorium series, and the potassium isotope, K^{40}. Nearly all sedimentary rocks contain traces of radioactive salts and, as a consequence, emit measurable radiation.

A gamma ray logging device of the ionization chamber (Geiger–Muller Counter) type is shown in Figure 6.10. Scintillation counter equipment is also in common use. However, the ionization chamber produces the simplest example of the principles involved. An ionization radiation chamber containing an inert gas at high pressure is penetrated by gamma rays. Some of these rays collide with gas atoms, liberate electrons from the gas, and thereby ionize the gas. The current resulting from this liberation of electrons is automatically amplified at the surface and recorded as a function of depth. Its magnitude is directly related to the intensity of the gamma radiation at any level. Gamma radiation is not emitted at a constant rate. Consequently, for a given length of ionization chamber (commonly 3–4 ft) the logging speed has to be adjusted to obtain a true statistical picture of radiation in different localities.

The log normally reflects the shale content in sedimentary formations because the radioactive elements tend to concentrate in clays and shales. Clean formations usually have a low level of radioactivity, unless radioactive contaminant such as volcanic ash or granite wash is present or the formation waters contain dissolved radioactive salts. The GR log can be recorded in cased wells, which makes it useful as a correlation curve in completion and workover operations. It is frequently used to complement the SP log and as a substitute for the SP curve in wells drilled with salt mud, air, or oil-based muds. In each case, it is useful for location of shales and nonshaly bed and, most importantly, for general correlation.

The GR log is particularly useful for defining shale beds when the SP is distorted (in very resistive formations), or when the SP is featureless (in freshwater-bearing formations or in salt mud), or when the SP cannot be recorded (in nonconductive mud, empty or air drilled holes, cased holes). The bed boundary is picked at a point midway between the maximum and minimum deflection of the anomaly.

The GR log reflects the proportion of shale and, particularly, can be used quantitatively as a shale indicator. It is also used for the detection and evaluation of radioactive minerals, such as potash or uranium ore. Its response, corrected for borehole effect, is practically proportional to the K_2O content, approximately 15 API units per 1% of K_2O. The GR log can also be used for delineation of nonradioactive minerals.

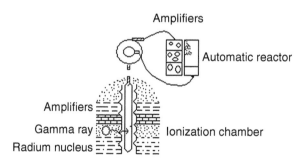

Figure 6-10   Ionization chamber device for radiation logging. Redrawn from Gatlin (1960).

6.3.1.5 Gamma ray absorption logs

The first commercial tool to use the physical phenomena of gamma ray scattering and absorption was introduced in the early 1950s (Bassiouni 1994). The tool, known as the gamma-gamma, was developed initially to measure bulk density *in situ*, as an aid to geophysicists in gravity-meter data interpretation. The tool consisted of a gamma ray source and a detector. The sources used are cesium-137 and cobalt-60. The cesium, preferred because of its stability, decays with a half-life of about 30 years, emitting gamma rays of about 0.66 MeV. The gamma rays emitted by the source, held in a skid in contact with the borehole wall, are directed toward the formation. Some of these gamma rays are absorbed and some are scattered away from the detector, but others are scattered into the detector and counted (Bassiouni 1994).

6.3.1.6 Neutron logs

Neutrons exist in the nuclei of all elements except hydrogen. They have approximately the same mass as hydrogen atoms, but with no charge. When emitted from fissionable material, they possess very high velocities but are rapidly slowed down by collisions with other atoms. Atoms of nearly the same mass as the neutron are most effective in reducing neutron velocity. The neutron is greatly slowed by collisions with hydrogen atoms. Since fluids (water, oil, and gas) contain a much higher hydrogen content than rocks, it is apparent that the behavior of emitted neutrons affords a means of evaluating the hydrogen (and hence fluid) content of a formation.

Inelastic scattering, elastic scattering, and absorption are the basic phenomena that occur after a fast neutron is introduced into a formation. The fast neutrons are slowed first by inelastic scattering (which takes place at a high neutron energy level) and then by elastic scattering. The neutrons eventually slow to a level of energy at which they coexist with the formation nuclei in thermal equilibrium. Neutrons in this state are called thermal neutrons. Thermal neutrons continue to scatter off the formation nuclei elastically and diffuse throughout the formation. Each thermal neutron eventually is captured by one of the nuclei. The nucleus instantaneously emits gamma rays, called capture gamma rays. Neutrons that have slowed almost to the thermal energy level yet are still energetic enough to avoid capture are known as epithermal neutrons. Several logging tools are based on these phenomena.

The neutron porosity log is based on the elastic scattering of neutrons as they collide with nuclei in the formation. Each neutron scatters off a nucleus with less kinetic energy. Energy and momentum conservation in elastic collisions dictates that the presence of hydrogen in the formation dominates the slowing process. The reason for this is that the mass of the hydrogen nucleus is approximately equal to that of the incident neutron. Consequently, at a point sufficiently removed from the neutron source, formations with high hydrogen content display low concentrations of epithermal and thermal neutrons and capture gamma rays. Inversely, formations with low hydrogen content display high concentrations of epithermal and thermal neutrons and capture gamma rays.

To provide a standard unit for neutron log measurements, the American Petroleum Institute (API) adopted the "API neutron unit". One API neutron unit is defined as 1/1000 of the difference between instrument zero (i.e., tool response to zero radiation) and log deflection opposite a 6-ft zone of Indiana limestone in a neutron calibration pit at the University of Houston (Figure 6-11). The pit is 24 ft deep, with a 15-ft rathole. The pit contains three 6-ft-thick limestone zones. Each zone is made up of six 1-ft-thick octagonal blocks. The zones consist of Carthage, Indiana, and Austin limestones displaying average porosities of 1.9% and 25%, respectively. The rock is saturated with fresh water. A 6-ft layer of fresh water atop the rock blocks provides a 100% porosity reference point. A $7^7/_8$- inch borehole extends through the pit center.

6.3.1.7 Sonic porosity logs

Sonic devices were first introduced for seismic velocity determination. These "continuous velocity logs" were widely used in petroleum exploration and development once it was discovered that a reliable formation

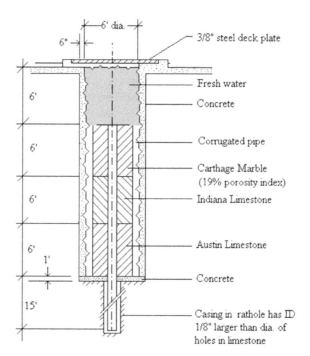

Figure 6-11   Neutron log calibration pit. Redrawn from Bassiouni (1994).

porosity value could be extracted from log response. Conventional sonic tools measure the reciprocal of the velocity of the compressional wave. This parameter is called interval travel time or slowness. It is expressed in microseconds per foot. Porosity of the consolidated formation is expressed by Wyllie's Equation (Bassiouni 1994).

Sonic tools can be classified as single-receiver systems or dual-receiver systems. A duel-receiver system is shown in Figure 6-12. This system is introduced to remove the mud-path contribution from the response of sonic tools. The figure shows a schematic of one of the first tools that incorporated the two-receiver system. The tool consists of a transmitter and three receiver located 3, 4, and 6 feet from the transmitter. The transmitter emits acoustic waves at 10 waves/s. The first arrival of acoustic energy at each receiver triggers its response system. A two-receiver system can be viewed as an accurate stopwatch. The stopwatch starts when the acoustic energy arrives at the first receiver and stops when it arrives at the second receiver. Indicated by the stopwatch is the time required for the sound wave to traverse a length of the formation equal to the spacing between the two receivers.

6.4   Current Practices of Core Analysis

The main goal of core analysis is to reduce uncertainty in reservoir evaluation by providing data representative of the reservoir at *in situ* conditions. Core analysis is important for the extensive use of reservoir simulation in the evaluation, development, and management of oil fields. The importance is increased due to the correct use of results from special core analysis, particularly those from relative permeability tests. Correct use of these data requires knowledge of the history of the core, drilling mud conditions during coring, process of extraction from ore, and measurements in the laboratory. It is also important to know the coring operations and core preservation techniques. Laboratory measurement methods also play a role because results are affected by the techniques used and the test conditions. It should be noted that results that are considered valid require adjustment and refinement. As a result, an understanding of laboratory techniques is needed because each method has its own strengths and weaknesses.

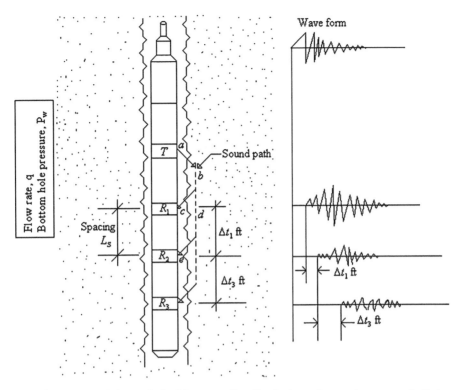

Figure 6-12   Schematic of the sonic tool incorporating the dual-receiver system concept. Redrawn from Bassiouni (1994).

6.4.1   Correlation with field

Core analysis has to be integrated with field and production data to minimize reservoir uncertainties that cannot be addressed from other data sources, such as well logging, well testing, or seismic. These requirements define the coring objectives, core handling, and core analysis schedule. These objectives cannot be achieved by coring a single well. The coring program is thus an integral part of the reservoir history cycle. The post-1980s economics of the petroleum industry expressed a need for evermore cost-effective technologies, combined with the need to evaluate thin bed and non-conventional reservoirs by means of vertical and horizontal wells. It serves as both the controlling factor and driving force, respectively and is behind the development of new techniques of coring and core analysis.

Techniques are constantly being improved or new ones introduced. In proper core analysis, the concept of automatic geological core description is growing with the use of the mini-permeameter and the creation of sophisticated analysis methods such as SEM/EDX, X-ray, CT, NMR, and PIA. These hi-tech methods provide a wealth of micro-structural and microscopic information, to enhance the reservoir description and evaluation processes. Eventually the main challenges in core analysis and recent trends affecting future tool developments will be provided. Traditional structural analysis of the core attempts to list, as accurately as possible, every fracture on the core and its geometry, but often with no connection to the other core analyses performed. As a result, the structural data is often difficult to extrapolate from the borehole to the reservoir scale, and therefore difficult to integrate into the reservoir model.

6.4.2   Background information

The material of which a petroleum reservoir rock may be composed can range from very loose and unconsolidated sand to very hard and dense sandstone, limestone, or dolomite. The grains may be bonded together with a number of materials, the most common of which are silica, calcite, or clay. Knowledge of the physical properties of the rock and the existing interaction between the hydrocarbon system and the formation is essential in understanding and evaluating the performance of a given reservoir. Rock properties are determined by performing laboratory analyses on cores from the reservoir to be evaluated. The cores are removed from the reservoir environment, with subsequent changes in the core bulk volume, pore volume, reservoir fluid saturations, and, sometimes, formation wettability. The effect of these changes on rock properties may range from negligible to substantial, depending on characteristics of the formation and property of interest, and should be evaluated in the testing program. There are basically two main categories of core analysis tests that are performed on core samples, regarding physical properties of reservoir rocks. These are:

Conventional core analysis:

- Porosity
- Permeability
- Saturation

Special core analysis:

- Overburden pressure
- Capillary pressure
- Relative permeability
- Wettability
- Surface and interfacial tension

The above rock property data are essential for reservoir engineering calculations as they directly affect both the quantity and the distribution of hydrocarbons and, when combined with fluid properties, control the flow of the existing phases (i.e., gas, oil, and water) within the reservoir.

6.4.2.1   Conventional core analysis

Porosity: The porosity of a rock is a measure of the storage capacity (pore volume) that is capable of holding fluids. Quantitatively, the porosity is the ratio of the pore volume to the total volume (bulk volume). This important rock property is determined mathematically by the following generalized relationship:

$$\phi = \frac{pore\ volume}{bulk\ volume}$$

where, ϕ = porosity.

As the sediments were deposited and the rocks were being formed during past geological times, some void spaces that developed became isolated from the others by excessive cementation. Thus, many of the void spaces are interconnected while some are completely isolated. This leads to two distinct types of porosity, namely:

1. absolute porosity
2. effective porosity

Absolute porosity: The absolute porosity is defined as the ratio of the total pore space in the rock to that of the bulk volume. A rock may have considerable absolute porosity and yet have no conductivity to fluid for lack of pore interconnection. The absolute porosity is generally expressed mathematically by the following relationships:

$$\phi_a = \frac{total\ Pore\ volume}{bulk\ volume}$$

$$\phi_a = \frac{bulk\ volume - grain\ volume}{bulk\ volume}$$

where, ϕ_a = absolute porosity.

Effective porosity: The effective porosity is the percentage of interconnected pore space with respect to the bulk volume, or

$$\phi = \frac{interconnected\ pore\ volume}{bulk\ volume}$$

where, ϕ = effective porosity.

The effective porosity is the value that is used in all reservoir engineering calculations, because it represents the interconnected pore space that contains the recoverable hydrocarbon fluids. Porosity may be classified according to the mode of origin as originally induced. The original porosity is that developed in the deposition of the material, while induced porosity is that developed by some geological process subsequent to deposition of the rock. The intergranular porosity of sandstones and the intercrystalline and oolitic porosity of some limestones typify original porosity. Induced porosity is typified by fracture development as found in sandstone and limestone, and by the slugs or solution cavities commonly found in limestones. Rocks having original porosity are more uniform in their characteristics than rocks in which a large part of the porosity is included. For direct quantitative measurement of porosity, reliance must be placed on formation samples obtained by coring.

Since effective porosity is the porosity value of interest to the petroleum engineer, particular attention should be paid to the methods used to determine porosity. For example, if the porosity of a rock sample was determined by saturating the rock sample with 100% of fluid of known density and then determining, by weighing, the increased weight due to the saturating fluid, this would yield an effective porosity measurement because the saturating fluid could enter only the interconnected pore spaces. However, if the rock sample were crushed with a mortar and pestle to determine the actual volume of the solids in the core sample, then an absolute porosity measurement would result because the identity of any isolated pores would be lost in the crushing process.

Permeability: Permeability is a property of the porous medium that measures the capacity and ability of the formation to transmit fluids. The rock permeability, k, is an important rock property because it controls the directional movement and flow rate of the reservoir fluids in the formation. This rock characterization was first defined mathematically by Henry Darcy in 1856, so the equation that defines permeability in terms of measurable quantities is called Darcy's Law.

Darcy developed a fluid flow equation that has since become one of the standard mathematical tools of the petroleum engineer. If a horizontal linear flow of an incompressible fluid is established through a core sample of length L and a cross-section of area A, then the governing fluid flow equation is defined as

$$v = -\frac{k}{\mu}\frac{dp}{dL} \tag{6.31}$$

where

v = apparent fluid flowing velocity, cm/sec

k = proportionality constant, or permeability, Darcys

μ = viscosity of the flowing fluid, cp

dp/dL = pressure drop per unit length, atm/cm

The velocity, v, in Equation 6.31 is not the actual velocity of the flowing fluid but is the apparent velocity (Darcy velocity) determined by dividing the flow rate by the cross-sectional area across which fluid is flowing. Substituting the relationship, q/A, in place of v in Equation 6.31 and solving for q results in

$$q = -\frac{kA}{\mu}\frac{dp}{dL} \qquad (6.32)$$

where

q = flow rate through the porous media, cm³/sec

A = cross-sectional area accross which flow occurs, cm²

With a flow rate of 1 cm³/s across a cross-sectional area of 1 cm², with a fluid of one centipoise viscosity and a pressure gradient at one atmosphere per centimeter of length, it is obvious that k is unity. For the units described above, k has been arbitrarily assigned a unit called a Darcy. Thus, when all other parts of Equation 6.32 have values of unity, k has a value of one Darcy. One Darcy is a relatively high permeability as the permeabilities of most reservoir rocks are less than one Darcy. In order to avoid the use of fractions in describing permeabilities, the term millidarcy is used. As the term indicates, one millidarcy (1 md), is equal to one-thousandth of one Darcy or

$$1 \text{ Darcy} = 1000 \text{ md}$$

The negative sign in Equation 6.32 is necessary as the pressure increases in one direction while the length increases in the opposite direction. Equation 6.32 can be integrated when the geometry of the system through which the fluid flows is known. For the simple linear system shown in Figure 6-13, the integration is performed as

$$q\int_{0}^{L} dL = -\frac{kA}{\mu}\int_{p_1}^{p_2} dp$$

Integrating the above expression yields

$$qL = -\frac{kA}{\mu}(p_2 - p_1)$$

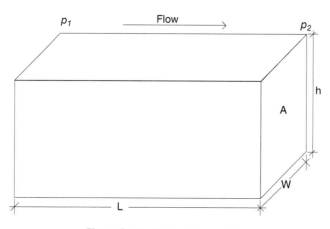

Figure 6-13   Linear flow model.

It should be pointed out that the volumetric flow rate, q, is constant for liquids because the density does not change significantly with pressure. Since p_1 is greater than p_2, the pressure terms can be rearranged, which will eliminate the negative term in the equation. The resulting equation is

$$q = \frac{kA}{\mu L}(p_1 - p_2) \tag{6.33}$$

Equation 6.33 is the conventional linear flow equation used in fluid flow calculations. Standard laboratory analysis procedures will generally provide reliable data on permeability of core samples. If the rock is not homogeneous, the whole core analysis technique will yield more accurate results than the analysis of core plugs (small pieces cut from the core). Procedures that have been used for improving the accuracy of the permeability determination include cutting the core with an oil-based mud, employing a pressure-core barrel, and conducting the permeability tests with reservoir oil. Permeability is reduced by overburden pressure, and this factor should be considered in estimating permeability of the reservoir rock in deep wells because permeability is an isotropic property of porous rock in some defined regions of the system, that is, it is directional. Routine core analysis is generally concerned with plug samples drilled parallel to bedding planes and, hence, parallel to direction of flow in the reservoir. These yield horizontal permeabilities (k_h). The measured permeability in plugs that are drilled perpendicular to bedding planes are referred to as vertical permeability (k_v). The factors that must be considered as possible sources of error in determining reservoir permeability are:

- Core sample may not be representative of the reservoir rock because of reservoir heterogeneity;
- Core recovery may be incomplete;
- Permeability of the core may be altered when it is cut, or when it is cleaned and dried in preparation for analysis. This problem is likely to occur when the rock contains reactive clays;
- Sampling process may be biased. There is a temptation to select the best parts of the core for analysis.

Permeability is measured by passing a fluid of known viscosity, μ, through a core plug of measured dimensions (A and L) and then measuring flow rate, q, and pressure drop, Δp. Solving Equation 6.33 for the permeability gives

$$k = \frac{q\mu L}{A\Delta p}$$

where

L = length of core, cm

A = cross-sectional area, cm^2

The following conditions must exist during the measurement of permeability:

- Laminar (viscous) flow
- No reaction between fluid and rock
- Only single phase present at 100% pore-space saturation.

This measured permeability at 100% saturation of a single phase is called the absolute permeability of the rock.

In order to overcome some of the difficulties associated with coring technique, Belhaj et al. (2006a, 2006b) have introduced a new line of permeameters that can eventually eliminate the need for coring, as measurements can be done *in situ*.

Saturation: Saturation is defined as that fraction, or percent, of the pore volume occupied by a particular fluid (oil, gas, or water). This property is expressed mathematically by the following relationship:

$$\text{Fluid saturation} = \frac{\text{total volume of the fluid}}{\text{pore volume}}$$

Applying the above mathematical concept of saturation to each reservoir fluid gives

$$S_o = \frac{\text{volume of oil}}{\text{pore volume}}$$

$$S_g = \frac{\text{volume of gas}}{\text{pore volume}}$$

$$S_w = \frac{\text{volume of water}}{\text{pore volume}}$$

where

S_o = oil saturation

S_g = gas saturation

S_w = water saturation

Thus, all saturation values are based on pore volume and not on the gross reservoir volume. The saturation of each individual phase ranges from 0 to 100%. By definition, the sum of the saturations is 100% (no void), therefore,

$$S_o + S_g + S_w = 1.0$$

The fluids in most reservoirs are believed to have reached a state of equilibrium and so will have become separated according to their density, i.e., oil overlain by gas and underlain by water. In addition to the bottom (or edge) water, there will be connate water distributed throughout the oil and gas zones. The water in these zones will have been reduced to some irreducible minimum. The forces retaining the water in the oil and gas zones are referred to as capillary forces because they are important only in pore spaces of capillary size. Connate (interstitial) water saturation, S_{wc}, is important primarily because it reduces the amount of space available between oil and gas. It is generally not uniformly distributed throughout the reservoir but varies with permeability, lithology, and height above the free water table. Another particular phase saturation of interest is called the critical saturation, associated with each reservoir fluid. The definition and the significance of the critical saturation for each phase is described below.

6.4.2.2 Special core analysis

Overburden Pressure: Overburden or lithostatic pressure is a term used in geology to denote the pressure imposed on a stratigraphic layer by the weight of overlying layers of material. It can also be described as the pressure regime in a stratigraphic unit that exhibits higher-than-hydrostatic pressure in its pore structure. If a given layer is in hydrostatic equilibrium, the lithostatic pressure at a depth, z, is given by

$$P(z) = P_O + g\int_0^z \rho(z)dz \tag{6.34}$$

where

$P(z)$ = overburden pressure at a depth z

$\rho(z)$ = density of the overlying rock at depth z

g = acceleration due to gravity

z = depth in meters

P_O = datum pressure, e.g., pressure at the surface

It is common for parts of stratigraphic layers to be isolated, or sealed, such that they are no longer in hydrostatic equilibrium. In such cases, the layer, or part of layer, is said to be in a condition of overpressure (if local pressure is greater than hydrostatic) or under pressure (if local pressure is less than hydrostatic). The physical process by which sediments are consolidated results in the reduction of pore space as grains are packed closer together. As layers of sediment accumulate, the ever increasing overburden pressures during burial cause compaction of the sediments, loss of pore fluids, and formation of rock as grains are welded or cemented together. Figure 6-14 shows the variation of pressure with depth.

Causes of overpressure: Overpressure in stratigraphic layers is fundamentally caused by the inability of connate pore fluids to escape, as the surrounding mineral matrix compacts under the lithostatic pressure caused by overlying layers. Fluid escape may be impeded by sealing of the compacting rock by surrounding impermeable layers (i.e., evaporites, chalk, and cemented sandstones). Alternatively, the rate of burial of the stratigraphic layer may be so great that the efflux of fluid is not sufficiently rapid to maintain hydrostatic pressure. It is extremely important to be able to diagnose overpressured units when drilling through them, as the drilling mud-weight must be adjusted to compensate. If it is not, there is a risk that the pressure difference down-well will cause a dramatic decompression of the overpressured layer and result in a blow-out at the well-head with possibly disastrous consequences. Details are found in Islam (2002)

Capillary Pressure: The capillary forces in a petroleum reservoir are the result of the combined effect of the surface and interfacial tensions of the rock and fluids, the pore size and geometry, and the wetting characteristics of the system. The mere presence of a surface indicates change in pressure. This pressure change is due to change in density, viscosity, interfacial tension (which itself is influenced by viscosity), gravity (magnitude depending on density), size of the pore (related to permeability), and others. Any curved surface between two immiscible fluids has the tendency to contract into the smallest possible area per unit volume. This is true whether the fluids are oil and water, water and gas (even air), or oil and gas.

When two immiscible fluids are in contact, a discontinuity in pressure exists between the two fluids, which depend upon the curvature of the interface, separating the fluids. This difference in pressure is called the capillary pressure and is referred to as P_C. Thus capillary pressure can be defined as the pressure difference

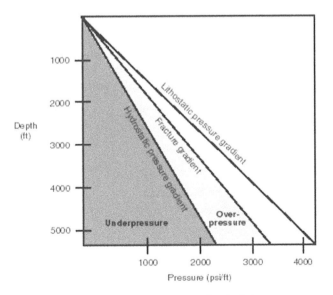

Figure 6-14 Pressure *vs* depth plot.

existing across the interface separating two immiscible fluids. It is related to wettability. The displacement of one fluid by another in the pores of a porous medium is either aided or opposed by the surface forces of capillary pressure. As a consequence, in order to maintain a porous medium partially saturated with nonwetting fluid and while the medium is also exposed to wetting fluid, it is necessary to maintain the pressure of the nonwetting fluid at a value greater than that in the wetting fluid. Denoting the pressure in the wetting fluid by P_W and that in the nonwetting fluid by P_{nw}, the capillary pressure can be expressed as

Capillary pressure = (pressure of the nonwetting phase) − (pressure of the wetting phase)

Mathematical derivations of capillary pressure are described below.

i) Capillary flow in a tube

Figure 6-15(a) shows a two-phase flow in a capillary tube with length, L and radius, r. The nonwetting fluid (colored black), covers a fraction, x/L of the tube while the wetting fluid (colored white) fills the rest of it. The pressures at the left- and the right-hand sides of the tube are P_i and P_j, respectively. P_{nw} and P_w define the capillary pressure due to the interface, i.e., $P_c = P_w - P_{nw}$. Figure 6.15(b) is a geometric representation of the curved interface with a wetting angle ϕ and principal curvature R.

Assume that a single pore can be treated as a thin cylindrical tube of fixed radius. On the basis of Darcy's equation and the knowledge of capillary pressure, we are able to derive a useful formula giving the local flow velocity. Consider horizontal laminar flow through a straight cylindrical tube of radius, r and length, L. The flow velocity, u through the tube is given by Darcy's equation (also analogous to Poiseuille law):

$$u = -\frac{k}{\mu}\frac{\Delta P}{L} \tag{6.35}$$

where,

$k = \dfrac{r^2}{8}$ = permeability of the tube (open channel equation)

μ = viscosity of the fluid

$\dfrac{\Delta P}{L}$ = pressure gradient across the tube

Suppose the tube contains two immiscible fluids, one wetting and the other nonwetting. Between the fluids there is a well-defined interface whose curvature gives rise to capillary pressure. The situation is shown in Figure 6-15 (a), where the pressure across the nonwetting and wetting fluids is defined as $\Delta P_{nw} = P_{nw} - P_i$ and $\Delta P_w = P_j - P_w$, respectively. The flow, u, constant through the whole tube and Darcy's equation and applied separately to the two liquids, gives (Lenormand et al. 1983, 1988)

$$u = -\frac{r^2}{8\mu_{nw}}\frac{\Delta P_{nw}}{x} \tag{6.36}$$

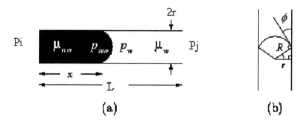

Figure 6-15 Capillary flow in a tube and its wettability. Redrawn from Lenormand et al. (1983).

$$u = -\frac{r^2}{8\mu_w}\frac{\Delta P_w}{L-x}$$ (6.37)

Here x denotes the position of the interface inside the tube and μ_{nw} and μ_w refer to the viscosities of the nonwetting and wetting fluid, respectively. The total pressure across the tube is given as $\Delta P = P_j - P_i$, where P_i and P_j are the pressures at the inlet and outlet (Figure 6.15(a)). P_{nw} and P_w are the pressures on the non-wetting and wetting side of the interface and the pressure difference across the interface defines the capillary pressure, P_c:

$$P_c \equiv P_w - P_{nw}$$ (6.38)

Now, the total pressure, ΔP across the tube can be written as

$$\Delta P = \Delta P_{nw} + \Delta P_w - P_{nw} + P_w$$
$$\Delta P = \Delta P_{nw} + \Delta P_w + P_c$$ (6.39)

Now substituting the value ΔP_{nw} and ΔP_w from Equations 6.38 and 6.39, we get, for the two liquids:

$$\Delta P = -\left[\frac{8\mu_{nw}xu}{r^2} + \frac{8\mu_w(L-x)u}{r^2}\right] + P_c$$

$$\Delta P = -\frac{8}{r^2}[\mu_{nw}x + \mu_w(L-x)]u + P_c$$

$$\Delta P - \Delta P_c = \frac{8L}{r^2}\left[\frac{\mu_{nw}x}{L} + \mu_w\left(1 - \frac{x}{L}\right)\right]u$$ (6.40)

$$u = -\frac{r^2}{8\mu_{eff}}\frac{\Delta P - P_c}{L}$$

where

$$\mu_{eff} = \frac{\mu_{nw}x}{L} + \mu_w\left(1 - \frac{x}{L}\right) = \text{effective viscosity due to the amount of the two fluids inside the tube.}$$

Equation 6.40 is often called the Washburn equation (Washburn 1921). The capillary pressure, P_c, defined in Equation 6.38, can be found exactly by simple geometrical calculations (Figure 6-15(b)). Let R denote the principal radius of curvature of the interface and ϕ indicate the wetting angle between the interface and the cylinder wall. Then $R = \frac{r}{\cos\phi}$, where r refers to the radius of the cylinder. Thus, the Young–Laplace equation with $R_1 = R_2 = R$ gives

$$P_c = \frac{2\gamma}{r}\cos\phi, \text{ where } \gamma = \text{interfacial tension}$$ (6.41)

The capillary pressure is derived under the assumption that the fluids are in static equilibrium, i.e., there is no flow in the tube and $\Delta P = P_c$. The dynamics of the moving interface is still unsolved theoretically, but at low velocities, when no turbulence occurs, the above expression is a sufficient approximation.

ii) Flow in a single tube

Figure 6-16 shows a typical situation for a single tube in the lattice. The capillary flow in a single tube contains a pore-interface. P_i and P_j denote the pressure at the ends, while P2 and P1 indicate the pressure at the left- and the right-hand side of the interface between the two fluids, respectively. ϕ denotes the wetting angle between the nonwetting fluid and the cylinder wall.

The volume flux q_{ij} from node i to node j in the tube are found from the Washburn equation (Washburn, 1921) for capillary flow with two fluids present:

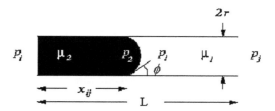

Figure 6-16   Capillary flow in a single tube.

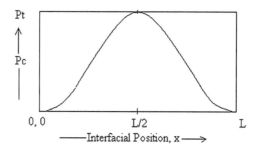

Figure 6-17   Capillary pressure dependence on interfacial position.

$$q_{ij} = -\frac{\pi r_{ij}^2 K_{ij}}{\mu_{eff}} \frac{(\Delta P_{ij} - P_c)}{L} \qquad (6.42)$$

where

$K_{ij} = \dfrac{r_{ij}^2}{8}$ = permeability of the rock

$\mu_{eff} = \mu_2 x_{ij} + \mu_1(1 - x_{ij})$ = weighted effective viscosity due to the two fluids

$\Delta P_{ij} = P_j - P_i$ = pressure difference between node i and j

$P_c = P_1 - P_2$ = capillary pressure in the tube ij

xij = position of the pore-interface in the tube

r_{ij} = radius of the tube

The position of the pore-interface is a continuous function in the range [0, 1] and the flow direction is given by Equation 6.42 (Figure 6-17).

This figure shows the capillary pressure P_c as a function of interfacial position x in that tube (Figure 6-17). In the middle of the tube, at x = L/2, the capillary pressure becomes equal to the threshold pressure P_t. For straight cylindrical tubes, the capillary pressure P_c is assumed to be constant. Let ϕ refer to the wetting angle (Figure 6.17), then according to Equation 6.41

$$P_c = \frac{2\gamma}{r} \cos \phi$$

where, r is the radius of the tube corresponding to the radius of curvature to the pore-interface. In a real porous medium consisting of pores with curved walls, the radius of curvature is a complicated function of the interfacial position. Thus, P_c depends on the local position of the pore-interface inside a single pore or pore throat. This variation in the capillary pressure results in perturbations in the total pressure across the sample, as the invading fluid covers new pores or even retires from invaded regions. To study the local pressure fluctuations due to the capillary pressure and the dynamics of the fluid movements, we apply a dependency in P_c and allow the pore-interfaces to move both in forward and backward directions. Let P_c

vary with the position of the pore-interface in each tube and assume perfect wetting ($\phi = 0$), then it can be defined (without index notations):

$$P_c = \frac{2\gamma}{r}\left[1 - \cos\left(2\pi\frac{x}{L}\right)\right] \tag{6.43}$$

Here r denotes the radius of the actual tube and x is the interfacial position in that tube (Figure 6-18). The definition sets the capillary pressure equal to the threshold pressure, P_t when the pore-interface is in the middle of the tube, i.e., $P_t = 2\frac{2\gamma}{r}$. The threshold pressure is the minimum capillary pressure required to allow the nonwetting fluid to invade the tube. The advantage of including a local dependency in the capillary pressure is to measure the fluctuations in the pressure evolution and study the interaction between viscous and capillary forces. In addition, allowing the fluids to move in both forward and the backward directions, gives a model physical properties that are closer to those of a real porous medium. Compared to the model of Lenormand et al., where each tube is assigned a constant capillary pressure and the fluids are only allowed to move in the direction of the displacements, this model becomes much more realistic.

Relative Permeability: Relative permeability is defined as the ratio of the effective permeability to a fluid at a given saturation to a base permeability. The base permeability is commonly taken as the effective permeability to the fluid at 100% saturation (absolute permeability) or the effective nonwetting phase permeability at irreducible wetting phase saturation. In the case of two or more fluids flowing simultaneously through a porous medium, a relative permeability for each of the fluids can be defined. It describes the extent to which one fluid is hindered by the other. The relative permeability is defined by setting up the Darcy equation individually for each phase, i, that flows in the pore space:

$$q_i = \left(\frac{Kk_{ri}}{\mu_i}\right)A\frac{\Delta P_i}{\Delta x} \tag{6.44}$$

where

q_i = flow rate of phase i

K = permeability of the porous system

k_{ri} = relative permeability of phase i

μ_i = viscosity of phase i

ΔP_i = pressure drop within phase i

$\frac{Kk_{ri}}{\mu_i}$ = called as "mobility" of phase i.

Relative permeability is one of the most important petrophysical parameters in describing multiphase flow through porous media. With increasing interest in use of thermal oil recovery methods, an understanding of the temperature effects on relative permeability curves is essential in modeling these types of recovery processes. Many researchers have investigated the temperature effects on relative permeabilities of high tension systems. However, many of the reported results are contradictory. Only one steady-state relative permeability result has been presented at elevated temperatures. Work on the isolation of temperature as a factor on relative permeability started in the early 1960s.

Edmondson (1965) performed hot waterflood tests in Berea Sandstone using both crude and refined oils. His results showed a change in the relative permeability ratio accompanied by a decrease in residual oil saturation with increasing temperature. He found that the changes in relative permeability ratios were different for different oils and assumed that the irreducible water saturation did not change with temperature. No data was given for water saturations below 40%.

Davidson (1969) investigated the temperature dependence of relative permeability ratios by conducting isothermal displacement of white oil from a sand pack by distilled water, nitrogen, or steam. His results indicated that the ratio is temperature dependent at low and high water saturations and insensitive in the mid-saturation range. A reduction in S_{or} with temperature was also observed in his study. He explained his results through changes in interfacial properties.

Poston et al. (1970) conducted a series of dynamic displacement relative permeability measurements on unconsolidated sands at elevated temperatures. Their results showed an increase in S_{wir} with temperature increase accompanied with a decrease in S_{or}. They also found that relative permeabilities to both oil and water increased at higher temperatures.

Ehrlich (1970) studied the effect of temperature on relative permeability curves. He reported that with an increase in temperature water wettability increased as oil-water-rock contact angle decreased to zero. His model showed that the water-oil relative permeability ratio at a given saturation increased with temperature for low residual oil porous media (unconsolidated) and decreased with temperature for high residual oil materials (consolidated).

Lo and Fiungan (1973) reported the only steady-state relative permeability data at elevated temperatures. Their results on consolidated porous Teflon and Berea Sandstone indicated that the temperature effects were similar in both oil-wet and water-wet systems. They suggested that the change in viscosity ratio with temperature may be responsible for the observed changes in relative permeability curves.

Weinbrandt et al. (1975), working with consolidated sands, obtained similar results to those of Sinnokrot et al. (1971). Their relative permeability curves shifted toward increasing water saturation with increasing temperature. Increase in water wetness of the porous medium with temperature was given as the main reason for observed changes by most of these researchers.

Nakornthap et al. (1982) presented a mathematical model to describe the changes in relative permeability with temperature in a water-oil system based on previous findings.

Sufi et al. (1982) found that oil and water in unconsolidated clean Ottawa Sand were independent of temperature between 70°F and 186°F. They concluded that the decrease in "practical" S_{or} was due to the reduction in the viscosity ratio and the "apparent" increase in S_{wi} was the result of a decrease in the viscous force provided by oil which, in turn, is caused by the viscosity reduction with temperature. The results suggest that temperature effects may be a result of measurement difficulties and laboratory-scale phenomena rather than the actual flow behavior.

Maini et al. (1983) investigated the effect of temperature on horizontal and vertical relative permeabilities of heavy oil-water system. Their results showed an increase in S_{wir}, a decrease in S_{or} to a minimum value, and changes in relative permeability curves with increasing temperature.

Torabzadeh and Handy (1984) investigated the effects of temperature and interfacial tension on relative permeability. Their experimental results indicate that relative permeability curves are affected by temperature, especially at low interfacial tension. For the high-tension system, relative permeability to oil increased and relative permeability to water decreased at a given saturation, while residual oil saturation decreased and irreducible water saturation increased with increasing temperature. The water/oil relative permeability ratio decreased with temperature at a given saturation. These results suggest an increase in the preferential water wettability of sandstone with temperature.

Miller and Ramey (1985) performed dynamic displacement tests on unconsolidated and consolidated sand cores using water and a refined oil and observed no changes with temperature in either residual saturations or relative permeability curves. They did not present any relative permeability curves for consolidated Berea Sandstone. Only end point relative permeabilities and saturations were compared at 97°F and 199°F, which in fact showed some temperature effects in oil relative permeabilities at irreducible water saturations. Miller and Ramey (1985) postulated that the results of the previous researchers may have been affected by viscous instabilities, capillary end effects, and/or material balance measurement difficulties.

Watson and Ertekin's (1988) experimental results indicate that the irreducible water saturation increased and the residual oil saturation decreased with average temperature. As the injection temperature was increased, the computed values of oil and water relative permeability decreased. The difference in the temperature gradient resulted in variations in both the irreducible water saturation and residual oil saturation and suggested that changes in wettability occurred. Both oil and water relative permeability decreased at a higher rate with an increasing temperature gradient.

Polikar et al. (1990) conducted an experimental study on Athabasca bitumen/water. They found that Athabasca bitumen/water relative permeabilities revealed little or no temperature effect on the relative permeabilities to water and bitumen over a range of 100 to 250°C (212 to 482°F). It was determined that the oil-phase relative permeability curve was convex. Measured curves were also compared with those obtained by history matching.

Schembre et al. (2005) studied the thermal recovery processes for the effects of temperature on relative permeabilities. They used CT scanning technology and B-spline coefficients. They found that relative permeabilities are linked to the effect of temperature on surface forces and ultimately to rock-fluid interactions. The residual oil saturation and water relative permeability endpoint decreases as temperature increases. These results confirm that temperature increases the water wettability.

Lion et al. (2005) conducted their experimental program to characterize the effects of increasing temperature upon the hydraulic properties of cement-based materials. They generated sets of permeability data, varying temperature for a constant pressure.

Huang and Rudnicki (2006) studied the migration of deeply buried groundwater by using a numerical simulation based on laboratory test results. They showed the variation of permeability with pressure and temperature. They identified a graphical relationship between permeability and temperature (Figures 6-18 and 6-19). These figures show the permeability variation *vs* pressure and temperature.

The permeability decreases with increasing confining pressure for a fixed temperature (Figure 6-18). This decrease can be expressed in terms of a negative exponential function:

$$k = a_k k_o e^{-b_k(p'-p_o)}$$

where

a_k & b_k = simulation constant

p' = confining pressure

P_o = initial pressure

K_o = initial permeability, 1 std atm pressure

Figure 6-18 also shows that curves of k *vs* P_o tend to become flat with increasing temperature, indicating that the increase of temperature diminishes the effect of pressure on the sample's permeability.

Figure 6-19 demonstrates that sandstone permeability decreases with increasing temperature at fixed confining pressure, and confirms that increasing pressure reduces the permeability. This behavior indicates that the increasing pressure dominates the effect of temperature on the sandstone's permeability. These two characteristics can be attributed to two physical processes (Figures 6-18 and 6.19). However, the solid skeleton of the sandstone undergoes thermal expansion with increasing temperature. Thus the seepage passageways between pores are compressed, and consequently the sandstone permeability decreases with increasing temperature. Also, increasing pressure causes the compressibility of pores to decrease and flow passages to become more restricted. As a result, the compressive effect of pressure and thermal expansion on pores is diminished.

An empirical correlation has been derived based on the experimental data of Lion et al. (2005). Figure 6.20 shows the variation of permeability with time. Zharikova et al. (2003) also found the same pattern in permeability and temperature plotting.

The empirical model based on experimental data is shown by Equation 6.45, describing the relation between permeability and temperature:

$$k = 8.33 \times 10^{-22}T^2 - 9.33 \times 10^{-19}T + 1.3 \times 10^{-17} \tag{6.45}$$

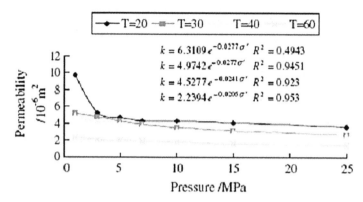

Figure 6-18   Sandstone permeability variation with pressure under different temperature levels. Redrawn from Huang and Rudnicki (2006).

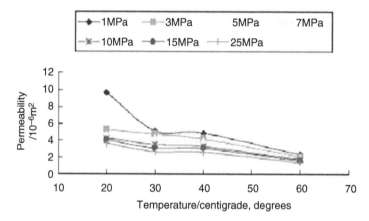

Figure 6-19   Sandstone permeability variation with temperature under different pressure level. Redrawn from Huang and Rudnicki (2006).

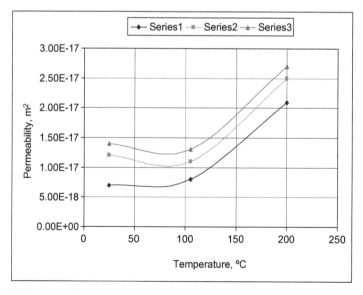

Figure 6-20   Dependence of permeability with time. Redrawn from Lion et al. (2005).

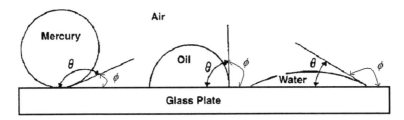

Figure 6-21   Illustration of wettability.

It is known that relative permeability can be defined as the ratio of effective permeability to absolute permeability. So, from the definition we can write

$$k_r = \frac{k_{eff}}{k}$$

(6.46)

Correlating both equations for relative permeability yields

$$\frac{k_{eff}}{k_r} = 8.33 \times 10^{-22} T^2 - 9.33 \times 10^{-19} T + 1.3 \times 10^{-17}$$

(6.47)

Wettability: Wettability is defined as the tendency of one fluid to spread on or adhere to a solid surface in the presence of other immiscible fluids. Wettability describes the relative preference of a rock to be covered by a certain phase. It also refers to the interaction between fluid and solid phases (Figure 6-21).

Small drops of three liquids, mercury, oil, and water, are placed on a clean glass plate. The three droplets are then observed from one side, as illustrated in Figure 6-21. It is noted that the mercury retains a spherical shape, the oil droplet develops an approximately hemispherical shape, but the water tends to spread over the glass surface. The tendency of a liquid to spread over the surface of a solid is an indication of the wetting characteristics of the liquid on the solid. This spreading tendency can be expressed more conveniently by measuring the angle of contact at the liquid-solid surface. This angle, which is always measured through the liquid to the solid, is called the contact angle, θ and the wetting angle, ϕ between the horizontal layer and the droplet interface defines the wettability of the liquid. The contact angle, θ has achieved significance as a measure of wettability.

As shown in Figure 6-21, as the contact angle decreases, the wetting characteristics of the liquid increase. Complete wettability is evidenced by a zero contact angle, and complete nonwetting by a contact angle of 180°. There have been various definitions of intermediate wettability but, in much of the published literature, contact angles of 60° to 90° will tend to repel the liquid. The wettability of reservoir rocks to the fluids is important in that the distribution of the fluids in the porous media is a function of wettability. Because of attractive forces, the wetting phase tends to occupy the smaller pores of the rock and the nonwetting phase occupies the more open channels.

Mathematically, the wettability of a liquid is defined as the contact angle between a droplet of the liquid in thermal equilibrium on a horizontal surface. Depending on the type of surface and liquid, the droplet may take a variety of shapes (Figure 6-21). The wetting angle, ϕ is given by the angle between the interface of the droplet and the horizontal surface. The liquid is seen to be wetting when $90 < \phi < 180°$ and nonwetting when $0 < \phi < 90$. $\phi = 0, 180°$ corresponds to perfect wetting and the drop spreads forming a film on the surface. The wetting angle, ϕ is a thermodynamic variable that depends on the interfacial tensions of the surfaces.

Let γ_{lg} denote the interfacial tension due to the liquid-gas surface, γ_{sl} refer to the interfacial tension due to the solid=liquid surface, and γ_{sg} indicate the interfacial tension of the solid-gas surface. In thermodynamic equilibrium, the wetting angle, ϕ is given by Young's law:

$$\gamma_{sg} = \gamma_{sl} + \gamma_{lg} \cos \phi \tag{6.48}$$

Basic reservoir properties such as relative permeability, capillary pressure, and resistivity depend strongly on wettability. Therefore, it is important that laboratory experiments in which these properties are measured are carried out on samples whose wettability is representative of the reservoir from which they are taken.

Surface and interfacial tension: In dealing with multiphase systems, it is necessary to consider the effect of the forces at the interface when two immiscible fluids are in contact. Consider two immiscible liquids – at the interface between the liquids there will be interactions between molecules of different types, with interfacial tension arising due to the attractive forces between the molecules in different fluids. Generally, the interfacial tension of a given liquid surface is measured by finding the force across any line on the surface divided by the length of the line segment. Thus, the interfacial tension becomes a force per unit length that is equal to the energy per surface area. When these two fluids are liquid and gas, the term surface tension is used to describe the forces acting on the interface. When the interface is between two liquids, the acting forces are called interfacial tension. Surfaces of liquids are usually blanketed with what acts as a thin film. Although this apparent film possesses little strength, it nevertheless acts like a thin membrane and resists being broken. This is believed to be caused by attraction between molecules within a given system. All molecules are attracted to one another in proportion to the product of their masses and inversely as the square of the distance between them.

The pressure difference across the interface between points 1 and 2 is essentially the capillary pressure (Figure 6-22), i.e.:

$$P_c = P_1 - P_2 \tag{6.49}$$

The nonwetting fluid will flow forward into the tube until the total force acting to pull the liquid forward is balanced by the weight of the column of liquid being supported in the tube. Assuming the radius of the capillary tube is r, the total forward force $F_{forward}$, which holds the liquid up, is equal to the force per unit length of surface times the total length of surface, or

$$F_{forward} = (2\pi r)(\sigma_{nww})(\cos \phi) = 2\pi r \sigma_{nww} \cos \phi \tag{6.50}$$

where

σ_{nww} = surface tension between nonwetting and wetting fluid, dynes/cm

ϕ = wetting angle

r = radius, cm

The forward force is counteracted by the weight of the fluids, which is equivalent to a backward force per unit length of mass times acceleration, or

$$F_{backward} = (\pi r^2)(\rho_w - \rho_{nw})g = \pi g r^2(\rho_w - \rho_{nw}) \tag{6.51}$$

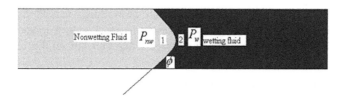

Figure 6-22   Pressure relation for two immiscible fluids.

where

g = acceleration due to gravity, cm/sec^2

ρ_{nw} = density of nonwetting fluid, gm/cm^3

ρ_w = density of wetting fluid, gm/cm^3

Now, equating Equation 6.50 with Equation 6.51

$$2\pi r \sigma_{nww} \cos \phi = \pi g r^2 (\rho_w - \rho_{nw})$$

or

$$\sigma_{nww} = \frac{g r (\rho_w - \rho_{nw})}{2 \cos \theta} \tag{6.52}$$

It is known that the backward force of the two fluids per unit area is the pressure difference between the nonwetting and wetting fluids. So equating Equations 6.49 and 6.51, one dotains:

$$P_c = -\frac{F_{backward}}{Area} = -\frac{\pi g r^2 (\rho_{nw} - \rho_w)}{\pi r^2} = g(\rho_w - \rho_{nw}) \tag{6.53}$$

Putting the value of $g(\rho_w - \rho_{nw})$ from Equation 6.52 into Equation 6.53:

$$P_c = \frac{2\sigma_{nww} \cos \theta}{r} \tag{6.54}$$

This is the fluid–fluid interface model to characterize the fluid behavior. Many researchers/authors (i.e., Hill 1952; Green and Willhite 1998; Boyd et al. 2006) have used the following mathematical model to describe the fluid–fluid interface in porous media. The pressure in each phase can be determined by application of Darcy's equation across the interface (Figure 6-23):

$$p_d - p_D = g(\rho_d - \rho_D)\Delta L \sin\alpha - \left(\frac{\mu_d}{k_d} - \frac{\mu_D}{k_D}\right)u\Delta L \tag{6.55}$$

where

P_d = displaced fluid pressure

P_D = displacing fluid pressure

g = gravity constant

α = inclination of flow relative to the horizontal axis (dip angle)

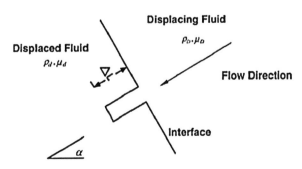

Figure 6-23 Model for determination of interface stability criterion in dipping reservoir.

ρ_d = displaced fluid density

ρ_D = displacing fluid density

$k_d = k_D$ = effective permeability

μ_d = displaced fluid viscosity

μ_D = displacing fluid viscosity

u = fluid velocity

ΔL = length of the perturbation

A finger remains stable at the interface when the pressure in the displaced phase is greater than the displacing phase, or $P_d > P_D$. If $P_d > P_D$, the interface remains stable. If $P_d < P_D$, the interface is considered unstable and the magnitude of the pressure difference indicates the potential for fingering. For miscible displacement, effective permeability is equal to intrinsic permeability, k, such that $k_d = k_D = k$. For horizontal flow, ($\alpha = 0$), the stability criterion is not influenced by gravity but determined by viscous forces.

Now, consider a spherical water droplet in equilibrium with a surrounding vapor when gravitational forces are neglected – the curved surface leads to a pressure difference between the water phase and the gas. The pressure inside the droplet on the concave side of the surface is expected to exceed the pressure on the convex side. If γ denotes the interfacial tension of the droplet surface, then the free energy of the surface is given by $4\pi R^2 \gamma$, where R is the radius of the droplet. Assume that the radius of the droplet is increased by dR because of molecules condensing on the surface. The corresponding surface free energy increases with $4 * 2\pi R\gamma * dR = 8\pi R\gamma dR$. This energy has to be balanced by the pressure forces between the droplet and the vapor. Let $*DP > 0$ denote the pressure difference between the droplet and the vapor. Then the work required to increase the size of the droplet by dR becomes $\Delta P * 4 * \pi R^2 dR = 8\pi R\gamma dR$. Thus, the pressure between the surface of the droplet and the gas becomes

$$\Delta P = \frac{2\gamma}{R} \tag{6.56}$$

This is called the Young–Laplace equation and has the general form:

$$\Delta P = \gamma\left(\frac{1}{R_1} + \frac{1}{R_2}\right) \tag{6.57}$$

where

R_1, R_2 = principal radii of curvature of the interface

In the above example, the droplet was assumed to be a perfect sphere, with $R_1 = R_2 = R$. For two-phase flow in porous media the interfacial tension of curved pore interfaces gives rise to a capillary pressure, P_c between the two liquids. At pore level, the curvature of the interface is often assumed to be equal to the pore size, denoted by a. Thus, the capillary pressure between the fluids in a pore of size, a is approximately $\frac{2\gamma}{a}$.

6.5   Practical Guidelines

6.5.1   *Core analysis*

The problem in core analysis has to do with the sample size. Much has been discussed regarding the representative elemental volume (REV). However, little has been done to address the problem of inadequate sample size. Islam (2002) pointed out the nature of the problem and suggested that the problem is particularly acute for fractured formations. Figure 6-24 shows how REV of a homogeneous formation (denoted as REV_{hom}) is much smaller than that of a fractured formation (denoted as REV_{frac}). This means the cores for fractured formations should be much larger than those extracted from homogeneous formations. Apart from

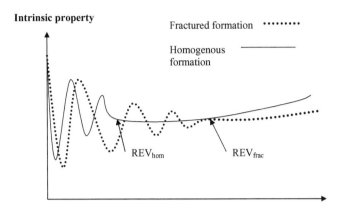

Figure 6-24 The representative elemental volumes (REV) are not the same for homogenous and fractured formations.

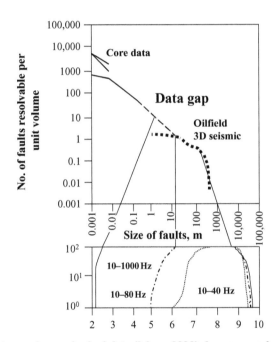

Figure 6-25 Data gap in geophysical data (Islam, 2002). Longer core is not the solution.

the fact that most reservoirs are heterogeneous (some fractured), it is important to note that the petroleum industry does not have the protocol of extracting larger samples from heterogeneous formations. Some researchers have attempted to address this problem by using longer samples. Unfortunately, longer samples are harder to extract, especially from fractured formations and, more importantly, are not the solution to the problem of REV. It is important to understand that REV cannot be achieved by merely increasing one dimension. All three dimensions have to be included in the elemental volume.

Figure 6-25 shows how a data gap exists in petrophysical data. However, few researchers realize that the problems within core samples are rarely addressed. If longer cores are not the solution, investigations should focus on measuring petrophysical properties *in situ*. An entire line of research work has been launched by our research group in order to develop a number of tools capable of measuring *in-situ* properties (Frempong

Figure 6-26   Experimental set-up for measuring rock fluid properties.

2005; Frempong et al. 2004; Nouri et al. 2003). Initial studies involve the study of wave propagations with a variety of cores under different pressure conditions, using a triaxial device (Figure 6.26).

In another series of experiments, we investigated the possibility of using different types of energy sources to determine the rock/fluid properties *in situ* (Zaman et al. 2004). In the following section, a new line of permeameters are introduced. These are likely to be operating *in situ* in the near future. Note that measuring rock fluid properties is the best remedy to the problems encountered in core analysis.

The inescapable issue of the Representative Elemental Volume (REV) is, "how big is enough?" This issue is not new, as all tested samples are subjected to this crucial definition of the REV, but in the case of permeability measurement using the Three-Dimensional Spot Gas permeameter, it brings special concern. When the sample is not homogeneous, a common situation, the larger the sample the more representative is the permeability measurement. However, when the sample gets larger than a critical size, with a small pressure-drop imposed by the permeameter, we know that the sample characteristics beyond the critical size are not making any difference. Therefore, the measurement is mainly reflecting the properties of the critical size of the sample. Of course, the characteristics of the media in general, specifically permeability and porosity, play a major role in defining the critical size, which is mainly governed by its particular properties. Recently, it has been reported that critical sample size can be affected by how far the sample boundaries are from the probe location. Boundaries located within 2.5 to 4.0 times the inner radii of the probe tip may skew its response by as much as 10%.

6.5.2.   A novel permeameter

This section introduces new permeability equipment and presents a new understanding in reservoir permeability that reflects the true behavior of fluid flow around producing wells. The traditional use of horizontal, vertical, and directional permeabilities to reflect the conductivity of a formation to fluid flow is often misleading. The flow comes from everywhere in the reservoir and reduces to the wellbore, and in many cases ends at the perforations. The flow pattern takes the shape of a cone where the base is at the boundary and the head is at the wellbore or the perforation opening. Thus, this flow pattern produces a conical or "tapering" permeability. This new 3D permeability term should enhance the accuracy of models used to represent fluid flow in porous media.

A 3D permeability term along with a device and techniques for measuring it is introduced. A 3D spot gas permeameter has been constructed in the laboratory. This device measures gas permeability with ease and precision, at any spot on the surface of the sample, regardless of the sample's shape and size.

The issues of probe sealing and gas slippage have been resolved by introducing a rubber backer at the tip of the probe and allowing low pressure injection. A new mathematical model has been derived to describe the kind of flow pattern taking place when measuring gas permeability using the proposed gas permeameter. The proposed mathematical model, along with the numerical solution presented, is expected to be used beyond the gas permeameter cases, as they prove more relevant to the reservoir behavior. This new 3D (tapering) permeability term should enhance the accuracy of the models used to represent fluid flow in porous media, with the 3D Spot Gas Permeameter and its methods of measurement likely to enjoy wide applications.

6.5.2.1 3D spot gas permeameter structure and testing procedure

The permeameter is simple in construction (Figure 6-27). The main component of the equipment is the X-Y-Z table, which allows the mobile head that contains the gas chamber and attached to the probe, to move forward/backward and upward/downward along the table where the samples are laid out. The gas chamber, which feeds the probe steadily with gas, is connected to a gas supply. The probe has a rubber end-tip to ensure it is leak-free of gas when the probe positioned on the surface of the sample and a designated force is applied. The opening diameter of the probe is 1.25 mm. A camera is installed with the probe to take a snapshot of the tested surface of the sample. Computer with specific software is part of the permeameter, mainly to control all movements and record all data automatically.

The sample can be of any shape and size. If the exposed surface of the sample is enough to place the probe (diameter 1.25 mm), then a permeability measurement can be performed. The sample has to be dry and free from any contaminants. Depending on the shape of the sample, it can be placed on the grooves made in the table, mostly to host cylindrical-shaped samples, or it may laid flat on the table for other shapes. The head can be moved in all directions and the camera is able to capture the image in realtime so that the spots where the permeability is required to be measured can be determined and the computer can record this information. Once this step is done, the permeameter can start operating. The head will move to the specified spots on the surface of the sample automatically and place the probe in the required position, and then apply a pre-set

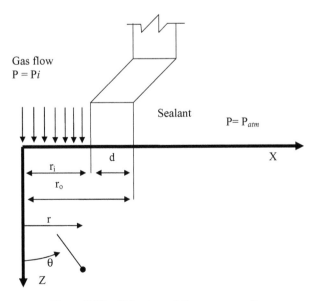

Figure 6-27   Side view of the permeameter.

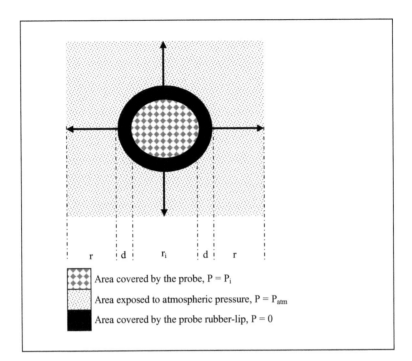

Area covered by the probe, $P = P_i$

Area exposed to atmospheric pressure, $P = P_{atm}$

Area covered by the probe rubber-lip, $P = 0$

Figure 6-28   Schematics of different flow-boundary assumptions made for the mathematical modeling.

force on the surface – enough to spread and seal the rubber to the sample surface, while not exceeding a certain damaging-stress on the sample – and then the gas will be released through the sample (Figure 6-28). The gas will flow into the sample until steady-state is reached. Only then can the mass flow rate and the mass flow pressure of the gas, measured by the permeameter, be recorded. Before the probe touches the sample, an image of the spot will be taken and stored as well.

The input mass flow rate and the mass flow pressure are the main parameters used to calculate permeability (Figure 6-28). Unlike the well-known Darcy permeability testing, this procedure does not depend on a specific sample geometry, length, or cross-sectional area. Rather, the minimum cross-sectional area is the size of the probe opening (radius 1.25 mm) and the maximum is the area of the end sides of the sample where the flow converges to atmospheric pressure. The flow is hemispherical in pattern initiated at the surface of the sample as a small spot (probe size) and dissipated in all directions across the sample. By determining the inlet mass flow rate and inlet mass flow pressure, the permeability can be calculated. The outlet pressure is always atmospheric. The cross-sectional area can be taken as the average area between the size of the probe and the end size of the sample.

6.5.2.2   Mathematical formulation

In order to formulate a suitable mathematical model to describe flow behavior within a rock sample, it is necessary to understand the specifics of such flow. Three main equations, Darcy law, the continuity equation, and the equation of state, will set the cornerstone of the 3D Gas-Spot Permeameter model (Figure 6-29). A plane view configuration of different samples' surface areas is shown in Figure 6-30.

To date, the empirical correlation introduced by Darcy is still used to demonstrate fluid flow in porous media. In its simplest form, this equation takes the form:

$$V = -\frac{k}{\mu}\nabla P \qquad (6.58)$$

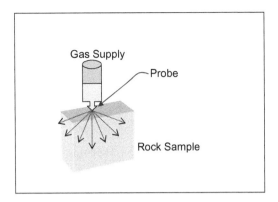

Figure 6-29   Schematic diagram of the permeameter main components.

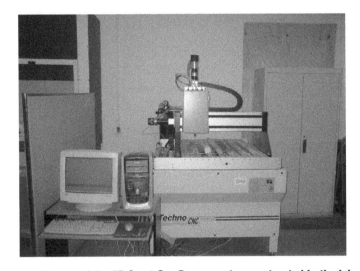

Figure 6-30   Picture of the 3D Sport Gas Permeameter constructed in the laboratory.

P is the pressure and V is the Darcy velocity. K and μ are the rock permeability and the gas dynamic viscosity, respectively.

We need the conservation of mass equation (continuity equation):

$$\nabla \cdot (\rho V) = -\phi \frac{\partial \rho}{\partial t} \tag{6.59}$$

ρ and ϕ are the gas density and the sample porosity, respectively. Also, the equation of state is required:

$$\frac{P}{P_o} = \left(\frac{\rho}{\rho_o} \right)^{\gamma} \tag{6.60}$$

P_o and ρ_o are a reference pressure (initial pressure) and reference density (density at initial pressure). The term denoted by γ is the ratio of the specific heat (c_p/c_v).

We can eliminate the velocity term by substituting Equation 6.58 into Equation 6.59:

$$\nabla \cdot \left[\frac{k}{\mu\phi} \rho \nabla P \right] = \frac{\partial \rho}{\partial t} \tag{6.61}$$

Substituting the value of P from Equation 6.60 into Equation 6.61, we get

$$\nabla \cdot \left[\frac{k\gamma P_o}{\mu\phi} \left(\frac{\rho}{\rho_o} \right)^{\gamma} \nabla \rho \right] = \frac{\partial \rho}{\partial t} \tag{6.62}$$

Equation 6.62 is a form of nonlinear diffusivity equation. Equation 6.62 can be rewritten as

$$\rho = \rho_o \left(\frac{P}{P_o} \right)^{1/\gamma} \tag{6.63}$$

Taking the derivative of Equation 6.63 with respect to time "t":

$$\frac{\partial \rho}{\partial t} = \frac{\rho_o}{P\gamma} \left(\frac{P}{P_o} \right)^{1/\gamma} \frac{\partial P}{\partial t} \tag{6.64}$$

Substituting Equation 6.63 into Equation 6.64 and rearranging:

$$\nabla \cdot \left[\frac{k}{\mu\phi} P^{1/\gamma} \nabla P \right] = \frac{1}{\gamma} P^{(1/\gamma - 1)} \frac{\partial P}{\partial t} \tag{6.65}$$

We assume an isothermal system action, which is a good assumption, since the laboratory temperature is almost constant for a particular test, and reservoir temperature is always considered unchanged unless thermal recovery is imposed. Validation of this assumption will make the term $\gamma = 1$, and consequently Equation 6.65 becomes

$$\nabla \cdot \left[\frac{k}{\mu\phi} P \nabla P \right] = \frac{\partial P}{\partial t} \tag{6.66}$$

Equation 6.9 may be rewritten in the form:

$$\nabla \cdot \left[\frac{k}{\mu\phi} \nabla P^2 \right] = \frac{1}{P} \frac{\partial P^2}{\partial t} \tag{6.67}$$

Equation 6.67 is the final form of the model used to demonstrate gas flow through the rock sample measured by the proposed permeameter. It is a nonlinear partial differential equation, so mathematical solution is difficult to achieve, therefore a numerical simulation scheme is used to solve this equation. The gas is assumed to behave as an ideal and the flow velocity is low enough to be in a range of laminar flow and also does not cause a significant slippage effect.

Figure 6-30 shows a configuration of the permeameter. The gas injection pressure is kept constant (after reaching steady-state). We can use the following transformation:

$$\Phi = \frac{P^2 - P_{atm}^2}{P_i^2 - P_{atm}^2} \tag{6.68}$$

Now, Equation 6.66 can be simplified:

$$\nabla \cdot \left[\frac{k}{\mu\phi} \nabla \Phi \right] = \frac{1}{P} \frac{\partial \Phi}{\partial t} \tag{6.69}$$

The term, ϕ is a non-dimensional potential. The pressure distribution within the system can be obtained as

$$P = P_{atm}\sqrt{\Phi(\psi_i^2 - 1) + 1} \tag{6.70}$$

where; $\psi_i = \dfrac{P_i^2}{P_{am}^2}$

The porous media is considered homogenous and the viscosity of the gas is assumed a function of temperature only.

Now, the spherical coordinate system may be applied, which is found to be representative of gas flow from the probe into the rock sample. The porons medium is considered homogeneous, therefore there are no change in characteristics in the azimuthal direction. Equation 6.70 takes the following form in spherical coordinates:

$$\frac{k_r}{\mu\phi}\frac{1}{r^2}\frac{\partial}{\partial r}\left(r^2\frac{\partial \Phi}{\partial r}\right) + \frac{k_\theta}{\mu\phi}\frac{1}{r^2\sin\theta}\frac{\partial}{\partial \theta}\left(\sin\theta\frac{\partial \Phi}{\partial \theta}\right) = \frac{1}{P_{atm}\sqrt{\Phi(\psi_i^2 - 1) + 1}}\frac{\partial \Phi}{\partial t} \tag{6.71}$$

In an effort to transform Equation 6.71 to a dimensionless form, we define the following quantities (Figure 6-27):

$$r^* = \frac{r}{r_i}, \quad \varepsilon = \frac{r_o}{r_i}, \quad t^* = \frac{t}{\tau}, \quad \text{and} \quad \kappa = \frac{k_\theta}{k_r}$$

where $\tau = \dfrac{\mu r_i^2}{P_{atm}k_r}$

Now, Equation 6.71 may be written in a dimensionless form as

$$\frac{1}{r^{*2}}\frac{\partial}{\partial r^*}\left(r^{*2}\frac{\partial \Phi}{\partial r^*}\right) + \frac{\kappa}{r^{*2}\sin\theta}\frac{\partial}{\partial \theta}\left(\sin\theta\frac{\partial \Phi}{\partial \theta}\right) = \frac{\varphi}{\sqrt{\Phi(\psi_i^2 - 1) + 1}}\frac{\partial \Phi}{\partial t^*} \tag{6.72}$$

r^* and t^* are the dimensionless radius and dimensionless time, respectively. In order to solve our final model, Equation 6.72, the following initial and boundary conditions are recognized:

$$\Phi(r^*, \theta, 0) = 0 \quad for \quad r^* > 0 \quad and \quad 0 \leq \theta \leq \frac{\pi}{2}$$

$$\Phi(r^*, \theta, t^*) = 1 \quad for \quad 0 \leq r^* \leq 1, \quad \theta = \frac{\pi}{2} \quad and \quad t^* > 0$$

$$\frac{\partial \Phi(r^*, \theta, t^*)}{\partial \theta} = 0 \quad for \quad 1 < r^* < \varepsilon, \quad \theta = \frac{\pi}{2} \quad and \quad t^* > 0$$

$$\Phi(r^*, \theta, t^*) = 0 \quad for \quad r^* \geq \varepsilon, \quad \theta = \frac{\pi}{2} \quad and \quad t^* > 0$$

$$\Phi(r^*, \theta, t^*) = 0 \quad as \quad r^* \to \infty, \quad 0 \leq \theta \leq \frac{\pi}{2} \quad and \quad t^* > 0$$

$$\frac{\partial \Phi(r^*, \theta, t^*)}{\partial \theta} = 0 \quad for \quad 0_i < r^* \leq \infty, \quad \theta = 0 \quad and \quad t^* > 0$$

6.5.2.3 Some results

Evaluation of any hydrocarbon accumulation in petroleum reservoirs is based on rock and fluid properties estimated directly or indirectly from those reservoirs. The quantity and the quality of collected data depend on the time of measurement, tool used, and testing cost. Permeability expresses the fluid conductivity of a certain rock and is among the important parameters used in reservoir evaluation and simulation studies. Laboratory core analysis, well-testing and, to a certain extent, well logging, are the traditional sources of permeability measurement.

Permeameters, known since the early days of the petroleum industry, have been used to determine permeability using either liquid or gas as flowing fluid. However, gas permeameters are the most commonly used and more favorable for a number of reasons. Among these reasons, gases are inexpensive, testing is easy, and more importantly the method is non-destructive. Rock sampling is a tedious and costly practice, and often obtained samples are multipurpose, so keeping them in a good shape (original status) assists in optimizing their use.

Only steady-state permeability measurements have been obtained by using the 3D gas permeameter, and the efforts of testing unsteady-state conditions is ongoing. So far, unsteady-state permeability measurements have not been conducted before using such novel equipment.

Goggin (1988) and Goggin et al. (1988) presented a technique, complete with numerical simulation to appropriately calculate permeability from the flow rate, the pressure in the probe, dimensions of the probe, air viscosity, and the atmospheric pressure at the outlet. Probe gas permeameters were first introduced by Dykstra and Parsons in 1950 to measure permeability at any spot surface of a rock sample. However, Goggin et al. (1988) formulated a mathematical model for such an application. Unlike Darcy's modeling law, the "pressure-volume gradient" is considered instead of the conventional pressure gradient. In fact, the pressure is more sensitive to the volume of the rock material that it passes through than to the 1D distance it traverses. If we can model Darcy's Law in such a fashion, then we can determine the permeability at the spot using the 3D Gas Permeameter. Furthermore, the measured permeability would reflect a 3D conductance (ability to allow fluids within) of the rock sample. This is not just the conventional horizontal, vertical, or any other directional permeability. This permeability is a transmissibility parameter expressing the flow ability of the media from all parts of the reservoir to the wellbore. The experimental aspect of this work focuses on measuring the permeability of core samples at a certain spot, thus the support volume was defined by the sensitivity of the geometric factor to the sample size. Specifically, the support volume was determined by numerically studying the convergence of the geometric factor for samples of increasing size to the geometric factor for the infinite half-space. Young (1989) tested the anisotropy effects on permeability measurements and defended the idea of the support volumes.

Suboor and Heller (1995) investigated the support volume of the mini-permeameter experimentally by conducting a series of measurements over a large sample of Berea Sandstone. An interesting part of this research considered the influence of both permeable and impermeable boundary conditions in an effort to emulate the influence of heterogeneities. Belhaj et al. (2006a, 2006b) considered the infinite boundary (permeable boundary) and a homogeneous media, since the permeameter is supposed to measure permeability at that spot. A homogeneous assumption is valid in such a case. The permeability is normally measured at many spots over the surface of a sample, and then a comprehensive permeability interpretation study takes place to evaluate the overall permeability of the sample. The infinite boundary assumption is also valid because the sample is usually large compared to the small volume surrounding the permeameter probe. Permeability will be much affected by that small volume rather than the whole sample volume. In the field, the reservoir boundaries are far apart within a specific well drainage area.

Figure 6.31 shows the pressure behavior with respect to the radius, r at different angles, "θ". The maximum pressure is at the centre (within the probe radius) and going outward from the probe. Inside the sample, the pressure declines until it reaches a minimum (atmospheric pressure) at the boundary. Since the sample is assumed homogenous, the maximum pressure profile is in the middle of the sample ($\theta = 0$). With time, the

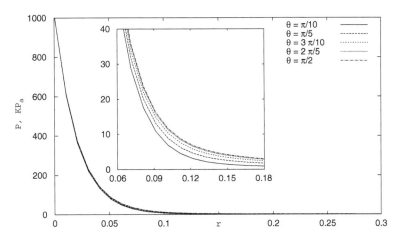

Figure 6-31   Pressure profile as predicted by the proposed model.

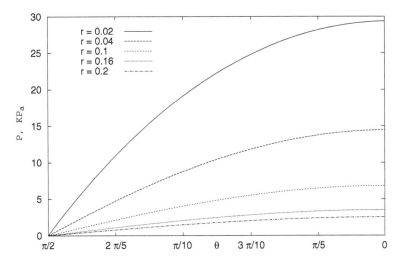

Figure 6-32   Pressure profile as predicted by the proposed model.

profile will eventually become a straight-line, extended from the maximum pressure at the probe (inlet pressure) to the far minimum pressure at the boundary (outlet pressure). This condition marks the steady-state action.

As seen in Figure 6.32, $\theta = \pi/2$, and the pressure is atmospheric at any radius, r, indicating a spot on the surface of the sample that indicates logical behavior and the maximum pressure happens at $\theta = 0$. The predictions of the model reflect the true behavior of fluid flow within the sample using the 3D Gas Spot Permeameter. Thus, it is safe to use the model to evaluate the permeability of the sample. The permeability parameter is an indication of the sample conductivity to fluid, and is neither vertical nor horizontal permeability. It is rather the conical or "tapering" conductivity of the media, because the flow is taking on a conical profile. This kind of flow scheme is more applicable to fluid flow from the reservoir into the wellbore.

The series of figures shown in Figure 6-33 show the dimensionless pressure distribution within the core.

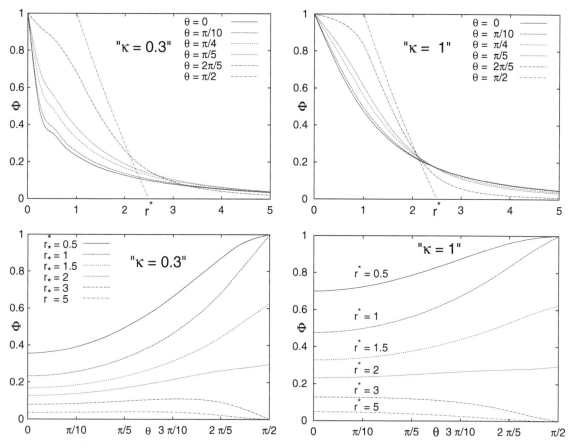

Figure 6-33   Dimensionless pressure as distributed in different sections of the core.

6.5.3   *Relative permeabilities that make sense in the field*

For decades, researchers have struggled with relative permeability data that always led to more optimistic results as compared to those in the field. Bentsen (1985) argued that such discrepancies can be explained through instability analysis. For the horizontal case, his instability number is given by

$$I_{sr} = \frac{\mu_w v(M-1)}{k_{wr}\sigma_e} \times \frac{M^{5/3}+1}{(M+1)(M^{1/3}+1)^2} \frac{4L_x^2 L_y^2}{L_x^2 + L_y^2} \tag{6.73}$$

Because of the dimension, if the mobility ratio, M is not less than 1, probably the instability number in the field will be large. This would put the field on a different scale when it comes to instability (Figure 6.34). Bentsen argued that for the same reason relative permeability graphs that are measured in the laboratory fall under the stable steady state regime as opposed to the pseudo-stable regime for the field. It becomes a problem as how to interpret laboratory results not representing the field regime. Islam and Bentsen (1986, 1987) also pointed out that all laboratory methods (until 1985) were using flow rates that would invariably make the flow regime stable and steady state. This, they argued, made it possible to use conventional steady-state and unsteady-state methods that required the flow regime to be both stable and stabilized. However, the dilemma arises as to the validity of the relative permeability curves in a different flow regime.

In order to answer this question, Islam (1985) conducted a series of laboratory experiments and measured relative permeability directly from the saturation profiles that were measured using microwave attenuation.

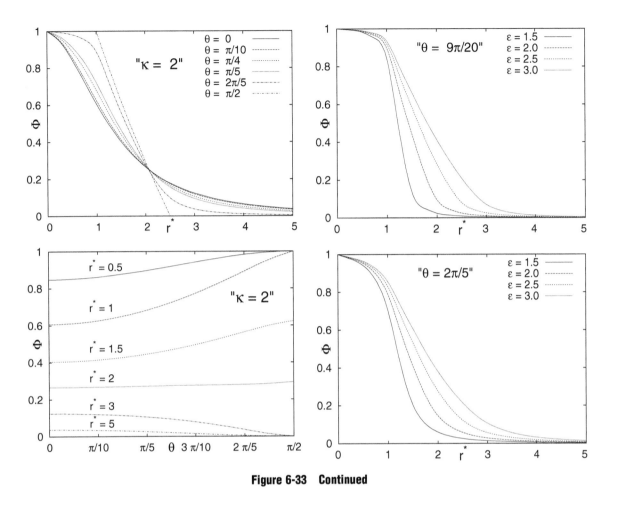

Figure 6-33   Continued

This allowed the calculation of relative permeabilities, even for unstable flow regimes. Islam showed that under unstable flow conditions, effective permeability to water is not affected. However, that of oil is affected significantly, depending on the instability number. Even though the limitations of this experimental set-up (high pressure could not be allowed) restricted the process within mildly unstable regimes, he postulated that with the newly introduced technique (reported as a dynamic method by Islam and Bentsen (1986), we can eventually determine relative permeabilities under pseudo-stable conditions. These would represent field conditions and modifications of the relative permeability graph during which further analysis (e.g., a problematic practice in history matching) would not be necessary. This relative permeability measuring technique was later used by Sarma and Bensen (1987), who further verified it. Later in 1992, Islam et al. demonstrated that the same graph is not applicable to gas oil relative permeability. For gas-oil, the pseudo-steady state was never reached (Figure 6-35). This required a renewed call to find new techniques for measuring relative permeabilities *in situ*.

Recently, our research group has developed a new permeameter that can be characterized as the first step toward developing *in situ* relative permeameters. Figure 6-36 shows the schematic of the set-up. Instead of using microwaves (Islam 1985), we have introduced ultrasonography. Because 3D ulstrasonography is now available, this set-up can be coupled with the 3D permeameter outlined above (Figure 6-36).

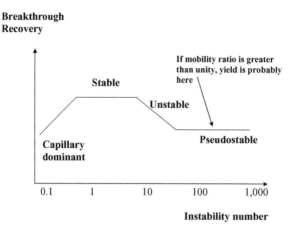

Figure 6-34   Breakthrough recovery *vs* instability number. Conventional analysis assumes that the yield and the laboratory fall under the same flow regime.

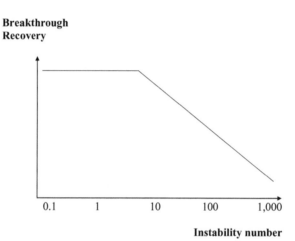

Figure 6-35   Breakthrough recovery during gas flooding. The pseudo-stable regime is never reached.

In-Situ Relative Permeameter

Index	Components
A	Ultrasound transducer
B	Signal amplifier
C	Ultrasound display
D	Well-casing
E	Pressurized well-cap with pressurized through fittings
F	Ultrasound co-axial cable
G	Milton Roy metering pump model RS11 12FRSESEALNN rated for 40 l/hr @ 1200 psi
H	Pulsation dampener
I	Pressure valve

Note: *Not to scale*

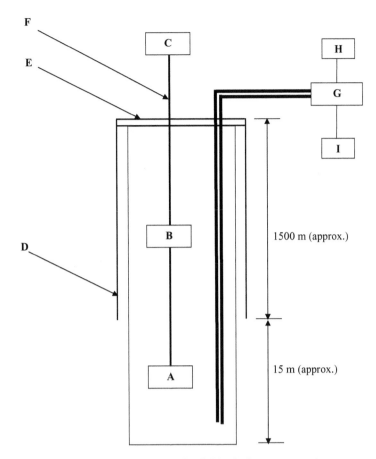

Figure 6-36 Schematic of a field relative permeameter.

6.5.4 Special guidelines for heterogeneous formations

It is well known that heterogeneous formations offer special challenges and that conventional reservoir engineering tools do not apply. Fluid flow is governed by permeability, which has a dimension of L^2, explaining why any heterogeneity affects fluid flow so drastically. This aspect is highlighted in Figure 6-37. Permeability is an even stronger function of a space variable, such as fracture frequency. This is because the presence of fractures adds to permeability without adding to porosity. It is the surface area that is most affected, increasing permeability unproportionately. However, such trends cannot continue as the frequency is increased to a higher value (Figure 6-38). The following steps are recommended for creating the database of a heterogeneous formation:

1. Decide on the type of average permeability to be used. It is important to have a clear geological picture of the formation. Depending on the origin of the formation, the direction of flow (in geological time), and orientation of fractures and fissures, different averages should be used. There are three options available, namely, arithmetic average, harmonic average, and geometric average, given below. A rule of thumb is to compare $hk_{well\ test}$ with hk_{core}. If they are comparable in general, use k_{arith}.

2. If hk_{core} is smaller than $hk_{well\ test}$, the formation is dominated by open fractures. Determine the following from available cores:

    a) fracture frequency

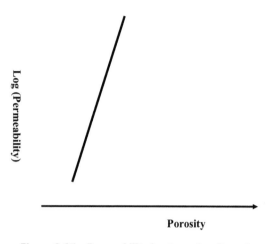

Figure 6-37   Permeability is strong function of space variables.

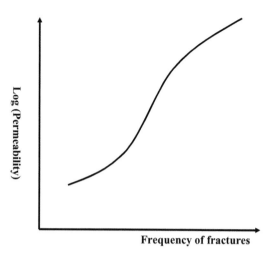

Figure 6-38   Permeability is a complex function of fracture frequencies.

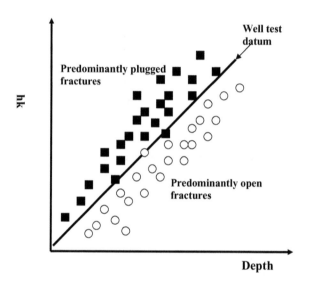

Figure 6-39   The location of core hk *vis a vis* well-test datum will indicate the nature of fractures.

b) fracture orientation
c) develop Rose diagram

$$k_{arith} = \frac{\sum\limits_{i=1}^{n} k_i h_i}{\sum\limits_{i=1}^{n} h_i}$$

$$k_{harm} = \frac{\sum\limits_{i=1}^{n} h_i}{\sum\limits_{i=1}^{n} h_i / k_i}$$

$$k_{geom} = \sqrt[n]{k_1 k_2 k_3 \ldots k_n}$$

3. If hk_{core} are generally larger than $hk_{well\ test}$, the formation is dominated by plugged fractures. Determine the following from available cores:
    a) plugged fracture frequency
    b) fracture orientation
    c) shale breaks and their orientation
    d) develop Rose diagram
4. Based on core data and fracture and/or shale break frequency, develop individual filters so that the resulting permeability data from cores correlate with those of the well test.
5. Use permeability magnitude of filtered data. Use direction of the Rose diagram in order to determine directionality of permeability tensor.
6. For very tight formations, consider using the Brinkman equation in cases of liquid-only flow, and the Forchheimer equation in cases of gas or oil-gas simultaneous flow.

Chapter 7

Enhanced Oil Recovery (EOR) Operations

7.1   Introduction

Historically, Enhanced Oil Recovery (EOR) refers to human intervention beyond waterflood or pressure maintenance gas injection. However, scientifically speaking, practically all recovery schemes beyond natural drive (e.g., solution gas, gravity drainage, aquifer-drive) should be considered as part of EOR. Because natural drive leads to steady decline, there has been a need to intervene in order to maintain the *status quo*. Natural flow also is called the *primary recovery* scheme. Historically, production decline can be contained by either increasing the number of wells or by introducing mass or energy into the reservoir in order to unlock additional oil. Water, being the most abundant fluid, was the first choice of "mass injection" to maintain pressure in the reservoir, pressure being the main drive for oil flow. This scheme is called the *secondary recovery* scheme, in conventional terms. Because water is also the most potent solvent, every waterflood scheme is also a chemical flooding process. Unaware of this fundamental fact, scientists introduced the concept of adding numerous chemicals in order to render waterflood into a chemical flood scheme. Because the word waterflood was synonymous with secondary recovery, the chemical flood scheme is called the *tertiary recovery* scheme.

The erosion of tax credits for tertiary recovery schemes made it difficult to justify from an economic standpoint, hence the term *Enhanced Oil Recovery* (EOR) was introduced. At the same time, the term tertiary did not make much sense when it came to thermal recovery that was successful, even though not necessarily preceded by a secondary recovery scheme. Thermal recovery schemes also were different from other schemes in the sense that they introduced the concept of injecting energy, often along with mass (steam injection as in steamflood, air injection as in *in-situ* combustion). Throughout the 1990s, because of the unprecedented dip in oil prices as well as the erosion of tax credits, EOR lost it economic appeal and the term *Improved Oil Recovery* (IOR) was coined. However, from a scientific point of view, there is no difference between secondary, tertiary, EOR, or IOR. They are all mass or energy introduction into the reservoir in order to increase oil production.

Islam (2000) defined EOR as a recovery scheme that uses the injection of fluids not normally present in the reservoir. For instance, chemical injection, steam injection, *in-situ* combustion, or even microbial enhanced recovery are considered as EOR. Three major categories of EOR have been commercially successful to varying degrees:

1. Thermal recovery, involves the introduction of heat, such as the injection of steam to lower the viscosity or thin the heavy viscous oil, and so improve its ability to flow through the reservoir. Thermal techniques account for over 50% of US EOR production (DOE 2004).

2. Gas injection uses gases such as natural gas, nitrogen, or carbon dioxide that expand in a reservoir to push additional oil to a production wellbore, or other gases that dissolve in the oil to lower its viscosity and improve its flow rate. Gas injection accounts for nearly 50% of US EOR (DOE 2004).

3. Most EOR techniques are based on oil viscosity reduction and/or improvement of the mobility ratio by increasing the displacement phase viscosity or by reducing oil viscosity and/or the interfacial tension between injected fluid and oil.

From the beginning of oil recovery, scientists have been puzzled by the huge amount of oil left following primary recovery. In most cases, recovery declines rapidly as viscosity of oil increases. For instance, primary recovery is less than 5% when oil viscosity exceeds 100,000 cp. It is well known that room for improvement in oil recovery is the greatest for heavy oil reservoirs (Islam 1999). Even though EOR has not been synonymous with heavy oil recovery, its history shows that thermal recovery schemes, predominantly EOR techniques for heavy oil recovery, have dominated incremental oil recovery attributed to EOR (Moritis 1998).

Even though thermal recovery would include several methods in addition to steam flooding, it remains by far the most widely successful thermal EOR technique. Recently it was realized that mobility control could be achieved by using surfactant with steam when steam /foam is generated. Ever since this discovery, most emerging technologies in steam flooding involve some kind of surfactant application. EOR in light oil reservoirs has mainly focused on surfactant and/or polymer injection.

Hundreds of patents have been issued for different forms of surfactant and polymer injection (in the form of surfactant, water flood, micellar flood, surfactant/polymer enhanced water flooding, etc.). Since the drop in oil prices in 1982, chemical EOR has been marked as too expensive. However, surfactants continue to play an important role in virtually all forms of successful EOR, be it in the form of foam mobility control in gas injection, steam foam mobility control in steam flooding, micellar, alkaline polymer flooding, or others.

Recent reports show that the contribution of EOR in total oil production has increased steadily throughout the last two decades, despite fluctuating oil prices (Figure 7-1A). Also the number of EOR projects has steadily declined since their peak in 1986. This shows more efficiency of EOR projects, indicating a trend that efficiency is the focus of future EOR schemes. This trend continued throughout the 1990s, mainly because the oil price was low. The previous forecast had been US $99/bbl for oil by 1999, but in reality, it was US $10/bbl. During this time, oil industries could not afford to experiment with EOR schemes that did not show immediate improvement in oil production. While this pressure to produce instant results emerges from a culture that finds its root in what has been termed "aphenomenal" by Zatzman and Islam (2006), it is also true that Enron-type models have been very much in control of the petroleum operation management. Oil and gas research were no exception and have benefited from research in both thermal and chemical processes (Butler and Mokrys 1991). Table 7.1 shows the profitability of various EOR options (Taber 1994).

Steamflood projects have enjoyed the most profitability. This trend continued even when the number of steamflood projects declined in the late 1990s and early 2000s. Steamflood projects are bound to be profitable because there is no other technique that can recover the very viscous oil. If a project is not profitable with steamflood, the oilfield is unlikely to be developed. The amount of *in-situ* combustion has declined drastically, but the profitability remains high. This is mainly because initial screening is only carried out with

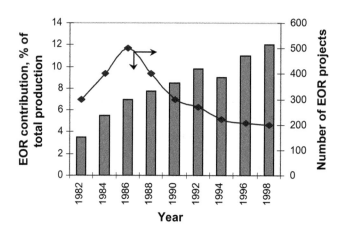

Figure 7-1A   Contribution of EOR as a percentage of US oil production. Data from Moritis (1998, 2006).

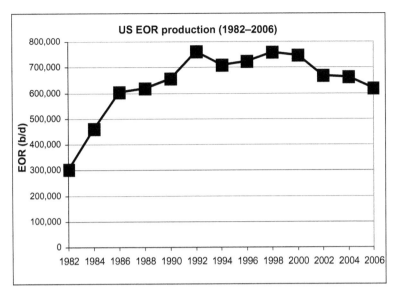

Figure 7-1B   Updated data on the contribution of EOR as percentage of US oil production. Data from *Oil and Gas Journal* **(2006).**

Table 7-1   Profitability of various EOR options (Taber 1994)

Method/Year	% Reported as Profitable			
	1982	1988	1990	1994
Steam	86	95	96	96
Combustion	65	78	88	80
Hot water	–	89	78	100
CO$_2$	21	66	81	81
Hydrocarbon	50	100	100	100
Nitrogen	100	100	100	100
Flue gas	100	100	100	–
Polymer	72	92	86	100
Micellar/Polymer	–	–	–	–
Alkaline/Surfactant	40	100	1 success	100

extreme caution. There are very few hot waterflood projects in place, which would explain why the profitability is so high. Islam et al. (1989) showed that many steamflood and even *in-situ* combustion projects are equivalent to hot water injection. This is particularly true for cold regions, some with permafrost (Wadadar and Islam 1994). Table 7-1 also shows the profitability of CO$_2$ injection has been increasing steadily. This trend will also continue in later years, even when the number of CO$_2$ projects continues to increase. The profitability of chemical floods is still high, mainly because there are only a few of them active.

Figure 7-1B shows the growth of different EOR schemes, drawing a particularly grim picture of the contribution of chemical flooding schemes. We must realize that this figure does not consider the chemical applications in thermal and other EOR schemes and that the scenario would probably change if they were so considered.

Figure 7-2 shows US EOR average daily production from January 1, 1984 (for 1983) through January 1, 2000 (for 1999). In 1999 (reported as of January 1, 2000), EOR processes contributed 748,000 barrels per

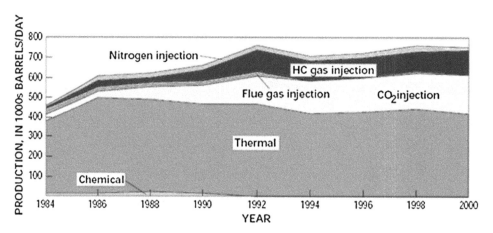

Figure 7-2   US EOR production using following methods: chemical, thermal, CO_2 injection, flue gas injection, miscible/immiscible hydrocarbon gas injection and nitrogen injection processes. Data from Moritis (2000, 2004).

day to US oil production or about 12% of US oil production, a percentage that is consistent with 1997 levels (reported as of January 1, 1998) (Moritis 2000).

By far the most important EOR scheme in the United States, and the world, has been thermal EOR. Obviously, the advantage gained by the exponential decrease in oil viscosity, due to the linear increase in temperature, has been the focal point of all thermal EOR schemes. Among thermal methods, steam injection has been the most dominant EOR scheme. The simplicity of the scheme and the unique latent heat properties of water are the major reasons why the oil industry has been active in steam flooding. With primary oil recovery being practically impossible, only huge heavy oil reserves are left as the target of the steam injection scheme. *In-situ* combustion, despite being popular in the 1960s, has lost its appeal in recent years. This is probably due to the fact that *in-situ* combustion is a better scheme for reservoirs that do not need *in-situ* combustion for improving oil recovery. Thermal recovery schemes continue to be prime targets for industry and academic research.

Despite the fact that the so-called chemical EOR scheme (micellar-polymer, polymer, surfactant, or caustic-alkaline process) has lost popularity in the United States, this scheme has enjoyed a great deal of industrial and academic research efforts in the last 30 years. It is likely that chemical EOR techniques will enjoy another 30 years of popularity throughout the rest of the world, as the light oil reserves of the Middle East become available to EOR applications. In fact, chemical EOR has been the major scheme among those suitable for light oil recovery in the United States. If chemical phenomena introduced in thermal recovery schemes (e.g., emulsification, froth, and foam generation) are considered, the impact of chemical EOR will be viewed as far more important than usually given credit for.

Emulsion flow in porous media is of interest in almost all EOR processes. Emulsions have been used as selectively plugging agents to improve oil recovery in water flooding, as well as chemical and steam flooding operations. It is also believed that emulsion flow occurs naturally in thermal processes such as *in-situ* combustion, steam flooding, chemical flooding, water flooding, or even in primary depletion (Doscher 1967). Natural porous media often provide enough energy to generate emulsions *in situ*. McAuliffe (1973a, 1973b) is one of the first researchers to report the use of oil-in-water emulsions in improving oil recovery during waterfloods. Broz et al. (1985) reported laboratory results in the development of an emulsion blocking technique for the correction and control of steam override and channeling.

Among gas injection schemes, by far the most important EOR scheme in the United States has been the miscible/immiscible gas injection. This scheme is followed in importance by carbon dioxide miscible flooding, which has started to gain popularity in some heavy oil reservoirs. If this trend continues, immiscible

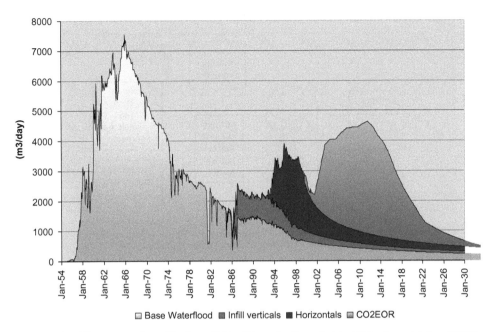

Figure 7-3   Oil recovery from Weyburn field of Canada. Reported by Mulcahy and Islam (2006), with data from EnCana.

carbon dioxide could well become the most dominant nonthermal EOR technique for heavy oil reservoirs. In 2000, Canada started a landmark project, the "Weyburn billion-dollar" miscible CO_2 injection project. This project has become the center of international attention, as it is the world's largest CO_2 storage project. This is also the first initiative that couples EOR with CO_2 sequestration as evidence of increased environmental awareness. The usual slogan of this project is:

> Not only does the CO_2 injected underground aid in oil recovery, but it is also equivalent to taking about 6.7 million cars off the road for one year over its life. Through its life, approximately 30 million tonnes of CO_2 will be stored underground, with some 8 million tonnes already injected.

However, the CO_2 is not collected from cars, but purchased from the Dakota Gasification Company (DGC) of in the United States, at an estimated \$35/ton, although the contract documents were unpublished (Mulcahy and Islam, 2006). Some 95 MMscf/d of CO_2 is transported by a 323-km pipeline, owned by DGC, to the Weyburn field in Southern Saskatchewan. The volume was expected to increase to 125 MMscf/d in 2006. Purchased CO_2 is combined with recycled CO_2 from the producing formation, and then injected into the formation. EnCana currently is using a 4D approach to measure the movement of CO_2 through the reservoir (Figure 7-3). In an internal report produced by Islam, scaled model studies showed that the breakthrough would occur much sooner than predicted. Production data on the field are not available, but the future of the project seems secure, mainly due to unexpectedly high oil prices. The original calculations were based on \$10/bbl oil prices.

Even though microbial EOR has received extraordinary attention from researchers around the world, this scheme has failed to receive any credit for any successful field application. Most field trials have been designed by microbiologists or other non-petroleum engineers. There are instances when nothing has been learned from a pilot test, mainly due to lack of proper engineering design (Jack et al. 1991). Even though successful microbial EOR results keep surfacing from European oilfields, the process cannot have any future unless petroleum engineers become more involved in MEOR research and development.

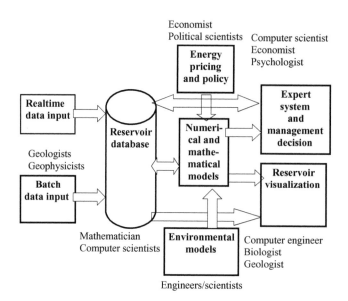

Figure 7-4   Various tasks and experts involved in a novel petroleum management project.

7.2   Contributions of Different Disciplines

So far, most research in EOR has been carried out by petroleum engineers. However, petroleum engineers with true petroleum engineering backgrounds have tended to take the so-called engineering approach, focusing mainly on macroscopic properties and overall improvement in productivity. While the credit for improving oil recovery goes mainly to petroleum engineers, so-called fundamental research has been dominated by chemical engineers and chemists. This is mainly due to the petroleum industry's impatience and the pressure on petroleum engineers to come up with quick solutions to problems facing the industry at any given time. This is the product of the Enron culture that promoted the concept of instant success. For some 30 years, Enron was considered the most creative energy company and until their collapse in 2001, they were considered to be the ideal of energy management. It is no surprise that virtually all new theories of fluid flow have been developed by chemical engineers, while petroleum engineers continued to excel in developing innovative ideas in field implementations. This aspect of numerical simulation of EOR schemes has been single-handedly addressed by petroleum engineers. In this regard, the role of chemical engineers has been similar to that of scientists in any discipline, whereas the petroleum engineers have carried the burden of engineering of EOR. Geologists and civil engineers have played only a small role in developing EOR schemes.

Over the last two decades, geologists seem to have been preoccupied with reservoir characterization, and their contribution appears to be in line with that of mathematicians and geo-statisticians. Civil engineers, on the other hand, have worked on material characterization in heavy oil recovery and environmental aspects of petroleum engineering. Once again, even with environmental aspects of EOR, petroleum engineers had to play a major role in process development and the economics of oil production. Microbiologists have contributed in the areas of microbial EOR. However, their contribution did not get past laboratory studies and microscopic studies of bacterial growth, leaving the entire engineering, including numerical simulation, to petroleum engineers.

Recently, the authors were involved in developing a comprehensive management tool for the petroleum industry. This was one of the most heavily funded and human resource-intensive project ever undertaken by a university researcher. The project had the objective of developing "the ultimate reservoir management tool" and employed over 30 researchers during its five-year execution time. The researchers came from all disciplines and the project generated numerous publications, about a dozen patent disclosures, and five books. The project graduated 10 PhDs and over 20 Masters, all completing their work as part of this project. The project aimed at developing new models in practically all aspects of petroleum engineering. The interdisciplinary nature of the project is in line with that of petroleum engineering itself. The project aimed at (original proposal written in 2000) replacing the Enron model with the so-called "Knowledge" model. It involved the solution of nonlinear equations with novel techniques to introducing attitude adjustment. It was postulated that by "humanizing the environment", short-term objectives would be covered while long-term would be secured. This five-year research period saw the collapse of Enron, World dot com, and a new world order that puts environmental considerations before short-term financial gains. With hundreds of publications and some half dozen books, it is not an exaggeration to state that the project was an astounding success, attributed to the interdisciplinary nature of the research group as well as the research theme. For petroleum engineering, thinking interdisciplinary has to come naturally.

7.3 Different EOR Techniques

Figure 7-5 shows the different EOR methods. In this classification, EOR methods are categorized into two major groups, one thermal and one non-thermal. Thermal categories are divided into four sub-categories, which include steam injection, hot waterflooding, *in-situ* combustion, and electrical heating. Steam injection and *in-situ* combustion are classified into other sub-groups. Nonthermal methods are chemical floods, miscible displacement, and gas drives. Each of the nonthermal groups is categorized into many sub-groups. All sub-categories are presented in the Figure 7-5. Figure 7-6 shows the contribution of various EOR recovery

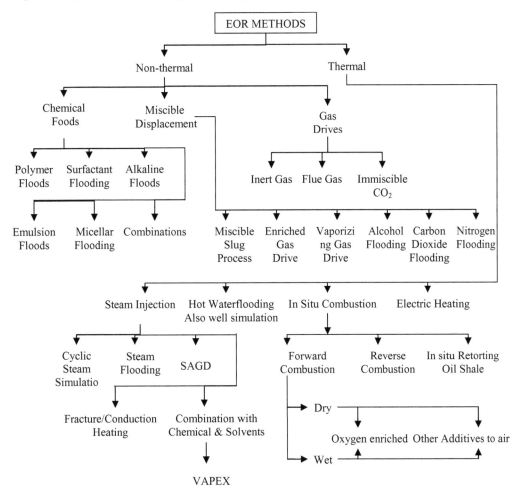

Figure 7-5   Different EOR methods (Farouq Ali 2004).

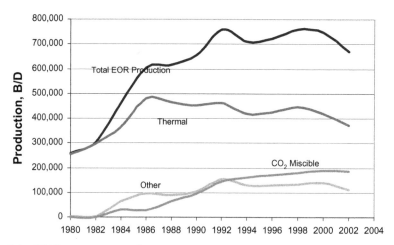

Figure 7-6   US EOR production (US oil production 7.9 × 106 B/D oil imports 12.4 × 106 B/D).
Source: Farouq Ali (2004).

schemes in US oil production. Note that the EOR contribution continued to increase until 1992. During this time, thermal recovery was very high, peaking in 1986. The following decline in thermal contribution did not trigger an overall decline because CO_2 contribution continued to rise. Throughout this period, oil prices remained much lower than predicted, and contributed to closure of new EOR schemes. Some of thermal projects reached maturity and did not show appreciable results despite the use of surfactant/steam and other improved schemes. Because CO_2 miscible also reached a plateau soon after the Enron collapse, the overall EOR took a dive after 2001. With increased oil prices, oil companies are struggling to put together their EOR research schemes again. This is bound to increase overall contributions. In the meantime, large tar sand schemes, such as in Alberta, Canada would definitely increase the overall world production that is attributed to EOR activities. The same applies to Canada's Weyburn CO_2 miscible flood that produced unexpectedly good results, mainly due to high oil prices.

7.3.1   Thermal EOR

Thermal recovery, CO_2 gas injection, and miscible/immiscible hydrocarbon gas injection are the EOR techniques that have contributed the most to field growth. At the end of 1999, 86 thermal recovery, 64 CO_2 injection, and 6 hydrocarbon miscible/immiscible gas injection projects were in operation in the United States (Moritis 2000). It is important to note that despite the decline in the number of EOR projects and the low price of oil during 1998–1999, EOR production was still a significant portion of the US total oil production (about 12%). Among the various thermal recovery techniques, steam flooding is the most widely used method. Of the various gas injection processes, the miscible CO_2 injection process has been a preferred option because it is usually more efficient in displacing oil than any other gas injection process. Miscible/immiscible hydrocarbon gas injection is technically complex and relatively more expensive. In the miscible gas injection process, the injection gas and the reservoir oil form a single phase, resulting in significantly higher oil recoveries. In the immiscible gas injection process, the injection gas and reservoir oil remain as two separate phases and the oil recoveries are not as high as in the miscible gas injection process.

Thermal recovery and CO_2 injection processes are technically and economically the most successful EOR techniques and will continue to play an important role in the field growth phenomenon. The key to expanding the use of the technically proven EOR processes is the reduction of implementation costs. One example is an innovative thermal recovery (steam-cogeneration) project in the Wilmington field, California, where steam was used first to produce electricity generating extra revenue and then was injected into reservoirs to mobilize viscous oil for better recovery (Moritis 1998).

Table 7-2 Canadian oil reserves. Source: NRCan (1997)

Source	Remaining Established		Remaining Potential	
	Reserve (million barrels)	Reserve/Production (Years)	Reserve (million barrels)	Reserve/Production (Years)
Conventional areas	3,710	7.6	12,600	26
Frontier	1,200		20,100	
Oil sands	3,615		306,520	

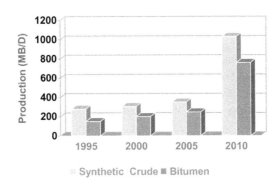

Figure 7-7 Synthetic crude and bitumen production from Canada's tar sand. Source: NRCan (1998).

Table 7-2 presents the estimation of Canadian oil reserve. While oil from conventional areas is not as plentiful, when frontier areas and oil sands are added to the reserve, Canada's oil potential becomes enormous. The following are the projections of energy production growth from Canada's Energy Outlook, 1996–2020 by NRCan (Natural Resources Canada) (NRCan 1997). Total oil production is forecast to increase from 1960 thousand barrels per day (MB/D) in 1995 to 2350 Mbbl/d by 2010. According to Canada's Energy Outlook, synthetic crude production from oil sands is anticipated to increase from 280 Mbbl/D in 1995 to 380 Mbbl/D in 2010, reaching 450 Mbbl/D by 2020. Similarly, bitumen production is expected to increase from 260 Mbbl/D in 1995 to 295 Mbbl/D in 2010. However, recently announced oil sands projects, if implemented, can potentially increase synthetic crude oil production to 1030 MB/D by 2010, an increase of 360% (Editorial 1999a). Similarly, bitumen production can potentially increase to 760 MB/D by 2010, another major increase (Editorial 1999b). Recent mega-projects in Alberta (Canada) involving tar sand have made sure that these expectations are met or even exceeded. Figure 7-7 shows the oil recovery projections from tar sand sources.

Canada's future in fossil fuel clearly lies in heavy oil and tar sand reserves (Islam et al. 2000). We can only imagine how beneficial it can be if techniques are developed to maximize oil recovered from fossil fuel sources that produce less than 5% through currently used technologies. Thermal recovery techniques are equally useful for producing from tar sand and bitumen. A number of commercially successful projects have been reported. The most prominent recovery technique is the so-called Steam-Assisted Gravity Drainage (SAGD), which has been synonymous with thermal recovery of tar sand and bitumen (Edmunds 1999). This process can be described as the main prize in Canadian heavy oil and tar sand research. Several varieties of SAGD have been tested, including the so-called VAPEX, best characterized as the solvent extraction method (Butler and Mokyr 1991). The greatest shortcoming of VAPEX, or any other chemical-dependent technique, continues to be the loss of chemicals that can be very costly. Even after recovering 99% of the solvent, the technique can be an economic failure (Farouq Ali 1997).

7.3.2   Steam injection

The steam injection process involves conversion of scale-free water to high-quality steam of about 232°C temperature and at a pressure higher than the corresponding saturation pressure (DeMirji 1978). Generally, by using directly fired heaters, water is converted to steam. By using insulated distribution lines, steam is transported to various injection wells. Steam injection can be carried out by two different methods, steam stimulation and steam displacement.

In the stimulation method, a predetermined volume of steam is injected into the well, which is then closed to stimulate the wellbore area. After a few days of closure, the well starts to produce. If necessary, the stimulation process is repeated. In the steam displacement process, continuous injection of steam is usually applied at lower rates. The steam is injected in place, as to distance and direction from production wells. Steam injection is a highly sophisticated process and requires extensive engineering and analytical inputs.

It is estimated that there are 85 to 110 billion barrels of heavy oil reserves in the United States. Since the 1960s, steam has become the predominant EOR method for these high viscosity, heavy oil reservoirs world-wide. However, factors such as steam-channeling, gravity segregation, and reservoir heterogeneity often result in poor contact of the heavy oil formation by the injected steam, leading to low recoveries. One method of conformance control that has received considerable attention is the use of surfactant foams that reduce steam mobility. Numerous laboratory and technically successful field studies have been reported (Hirasaki 1989; Castanier 1989; Mohammad 1989).

For a long time, steam has been used as the driving fluid in heavy oil reservoirs. The steam injection scheme has been popular because of its simplicity. However, steam injection leads to an unfavorable mobility ratio in most applications. Besides, gravity lay over is a problem with most reservoirs with little or no dip. Injected steam, because of its low density, rises to the top of the reservoir and tends to form a channel beneath the cap rock to the production well. Early steam breakthrough can occur at producing wells owing to gravity override, channeling, and viscous fingering, resulting in low oil recovery efficiency. Because of high steam mobility, there is little pressure differential between injector and producer once steam breakthrough occurs. The majority of subsequently injected steam follows this established path of least resistance and the process efficiency is impaired. Injecting surfactants to generate foam *in situ* can reduce steam mobility and improve the volumetric sweep efficiency in oil reservoirs. There have been many examples of increased oil production in Californian heavy oil reservoirs when steam/foam is used.

During the last two decades, there have been many attempts to improve steam injection efficiency by the use of additives. Among the many additives tried, the aqueous surfactant solution appears to be the most promising. The objective of such surfactant injection is either to increase the pressure gradient across the region of interest by generation of foam or to use the surface-active properties of the surfactant to reduce the oil–water interfacial tension and to alter the relative permeability curve. Following is a list of research areas on this topic.

7.3.2.1   Surfactant selection criteria for steam flooding

In selecting surfactants for application in thermal recovery, two criteria are set, namely, the resistance of surfactants to hydrolytic degradation and to thermal degradation. It is a common practice to study surfactants at elevated temperatures exterior to porous media. The foam tube test is the most commonly applied technique for determining foam stability exterior to porous media (DeVries 1958). Some studies found foam stability outside of porous media to be an important indicator of potential mobility reduction within porous media (Doscher and Hammershaimb 1981). However, in other studies, no such correlation was found (Dilgren et al. 1982). It is likely that the tube test represents foam behavior in large pore throats and may not represent foam stability in a confined case as in a real porous medium. This observation has been confirmed by Zhong et al. (1999).

Handy et al. (1982) indicated that thermal stability is a critical factor in the choice of foaming agent for thermal EOR processes. Many studies have demonstrated that foam can be used for flow diversion in a steam flood process. Recently, Djabbarah et al. (1990) reported thermal stability of several surfactants. Despite many disjointed efforts, a comprehensive selection criterion, applicable to steam flooding, has not yet been developed (Zhong et al. 1999).

7.3.2.2   Microscopic behavior of surfactant-steam flooding

It is important to understand the microscopic behavior of a system before a field application can be recommended. In steam flooding research, little effort has been made to study microscopic behavior and extending that observation to the scaled-up version. Several theories have been proposed to explain surface phenomena in a surfactant-steam system (Ransohoff and Radke 1988; Falls et al. 1988, 1989; Hirasaki and Lawson 1985). However, little agreement among researchers exists and fundamental questions, such as the role of gas rate on foam apparent viscosity, mechanism of bubble generation, effect of surfactant concentration, or the effect of temperature on foam flow, cannot be answered without some degree of ambiguity.

7.3.2.3   Role of residual oil on foam

This fundamental aspect of the steam/foam process has not been properly addressed. Most papers on the topic claim to offer different solutions. One possible way to address this process is to conduct research on the microphysical aspect of the process (Zhong et al. 1999; George and Islam 1998).

7.3.3   In-situ *combustion*

The *in-situ* combustion process generates the required thermal energy in the reservoir. Combustion takes place by burning a portion of in-place crude oil by injecting compressed air. *In-situ* combustion is also known as fireflooding. The application method is complex in comparison to other EOR methods. An *in-situ* combustion technique has been applied in many conditions and it is hard to achieve success in all conditions. DeMirji (1978) reported that *in-situ* combustion is only suitable for a narrow range of reservoirs. Some of the important variations of fireflooding are forward combustion, reverse combustion, and waterflooding. The most commonly used form of *in-situ* combustion is forward combustion. In this process, ignition of the reservoir starts near an injection well and the burning front moves rapidly from this point to the production wells. According to DeMirji (1978), under the most ideal reservoir conditions represented by homogeneous rock and fluid properties, the product of aerial and vertical sweep efficiencies could approach 60%.

The reverse combustion process is used to overcome the problem of forward combustion, such as fluid blocking, ignition, and injected air flow. In this process, oil moves though the heated region of the reservoir toward the producing well. This technique is good for viscous crude oil reserves, where recovery is very low.

In the forward combustion and waterflooding processes, water is injected simultaneously with the air. Water is injected after ignition of the formation. The injected water acts like a steam injection because due to burning, hot water turns to steam and joins the combustion gases to flood the cold regions. Economically, this process is better than all other combustion process, efficiencies being similar to the forward combustion process.

There are several reasons why *in-situ* combustion may fail. One is that low permeability does not allow the injection of compressed air or steam. High thermal efficiency or economic efficiency cannot be achieved without sufficiently high permeability. A second parameter effecting flow is the available pressure to move the extraneous fluids through the formation. Another factor that might have an effect is the initial oil saturation.

7.3.4   *Electromagnetic heating*

This process holds great promise for recovery of heavy oil from cold regions. This technique eliminates one of the larger technical problems with conventional steam injection in regions where permafrost exists.

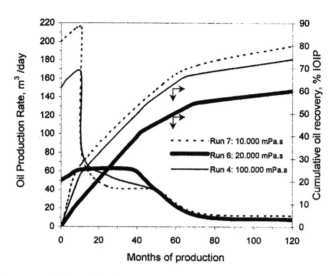

Figure 7-8   Oil recovery with gas injection and electromagnetic heating (Islam and Chilingar 1995).

Several options have been tried, including the use of electricity to heat oil reservoirs. These methods can be classified according to the mechanisms of thermal dissipation that dominate the recovery process (Pizzaro and Trevisan 1990). They range from dielectric heating with a high frequency range to radio frequency in the microwave range. Wadadar and Islam (1994) have recently investigated the possibility of using electromagnetic heating with horizontal wells. Islam and Chilingar (1995) reported the results of a series of numerical simulation tests based on the method originally proposed by Islam and Chakma (1992). They showed that by coupling electromagnetic heating with other EOR schemes, the recovery of heavy oil or tar sand could be significantly increased. Figure 7-8 shows that as much as 80% of the oil in place can be recovered, if the process is combined with gas injection from the top horizontal well. However, this process has never been tested in a field. The company in charge of promoting electromagnetic heating (Electromagnetic Oil Recovery, EOR, Calgary) has not been active for some time.

7.4   Gas Injection

Even though the gas injection technique has been prominent in light oil reservoirs for several decades, it has been considered for heavy oil reservoirs only recently (Malik and Islam 1999). Recently, immiscible carbon dioxide has started to gain popularity in some of the heavy oil reservoirs. If this trend continues, immiscible carbon dioxide could become the most dominant nonthermal EOR technique for heavy oil reservoirs.

Even though the oil recovery in the United States from other EOR processes has decreased since 1992, EOR recovery from CO_2 flooding has continued to increase (Islam 1999). In 1996, the oil production from CO_2 processes was six times that of the same process in 1986 (Grigg and Schechter 1997). Of all the recovery techniques, CO_2 flooding shows the widest applicability with a potential of recovering additional 7–14% of the initial oil in place (Klins 1984). For heavy oil reservoirs, this incremental recovery can be significantly higher, especially if operating parameters are optimized or CO_2 injection is combined with other recovery schemes (Islam and Chakma 1992; Islam et al. 1994).

Numerous laboratory studies have shown that immiscible CO_2 in heavy oil formations can recover billions of barrels of additional oil in Alberta and Saskatchewan alone. Before such a process can be implemented in the field, operating parameters need to be optimized. This is a difficult task because CO_2 injection in heavy oil reservoirs involves many non-equilibrium phenomena. These non-equilibrium phenomena are often manifested through sensitivity to flow rate. For instance, Figure 7-9 shows the sensitivity of oil recovery (expressed

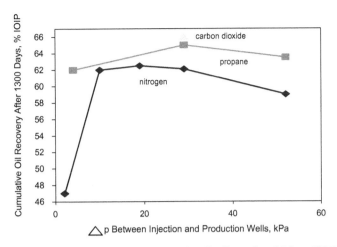

Figure 7-9   Gas injection in horizontal wells (Bansal and Islam 1994).

as a percentage of the initial oil in place, IOIP) to flow rates (due to difference in total pressure drops). This dependence is not due to viscous instability as all runs were for gravity stabilized flow. These results clearly show that flow rate must be optimized, even when operating under stable flow regime.

Gas injection can lead to partial miscibility (Frauenfeld et al. 1998) with heavy oil. However, even under miscible displacement conditions, poor sweep, gravity segregation, and viscous fingering can result in low oil recovery during miscible gas injection of a heavy oil formation. Gravity stabilization has been recommended by researchers in order to improve oil recovery under unfavourable mobility ratio conditions (Islam et al. 1994).

Srivastava and Huang (1995), in their feasibility study of miscible or near-miscible flooding with CO_2 on heavy crude, reported an initial rapid decrease in fluid viscosity followed by slow decrease at higher concentrations of CO_2. The bubble point-pressure, gas-oil ratio, swelling, and formation volume factors for the reservoir fluid-CO_2 mixtures increased smoothly with CO_2 concentrations. Deposition of asphaltene, during a miscible displacement process in the presence of CO_2 as an EOR agent, can cause numerous problems with a negative effect on oil recovery. Srivastava and Huang (1997), in their laboratory study, addressed the deposition of asphaltenes and other heavy particles. They concluded that asphaltene precipitation depends on the CO_2 concentration and is independent of the operating pressure. Ali and Islam (1998) mentioned that the deposited asphaltenes can be removed mechanically with increased flow rate, whereas adsorbed asphaltene can only be removed through desorption, the rate of which is much lower than that of adsorption. They concluded that asphaltene plugging is dependent on the flow rate, leading to greater deposition near the wellbore.

Loss of miscibility during a miscible CO_2 injection process can have severe consequences. Because most reservoir simulators are unsuitable for modeling the transition between miscible and immiscible displacements, CO_2 injection design can be seriously flawed. Figure 7-10 shows numerical simulation results of recovery performance with miscible and immiscible displacement processes (Islam and Chakma 1993). Note that both these recovery processes have stable displacement fronts and, therefore, viscous fingering is not a factor.

Contrary to miscible CO_2 cases, few applications have been reported on immiscible CO_2 injection. Only recently, immiscible CO_2 injection has been introduced to heavy oil reservoirs in the context of nonthermal EOR (Rojas et al. 1995; Lozada and Farouq Ali 1988; Islam et al. 1994, Srivastava et al. 1993, 1995). CO_2 can also be useful in the heavy oil reservoirs where thermal methods are difficult to implement. CO_2-saturated crude oils exhibit moderate swelling, leaving fewer stock tank barrels of residual oil in place, and reduce oil viscosity to a point at which mobility ratios are drastically affected.

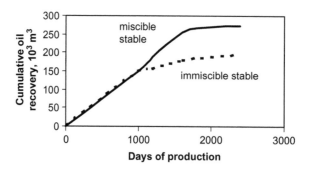

Figure 7-10   The role of immiscibility in oil recovery with gas injection (Islam and Chakma 1993).

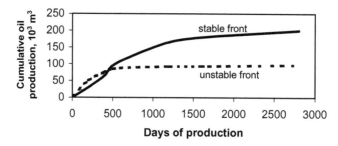

Figure 7-11   Role of stability during heavy oil recovery with CO_2 9 (Islam and Chakma 1993).

Dyer et al. (1994) studied the phase behavior and scaled model behavior of a Saskatchewan reservoir, to investigate displacement mechanisms associated with immiscible gas (CO_2) injection processes. They reported that nonthermal EOR techniques show good potential for recovering oils from the thin and shaly heavy oil reservoirs of Saskatchewan. Among the nonthermal processes, immiscible CO_2 injection holds the most promise of accessing these reservoirs. Dyer et al. (1994) concluded that the process is proven and is applicable to Saskatchewan reservoirs. They further suggested that approximately 90% of the total oil-in-place could be accessed from the Saskatchewan reservoirs with a pay thickness of three to seven meters.

Several recent studies showed the importance of gravity stabilization during immiscible CO_2 injection, especially in a heavy oil formation (Islam et al. 1994; Islam et al. 1992). In heavy oil reservoirs, an unstable displacement front can lead to significant loss in oil recovery. Figure 7-11 shows one such case for which the oil recovery is several times smaller during unstable flow than during stable flow. The reason for such dependence is attributed to the presence of viscous fingering that decreases the gas breakthrough time significantly.

Figure 7-12 shows that the instability number (that lumps flow rate, geometry, and other factors into a dimensionless group, see Chapter 6 for details) affects breakthrough recovery during gas injection for which viscous fingering occurs. In contrast to water flooding, that shows the existence of a pseudo-stable regime, gas breakthrough appears to decline continuously as a function of instability number.

Numerous field reports in Canada (Amoco, Shell, Imperial, and Husky Oil) show that horizontal wells can be used to successfully conduct stable miscible displacement of both light and heavy oil. The key to success in miscible or immiscible gas injection appears to be the accurate prediction of the frontal stability, which is very sensitive to reservoir heterogeneity (Islam et al. 1994). Another method of reducing viscous fingering during gas injection is the use of foam. Foam increases the viscosity of the displacing gas phase to the extent that an otherwise unstable front (with gas only) can become stable (Islam et al. 1989). Foam also has a

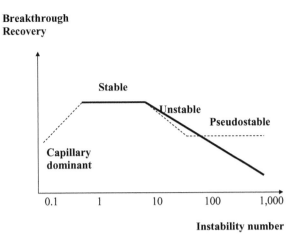

Figure 7-12 The role of instability number in breakthrough recovery (the solid line is for gas injection and the dotted line for liquid injection). Data combined from Islam and Bentsen (1987) and Islam et al. (1994).

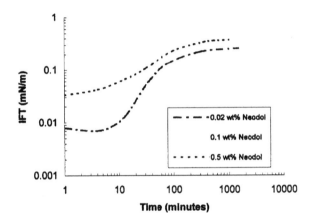

Figure 7-13 Transient interfacial tension between a Canadian crude and 1% wt Na_2CO_3 solution (Taylor et al. 1990).

homogenizing effect when the displacement front encounters a heterogeneous spot in the reservoir. Even though no study has yet been reported on frontal stability in a heterogeneous medium, foam is likely to eliminate some of the problems associated with frontal instability due to heterogeneity.

7.5 Chemical EOR

Recently, the focus in chemical EOR has been to reduce costs. One way to reduce costs is to use surfactants that are less expensive or, better yet, to use cosurfactants with inexpensive chemicals, such as sodium carbonate (Taylor et al. 1990). Figure 7-13 shows how the use of only 200 ppm cosurfactant can decrease the interfacial tension to an ultra-low value that would otherwise require more than 1% of conventional surfactant (Sundaram and Islam 1994). Islam and Chakma (1990) reported that similar advantages can be maintained even in the presence of polymer in the system. This mixture becomes less expensive than the conventional surfactant, despite maintaining an ultra-low interfacial tension. However, with acidic crude, the interfacial tension value increases following a minimum. Fortunately, it is not necessary to maintain an ultra-low value throughout the displacement. As long as the displacement front encounters an ultra-low

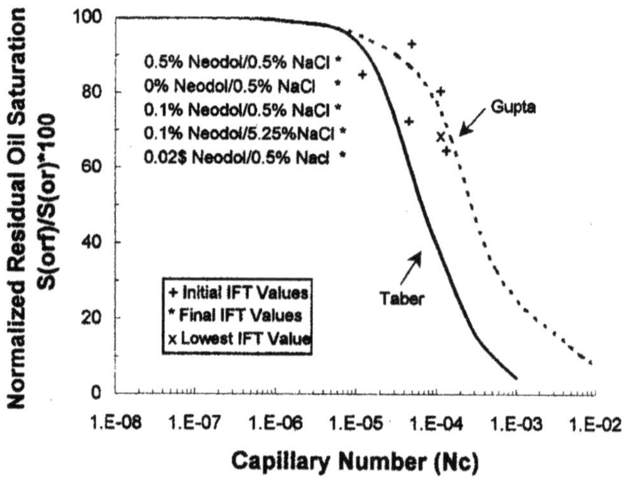

Figure 7-14   Capillary numbers for different 1% wt NaCO₃ and a Canadian crude (Taylor et al. 1990).

value for a reasonable time, the recovery is likely to be high (Islam and Chakma, 1990). Figure 7.14 clearly shows how the oil recovery corresponds to the lowest interfacial tension value. Historically, researchers have ignored the transient values of interfacial tension, considering them irrelevant. Until Borwankar and Wasan (1986), all IFT measurements were reported for steady states. They identified the relevance of transient IFTs, indicating that the transient values were not experimentally spurious and should not be ignored in recovery analysis. Taylor et al. (1990) showed that it is the lowest IFT value that corresponds to the additional oil recovery (Figure 7-14).

The addition of these cosurfactants with polymer will ensure the maintenance of an ultra-low interfacial tension (IFT) while improving the mobility ratio. The IFT properties are likely to change in the presence of polymer and need to be investigated in more detail (Islam and Chakma 1988; Islam and Chakma 1991; Sundaram et al. 1995). Islam and Chakma (1991) presented a model for alteration of IFT in the presence of polymer. This model did not include salinity. It is recommended that, in any field study, transient IFT be determined in presence of the aqueous phase that has similar composition to the reservoir water and the chemicals that are being used to bolster the interfacial action.

7.6   Microbial EOR

Microbial Enhanced Oil Recovery (MEOR) is a biological technique used for enhancing the oil production from reservoir. In this technique, selected natural microorganisms are introduced into oil wells to produce harmless by-products, such as surface-active natural substances or gases, all of which help propel oil out of the well (Bio Basic 2006). These processes help to mobilize the oil and facilitate oil flow, and so allow a greater recovery from the well.

A special type of microorganism is needed to carry out EOR processes, because the physiochemical condition of a well is exceptional. For the microorganisms to be able to survive in such a harsh environment, nutrients and oxygen need to be introduced. MEOR also requires that water be present. Microbial growth changes the reservoir conditions, helping to produce more oil. Some of the basic mechanisms, which help to enhance oil production, are (Bio Basic 2006):

- *Reduction of oil viscosity*: Oil is a thick fluid that is viscous, meaning that it does not flow easily. Microorganisms help break down the molecular structure of crude oil, making it more fluid and easier to recover from the well.

- *Production of carbon dioxide gas*: As a by-product of metabolism, microorganisms produce carbon dioxide gas. Over time, this gas accumulates and displaces the oil in the well, driving it up and out of the ground.

- *Production of biomass*: When microorganisms metabolize the nutrients they need for survival, they produce organic biomass as a by-product. This biomass accumulates between the oil and the rock surface of the well, physically displacing the oil and making it easier to recover from the well.

- *Selective plugging*: Some microorganisms secrete slimy substances called exopolysaccharides to protect themselves from drying out or falling prey to other organisms. This substance helps bacteria plug the pores found in the rocks so that oil may move past rock surfaces more easily. Blocking rock pores to facilitate the movement of oil is known as selective plugging.

- *Production of biosurfactants*: Microorganisms produce slippery substances called surfactants as they break down oil. These biosurfacants, being naturally produced by biological microorganisms, are referred to as biosurfactants. Biosurfactants act like slippery detergents, helping the oil move more freely away from rocks and crevices so that it may travel more easily out of the well.

Islam (1999) reported that MEOR has received a great deal of attention in recent years. However, field application in the United States has not been successful and, therefore, little oil production has been attributed to this method. In this regard, Russia has implemented several MEOR projects with more success. Canada has reported the use of microbes for plugging water-producing zones (Jack et al. 1991).

MEOR is one such technique that is gaining popularity, but has not yet reached commercial use. More fundamental research needs to be done in order to understand the scaling up of laboratory events to field conditions. MEOR uses microorganisms and their metabolic products to improve the recovery of crude oil from reservoir rock. The application of MEOR techniques involves three main mechanisms by which microorganisms sent into oil reservoirs can improve oil recovery, classified as:

1. Splitting of heavier fractions of the crude oil;
2. Generation of gases (CO_2, N_2, H_2, CH_4); and
3. Production of chemicals (surfactants, solvents, acids, biopolymers).

The first publication, suggesting the use of bacteria to increase petroleum recovery from oilfields, is that of Beck (1947). This work dealt with viscosity reduction of mineral oil. During the 1940s, ZoBell and his colleagues observed that certain enriched cultures of sulfate-reducing bacteria, growing in the presence of Athabaska tar sands, appeared to release oil from the solid sediment, causing it to rise to the surface of the surrounding medium (Updegraff 1990). This led ZoBell to obtain a US patent on bacteriological processes for recovering oil, and to publish a series of journal publications on the subject. These papers led to the birth of MEOR and to further investigations by large oil companies such as Penn Grade Crude Oil association, Standard Oil of California, Texaco, and Mobil. The research carried out by these groups continued during the 1940s through to 1955, at which time research of MEOR virtually stopped in the United States. However, this work aroused considerable interest in Canada, Eastern Europe, the Soviet Union, Poland, Hungary, and Checkoslovakia (Jack 1990; Updegraff 1990). In the mid-1970s, with the dramatic increase in the price of oil, EOR such as MEOR became attractive once again. Research on the subject was restarted and continues to be a subject of interest to this day.

7.6.1 Advantages of MEOR

Chemically-synthesized surfactants have been used in the oil industry to aid the clean up of oil spills, as well as to enhance oil recovery from reservoirs for the past few decades. These compounds are not biodegradable and are toxic to the environment (Khan and Islam 2007). The reason that toxicity occurs is because synthetic surfactants are inherently toxic at any concentration (Khan 2006b). Biosurfactants, on the other hand, have been shown to have equivalent emulsification properties and are still be biodegradable. Biodegradability is essential to the acceptance of a chemical in any process, irrespective of how late it is released

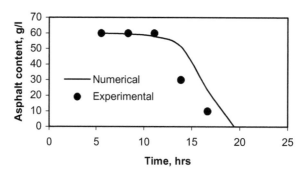

Figure 7-15   Degradation of asphaltenes with thermophilic bacteria at 80°C (Al-Maghrabi et al. 1998).

into the environment. The second advantage of MEOR is that it is inexpensive. It is recommended that local strains of bacteria are used, as they are most likely to be effective for a given region (Al-Maghrabi et al. 1999). MEOR offers major advantages over conventional EOR in that lower capital and chemical/energy costs are required (Sarker et al. 1989). This factor, coupled with the fact that bacteria grow exponentially in presence of food (crude oil in this particular case) makes a compelling case for inexpensive MEOR. Thus, there is increasing interest in the possible use of biosurfactants in mobilizing heavy crude oil, transporting petroleum in pipelines, managing oil spills, oil-pollution control, and MOER.

7.6.2   Applicability of MEOR

Early research on MEOR had already established that it could not be applied to all petroleum reservoirs. Beck (1947) found that bacteria would not penetrate the fine Bedford Sands. Updegraff (1990) studied the effect of permeability and pore size on the penetration of bacteria through reservoir rock and concluded that MEOR was not applicable to petroleum reservoirs where the average permeability exceeds 100 md. Deep reservoirs, where the temperature exceeds the maximum growth temperature of desired bacteria, were also excluded. One other reservoir condition that restricts MEOR applications is high salinity (over 12%). While microbes are known to grow under conditions outside of these environmental parameters, most MEOR field demonstration projects have been in reservoirs with the above conditions.

In the past, scientists considered high temperatures (above 60°C) untenable for microbial growth. This would mean that most of the light oil reservoirs would be unsuitable for microbial EOR. However, Al-Maghrabi et al. (1999) showed that there are bacteria that can thrive under high temperatures. These thermophilic bacteria have their optimum temperature at around 80C. Figure 7-15 shows degradation of asphaltenes with bacteria. Note that asphaltenes are considered difficult to breakdown. These bacteria are likely to be much more effective in the presence of lighter components of the crude oil. In Figure 7-15, the numerical model was a modified version of that first developed by Islam and Gianetto (1993) and later adapted by Kowalski et al. (1993). Using this modified version, Al-Maghrabi et al. (1998, 1999) were able to predict the impact of bacterial solution around the wellbore. Figure 7-16 shows that the bacteria can propagate slowly and can degrade heavy components, leading to improving oil flow.

Table 7-3 shows the bacterial growth rate for some of the thermophilic bacteria under various conditions. Corresponding equations that should be used in a simulation are also given. It is important to note that these values should be determined for each field before an MEOR scheme is carried out.

The above applications refer to MEOR specifically geared toward alteration of fluid properties. There are other applications of microbial use. For instance, bacteria can consolidate unconsolidated sands, thereby diminishing sanding problems (Zhong and Islam 1995). This technique would allow the use of bacteria in shallow formations that usually have sanding problems. Jack and his research group led a series of studies in order to use bacteria for sand consolidation (Jack et al. 1987).

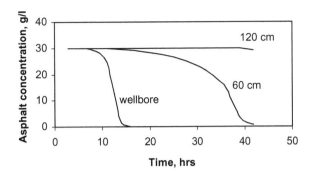

Figure 7-16   The effect of bacteria on asphaltene concentrations in and around the wellbore (Al-Maghrabi et al. 1998).

Table 7-3   Growth rates of bacteria in crude oil and wax at different concentration and temperatures

Concentration.	Temp.	Growth rate, μ	Complete equation
3% asphalt*	45°C	0.495/hr	$C_t = 75.81e^{(0.495t)}$
6% asphalt*	45°C	0.424/hr	$C_t = 99.53e^{(0.424t)}$
3% asphalt*	80°C	0.605/hr	$C_t = 56.00e^{(0.605t)}$
6% asphalt*	80°C	0.519/hr	$C_t = 77.16e^{(0.519t)}$
3% wax	80°C	0.0293/hr	$C_t = 48.038e^{(0.029t)}$
10% crude oil (0% salinity)	22°C	0.52/hr	$C_t = 10.67e^{(0.52t)}$
10% crude oil (10% salinity)	22°C	10.19/hr	$C_t = 12.45e^{(0.19t)}$

7.6.3   Type of bacteria to be selected for MEOR

When considering which microorganisms to use for MEOR, the varying conditions under which they are to be used must be given priority (Khire and Khan 1994). Typically, microorganisms injected into an oil well should endure high temperatures, pressures, and salinity, and be capable of growth under anaerobic conditions. Certain microorganisms are better adapted to particular environments, such as oil reservoirs, soil, or the ocean. Table 7-4 lists microorganisms, which produce biosurfactants, and their structures, as reported by Banat (1994).

7.6.4   Mechanisms of MEOR

The mechanisms of MEOR action *in situ* are due to the multiple effects of microorganisms on the environment and the oil itself. These mechanisms include gas formation and pressure increases, acid production and degradation of limestone matrices, reduction in oil viscosity and interfacial tensions by biosurfactants, solvent production, plugging by biomass accumulation or polymer formations, and degradation of large molecules in oil, resulting in decreases in viscosity (Jack 1988; Khire and Khan 1994).

The presence of different types of microorganisms with varying growth properties will have different effects on the reservoir environment. Thus, it is important to collect information at the level of interfacial phenomena. The study of the ability of microbial systems to lower interfacial energies and effectively aid in the release of oil from rock surfaces needs to be carried out. There are several strategies involving the use of biosurfactants in MEOR.

1.   The first involves injection of biosurfactant-producing microorganisms into a reservoir through the well, with subsequent propagation *in situ* through the reservoir rock (Bubela 1985).

Table 7-4   Various biosurfactants produced by microorganisms

Microorganism	Biosurfactant
Arthrobacter RAG-1	*hetropolysaccharides*
Arthrobacter M1S38	*lipopeptide*
Arthrobacter sp.	*Trehalose, sucrose and fractose lipids*
Bacillus lichenformis JF-2	*lipopeptides*
Bacillus lichenformis 86	*lipopeptides*
Bacillus subtilus	*surfactin*
Bacillus pumilus A1	*surfactin*
Bacillus sp. AB-2	*rhamnolipids*
Bacillus sp. C-14	*hydrocarbon-lipid-protein*
Candida antarctica	*mannan-fattyacid*
Candida bombicola	*sophorose lipids*
Candida tropicalis	*mannan-fatty acid*
Candida lipolyttica Y-917	*sophoros lipid*
Clostridium pasteurianum	*neutral lipids*
Corynebacterium hydrocarbolastus	*protein-lipid-carbohy*
Corynebacterium insidiosum	*phosholipids*
Corynebacterium lepus	*fatty-acids*
Strain MM1	*glucose, lipid and hydroxydecanoic acids*
Nocardia erthropolis	*neutral lipids*
Ochrobactrum anthropii	*protein*
Penicillium spiculisporum	*spiculosporic acid*
Pseudomonas aeruginosa	*rhamnolipid*
Pseudomonas fluorescens	*lipopeptide*
Phaffia rhodozzyma	*carbohydrate-lipid*
Rhodococcus erthropolis	*trehalose dicorynomycolate*
Rhodococcus sp. ST-5	*glycolipid*
Rhodococcus sp. H13-A	*glycolipid*
Rhodococcus sp. 33	*polysaccharide*
Torulopsis bombicola	*sophorose lipids*

Source: Banat (1994)

2. The second involves the injection of selected nutrients into a reservoir, thus stimulating the growth of indigenous biosurfactant-producing microorganisms

3. The third mechanism involves the production of biosurfactants in bioreactors *ex situ* and subsequent injection into the reservoir.

7.6.5   *Laboratory studies and scaling up*

Laboratory studies on MEOR have typically contained core samples or columns with the desired microorganisms. In one experiment, *B. subtilis* was injected through sand columns and a 35% release of residual oil was observed compared to 21% using the nutrient solution control. Similar experiments with *C. acetobutylicum* with molasses (4%) and 0.5% ammonium diphosphate yielded a 66% and 59% oil recovery in the presence of and absence of pyrophoshpate, respectively (Chang 1987).

Vibrio aspartigenicus strain GSP-1 and *Bacillus licheniformis* JF-2 were also tested for their ability to recover residual oil from crushed limestone cores (Adkins et al. 1992). *Vibrio aspartigenicus* was observed to recover 32–36% more oil from saturated cores than that of the control columns after three treatments, *and B. licheniformis* increased oil recovery by 27%.

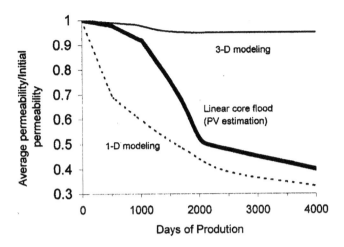

Figure 7-17   Effect of scaling up on microbial plugging (Chilingar and Islam 1995).

Jack and his colleagues conducted a series of laboratory studies dealing with microbial plugging in order to improve oil recovery. Despite extensive background laboratory work, the project was a failure in the field, because of scaling-up problems. It was not until later that it was found that linear model results could not be used directly to design a field project. Figure 7.17 shows how the same strain of bacteria reduces permeability in a linear core significantly, whilst making no difference in a 3D system. Microbial EOR has enjoyed significant scientific research and it is time for some of the engineering problems associated with the scheme to be addressed before field implementation.

One specific aspect of microbial EOR appears to be unexplored, the possibility of using thermophilic bacteria along with hot water injection. Hot water alone can reduce oil viscosity significantly (Islam et al. 1992). If thermophilic bacteria are added, additional improvement is observed due to decomposition of inorganic carbonates, evolution of viscosity-reducing gases, and IFT reduction with surface active bioagents. Al-Maghrabi et al. (1999) have conducted some preliminary studies with this scheme. This study also opens up prospects of using bacteria in combination with other EOR techniques, such as surfactant, water flooding, and others. In the combined process, it is important to ensure that the selected surfactant is compatible with the bacteria to be used (Sundaram et al. 1994). It is possible that the strain of bacteria will eventually use the surfactant as a nutrient and can add synergy to the system.

MEOR is the most environmentally friendly EOR process, eliminating the need to use toxic chemicals from oil recovery processes. Biotechnological advances helps to develop more effective bacteria. This progress makes MEOR inexpensive. Developing and growing MEOR bacteria are improving, thereby lowering production costs and making it a more attractive alternative to traditional chemical methods of tertiary oil recovery (Bio Basic 2006).

7.7   EOR in Marginal Reservoirs

Among marginal reservoirs, reservoirs with a bottomwater zone are considered the least likely candidates for any EOR application. Islam (1993) reported the prospect of using EOR methods in improving water flood performance for reservoirs with a bottomwater zone. Figures 7-18 and 7-19 show how recovery can be improved by various EOR schemes, for different toil-to-water zone capacity ratio and oil viscosities.

However, these results are derived through experimental modeling, and none of these techniques has been tested in reservoirs with a bottomwater zone (Islam, 1999). The technologies tested in the field are the anti-water-coning technology (AWACT) and electromagnetic heating. AWACT involves the injection of a gas (methane in most cases) to reduce the effective permeability of water (in a three-phase system) during the

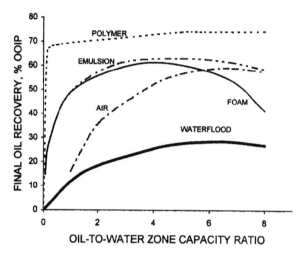

Figure 7-18   Oil recovery from bottomwater reservoirs (Islam 1993).

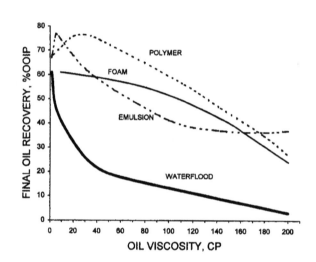

Figure 7-19   Effect of oil viscosity on different EOR schemes for bottomwater reservoirs (Islam 1993).

production phase. This process has given good results in some Canadian heavy oil reservoirs (Freeborn et al. 1989). Several reservoirs in Canada have been treated with electromagnetic heating with favorable results (Islam 1993). Similar results have been reported for some of the European oilfields (Rice et al. 1992)].

The possibility of using EOR methods for improving recovery from marginal reservoirs should be considered. Several EOR methods, along with horizontal wells, have good potential for future applications.

7.8   Scaling of EOR Schemes

According to Islam (1999), all EOR techniques should be tested in the laboratory prior to field tests. Unfortunately, if the scaling criteria are unknown, laboratory results only provide qualitative information about a specific technique. Farouq Ali (World Petroleum Congress 1967) was one of the first petroleum engineers to highlight the need of scaled model experiments for EOR applications. Recently, many papers have been published on the topic of scaling of EOR experiments. However, all these studies point toward the difficulties associated with the scaling process. It is generally accepted that all scaling groups cannot be scaled properly with a single model. Therefore, the idea is to identify major scaling groups that are considered to be crucial in a given EOR process and to recommend a model that scales all these groups properly.

Islam and Chakma (1992) discussed the scaling groups involved in modeling thermal EOR schemes (Kimber and Farouq Ali 1988; and Farouq Ali et al. 1987). Some of the most difficult to satisfy dimensionless groups involve chemically EOR schemes (Islam and Farouq Ali 1990). The most important observation is that to model dimensionless groups involving chemical reactions would require the use of a different rock fluid system in the laboratory from that of the field. This is also true for polymer or cosurfactant enhanced alkaline flooding for which the rock-fluid interaction is important in terms of adsorption. Similar difficulties persist with experimental modeling of *in-situ* combustion. Islam and Farouq Ali (1991) derived the scaling groups involved in modeling of an *in-situ* combustion process. They also reported that combustion tube experiments cannot be used to make quantitative observations. Islam and Gianetto (1993) showed that the difficulties in scaling up of microbial EOR can lead to erroneous judgments in the field. Chilingar and Islam (1995) later verified this observation through numerical simulation.

One aspect of modeling EOR processes is the modeling of unstable displacement fronts. When the displacement front is unstable due to viscous fingering in an immiscible system, viscous grading in a miscible displacement process, instability due to heterogeneity, etc., it is important that the displacement in the laboratory be unstable too. Thus, the degree of instability needs to be similar in the laboratory to that in the field (Islam and Bentsen 1986). Because the degree of instability depends on the dimension of the domain, having a similar degree of instability in the laboratory translates into having a much higher velocity in the laboratory than in the field. This is contrary to the common belief that laboratory experiments should match field velocities.

The competitive oil prices have driven the oil industry to look for inexpensive alternatives to commonly used EOR schemes, in the search for greater efficiency. As more reservoirs become depleted for primary production, the trend of EOR is likely to increase. More recent developments of expensive surfactants, electromagnetic heating, and possible use of foam and emulsion have added to the EOR arsenal. Recent developments in surface properties of chemicals and relatively stable chemical prices have opened up prospects of chemically based techniques. MOER has some appeal and a certain potential, but it requires more engineering studies before implementation in the field. New applications should focus on better planning, proper reservoir characterization, and minimizing costs. Marginal oil reservoirs also make good future candidates for EOR. The use of horizontal wells should be considered as an integral factor in designing EOR schemes.

7.8.1 Scaling unstable displacement

An EOR scheme is often accompanied by unstable displacement. As discussed in Chapter 6, if the mobility ratio is unfavorable, it is likely that the field dimension will make the process unstable. Under this condition, profuse viscous fingering occurs and the recovery efficiency can drop significantly (Figure 7-20). At present, few researchers have conducted scaled model studies of an unstable displacement process. Bansal and Islam (1994) were the first to attempt modeling unstable displacement in a scaled model. Their findings indicated that it is important to conduct experiments in the laboratory with a flow rate that would give rise to the same instability number. Only then, they argued, the flow regime in the laboratory would be the same as the one prevailing in the field.

Islam (PTRC 1998) conducted a series of displacement tests under unstable conditions. Figure 7-21 shows the recovery performance when the displacement becomes unstable for a gas injection project. It was noted that 3D modeling was essential, as with 1D cores such an unstable regime could not be established even with a velocity of 50 ft/day. Note that conventionally, 1 ft/day flow rate is used during laboratory core testing. This slow flow rate puts the flow regime invariably on the stable side, making it practically impossible to simulate unstable flow regimes in a laboratory. Basu (2005) used this technique to model even chemical flooding experiments. He showed reasonable agreement between numerical modeling and scaled modeling results (Figure 7-22).

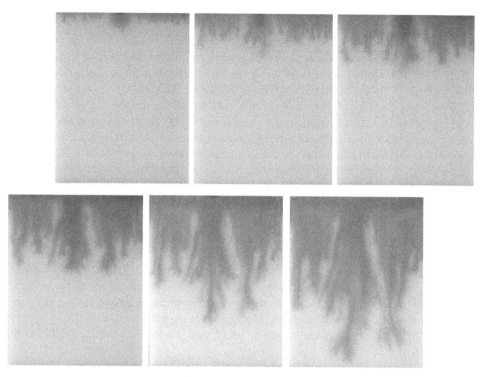

Figure 7-20   During unstable displacement, viscous fingering is dominant and the recovery dips down to a very low value (Mustafiz et al. 2006).

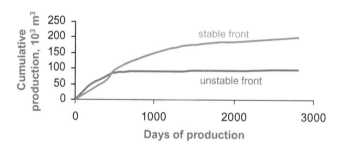

Figure 7-21   The role of stability in recovering oil during miscible displacement (scaled model results).

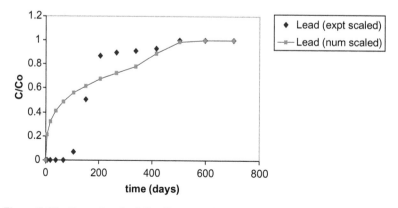

Figure 7-22   Even chemical flooding experiments should be scaled (Basu 2005).

For scaling an EOR process, the following steps are required:

1. Determine the end-point permeabilities.
2. Determine the capillary pressure curve.
3. Determine interfacial tensions.
4. Estimate flood pattern dimension.
5. Estimate frontal velocity from the injection well.
6. Calculate gravity number.
7. Calculate instability number.
8. If the instability number is less than π^2, follow conventional approach (velocity matched with that of the field as per the scaling requirement).
9. If the instability number is greater than π^2, calculate laboratory velocity such that the instability number in the laboratory matches with that of the field.

The instability number for a horizontal case is given by

$$I_{sr} = \frac{\mu_w v(M-1)}{k_{wr}\sigma_e} \times \frac{M^{5/3}+1}{(M+1)(M^{1/3}+1)^2} \frac{4L_x^2 L_y^2}{L_x^2 L_y^2} \tag{7.1}$$

For an inclined system, the instability number is given by

$$I_{sr} = \frac{\mu_w v(M-1-N_g)}{k_{wr}\sigma_e} \times \frac{M^{5/3}+1}{(M+1)(M^{1/3}+1)^2} \frac{4L_x^2 L_y^2}{L_x^2 + L_y^2}. \tag{7.2}$$

The gravity number, N_g is given by

$$N_g = \frac{\Delta\rho g k_{or} \cos\alpha}{\mu_o v} \tag{7.3}$$

In Equation 7.2, σ_e is the pseudo-interfacial tension. In order to calculate this parameter, we need to know the area of the capillary pressure *vs* saturation graph (Islam and Bentsen 1985). The actual expression for is given as

$$\sigma_e = \frac{A_c\phi(1-S_{wi}-S_{or})}{2/r_m} \tag{7.4}$$

where A_c is the area under the capillary pressure curve and r_m is the characteristic mean radius, usually assumed to be 1 (Islam and Bentsen 1987). In the above equation, mobility ratio and end-point mobility ratios are given by Equations 7.1 and 7.5, respectively:

$$M = \frac{\mu_o k_w}{\mu_w k_o} \tag{7.5}$$

$$M = \frac{\mu_o k_{wr}}{\mu_w k_{or}} \tag{7.6}$$

It is important to note that for gas injection the instability number continues to affect the recovery, meaning the higher the flow rate the less recovery there will be. This is because the pseudo-stable regime is never reached with gas (Islam et al. 1991). Gravity plays an important role during gas injection as the value of

Figure 7-23   For unstable displacement, scaled model tests should accompany the use of reservoir simulators that can handle viscous fingering.

the gravity number can stabilize a process. Islam et al. (1994) reported that by placing gas injection on top of the producer (e.g., in a dual horizontal well), gas displacement, even in a heavy oilfield, can be rendered stable. However, such phenomena cannot be simulated by the conventional approach. If the instability number is greater than π^2, viscous fingering model must be used, both for experimental and numerical modeling (Figure 7-23).

7.9   Environmental Consideration in EOR Operations

EOR in tar sands is the most important example of a scenario for which environmental impacts are important considerations. Such a scheme would require steam generation and bitumen upgrading facilities. Huge amounts of source water are need for steam generation. The thermal method also generates a large amount of produced water and recycling of this water would be required in order to reduce source and disposal volumes to acceptable levels. A hot water extraction process needs to use open pit mines. Sometimes large-scale tailing ponds are also required. There would be a relatively minor disposal problem for produced sand and fines.

Disposal water can be separated into two streams, the most offensive waste being disposed of underground and the safe stream discharged into a river system. In underground disposal of waste water, it is essential to ensure that groundwater sources are not contaminated. Noxious gas emissions into the atmosphere would be an issue of concern. Sulfur dioxide is the main pollutant and is produced by burning high sulfur crude or oil in the boilers. Injection of the flue gases along with steam into the reservoir may have some advantages in reducing atmospheric pollution.

In order to develop EOR schemes that are inherently environmentally friendly, technically effective, and socially responsible, the following steps should be taken (Khan and Islam 2006e):

1. Set up an interdisciplinary team (engineers, scientists, economists, and even social scientists).
2. The problems need to be openly discussed in the presence of top executives and policymakers before solutions can be addressed.
3. Ask each participant to propose his/her own solution to the problem. At least one solution per person is ideal. This should apply to every participant, including those who are from social science or other non-technical disciplines.
4. Document multiple solutions for each problem.
5. Evaluate objectively each solution, irrespective of who is proposing it.
6. Evaluate cost of the *status quo*.
7. Use the screening criterion of Khan and Islam (2006e) to evaluate the long-term benefit/cost of a particular solution.
8. List all the waste materials naturally generated in a particular project.
9. Select injection fluid from the waste products (Point 8).
10. If a particular solution is not fit for a specific field, investigate the possibility of using that solution in a different one.
11. Develop scaling criteria for each solution.
12. Conduct scaled model experiments using scaling groups that are most relevant.

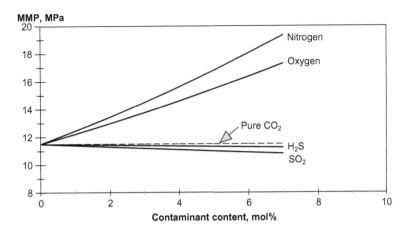

Figure 7-24 The role of contaminants in altering Minimum Miscibility Pressure (MMP). Data from Islam and Huang (1999).

13. Conduct numerical modeling.

14. Evaluate overall long-term cost and benefit.

15. Make sure global efficiency is high. This always means natural processes must not be violated, both in content and process.

In the above 15 steps, it should be clarified that the overall criterion for making any scheme environmentally friendly is to make sure it is natural, both in content and in process. Recently, Zatzman and Islam (2006) discussed what is natural. It essentially requires that the process have a real origin as well as a real pathway, reality implying "conformance with nature". This definition automatically fulfills Khan's (2006b) criterion of sustainability. If a process is natural, then not only it is environmentally friendly, it is also socially responsible and inherently sustainable (Khan and Islam, 2006a; Chhetri and Islam 2007).

7.9.1 Examples of humanization of the environment

The first example can be from miscible gas flood. A great deal of effort has been expended to keep the front miscible. Many synthetic chemicals are added in this process, so the minimum miscibility pressure is altered. These chemicals are expensive and toxic in the long term. Here, toxicity does not imply short-term reactions, but long-term harm to the environment. By analyzing the miscibility, the following scenario emerges (Figure 7-24). It is clear from this figure that air injection will not be suitable for miscible gas injection. The presence of pure CO_2 maintains a low MMP, so does the presence of H_2S or SO_2. Now that the injection fluid is to be selected, which one should be picked? Because most experiments are conducted under "controlled" environmental conditions with "pure" material, the temptation would be to recommend pure CO_2, which would be expensive. This will benefit marketing of the chemicals, but will not help the producer. However, there are numerous exhaust sites that are already producing CO_2, along with other gases. Even the oilfield itself is generating tremendous amounts of these gases (i.e., flaring accounts for 30% of total pollution through petroleum operations). Similar MMP can be achieved (if that is indeed the objective) using exhaust gas. If global efficiency is calculated (Chhetri, 2007), the efficiency of the latter system would be much higher because a number of steps (extraction, purification, etc.) are eliminated. In terms of economics, any time that waste material is used, it is a negative cost. In terms of environmental impact, the latter process is superior because it eliminates steps that generate other pollutants. This aspect of the overall efficiency calculation has recently been highlighted by Chhetri (2007). Figure 7-25 shows that global efficiency decreases as the system diverges from its natural state. However, local efficiency increases during the same transition. Local efficiency is the only one that is used in conventional calculations.

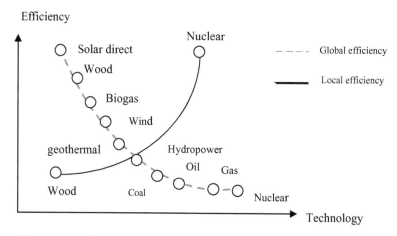

Figure 7-25   Global efficiency for various energy schemes (Chhetri 2007).

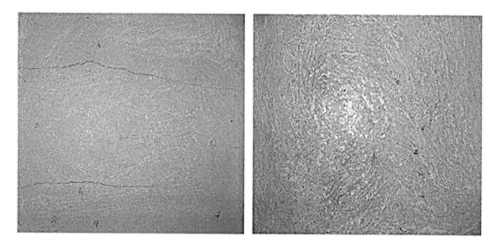

Figure 7-26   The addition of natural fibers can remediate micro-cracks in cement (left: no additive; right: natural fiber added). Data from Saeed et al. (2003).

The second example is from cementing operations. A number of cement additives are available. All these additives are synthetic and hence toxic to the environment in the long term (Khan 2006a). One popular additive is synthetic fiber, known to reduce the occurrence of micro-cracks in cement bodies. However, they are expensive, in addition to being toxic. Al-Darbi et al. (2007) and Saeed et al. (2003) reported the use of natural fibers, namely human hair, and showed that the addition of these fibers have similar effects on cement and concrete (Figure 7-26). Further analyses have indicated that there is no harmful effect of human hair waste, and there is nothing benign about synthetic polymer. By replacing synthetic polymer fibers with natural fibers that are just as effective, tremendous environmental and economic benefits have been added to the scheme.

The use of natural fibers was also found to be useful for other applications. For instance, they can separate oil from oil and water emulsion, and the recovered oil can be reused. If heating is introduced with direct solar heating, the entire extraction process can become self-sustaining (Khan et al. 2006a). Also, natural fibers are excellent for removing contaminants, especially at low concentrations for which it is usually difficult to remove the contaminants (Wasiuddin et al. 2002). The same principle can be applied for gas–gas separation, in which natural fibers, such as human hair, can replace hollow fibers that are toxic in the long term.

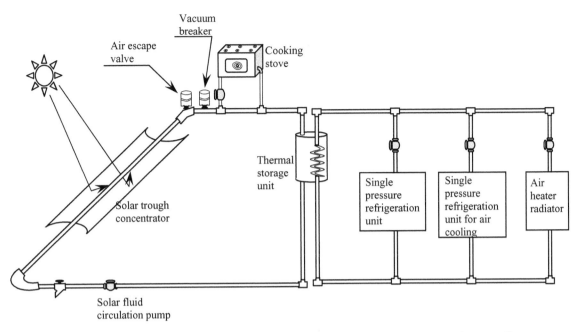

Figure 7-27   Direct solar heating can generate steam and run air conditioners at the same time (Khan and Islam 2006e).

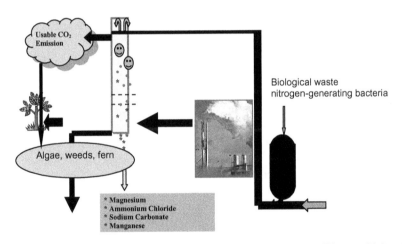

Figure 7-28   Schematic of an environmentally friendly desalination unit (Khan and Islam 2006e).

The next example is from thermal applications. One of the most expensive components of steam injection is steam generation. Usually combustion is used to heat water. Because this scheme requires fresh water, produced water cannot be used. If solar heating is used directly (e.g., solar trough to heat a reservoir of vegetable oil, as reported by Khan et al. 2006a), very high temperatures can be reached in the reservoir, which would be sufficient to run a steam generator. By using Einstein's absorption cycle, residual heat can be used to run a refrigerator or an air conditioner that would need no moving parts, hence eliminating the need of electricity. Figure 7-27 shows the schematic of the entire process. Note that this particular figure refers to a stove, easily modified into a steam generator. As for the fresh water needed, yet another scheme can be used in desalination. Figure 7-27 shows a recently developed desalination technique that replaces a current patent. The patent calls for pure CO_2 and ammonia, whereas the revised process uses exhaust as the

Figure 7-29   The use of olive oil clearly protected metal coupons from microbial corrosion (Al-Darbi et al. 2003).

CO_2 source and sewage water for ammonia source. As a result, both of these waste materials are converted into useful materials, while eliminating the high cost of chemicals, such as pure CO_2 and ammonia. This same process can also produce refined oil. Solar heating can be used to heat distillation columns and natural additives, such as zeolite and limestone.

One of the most important sources of toxicity comes from using catalysts, which should be replaced with natural alternatives. The economic benefit of such a process is obvious, but the environmental and social impacts are even more astounding.

The next example concerns corrosion. Billions of dollars of chemicals are injected into pipelines and down holes, to eliminate microbial corrosion. Expensive and highly toxic coatings are painted onto pipelines to prevent such corrosion. These applications are not only expensive but are also toxic. Al-Darbi et al. (2002, 2003) have demonstrated a number of naturally available chemicals that can efficiently destroy corrosive bacteria. For instance, olive oil, mustard oil, and fish oil all can protect metal from microbial corrosion. Figure 7-29 shows some metal coupons that are immersed in a culture of sulfate-reducing bacteria. The coupons immersed in olive oil are better protected than expected from either plastic or metallic coating (electroplated).

7.10   Alternative Technologies for EOR

7.10.1   *Environmentally friendly alkaline solution for EOR*

The history of chemical flooding dates back to the early 1920s, with its roots in the combination of reservoir engineering and chemistry. Alkaline or caustic flooding began in 1925 with the injection of a sodium carbonate solution in the Bradford area of Pennsylvania (Nultine et al. 1927; Mayer et al. 1983). The alkaline flooding process is simple when compared to other chemical floods, yet it is sufficiently complex to require detailed laboratory evaluation and careful selection of a reservoir for application. Caustic flooding is an economical option because the cost of caustic chemicals is low compared to other tertiary enhancement systems.

Alkaline water flooding is an old recovery process in which pH of the injected water is increased by the addition of relatively inexpensive chemicals. Many crude oils naturally contain a certain amount of organic acids (Thibodeau et al. 2003; Zhang et al. 2004). When this acidic oil becomes displaced by an alkaline solution, chemical reactions will occur at the interface. These reactions produce surface-active agents, which will in turn reduce the interfacial tension between the two fluids, emulsified oil and water (Cooke et al. 1974; Islam et al. 1990). The reduction in IFT between acidic oil and alkaline water depends on the pH of the water, the concentration and type of organic acids in the oil, and the concentration and type of salts in solution (Ramakrishnan and Wasan 1983; Elkamel et al. 2002).

Alkali and alkali/polymer solutions are well-known techniques for the chemical flooding application. For this scheme, synthetic high-pH alkaline solutions are commonly used. These solutions are not environmentally friendly and are expensive, and alkaline flooding has lost its appeal in last few decades. Recently, Rahman et al. (2006) developed low-cost, environmentally friendly alkaline solutions from local sources, such as wood ash.

7.10.2   Mechanisms of alkaline flooding

Some researchers (Johnson 1976; deZabala et al. 1980; Islam et al. 1989; Turksoy and Bagii 2000) reported that there are four common mechanisms that contribute to improved oil recovery with chemical methods:

1. emulsification and entrainment;
2. wettability reversal water-wet to oil-wet;
3. wettability reversal oil-wet to water-wet; and
4. emulsification and entrapment.

Other related mechanisms include emulsification with coalescence, wettability gradients, oil-phase swelling, disruption of rigid films, and improved sweep resulting from precipitates altering flow (Campbell 1982). All the postulated mechanisms have some superficial similarities. Laboratory experiments (Subkow 1942; Robinson et al. 1977) and field trials (Emery et al. 1970; Cooke et al. 1974) have shown that alkaline flooding performance will depend on:

1. water composition;
2. rock oil composition;
3. rock type and reactivity; and
4. alkaline concentration, especially how it interacts with the previously mentioned parameters.

7.10.2.1   Alkaline chemicals

The chemicals most commonly used for alkaline flooding are sodium hydroxide (NaOH), sodium orthosilicate (Na_4SiO_4), sodium metasilicate (Na_2SiO_3), sodium carbonate (Na_2CO_3), ammonium hydroxide (NH_4OH), and ammonium carbonate ($(NH_4)_2CO_3$ (Jennings 1975; Novosad et al. 1981; Larrondo et al. 1985; Burk 1987; Taylor and Nasr-El-Din 1996; Almalik et al. 1997). Due to reservoir heterogeneity and the mineral compositions of rock and reservoir fluids, the same alkaline solution might induce a different mechanism. A number of laboratory investigations, dealing with the interaction of alkaline solutions with reservoir fluids and reservoir rocks, have been reported (Ehrlich and Wygal 1977; Campbell and Krumrine 1979; Ramskrishnan and Wasan 1982; Trujillo 1983; Sharma and Yen 1983). Due to its higher pH value, sodium hydroxide is considered the most useful alkaline chemical for use in oil recovery schemes (Campbell 1977). The price comparison of the most common synthetic alkaline substances, between 1982 and 2006, is shown in Table 7-5. It shows that alkaline price has increased 5 to 12 times higher during the last 15 years.

The greatest challenge of any novel recovery technique is to be able to produce under attractive economic and environmental conditions (Islam 1999; Khan et al. 2006e). Due to the high cost of synthetic alkaline substances and the environmental impact, the alkaline flooding has lost its popularity (Figures 7-30 and 7-31). The graphs have been generated using data reported by Moritis (2004). However, cost effective alkali might recover its popularity in the recovery scheme. It has become a research challenge for the petroleum industry to explore the use of low-cost natural alkaline solutions for EOR during chemical flooding. Rahman

Table 7-5   Comparison of price and physical properties of most common alkalis (Mayer et al. 1983; Chemistry Store 2005; ClearTech. 2006).

Name of alkali	Formula	Molecular weight	pH of 1% solution	Na_2O (%)	Solubility (gm/100cm³)		Price range (dollar/ton) in 1988 (Mayer, et al. 1983)	Price range (dollar/ton) in 2006 (ClearTech 2006 Chemistry Store 2005)
					Cold water	Hot water		
Sodium hydroxide	NaOH	40	13.15	0.775	42	347	285 to 335	830
Sodium orthosilicate	Na_4SiO_4	184	12.92	0.674	15	56	300 to 385	1385
Sodium metasilicate	Na_2SiO_3	122	12.60	0.508	19	91	310 to 415	1340
Ammonia	NH_3	17	11.45	–	89	7.4	190 to 205	1920
Sodium carbonate	Na_2CO_3	106	11.37	0.585	7.1	45.5	90–95	1400

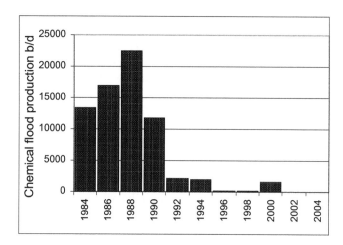

Figure 7-30   Total oil production by chemical flooding projects in the United States.

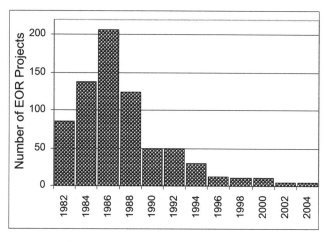

Figure 7-31   Chemical flooding field projects in the United States.

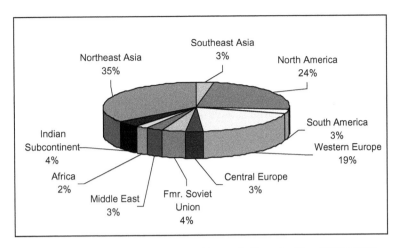

Figure 7-32   Distribution of global chloro-alkali production (CMAI 2005).

et al. (2006) used wood ash extracted solution was used as a low-cost natural alkaline solution. Several experiments were conducted to test the feasibility of this natural alkaline solution. They offered the following observations:

Toxicity of the alkalis: Alkali is one of the most commonly used chemicals in a variety of applications. It has a wide range of applications in different industries, such as petroleum refinery, pulp and paper mills, battery manufacturer, cosmetics, soap and detergent, leather processing industry, metal processing industry, water treatment plants, etc. The estimated worldwide demand of sodium hydroxide was 44 million tons expressed as NaOH 100% in 1999. The global demand is expected to grow by 3.1% per year (SAL 2006). CMAI (2005) reports that 62 million tons of alkalis were produced in 2005 alone (Figure 7-32). Alkalis are raw commercial products, and when transferred to other parts of the manufacturer's plant for use in further chemical processing, there is always the risk of leakage.

Every year huge amounts of synthetic alkalis are produced, all responsible for direct or indirect pollution of the environment. These alkalis also have significant adverse effects on human health. Inhalation of dust, mist, or aerosol of sodium hydroxide and other alkalis may cause irritation of the mucous membranes of the nose, throat, and respiratory tract. Exposure to the alkalis, in solid or solution, can cause skin and eye irritation. Direct contact with solid or concentrated solutions causes thermal and chemical burns, leading to deep-tissue injuries and permanent damage to such tissue (MSDS 2006; ATSDR 2006). Haynes et al. (1976) reported that a single dose of 1.95 grams of sodium hydroxide can cause death.

The chemistry of wood ash: Wood ash is a by-product of combustion in wood-fired power plants, paper mills, and other wood burning facilities. A huge amount of wood ash is produced annually worldwide and approximately three million tons is produced annually in the United States alone (SAL 2006). Wood ash is a complex heterogeneous mixture of all the non-flammable, non-volatile minerals that remain after the wood and charcoal have burned away. Because of the presence of carbon dioxide in the fire gases, many of these minerals will have converted into carbonates (Dunn 2003). The major components of wood ash are potassium carbonate "potash" and sodium carbonate "soda ash". From a chemical point of view, these two compounds are similar. In the United States, from the 1700s through to the early 1900s, wood was combusted to produce ash for chemical extraction. Wood ash was mainly used to produce potash for fertilizer and alkali for industry. On average, the burning of wood results in about 6–10% of ash. Ash is an alkaline material with a pH ranging from 9–13 and due to its high alkalinity characteristics, has various application in different sectors as an environmentally friendly alkaline substance.

Laboratory Study: Rahman et al. (2006) collected wood ash samples from a wood furnace and the ash samples were sifted with a sieve size 30 to remove as much of the charcoal as possible. Screened wood ash

Table 7-6   Physical properties of the crude oil

Sl No	Physical properties	Value
01.	Specific gravity	: 0.7 to 0.95
02.	Vapor pressure	: >0.36 Kpa at 20°C
03.	Vapor density	: 3 to 5 (approx),
04.	Freezing point	: −60°C to −20°C
05.	Viscosity,	: <15 centistokes at 20°C
06.	Solubility	: Insoluble
07.	Co-efficient of water/oil distribution	: <1

samples and synthetic sodium hydroxide were placed in beakers in differing amounts. Different concentrated alkaline solutions were prepared for laboratory testing and the alkalinity of the solutions was measured with a pH meter.

Natural alkaline solutions, at different concentrations, were placed on microscope slides and crude oil droplets were added with the help of a needle tip. The properties of the crude oil are shown in Table 7-6. The process of coalescences and flocculation of the oil droplets were observed by the Carl Zeiss light microscope attached with Axiovision 4.0 software and an AxioCam digital camera. Microscopic digital images of the crude oil were captured every five seconds and the images of the oil droplets were analyzed using image-processing software.

IFT measurement using Du Nouy ring method: Interfacial tension (IFT) is an important physical property. It characterizes interfaces between two immiscible liquids, frequently encountered, and has wide application in many industries (Alguacil et al. 2006). There are a number of techniques available for measuring IFT. The equipment price ranges from a few thousand to over hundreds of thousands dollars. However, the accuracy of measurement does not increase with increasing price. Most often equipment price is high, not because of the advanced physics used, but because of automatization, data display, and other matters, which are not at the core of the measurement. Automated instrumentation takes away the analytical thinking of researchers and makes them dependent on the manufacturers. Some of the most sophisticated measurements, based on simple phenomena, can be measured through inexpensive devices. One such method in IFT analysis is the Du Nouy ring method (Lecomte du Nouy 1919; Couper 1993). In this method, the liquid is raised until it is in contact with the surface. The sample is then lowered so that the liquid film produced beneath the ring is stretched. As the film is stretched, a maximum force is experienced that is recorded in the measurement. At the maximum of the force, the vector is exactly parallel to the direction of motion; at this moment the contact angle θ is 0 (Figure 7-33). The following illustration shows the change of force as the distance of the ring from the surface increases. In practice, the distance is first increased until the area of maximum force has been passed through. The sample vessel containing the liquid is then moved back so that the maximum point is passed through a second time. The maximum force is only determined exactly on this return movement and used to calculate the tension. The calculation is made according to the following equation (KRÜSS 2006):

$$\sigma = \frac{F_{\max} - F_{\mathrm{V}}}{L \cos \theta} \tag{7.5}$$

where

σ = surface or interfacial tension

$F_{\max}$ = maximum force

F_{V} = weight of volume of liquid lifted

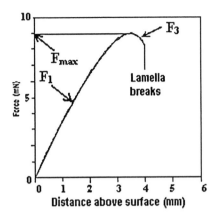

Figure 7-33 Change of force with ring distance. Redrawn from KRÜSS (2006).

Figure 7-34 Before filtration of alkaline solution (6% wood ash).

Figure 7-35 After filtration of natural alkaline solution (6% wood ash).

L = wetted length

θ = contact angle.

The contact angle, θ decreases as the extension increases and has the value $0°$ at the point of maximum force, meaning that the term $\cos \theta$ has the value 1.

Details of investigation: Figure 7-34 shows the wood ash solution preparation. Initially this solution looked turbid, but after extraction is clear (Figure 7-35). pH values of the natural alkaline solutions at different percentages (1%, 2%, 4%, and 6%) are presented in Table 7-7. The alkalinity of 6% wood ash solution is close to 0.5% synthetic sodium hydroxide solution. This value is also close to the pH of 0.75% Na_2SiO_3 solution (Green and Willhite, 1998). The alkaline solution of pH range 12–14 is a strong base. During the alkaline flooding, pH value of the synthetic alkaline is maintained in the range 11.5–13.5 as a common practice.

Table 7-7   Comparison of alkalinity between natural alkaline solution extracted from wood ash and synthetic sodium hydroxide solution at different concentrations.

	Synthetic sodium hydroxide solution	
	Synthetic NaOH solution (%)	pH value
Synthetic sodium hydroxide solution (NaOH)	1.5% NaOH solution	13.05
	1.0% NaOH solution	13.02
	0.5% NaOH solution	*12.35*
	0.2% NaOH solution	11.85
	Wood ash solutions	
	Percentage of wood ash solution	pH value
Wood ash solution	6% wood ash solution	*12.27*
	4% wood ash solution	11.94
	2% wood ash solution	11.61
	1% wood ash solution	11.22

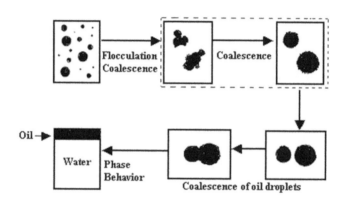

Figure 7-36   Schematic illustration of different steps in droplet growth during coalescence.

Therefore, the natural alkaline solution extracted from the 6% wood ash can be used instead of the 0.5% synthetic sodium hydroxide solution or the 0.75% synthetic sodium metasilicate solution during the chemical flooding scheme in an acidic reservoir. Burk (1987) reported that Na_2CO_3 solutions are less corrosive to sandstone than NaOH or Na_4SiO_4. Sodium carbonate (Na_2CO_3) buffering action can reduce alkali retention in the rock formation. The main components of wood ash is carbonate salts (Na_2CO_3 or soda ash and K_2CO_3 or potash). Carbonate salts offer an additional advantage upon contact with hard water. The resulting carbonate precipitations do not adversely affect permeability as compared to the precipitations of the hydroxides or silicate (Cheng 1986). Therefore, it is suggested that the use of carbonate buffer solution extracted from wood ash might result in longer alkali breakthrough times and increased tertiary oil recovery.

The microscopic study of oil–oil droplet interaction in wood ash extracted solution showed that the oil–water interface changes with time and its effects on oil–oil droplets coalescence. The natural alkaline solution reacted with the organic acids in the oil, producing a surfactant. This surfactant contained hydrophilic molecules and hydrophobic molecules that started to form a layer around the oil droplet called "micelles" and caused the smoothing of surfaces, resulting in less interfacial friction. Once the micelle is formed, mobility of the oil droplets increases and they move faster under the influence of the buoyancy force, which results in drainage of thin surfactant water film at the contact between flocculating oil droplets (Figure 7-36). This film reaches the critical thickness at which it ruptures and oil droplets coalesce to form a larger globule

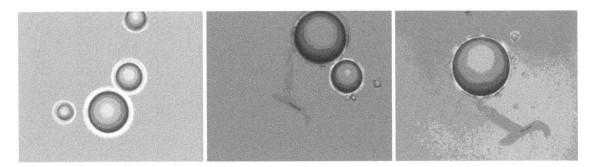

Figure 7-37   Coalescence of oil droplets in natural alkaline solution.

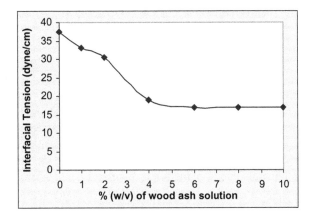

Figure 7-38   Interfacial tension vs different concentration of wood ash solution at 22°C.
(From Rahman et al. 2006).

(Figure 7-37). It was also found that two oil droplets were coalesced after 3.5 minutes in 6% wood ash solution that contain the same alkalinity of 0.5% NaOH and 0.75% Na_2SiO_3 solutions.

IFT measurements between a crude oil and an alkaline solution have generally been accepted as a screening tool to evaluate the EOR potential of the crude oil by alkali (Jennings 1975). Recently, Mollet et al. (1996) showed, in an experimental study, that a minimum IFT is unobserved when alkali is absent in the aqueous phase. From experimental studies, it was found that IFT gradually decreased with increasing concentration of natural alkaline solution (Figure 7-38). It was also observed that IFT decreased up to a certain limit with pH (Figure 7-39). The higher the concentration of the alkaline solution, the more the surface active agent reacts with organic acid in the crude oil and alkali in the aqueous phase. Hence, this surface active agent (petroleum soap) can cause the decrease of interfacial tension and increase the mobility of oil in the continuous water phase.

Based on the experimental results presented by Rahman et al. (2006), the following conclusions are reached. This study supports the basic idea of the applicability of natural alkaline solution for the EOR scheme. However, a wide range of experiments from different viewpoints may increase this appealing field application:

- The natural alkaline solution extracted from wood ash is highly alkaline and the alkalinity (pH value) of 6% wood ash solution is close to 0.5% synthetic sodium hydroxide and 0.75% synthetic sodium metasilicate solution.

- Wood ash extracted alkaline solution reduces the interfacial tension with crude oil, which helps to increase oil mobility in aqueous phase.

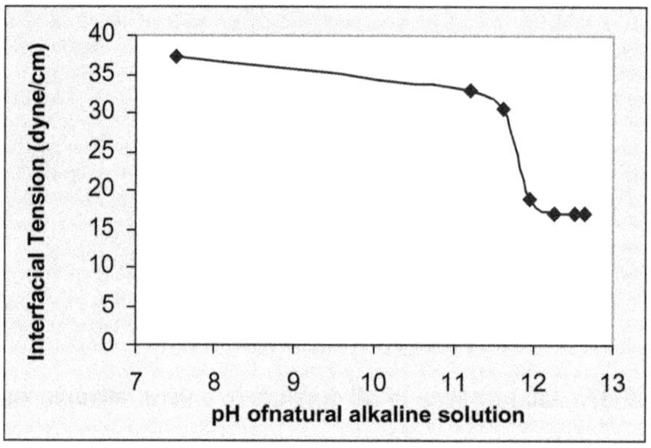

Figure 7-39   Interfacial tension vs pH of natural alkaline solution at 22°C.

- Coalescence time of oil droplets seem to be strongly influenced by the early micelle-forming stage. It is also dependent on the decrease in interfacial tension. From the study, it is observed that coalescence time of oil–oil droplets decreases with the increasing pH and two oil droplets are coalesced after 3.5 minute in 6% wood ash solution. Wood ash extracted alkaline solution, which contains mainly soda ash (Na_2CO_3) and potash (K_2CO_3), could be more advantageous than the alkaline solution of NaOH or Na_4SiO_4 for alkaline flooding, because the buffered slug would be less reactive with sandstone minerals due to reduction of hydroxyl ion activity.

- Wood ash extracted alkaline substances are environmentally friendly and naturally abundant, whereas injected synthetic alkaline solutions are cost effective and environmentally toxic and harmful.

7.11 CO_2 EOR Technology

7.11.1 Progress in carbon-dioxide EOR

Carbon dioxide started to gain popularity in 1990s (Islam 1990). At present, it is the fastest-growing EOR process in the United States (NETL 2004). While production volumes and the number of projects for thermal, chemical, and other EOR processes have fallen off sharply since the mid-1980s, the number of CO_2 projects has remained steady or increased slightly and CO_2 production volumes have jumped sevenfold (Figure 7-40). The CO_2 share of US crude oil production was estimated at almost 206,000 b/d in 2004 (EOR Survey 2004). That is about 4% of the national total. If this trend continues, immiscible carbon dioxide could well become the most dominant nonthermal EOR technique for heavy oil reservoirs (Islam 1990). The EOR technique that is attracting the most new market interest is carbon dioxide (CO_2)-EOR. Until recently, most of the CO_2 used for EOR has come from naturally occurring reservoirs. However, new technologies are being developed to produce CO_2 from industrial applications, such as natural gas processing, fertilizer, ethanol, and hydrogen plants in locations where naturally occurring reservoirs are not available.

Islam and Chakma (1992) introduced the concept of CO_2 injection in order to sequester CO_2. At that time, they also argued that most depleted reservoirs could be used as a sink for CO_2 sequestration. During this injection period, they anticipated a rise in reservoir pressure leading to additional oil recovery from wells that might have been abandoned previously.

7.11.2 Mechanism of CO_2 oil recovery

Miscible flooding with carbon dioxide or hydrocarbon solvents is considered one of the most effective EOR techniques applicable to light to medium oil reservoirs (Malik and Islam 2000). CO_2 has a viscosity similar

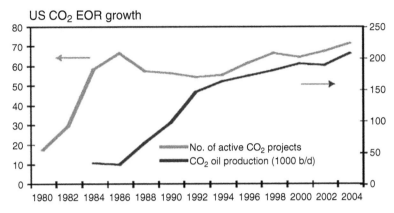

Figure 7-40   Carbon dioxide EOR growth in the United States (EOR Survey 2004).

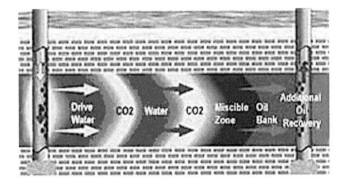

Figure 7-41   Graphic of CO₂ EOR. Courtesy of Occidental Petroleum Corp., DOE, 2004.

to hydrocarbon miscible solvents. Both types of solvents affect the volumetric sweep-out because of unfavorable viscosity ratio. However, CO_2 density is similar to that of oil. Therefore, CO_2 floods minimize gravity segregation compared with the hydrocarbon solvents. The miscible displacement between crude oil and CO_2 is caused by the extraction of hydrocarbons from the oil into the CO_2 and by dissolution of CO_2 into the oil. Light and intermediate molecular weight hydrocarbon fractions, as well as the heavier gasoline and gas oil fractions, are vaporized into the CO_2 front. Consequently, vaporizing-gas drive miscibility with CO_2 can occur with few or no C_2 to C_6 components present in the crude oil.

Figure 7-41 shows the CO_2 EOR method. When injected into the reservoir, CO_2 is dissolved in the oil. As a result, oil viscosity is reduced and mobility is increased. The efficiency of EOR depends on the pressure and thus on reservoir depth. Carbon dioxide is miscible with reservoir oil at high pressures and greater miscibility has cost benefits associated with increased oil recovery. The threshold pressure, above which miscibility occurs, is called the Minimum Miscibility Pressure (MMP). Miscible CO_2 displacement results in approximately 22% incremental recovery, while immiscible displacement achieves approximately 10% incremental recovery (Islam and Chakma 1992). Therefore, there is a greater payback for miscible displacement. For this, however, deeper reservoirs are preferred for which pressures are above the MMP. The minimum miscibility pressure depends on the composition of the oil, higher density and higher viscosity oils with more multiple aromatic ring structure having a higher MMP (Islam and Chakma 1992). Historically, CO_2 for EOR has only been used on oils with API gravity values greater than 22 and viscosities lower than 10 cp, because of greater miscibility and higher recovery efficiencies (Bergman et al. 1996).

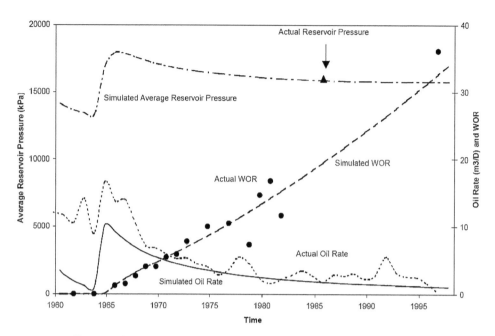

Figure 7-42   History match of studied reservoir (Malik and Islam 2000).

Recently, a billion-dollar CO_2 injection project was launched to increase oil recovery in the Weyburn field of Canada (Malik and Islam 2000). Initially, the project uses 5000 tonnes of CO_2 daily in 19 patterns. Malik and Islam (2000) find successful history matching, optimize secondary and tertiary recovery schemes, predict an incremental oil recovery through improved oil recovery, and give an estimate of the subsurface storage capacity of greenhouse gases (GHG). The validity of the models developed is evident from the agreement obtained and the conclusions drawn from these diverse data sources that enabled a successful history match of the fluid movement in the reservoir. The models are robust in nature and helped determine key parameters controlling miscible CO_2 flood. The effects observed due to the contaminated gas stream highlight several of the significant unexplained phenomena faced by the industry. They have the role of impurities in controlling mobility of the injected gas, effect of impurities in the GHG stream, effect of the loss of miscibility, effect of gas injection directly into the bottomwater zone, and the effect of bottomwater transmissivity on oil recovery.

A strategic conflict occurs toward the optimal operating conditions for simultaneous objectives of higher recovery and higher CO_2 storage. In this study, horizontal injection wells have proved to be efficient for CO_2 flooding processes to improve recovery while increasing the storage of anthropogenic CO_2. Twenty-one different scenarios for two different schemes were simulated and investigated simultaneously for storage and recovery. The incremental recovery is related to the flood injection operating strategies employed, introducing back pressure on the reservoir through injectors and producers. The CO_2 flood front is controlled through horizontal well adjusting pressure, simultaneously adjusting water injection in the offsetting vertical injection wells and holding back pressure on the associated production wells. Efficient back pressure is achieved by limiting high operating bottomhole pressure of the producers corresponding to those of injectors that helped maximize the vertical sweep. In addition, opting for the GOR strategy and location of the horizontal well, meant that optimal injection rates helped to achieve conformance within the reservoir, enabling the conflict of achieving the simultaneous objectives to be overcome.

A field-scale match is successfully achieved by matching 45 years of actual primary production and water flood history, by employing a 3D reservoir simulator. Due to lack of the detailed profiles of the reservoir, the primary production profiles for a typical well in the field is constrained to the cumulative production

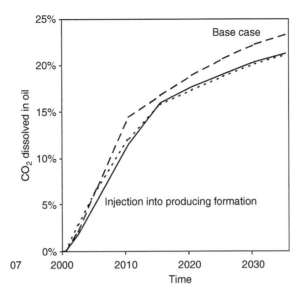

Figure 7-43   CO₂ Injection (Malik and Islam 2000).

profiles on a typical well basis being generalized for the field case. The criterion adopted is to history match different parameters simultaneously. The waterflood recovery from 1964–1998 is matched without any major fine-tuning of the model. Figure 7-43, shows the production profiles and the average reservoir pressures for primary and waterflood. The actual and simulated production profiles are in good agreement, representing a successful history match for the primary and secondary recovery mechanisms that provide a quality test to validate the simulated fluid movement within the reservoir over the years. Simulated average reservoir pressures are in excellent agreement with the monitored values, showing that the modeled reservoir phase behavior corresponds to the field conditions. Similarly, simulated water oil ratio (WOR) is in good agreement with the actual WOR. The same is true for GOR. The simulated GOR is within the acceptable range of the actual GOR, reported between 17 and 32 m^3/m^3.

Malik and Islam (2000) also simulated horizontal CO_2 flooding after primary for different scenarios, and reservoirs with bottomwaters have been simulated for two different schemes. CO_2 flood has responded differently to varying scenarios and impurities. These behaviors have been attributed to the probable development of asphaltenes, blockage of small pores, early miscibility obtained, better conformance achieved, which is related to fluid flow mechanics, and the development of different processes such as double displacement oil recovery process. Other attributions for CO_2 injection, with additional impurities into bottomwaters, have been made, similar to those mentioned by Islam and Farouq Ali (1989). They reported that in oil recovery, injection pressure, oil-to-water viscosity ratio, and permeability ratios appear to have a prominent role when injecting surfactants to develop flood, and similar attributions have been made for CO_2 flooding. One of the exhibits is attributed to the injection of CO_2 into water flooded oil zones that results in double displacement oil recovery processes, which is a response to injecting gas into the water flooded oil zones.

Figure 7.43 delineates a plot of CO_2 dissolved in the oil phase during the tertiary recovery process for two scenarios and schemes resulting in improved oil recovery. In Scheme-2, Case-2, the initial percentage of CO_2 dissolved in bottomwater is higher than in Case-1 of the same scheme and is marginally lower in the later part of the flood. The percentage of CO_2 dissolved in Scheme-2 corresponds to the production. Percentage of CO_2 dissolved in Scheme-2, Case-1 is unexpectedly low initially. This may be attributed to the location of a horizontal well that is in the producing formation adjacent to bottomwater layers for horizontal floods.

7.11.3 Improved oil recovery for post-primary CO_2 injection

Figure 7-44 shows cumulative oil produced for primary and post primary CO_2 flood adopting simultaneous straight gas injection technique. The two case scenarios, comparing base case to Scheme-3, cumulative oil produced varies significantly. It is interesting to note that Scheme-3 results in an increase of cumulative oil produced as a result of post primary CO_2 flood compared to post waterflood injection (Scheme-2). This is because CO_2 helps oil to swell, decreases viscosity, and helps mobilize the oil while simultaneous waterflood helps to drive swelled oil toward the producers. The higher cumulative oil production for base case is expected. The presence of the bottomwaters affects oil production due to coning effects that result in relatively low oil production (Stright et al. 1976). Having achieved early miscibility and conformance, transmissivity between the layers plays a significant role. Injection of CO_2 into bottomwaters results in efficient solubility, because the unique low viscosity of CO_2 inhibits even smaller pores. The presence of the interaction between the layers, for Scheme-3, Case-2, results in a marginal increase in oil production. This optimal response is only possible when an effective differential pressure within the reservoir is achieved. Injection of CO_2 in the bottomwaters would create three-phase flow, reducing the effective permeability to water (Islam and Farouq Ali 1989). In addition, the injection of gas creates a drive that helps mobilize oil from the oil zone.

An aggressive CO_2 injection scenario was investigated, in order to improve production and storage potential, which should increase rate of return for the investors. However, results of this aggressive injection into bottomwaters has not been encouraging. Simulation aspect is also to investigate an alternate way of controlling emissions of the anthropogenic CO_2 from the environment by injecting into the oil reservoirs. Aggressive injection did not result in higher storage volumes in the bottomwaters.

Figure 7-45 shows the simulated average reservoir pressures for primary and post-primary CO_2 recovery process as fluid flow mechanics respond differently and correspond to the average reservoir pressure for reservoirs with and without bottomwaters. Initially reservoir pressure drops due to primary production. For the bottomwaters, pressures observed are comparatively lower than the base case. CO_2 flood response is prompt. The average reservoir pressure increases sharply for the base case as compared to Scheme-3. This is evident to a point where it stabilizes over the years and later breakthrough occurs. As a result of this breakthrough, pressure drops sharply over the lifetime of the CO_2 flood. This observed behavior is the same for all the scenarios. In response to aggressive simultaneous straight gas injection (SSGI) into bottomwaters, CO_2 flood has not differed from non-aggressive injection. SSGI for the base case energizes the reservoir to a maximum average reservoir pressure above 20 MPa. The instantaneous response to increasing pressure is due to first contact miscibility achieved. The breakthrough is in several years' time because the average

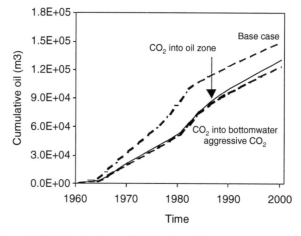

Figure 7-44   Post primary CO_2 injection (Malik and Islam 2000).

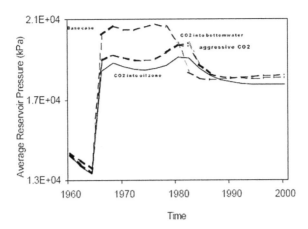

Figure 7-45   Post Primary CO$_2$ Injection (Malik and Islam 2000).

reservoir pressure has been maintained above MMP for the CO$_2$ gas stream. CO$_2$ injection for Scheme-3, Case-2, energizes the reservoir to an average reservoir pressure that is significant compared to Scheme-2. This is probably due to efficient solubility of CO$_2$ into bottomwaters and an inefficient transfer of CO$_2$ from water phase to oil phase.

As mentioned earlier, injection of CO$_2$ into the bottomwaters will create a three-phase flow, reducing the effective permeability to water (Islam and Farouq Ali 1989). In addition, the injection of gas creates a drive that helps mobilize oil from the oil zone. It is evident from the behavior exhibited that miscibility increases with increasing pressure to a point where breakthrough occurs in several years. The subsequent percolation of the CO$_2$ up through the oil zone provides maximum contact efficiency and keeps the reservoir energized.

According to Malik and Islam (2000), nitrogen contamination in the gas stream has an impact on tertiary recovery. The presence of contaminants decreases the solubility and diffusivity of carbon dioxide into oil consequently, leading to reduction in swelling of oil by carbon dioxide. The molecular diffusion, considered to be controlling mechanism in the carbon dioxide flood, is affected by an impure CO$_2$ gas stream that affects the diffusion rate of carbon dioxide into oil. The negative impact on production is due to the presence of the nitrogen content that tends to form a stagnant phase between the oil and carbon dioxide, through which carbon dioxide has to diffuse before contacting oil. The presence of nitrogen also increases the viscosity of carbon dioxide-oil mixture, resulting in an inefficient displacement from the pores. The low cumulative oil production, due to injection of an impure CO$_2$ gas stream into bottomwaters, is attributed to an inefficient transfer of CO$_2$ from the water phase to the oil phase. On the basis of the simulation results of Malik and Islam (2000), it is evident that bottomwaters have a significant impact on oil production and CO$_2$ storage capacity.

7.11.4   *Cost and environmental benefits*

Carbon dioxide flooding is found in the many projects in the United States and Canada (NETL 2004; Malik and Islam 2000). According to a recent DOE (2004) analysis, CO$_2$ EOR oil production in the United States could double by 2010 and quadruple by 2020, with CO$_2$ incentives. The CO$_2$ flooding methods can be replaced by other gas flooding processes, such as hydrocarbon injection for EOR. NETL (2004) reported over 300,000 b/d of oil is being produced by gas injection in the United States, which is replaceable by CO$_2$.

The benefits of CO$_2$ will include the following benefits:

- Extending the life of the petroleum reservoir (thus maintaining or increasing industry employment), increasing oil recovery, and expanding the range of reservoirs amenable to CO_2 flooding.

- Minimizing the other industries' greenhouse gas emission by using an industrial source of CO_2. However, currently available technology costs more to use industrial CO_2 than natural CO_2.

- Expanding the use of industrial CO_2, in the huge fossil fuel sector, could expand the applicability of CO_2 EOR while "closing the carbon cycle" (NETL 2004).

- Reducing of chemical cost by optimizing oil saturation tolerance of foam, decreasing primary foaming agent adsorption, and decreasing required primary foaming agent concentration.

- Oil industries are likely to get incentives, such as fiscal relief or greenhouse emission credit by using CO_2.

- This injection process might help delaying production of CO_2 and increasing retention of CO_2 in the reservoir (sequestration).

- Uses of CO_2 is an expensive process, because the gas is readily available at low cost. Kinder Morgan estimates total operating expenses exclusive of CO_2 costs at $2–3/bbl (NETL 2004). For the Weyburn field of Canada, cost was estimated as $35/tonne (Malik and Islam 2000).

7.12   Electromagnetic Heating for EOR

Electromagnetic heating is a technology that uses microwave energy and radio frequency to directly heat the reservoirs, in order to reduce the viscosity of oil. The application of electromagnetic heating includes radio frequency and microwave energy ranging broadly from 1 MHz to 30 GHz. Electromagnetic energy is transformed in the form of heat energy in dielectric, inductive, and resistive conductor materials. Dielectric properties of materials affect the way transmitted waves react under reservoir conditions, which depend on the frequency of operation, operating temperature, electric properties of material, concentration, and materials in the pathway of transmission. The dielectric properties of crude oil, water, and mixture also play an important role in the electromagnetic heating. The crude oil has low dielectric constant (2–3) and is weak in absorbing microwaves (Jackson 2002).

In cases where other EOR methods are not applicable, due to permafrost and other environmental constraints, electromagnetic heating may be the best alternative option (Sahni et al. 2000). Heating the reservoirs with electrical energy can be done in two ways: i) high frequency (radio frequency (RF) and microwave range); and II0 resistive or inductive heating. At high frequencies, dielectric heating prevails, where the dipoles formed by molecules tend to align themselves, results in rotational movement with velocity proportional to the frequency of alteration that generates heat. In the case of resistive heating, a low frequency alternating current is used where ohmic loss, $P_{dissipated} = I^2R$. Figure 7-46 is the schematic of resistive heating where two electrodes are placed close to the producing well. When the current is forced to flow in the reservoir between the electrodes, heat is generated in the reservoir due to ohmic losses. The electrodes are located in direct contact with the oil formation.

In microwave heating (Figure 7.47), an antenna is placed in a drilled hole close to producing well. The microwave energy heats the fluid in the reservoir moving toward the producing well. The power dissipated by high frequency electromagnetic heating can be expressed by Maxwell's equation, $p = \sigma E^2$, where p is power dissipated, σ is conductivity of medium, and E is the root mean square electric field intensity in volts per meter.

7.12.1   *Current electromagnetic heating practices*

Abernethy (1976) developed a mathematical model to evaluate the temperature distributions and other physical effects resulting from the radiation of electromagnetic energy in an oil reservoir. He compared the flow performance to the unheated case, on the basis of steady state temperature distribution derived from the mathematical model. He reported that under a wide variety of operating circumstances, the flow rate was

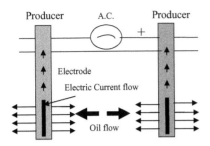

Figure 7-46 Schematic of resistive heating.

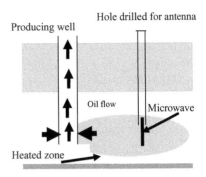

Figure 7-47 Schematic of electromagnetic heating.

doubled within four weeks of radiation heating. He also reported an increase of flow rate by 300% under specific conditions under direct electrical heating.

Sresty et al. (1986) performed the laboratory test as well as a pilot plant scale test for the recovery of bitumen from tar sand deposits, with the radio frequency process for Utah tar sands. They reported that after the *in-situ* heating of the tar sand deposit with radio frequency, 50–80% of the bitumen was recovered depending upon the temperature of heating and recovery method used. The recovery methods applied were gravity drainage, gas drive, or fluid replacement methods. The ratio of energy recovered to energy input ranged from 3 to 12, depending on the total quantity of bitumen recovered. However, the study does not consider the global efficiency of the system. Several environmental effects, due to the radio frequencies, have been overlooked when calculating the net energy out ratio. Pizarro and Trevisan (1989) presented low frequency electrical heating data from a field test performed at the Rio Panan oil field in Brazil. They reported that the production increases from 1.2 Bbls/day to 10 Bbls day after 70 days of applying 30 kW power across the producing wells, which were 328 feet apart.

Fanchi (1990) carried out a feasibility test with a numerical simulator for using electromagnetic irradiation as a stimulation tool. The study concluded that the electromagnetic power attenuates exponentially in a linear, homogenous, and dielectric medium. It was reported that the results of simulation were sensitive to input data such as electrical conductivity, water saturation, and relative electric permittivity. However, the assumption of attenuation of electromagnetic power exponentially in linear, homogenous, and dielectric medium is not practical as it is impossible to obtain such ideal medium. Moreover, this model did not account for the inter-phase mass transfer. As the microwave heating continues, *in-situ* water vaporizes, which lowers the electrical conductivity of the system.

McGee and Vermeulen (1991) reported that in a horizontal well, the heat transfer to the adjacent reservoir by conduction is more significant than heat transfer by convection and electrical heating. They also reported that the safe operating magnitude of the current in a horizontal well is limited by the cooling effects of the produced fluids. However, the calculation for heat transfer in the horizontal well was based on the assumption that the reservoir was homogenous and all physical properties such as thermal and electrical conductivity, permeability, porosity, reservoir thickness, and heat capacity were constant and do not change with temperature. Such a case can never be achieved in a real field situation.

Islam et al. (1991) investigated the EOR by electromagnetic heating along with various fluid injection methods in horizontal wells. They showed that the vertical fluid injector along with a horizontal well heated electromagnetically recovered 40% oil initially in place in both the upper and lower Ugnu tar sand formation of Alaska. This application was considered useful even if there is permafrost in the formations. Islam and Chakma (1992) studied the use of electromagnetic heating on recovery of heavy oil reservoirs containing a bottomwater zone. They reported that electromagnetic heating is practically limited to small radii but becomes effective in combination with horizontal wells. It was further reported that the recovery could be enhanced by coupling the electromagnetic heating with gas or water injection in order to create the favorable pressure gradient in the presence of bottom water. The model showed that a recovery as high as 77% of oil in place (OIP) was obtained. The study reported that the velocity of the gas–oil front near the wellbore was doubled by increasing the temperature up to 140°C. However, only reducing near-wellbore viscosity does not increase the frontal velocity to the same degree of temperature increment.

Chakma and Jha (1992) developed the laboratory model for heavy oil recovery from thin pay zones, by electromagnetic heating in combination with gas injection in horizontal wells. The study showed that the heat losses in the reservoir, in the case of electromagnetic heating, could be kept to a minimum by confining the heating to the oil bearing zone only. Water saturation, salinity of brine, and frequency of electromagnetic waves are the major factors affecting the heat transfer in the reservoir. The presence of water is essential for continuous propagation of waves. The study reported the yield of 45% of oil, initially in place of heavy oil of moderate viscosity, was recovered.

Kasevich et al. (1994) carried out a laboratory study of radio frequency heating with low permeability diatomic samples in a 55-gallon drum heated with 400 watts and 50.55 MHz power source, increased the temperature by 125°C after 49 min of heating. A field test was carried out using a mobile RF heating system with a 25 kW and a 13.56 MHz source. The RF antenna was 25 feet long, placed at a depth of 620 feet and enclosed in a 250-feet transparent composite liner. The borehole temperature, measured at 605 ft after 40 h of RF heating, rose to 220°F from 90°F of average formation temperature.

Soliman (1997) developed numerical as well as analytical model to study the effects of microwave technology for EOR. The study reported the solutions with and without considering the heat loss into the adjacent strata due to microwave heating. In the model, which considers no heat loss to adjacent strata, it was observed that 38% of the energy was absorbed at the front after the front was moved out by 5 m. The remaining 62% of the energy was dissipated ahead of the front to heat the formation. Where heat loss to adjacent strata was considered, the maximum energy dissipation decreased after 2.5 m distance from the front. However, this model has limitations. First, the thermal conductivity of the formation has been neglected, based on the assumption that heat flow due to fluid flow and front movement is more efficient than heat conduction in rocks. Second, the assumption for no heat loss at the adjacent boundary of the formation is not practical. Third, there may be a decrease in permeability to oil flow in the heated area due to the presence of the steam, which may decease the overall production efficiency. A laboratory study of frontal distance movement due to electromagnetic heating is presented in Figure 7-48 (Soliman 1997). Islam et al. (1998) have modeled microwave irradiation with a compositional simulator. With experimentally validated models, they constructed the compositional simulator and studied the effect of power (Figure 7-49). Recently, Mustafiz et al. (2006) investigated the role of ultrasonic and microwave irradiation on the separation of oil and water from a stable emulsion. This technique has the potential of increasing oil productivity when applied to a wellbore. The same technique can be applied to the recovery of oil from petroleum sludge (Bjorndalen

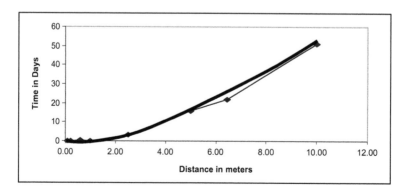

Figure 7-48   Frontal distance movement in the wellbore due to microwave heating (Soliman, 1997).

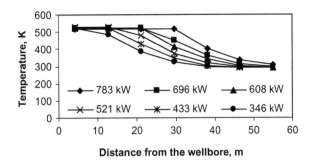

Figure 7-49   Heat propagation in the wellbore for various power (Islam et al. 1998).

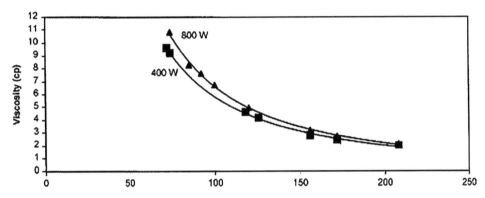

Figure 7-50   Alteration of crude oil with 8% asphaltene. Redrawn from Gunal and Islam (2000).

et al. 2004). Islam (1999) showed that the recovery of heavy oil or tar sands increases significantly if electromagnetic heating is used. The model showed that more than 8% of the recovery was possible, coupling with gas injection, especially in horizontal wells.

Gunal and Islam (2000) studied the application of electromagnetic heating as an effective option to recover heavy oil and gas from sandstone and carbonate reservoirs. The study investigated the extent of thermal alteration of crude oil properties due to electromagnetic or ultrasonic irradiation. Even though the efficiency of electromagnetic heating of reservoir was high, the presence of asphaltenes led to some irreversible alterations in crude oil rheology for comparable viscosity reduction (Figure 7-50). The viscosity reduction was not rapid, despite doubling the electrical power output. This study further reported that electromagnetic heating can be applied to carbonate formation as well as crude oil without the presence of water.

Vermeulen and McGee (2000) studied the *in-situ* electromagnetic heating for hydrocarbon recovery and environmental bioremediation. Increasing the reservoir temperature will not only reduce the viscosity of oil and bitumen but will increase the vapor pressure also. These effects continued to assist in sweeping the substances to be recovered from the formation. The study further reported that due to the heating, soil remediation was possible as the semi-volatile compounds were removed from soil faster than nonthermal removal of volatile compounds. In the case of bioremediation, every 10°C increase in soil temperature doubles the rate at which bio-matter populates and effectiveness of the bioremediation increases.

Sahni et al. (2000) modeled pre-heating of the reservoir, with thin sands separated by shale layers, with two horizontal electrodes operating at 300 V and 60 Hz (2 phase AC). They mentioned that the low temperature ohmic heating is a near wellbore effect with a temperature increase of 300°F near the two electrodes. At a distance of 100 feet from the well, the temperature increased by 75°F after 6 months of heating. Preheating

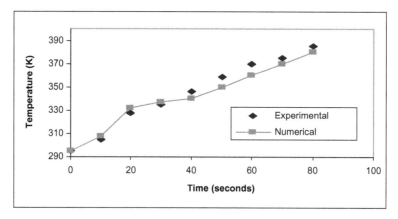

Figure 7-51   Experimental and numerical data for 50 ml crude oil under irradiation. Redrawn from Bjorndalen et al. (2003).

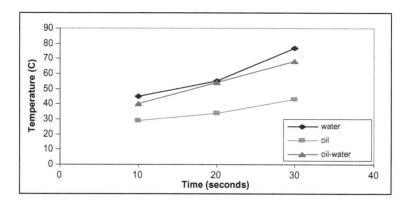

Figure 7-52   Temperature increment due to microwave heating (Chhetri et al. 2006).

provides a uniform distribution of heat in the reservoir followed by steam injection. Simulation results showed that electrical preheating significantly accelerated cumulative oil production compared to non-preheated case. The same study reported 80% increase in cumulative oil production over primary production when two 60 kW microwave sources were placed in the formation over 10 years.

Ovalles et al. (2002) carried out a study on the opportunities of downhole dielectric heating in Venezuela for medium, heavy, and extra heavy crude oil reservoirs. They evaluated three conceptual reservoirs, which involved medium crude oil (24°API), a shallow lake Maracaibo heavy oil (11°API), and a thin pay zone heavy oil (7.7°API). The simulation results showed an increase of 86% of cumulative oil production in a shallow heavy oil reservoir using 140 MHz radio frequency at 50 kW over a 10-year period. The net energy ratio calculated was also 8–20 times (output/input energy).

Jackson (2002) studied the effect of variable frequency in the case of heavy oil in North Carolina, United States. The major experimental parameters were time, frequency, and some chemical additives such as activated carbon, molybdic acid, and iron powders. Coke formation, due to heating, was reported as less than 2% by weight in the case of molybdic acid-iron powder combination, which was interpreted as low coke propensity due to selective heating.

Bjorndalen et al. (2003a) developed numerical models based on experimental results, and reported that the microwave irradiation can overcome the problem of asphaltene and paraffin precipitation out of the crude oil in horizontal wells (Figure 7-52). Another study reported on the effects of temperature rise due to microwave heating on various concentrations of crude oil, paraffin wax, bentonite, and gypsum under different exposure times (Bjorndalen et al. 2005). The study concluded that in an increase in density of the fluid, there is a decrease in the temperature change with time. The study further reported that the heterogeneous mixture in the reservoir can behave as a homogenous mixture, in which the crude oil content is about 10% with high water content.

Electromagnetic EOR is a relatively new method and has been of great interest in recent years. Recently, Chhettri and Islam (2006) conducted extensive study on electromagnetic heating of reservoirs for EOR. They found that the global efficiency of electromagnetic technology is low and suggested the use of sludge from the produced water or other waste material to produce biogas and then convert to electricity. The state of sustainability of electromagnetic EOR was also analyzed by Chhettri and Islam (2006) and showed that this EOR method is not sustainable for the long term.

7.12.2 Impact of irradiation on fluid and rock behavior

Chhettri and Islam (2006) carried out an extensive experiment with heating different fluids such as water, vegetable oil, and the mixture of water and vegetable oil with microwave heating. The heating temperature was recorded for 10, 20, and 30 seconds. Figure 7-52 shows the temperature raised due to microwave heating for these fluids. A mixture of oil and water reaches higher temperature compared to oil alone. The oil alone has least temperature rise because it has a low dielectric constant and is therefore weak in microwave absorption. This indicates that heating the oil in the reservoir is a complex phenomenon, as the reservoir fluid has a combination of several fluids of different physical, electrical, and chemical properties. As reservoirs generally have bottomwaters of different thickness, the absorption and dielectric properties of fluid as well as medium plays a significant role in the recovery mechanism.

Chakma and Jha (1992) reported that the presence of water is essential in the oil-bearing zone for continuous propagation of electromagnetic waves. The study further mentioned that the presence of higher salinity increases the conduction of the electromagnetic wave. However, a high thickness of bottom water may result in huge heat loss, causing inefficiency of the system. Islam and Chakma (1992) showed that the numerical model predicted higher cumulative oil recovery for 1 : 1 oil-water zone thickness compared to no bottomwater and higher bottomwater zone thickness. Henda et al. (2005) studied the microwave-enhanced recovery of nickel-copper ore and showed that ore concentrate recoveries and grade of nickel and copper are significantly enhanced after microwave treatment.

7.12.3 Efficiency of microwave heating

It is generally reported that microwave energy has high efficiency as it takes very little time to heat the materials. It may be true if only the local efficiency of a microwave unit is considered. However, if we consider the "global efficiency" from oil or gas burning in the combustion unit, the efficiency of the boiler to produce steam, the efficiency of the turbine and generator, the efficiency of transmission and distribution, this technology is not as efficient as projected. This concept of global efficiency was introduced by Khan et al. (2006e) and later expanded by Chhetri (2006). The electrical energy has to be converted to microwaves or radiofrequencies by using transformers and microwaves. Considering all the factors, the "total efficiency" of microwave is very low. Figure 7-53 shows the energy cycle for electromagnetic heating system.

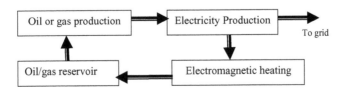

Figure 7-53   Energy flow in electromagnetic heating (Chhettri and Islam 2006).

The oil or gas produced from the reservoir is used again to heat the reservoir. The coal or oil boiler has an efficiency of about 35%, turbine efficiency is about 70%, generator efficiency is considered to be about 80%, and the transmission and distribution has efficiency of 90%. The efficiency of microwave energy itself is about 50%. The electricity production and transmission has significant environmental impacts. Thus, the overall or total system efficiency of the microwave system becomes less than 10%. From fuel to microwave, efficiency is greatly reduced.

7.12.4 Environmental impacts

Vermeulen and McGee (2000) proposed the use of microwaves for environmental remediation. They mentioned that the soil heating by microwaves can accelerate the removal of many contaminants such as gasoline and other volatile organic compounds. However, the volatiles removed from soil will go back into the environment and may have more impact than they do from the soil. Also, microwaves cause the cellular membranes of the trees to resonate and thereby interrupt the water circulation (Martinez, 2003). The balance of electrically-charged particles in the plant is also distorted. This also disturbs the mineral management of the trees and retards the effects of soil organisms.

Kakita et al. (1999) showed with an experiment that a cloth contaminated with *E-coli* and irradiated with microwaves (2450 MHz) rapidly killed bacteria in 20s exposure. This is a clear indication that microwaves severely affect the microorganism, plants, and environment. The biological effects of microwaves (Michaelson 1971) and radio frequency exposure (Ingram 1986) can be health threatening. Gunal and Islam (2000) reported that microwave heating in a reservoir with asphaltene irreversibly altered its properties. Microwave heating could change the entire pathway of heated substances, including toxic emissions, quality of heat, and ultimate products such as food, air, etc.

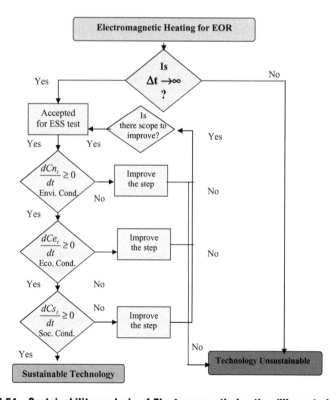

Figure 7-54 Sustainability analysis of Electromagnetic heating (Khan et al. 2006b).

7.12.5 *Sustainability analysis*

To evaluate the sustainability of electromagnetic heating, the model proposed by Khan et al. (2005d) is used (Figure 7-54). Any technology is considered sustainable if it fulfills the environmental, economical, and social conditions $(C_n + C_e + C_c)_t \geq$ constant for any time, t, provided that $dCn_t/dt \geq 0$, $dCe_t/dt \geq 0$, and $dCs_t/dt \geq 0$. The crude oil and natural gas refining processes use lots of synthetic chemicals and toxic catalysts. Burning of fossil fuel to produce electricity releases large amount of CO_2 and causes environmental pollution. The global efficiency of the electromagnetic energy system is very low. Conversion from electrical energy to electromagnetic energy needs costly transformers. The effect of electromagnetic radiation has several impacts on overall biodiversity. Based on these assumptions, this technology does not fulfill the economic criterion $(dCe_t/dt \geq 0)$, as the total change of economic benefit is not positive and hence is not sustainable for the long term. Similarly, the social criterion $(dCs_t/dt \geq 0)$ cannot be fulfilled as all fossil fuel-based technologies are developed based on top-down corporate approaches and users have no "say". Due to these reasons, the electromagnetic energy cannot fulfill the social criterion and fall within the unsustainable category.

A critical review of electromagnetic heating for EOR is carried out. Various studies show that electromagnetic heating holds some promise in EOR. However, the study to date is mostly limited to laboratory experiments and numerical models. A better understanding of the electrical properties of radiating materials and their degradation under temperature and pressure is necessary so that current leakage through the electrical insulation can be assessed. The global efficiency of electromagnetic technology is very low. Using the sludge from the produced water or other waste material to produce biogas and then convert to electricity would be one of the best options. The evaluation of sustainability of this technology showed that it is not sustainable for the long term (Chhettri and Islam 2006).

Chapter 8

Transportation, Processing, and Refining Operations

8.1   Introduction

After production, oil and natural gas are mainly transported through pipelines. For this purpose, complex piping systems are installed to transport and distribute the oil and gas to refineries or users (Figure 8-1). Such pipelines are necessary because the locations of oil/gas production sites tend to be far away from the refineries, processing plants, and end users. Because increasing pipe diameter would increase cost while decreasing energy requirement (to transport), the pipe diameter has to be optimized. This chapter deals with various aspects of transportation, processing, and refining operations, along with risk assessments.

There are two different phases of transportation operations (Figure 8-2). The first phase is the transportation of crude oil or gas from production facilities to oil refinery/gas processing facilities. The second phase is the distribution of refined products to end users. For conventional usage, petroleum products need to be purified. There are certain restrictions imposed on major transportation through pipelines. This involves the make up of the natural gas that is allowed into the pipeline, referred to as pipe "line quality" gas. This makes it mandatory that natural gas is purified before being sent through transportation pipelines. Gas processing is aimed at preventing corrosion, and environmental and safety hazards associated with transport of natural gas.

Because most natural hydrocarbons are a complex mixture of hundreds of different compounds, it is difficult to design a device that can make use of them directly. By refining, crude oil is fractionated into products such as gasoline, diesel fuel oil, light and heavy fuel oils, petrochemical feedstock, aviation fuels, asphalt, liquefied petroleum gas (LPG), lubricants, kerosene, and other products. In the refinery, many sophisticated technological processes are used to literally break down and then recombine the molecules of the original crude oil to produce the desired end products. Catalytic cracking and reforming, thermal cracking, and other secondary processes are used to achieve the desired product specifications. To accomplish this process, many toxic chemicals/catalysts are used. As a result, current oil refining and gas processing cause many environmental problems (Environment Canada 2005; Chhetri and Islam 2006a, 2006b).

This chapter discusses different aspects of oil and gas transportation and refining operations. Different risks associated with pipelines are investigated. A new dimension of risk assessment, due to flammability limit and lethal failure in the different accident scenarios, is discussed. These will help predict future outcomes with certainty and eliminate potential risks. In addition, this chapter examines the current practices of crude oil refining and gas processing methods. The use of toxic chemicals and catalysts is also investigated. Pathways of these chemicals used in gas processing along with their impacts to humans and the environment are presented. Some prospective natural materials for natural gas processing are also proposed. This chapter puts forward parameters to benchmark oil refinery companies based of the "Olympic" green supply chain that analyses operations, processes, materials design, and selection, according to environmental policies.

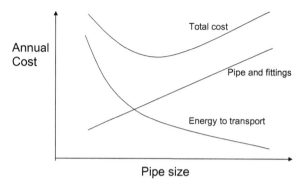

Figure 8-1   The need to optimize pipe size becomes clear when pipeline cost and energy to transport costs are considered.

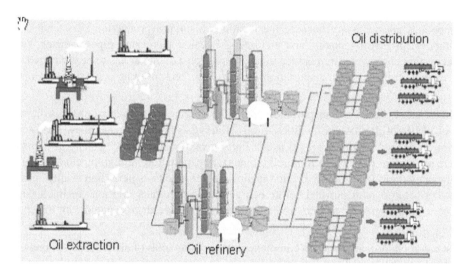

Figure 8-2   Schematic view of petroleum transportation and refining operations. Modified from Lakhal et al. (2007).

8.2   Pipelines and Risk Management

Oil and gas are primarily transported by a pipeline or series of pipelines, both above and below ground (Figure 8-3). Pipelines run between provinces, states, and countries. Gas moves through pipelines under pressure from large compressors. Some compressors run on natural gas from the pipeline itself, while some run on electricity. Crude oil is also pumped through pipelines that are often laid out in environmentally-sensitive areas.

The piping systems, being installed both underground and overground, can often be damaged by various activities. The most frequent cause is perforation of the pipe or complete fracture. Gas will be released into the environment at a flow rate depending on the hole diameter and the pressure within the pipe. Eventually the release will be stopped automatically by means of a regulator, as a reaction to excessive flow rate, or even manually. The failure of natural gas pipelines can occur due to natural or manmade disasters, such as earthquake, hurricane, sabotage, overpressure, flood, corrosion, or fatigue failures. Failure rate is also influenced by design factors, construction conditions, maintenance policy, technology usage, and environmental

Figure 8-3   Oil pipeline Alaska. Photo courtesy of _AP News_.

factors. All kinds of accidents in pipelines are determined by risk assessment and management (Ramanathan 2001). Risk assessment is the process of obtaining a quantitative estimate of a risk by evaluating its probability and consequences. Risk is generally referred to the potential for human harm. This risk represents a hazardous scenario, which is a physical or societal situation. If encountered, it could initiate a range of undesirable consequences.

The failure of pipelines are potentially hazardous events, especially in urban areas and near to roads. Therefore, people close to pipeline routes are subject to significant risk from pipeline failure. The hazard distance associated with a pipeline ranges from under 20 meters for a smaller pipeline at lower pressure, up to over 300 meters for a larger one at higher pressure (Jo and Ahn 2002). So it is essential to study the level of pipeline safety for better risk assessment and management.

Risk assessment addresses the safety, environmental protection, financial management, project or product development, and many other areas of business performance. In the pipeline sector, risk assessments are mainly considered on pipeline safety, needed for protecting human life, the environment, and property due to pipeline failure accidents. A pipeline can fail and release oil or natural gas into the environment and may cause many problems, including environmental degradation and loss of human life due to flammability.

The main reason for the risk assessment is to assess the likelihood of possible threats that could lead to failure at a particular location on the pipeline and what the consequences may be. This assessment is conducted by identifying the specific characteristics of the pipeline at any given location, along with the unique characteristics of the surrounding area. The susceptibility of the pipeline to failure and its impacts is dependent on numerous characteristics, such as the type and condition of the pipe coating, condition of soil around pipe, distance of pipeline from a specific locality, contents of pipeline, etc.

To determine the individual risk of an explosion hazard, flammability limits data are essential in a natural gas pipeline. Flammability limits are commonly used indices to represent the flammability characteristics of gases. The flammability limit criterion, and other related parameters, have been broadly discussed in the literature (Vanderstraeten et al. 1997; Kenneth et al. 2000; Kevin et al. 2000; Pfahi et al. 2000; Wierzba and Ale 2000; Mishra and Rahman 2003; Takahashi et al. 2003).

Hossain et al. (2005) studied the flammability and individual risk assessment for natural gas pipelines. They developed a comprehensive model for the individual risk assessment, for which the flammability limit and existing individual risk for an accidental scenario have been combined. Their model aims to determine the major accidental area within a locality surrounded by pipelines and for any natural gas pipeline risk assessment scenario. Hossain et al. (2005) also verified the model using available field data. However, they assume 10% accident risk due to flammability in natural gas pipeline accidents. An accident scenario may be any percentage within a limiting value. Hossain et al. (2007) applied the same model to verify different accidental scenarios. For a case study, 1%–20% accidental rates are considered in this chapter, which is a conservative figure in risk assessment.

In cases of risk assessment, Fabbrocino et al. (2005) reported that the assessment has always to be as conservative as possible. They also added that whatever the finality of the assessment, "worst case" should always be considered. When uncertainties are faced, the deterministic assessment, even within the framework of probabilistic safety assessment, should be taken in account. This approach is particularly effective, when late or early ignition assumption is considered in risk assessment.

Human health risk assessments determine how threatening a pipeline accident is to human health. The main objective of this assessment is to determine a safe level of contaminants or releases of toxic compounds, such as oil and natural gas from the pipeline. In the case of individual humans, this is a standard at which ill health effects are unlikely. It also estimates the current and possible future risks. This section examines individual risk of natural gas flammability on human health. The goal of this study is to manage risks to acceptable levels, and for risk managers to incorporate risk assessment information for planning and developing of pipeline networks.

To determine the individual risk of an explosion hazard, flammability limits data are essential in a natural gas pipeline. Flammability limits are commonly used indices that represent the flammability characteristics of gases. These limits can be defined as those fuel – air ratios within which flame propagation can be possible and beyond which flames cannot propagate. By definition there are two flammability limits, namely lower flammability limit (LFL) and upper flammability limit (UFL). LFL can be defined as the least fuel limit up to which the flame can propagate, the highest limit being the UFL (Liao et al. 2005). The flammability limit, criterion, and other related parameters have been broadly discussed in the literature (Vanderstraeten et al. 1997; Kenneth et al. 2000; Kevin et al. 2000; Pfahi et al. 2000; Wierzba and Ale 2000; Mishra and Rahman 2003; Takahashi et al. 2003).

8.2.1   Pipeline risk management

Although underground burial of pipelines is recommended, it does not prevent accidents from occurring, gas leakage and pipeline failure still being possible. A means for emergency isolation should be supplied at pipeline entries and exits from various facilities. For integrity assurances, pipelines should be checked regularly for failures and leakages at vulnerable locations, including weld joints and flange connections. These are usually checked using testing techniques, such as ultrasound, X-ray, and dye penetrating dyes.

The primary factor affecting pipeline hazardous incidents under normal conditions is corrosion. Therefore, it is important to take care of the pipelines by using proper anti-corrosion materials. Furthermore, pipeline failure can result from third-party activity, sabotage, or natural disasters. Figure 8-4 illustrates the risk management approach for natural gas pipelines, comprising the following steps:

- Piping system identification
- Operations information
- Risk assessment
- Strategy
- Actions
- Evaluation

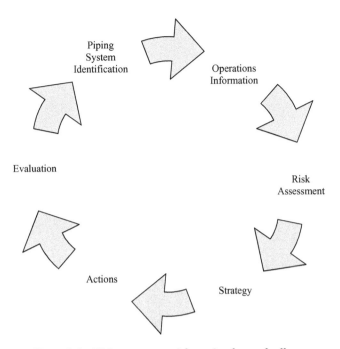

Figure 8-4   Risk management for natural gas pipelines.

8.2.2   Human health risk assessment

Different components are planning and scoping, exposure assessment, acute hazards, toxicity, and risk characterization. The main components of human health risk assessment are shown in Figure 8-4. For efficient risk assessments, "planning and scoping" of the information and data are needed. It should be done before the field investigations and site characterization.

The second step of human health risk assessment is "exposure assessment" (Figure 8-5). Exposure assessment refers to humans coming into contact with natural gas. This process considers the timing, duration, and frequencies of chemicals contact with humans in past, present, and future time periods. In the case of human risk assessment, the "acute hazards" mean the conditions that create the potential for injury or damage due to an instantaneous or short duration exposure to the effects of an accidental release. In this study, it is mainly flammability of natural gas.

"Hazard identification" is the process of determining whether exposure to natural gas can cause an increase in the incidence of a particular adverse health effect. Generally, it is done by the dose responses of particular chemicals. The "risk characterization" process is the synthesis of results of all other steps and determines how dangerous the accident is. It also considers the major assumptions and scientific judgments. Finally, there are the risk characterization estimates of the uncertainties embodied in the assessment.

8.2.2.1   Human health risk levels

Pipelines carry natural gas that contains methane, ethane, propane, iso-butane, normal butane, iso-pentane, normal pentane, hexanes plus, nitrogen, carbon dioxide, oxygen, hydrogen, and hydrogen sulfide. Sour gas contains a larger amount of hydrogen sulfide. In the case of a pipeline accident, all of these compounds are released. Due to flammability and exposure of all of these compounds, there are different levels of risk. Recently (May 2006) more than 150 people were killed due to flammability caused by pipeline accidents.

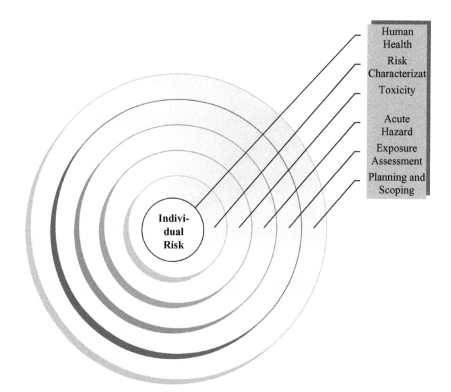

Figure 8-5   Different components of human health risks assessment.

It is reported that a ruptured fuel pipeline exploded and caught fire near Lagos, Nigeria (IRIN 2006). This pipeline transports fuel from a depot at the Lagos port for domestic use inland. Victims were inhabitants of poor fishing villages. Pipelines accidents are common in third-world counties, such as Nigeria, an oil-rich African nation. In 1998, it was also reported that more than 1000 people died due to a flammability accident in Jesse, near the oil town of Warri, Niger Delta (IRIN 2006).

In the above accident report, it is revealed that due to strong flammability the fate is certain death, but exposure to other components, such as hydrogen sulfide results in different risk levels. In Table 8-1, different risk levels caused by hydrogen sulfide are shown. This phenomenon should be considered seriously in the case of sour gas, where hydrogen sulfide concentrations are higher. Generally, the typical sulfur content is 5.5 mg/m^3, which includes the 4.9 mg/m^3 of sulfur in the odorant (mercaptan) added to gas for safety reasons.

8.2.2.2 Combustion properties of natural gas

As mention earlier, natural gas has an extreme risk of flammability due to its composition. To understand the flammability risk, its combustion properties are presented in Table 8-2. Note that the combustion properties depend on composition, but general estimations is shown in this table. The properties shown are an overall average of the Union Gas system (Union Gas 2006).

8.2.3   *Risk assessment*

In order to assess the risk regarding a natural gas pipeline, it is necessary to evaluate probable undesirable consequences resulting from any leakage or rupture.

Table 8-1   Human health risk levels

Risk levels	Concentration (ppm)	Effects
Negligible or no-risk	0.01–0.3	Odor threshold (highly variable)
Minimal risk	1–5	Moderate offensive odor, may be associated with nausea, eye irritation, headaches, or loss of sleep with prolonged exposure; healthy young male subjects experience no decline in maximum physical work capacity
Slightly moderate risk	108 h	Occupational exposure limit
Moderate risk	20–50	Ceiling occupational exposure limit and community evacuation level, odor very strong
Risk	100	Eye and lung irritation; olfactory paralysis, odor disappears
High risk	150–200	Sense of smell paralyzed; severe eye and lung irritation
Severe risk	250–500	Pulmonary edema may occur, especially if prolonged
Extremely risk	500	Serious damage to eyes within 30 min; severe lung irritation; unconsciousness and death within 4–8 h; amnesia for period of exposure; "knockdown"
Critical level	1000	Breathing may stop within one or two breaths; immediate collapse

Source: Guidotti (1994)

Table 8-2   Typical combustion properties of natural gas

Ignition Point:	59°C*
Flammability limits	4%–16% (vol. % in air)
Theoretical flame temperature (stoichiometric air/fuel ratio)	1960°C (3562°F)
Maximum flame velocity	0.3 m/s
Relative density (specific gravity)	0.585

Data source: Union Gas (2006)

The quantitative risk can be estimated from the flammability limit for a natural gas pipeline. Risks has been described as individual risk, societal risk, maximum individual risk, average individual risk of exposed population, average individual risk of total population, and average rate of death (TNO Purple Book 1999; Jo and Ahn 2002, 2005).

The failure rate of pipelines depends on various parameters such as soil conditions, coating type and properties, design considerations, and pipeline age. So, a long pipeline is divided into sections, due to significant changes of these parameters. Considering a constant failure rate, the individual risk can be written as (Jo and Ahn 2005):

$$IR = \sum_i \varphi_i \int_{l_-}^{l_+} p_i \, dl \qquad (8.1)$$

where

φ_i = Failure rate per unit length of the pipeline associated with the accident scenario, i, due to soil condition, coating, design, and age

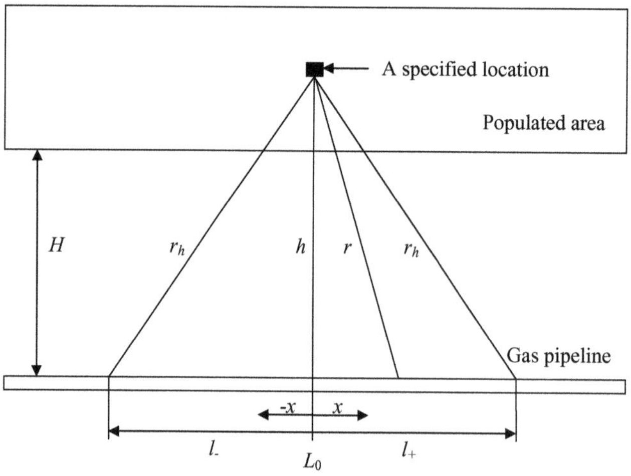

Figure 8-6   Individual risk variables.

l = Pipeline length

p_i = Lethality associated with the accident scenario, i

$l_\pm$ = Ends of the interacting section of the pipeline in which an accident poses a hazard to a specified location

The release of gas through a hole in the pipeline causes explosion and fire within the natural gas pipeline and the surrounding area. The affected section causes a hazard distance. The release rate of natural gas and hazard distance are correlated (Jo and Ahn, 2002):

$$r_h = 10.285\sqrt{Q_{eff}}\qquad(8.2)$$

where

Q_{eff} = Effective release rate from a hole in a pipeline carrying natural gas

r_h = Hazard distance

The hazard distance is the distance within which there is more than 1% chance of fatality due to the radiational heat of jet fire from pipeline rupture. Figure 8-6 shows the geometric relations among the variables in a specified location from a natural gas pipeline. From this figure, the interacting section of a straight pipeline, h from a specified location, is estimated by Equation 8.3 (Jo and Ahn 2005):

$$l_\pm = \sqrt{106Q_{eff} - h^2}\qquad(8.3)$$

Jo and Ahn (2005) show the different causes of failure based on hole size and other activities. The external interference by third-party activity is the major cause of key accidents related to hole size. Therefore, a more detailed concept is required to analyze the external interference. The third-party activity depends on several factors, such as pipe diameter, cover depth, wall thickness, population density, and prevention methods. The failure rate of a pipeline has been estimated by some researchers (Jo and Ahn 2005; John et al. 2001).

8.2.4   Effects of composition on flammability limit

An experimental study is usually conducted to investigate the effects of concentration or dilution in natural gas – an air mixture by adding CO_2, N_2 gas. The limit ranges are 85–90% of N_2 and 15–10% of CO_2 by

Table 8-3 **Flammability limit data (vol %) for methane-air and natural gas-air flames (quiescent mixtures with spark ignition)**

Mixture	Test Condition	LFL (vol %)	UFL (vol %)
NG-air	1.57 L chamber	5.0	15.6
	LeChatelier's rule	4.98	–
Methane-air	8 L chamber	5.0	–
	20 L chamber	4.9	15.9
	120 L chamber	5.0	15.7
	25.5 m³ sphere	4.9, 5.1 ± 0.1	–
	Flammability tube	4.9	15.0

volume. This is a practical consideration of natural gas stoichiometric combustion at ambient temperatures. Flammability experiments have been performed to simulate real explosions, in order to access and prevent hazards in the practical applications (Liao et al. 2005). Table 8-3 shows the flammability limit data for methane-air and natural gas-air flames (Liao et al. 2005).

LFL depends on the composition of fuel mixture in air. This value can be estimated by LeChatelier's rule (Liao et al. 2005):

$$LFL = \frac{100}{\sum (C_i/LFL_i)} \tag{8.4}$$

where

LFL = Lower flammability limit of mixture (vol. %)

C_i = Concentration of component, i in the gas mixture on an air-free basis (vol. %)

LFL_i = Lower flammability limit for component, i (vol. %)

The estimation of LeChatelier's rule is shown in Table 8-3. The reliance of the natural gas flammability limit upon ethane concentration has been studied by Liao et al. (2005) (Figure 8-7). It is shown that the flammability region is slightly extended with the increase of ethane content in natural gas. LFL is almost 5% in volume and UFL is about 15%. The flammability limits are 3% to 12.5% in volume for an ethane-air mixture. Their equivalent ratios are 0.512 and 2.506. The ratios are 0.486 and 1.707 with methane, respectively. It is noted that the increase of ethane content in natural gas extends the UFL equivalence ratio but there is no remarkable change in LFL. Liao et al. (2005) shows the effect of diluents ratio (ϕ_r) on the flammability ratio. The increase of diluents ratio decreases the flammability region. The reason is that the addition of diluents decreases the temperature of flames, which decreases the burning velocity. Thus, the flammability limit narrows. Normally, CO_2 is more influential than the addition of N_2. Shebeko et al. (2002) presented an analytical evaluation of flammability limits on ternary gaseous mixtures of fuel-air diluent.

8.2.5   *Individual risk based on flammability*

Figure 8-8 shows the incidental zone founded on basic fluid dynamics. The accident scenario represents this incidental zone. If an explosion occurs, the incidental zone will be covered by the projectile theory of fluid dynamics. This concept is the basic difference from the model of Jo and Ahn (2005), which is shown in Figure 8.8. An accident due to flammability is assumed here as the main cause of the incident. OB is the maximum distance covered by the flames within which a fatality or injury is likely to occur (Figure 8-8). BA and BC are the maximum distances traveled by the flames.

The velocity of the natural gas evolved through the hole can be written as

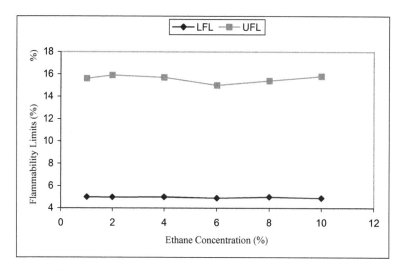

Figure 8-7   Dependence of NG flammability limits on ethane.

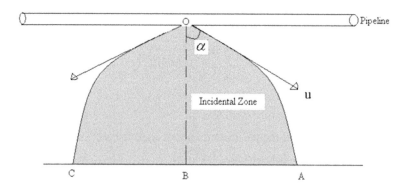

Figure 8-8   The relation of variables related with *IR*r.

$$u = 1.273 \frac{q_{\min}}{d^2_{hole}}$$ (8.5)

where

q_{min} = Minimum gas flow rate evolved through the hole that causes an explosion = $f(u,d_{hole})$

d_{hole} = Diameter of the hole through which gas passes.

Hazard distance or maximum distance covered by gas particles can be written as

$$h_{\max} = \frac{1}{2} ut \cos \alpha$$ (8.6)

where

$h_{\max}$ = Hazard distance

u = Velocity of gas

t = Travel time to reach the hazard distance

α = Angle between velocity of gas and hazard distance

Figure 8-8 shows the geometric relations among the variables in a specific location from a natural gas pipeline. From this figure, the interacting section of a straight pipeline, $l_{\pm}$ from a specific location, B, and the angle, α, are estimated by

$$l_{\pm} = \frac{1}{2}ut\sin\alpha \tag{8.7}$$

and

$$\alpha = \tan^{-1}\left(\frac{l}{h_{max}}\right) \tag{8.8}$$

The individual risk (IR_f) due to the flammability limit in a natural pipeline can be written as

$$IR_f = \sum_i \frac{\varphi_i}{100}\int_{-l}^{+l}\int_0^{h_{max}}(UFL_i - LFL_i)dhdl \tag{8.9}$$

where

φ_i = The failure rate per unit length of the pipeline associated with the accident scenario, i due to flammability

l = Pipeline length

UFL = Upper flammability limit

LFL = Lower flammability limit

$l_{\pm}$ = Ends of the interacting section of the pipeline in which an accident poses a hazard to the specific location.

Figure 8-9 shows the number of incidents with pipeline distance from the source of gas. The data have been collected from the US Office of Pipeline Safety incident summary statistics from 1986 to August, 2005 (Hossain et al. 2006a). The number of incidents show an oscillating pattern within the region of 67,775 and 259,136 miles. However, beyond this distance, the rate of incidents show an abnormal pattern. This might be caused by other factors, such as natural disaster, human activities, etc.

There is no available information that handles both the flammability limit and lethality for measuring individual risk. It is difficult to obtain the data due to flammability at the scene of the accident. In this study, 10% of accidental scenarios are assumed to be due to flammability. Using a 10% accidental scenario, the proposed model (Equation 8.9) is tested. Results show the individual risk due to flammability, with number of injuries (Figure 8-10). The normal trend of the curve is increasing with the increase in number of incidents, which leads to a separate scenario of accidents due to flammability. This chart also shows that there is a great impact of flammability on the accidental scenario.

Figure 8-10 shows the probability of individual risk due to flammability with pipeline distance, using Equation 8.9. Here it has been assumed that the UFL and LFL are 15.6 and 5.0, respectively. For this case study calculation, q_{min} is considered as 1 ft^3/sec, $\alpha = 45°$, $t = 1$ min, and $d_{hole} = 0.5$ ft. Available literature shows that the maximum value of h is 20 m and l is 30 m. Here the calculation shows that h is 80.5 ft and l is 129.93 ft. These values seem to be reasonable. The individual risk due to flammability is decreasing with pipeline distance from the gas supply center. However, the trend tends to be unpredictable and more frequent in an accident scenario within a pipeline range of 124,931 miles. This graph also shows that there is a great impact of flammability on the accidental scenario.

By combining Equations 8.1 and 8.9, a combined individual risk in a natural gas pipeline is obtained:

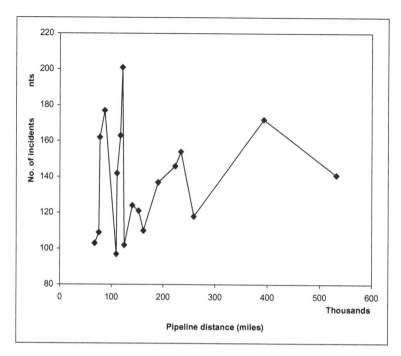

Figure 8-9   Incident related with pipeline distance. Redrawn from Hossain et al. (2006a).

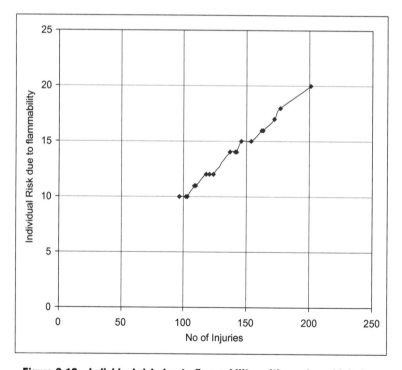

Figure 8-10   Individual risk due to flammability, with number of injuries.

$$IR_T = IR + IR_f \tag{8.10}$$

This represents a true scenario of an individual risk due to lethality and flammability of natural gas. The lethality of a natural gas pipeline depends on operating pressure, pipeline diameter, distance from the gas supply to pipeline, and the length of the pipeline from the gas supply or compressing station to the failure point.

Hossain et al. (2006a) have shown the concept of individual risk due to flammability at a locality with a dense population. Figure 8-8 has been redrawn from this reference where detail analysis has been presented. An accident due to flammability is considered here as the main cause of the incident. OB is the maximum distance covered by the flame within which a fatality or injury can take place. BA and BC are the maximum distances traveled by the flame.

The individual risk (IR_f), due to the flammability limit in a natural pipeline, can be written as

$$IR_f = \sum_i \frac{\varphi_i}{100} \int_{-l}^{+l} \int_0^{h_{max}} (UFL_i - LFL_i) dh dl \tag{8.11}$$

and the total individual risk can be written as

$$IR_T = IR + IR_f \tag{8.12}$$

where

φ_i = The failure rate per unit length of the pipeline associated with the accident scenario, i due to flammability

l = Pipeline length, ft

UFL = Upper flammability limit

LFL = Lower flammability limit

$l_\pm$ = Ends of the interacting section of the pipeline in which an accident poses hazard to the specified location, ft

Table 8-4 shows the different data for number of fatalities/injuries due to natrual gas flammability accidents in a pipeline from 1985 to 2005. The data have been collected from the US Department of Pipeline Safety.

Figure 8-11 has been generated using the data shown in Table 8-4. It shows the number of incidents with individual risk due to flammability, for different percentages of flammability risk at the pipeline. The data have been collected from the US Department of Pipeline Safety, incident summary statistics from 1986 to August, 2005. In this figure, the individual risk is increasing and exhibits a much steeper trend when human health hazard risk due to flammability injuries are increased. It means that the individual risk factor is influenced by the flammability risk factor within the contour locality.

At present, there are many models available to investigate individual risk (John et al. 2001; Jo et al. 2002, 2005; Fabbrocino et al. 2005). However, there is no model available that handles both the flammability limit and lethality for measuring individual risk for human health hazards. It is difficult to obtain data from the accidental scenario due to flammability. Based on available information and data dealing with this issue, the Hossain et al. (2006a) model can be easily used to verify any sets of data with confidence. In this study, 1–20% of accidental scenarios are considered to be due to flammability (Hossain et al. 2006). Using these data, the model (Equation 8.1) is tested and results are shown in Figures 8-12 and 8-13. Here it has been assumed that the UFL and LFL are 15.6 and 5.0, respectively. q_{min} is considered as $1 ft^3$/sec, $\alpha = 45°$, $t = 1$ min, and $d_{hole} = 0.5$ ft for the case study. Available literature shows that the maximum value of h is 66 ft and l is 99 ft (Hossain et al. 2006a). Here the calculation shows that h is 80.5 ft and l is 129.93 ft and these values seem to be reasonable in this case.

Table 8-4   Number of injuries and flammability data for different percentages (Hossain et al. 2006a)

Fatality/Injury	Fatality/injury due to natural gas flammability									
	1%	3%	6%	8%	10%	12%	14%	16%	18%	20%
97	0.97	2.91	5.82	7.76	9.7	11.64	13.58	15.52	17.46	19.4
102	1.02	3.06	6.12	8.16	10.2	12.24	14.28	16.32	18.36	20.4
103	1.03	3.09	6.18	8.24	10.3	12.36	14.42	16.48	18.54	20.6
109	1.09	3.27	6.54	8.72	10.9	13.08	15.26	17.44	19.62	21.8
110	1.1	3.3	6.6	8.8	11	13.2	15.4	17.6	19.8	22
118	1.18	3.54	7.08	9.44	11.8	14.16	16.52	18.88	21.24	23.6
121	1.21	3.63	7.26	9.68	12.1	14.52	16.94	19.36	21.78	24.2
124	1.24	3.72	7.44	9.92	12.4	14.88	17.36	19.84	22.32	24.8
137	1.37	4.11	8.22	10.96	13.7	16.44	19.18	21.92	24.66	27.4
141	1.41	4.23	8.46	11.28	14.1	16.92	19.74	22.56	25.38	28.2
142	1.42	4.26	8.52	11.36	14.2	17.04	19.88	22.72	25.56	28.4
146	1.46	4.38	8.76	11.68	14.6	17.52	20.44	23.36	26.28	29.2
154	1.54	4.62	9.24	12.32	15.4	18.48	21.56	24.64	27.72	30.8
162	1.62	4.86	9.72	12.96	16.2	19.44	22.68	25.92	29.16	32.4
163	1.63	4.89	9.78	13.04	16.3	19.56	22.82	26.08	29.34	32.6
172	1.72	5.16	10.32	13.76	17.2	20.64	24.08	27.52	30.96	34.4
177	1.77	5.31	10.62	14.16	17.7	21.24	24.78	28.32	31.86	35.4
201	2.01	6.03	12.06	16.08	20.1	24.12	28.14	32.16	36.18	40.2

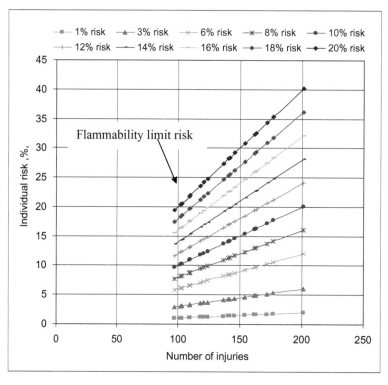

Figure 8-11   Individual risk due to flammability with number of injuries (Hossain et al. 2006a).

Table 8-5 Individual risk due to flammability with pipeline distance

Pipeline distance, (miles)	Individual risk due to flammability					
	1.0%	6.0%	10.0%	14.0%	18.0%	20.0%
5,320,616	8.72E-06	2.33E-05	3.49E-05	4.65E-05	0.005078	0.009638
3,928,390	0.00048742	0.002927	0.004879	0.00683	0.008781	0.538715
2,591,365	0.0005272	0.003163	0.005272	0.00738	0.009489	0.582692
2,339,883	0.00077257	0.004634	0.007723	0.010812	0.013901	0.853882
2,229,440	0.00081789	0.004908	0.00818	0.011452	0.014724	0.903967
1,905,511	0.00095825	0.005749	0.009582	0.013415	0.017248	1.059106
1,625,284	0.0009052	0.005431	0.009051	0.012672	0.016292	1.000471
1,534,665	0.00107209	0.00643	0.010716	0.015003	0.01929	1.184929
1,407,148	0.00129314	0.007748	0.012913	0.018078	0.023243	1.429245
1,249,316	0.00124893	0.007484	0.012474	0.017464	0.022453	1.380382
1,213,143	0.00258628	0.015525	0.025874	0.036224	0.046574	2.858489
1,173,612	0.00219945	0.0132	0.021999	0.030799	0.039599	2.430938
1,107,880	0.00215524	0.012905	0.021509	0.030112	0.038716	2.382075
1,095,067	0.00163577	0.00981	0.016351	0.022891	0.029431	1.807933
867,581	0.00373574	0.022426	0.037377	0.052328	0.067279	4.128928
776,574	0.00391258	0.0235	0.039167	0.054834	0.070501	4.324381
759,404	0.00282944	0.017002	0.028337	0.039672	0.051006	3.127236
677,750	0.00328259	0.019667	0.032778	0.04589	0.059001	3.628082

Table 8-5 shows the different individual risk due to flammability data for different pipeline distances. The flammability data have been calculated using Equation 8.11. The data showing fatalities/injuries from natural gas in pipeline accidents are from 1985 to 2005. The data has been obtained from the US Department of Pipeline Safety.

Figure 8-12 shows the individual risk due to flammability with pipeline distance. The normal trend of the curve is decreasing with the increase in pipeline distance, which leads to a separate scenarios of accidents due to flammability. The graph also shows that there is a great impact of flammability on accidental scenarios. An interesting point is that this model shows that the human health hazard risk due to flammability in individual risk assessments of natural gas is limited by 18% of the total risk factor (Figures 8-12 and 8-13). These figures have been generated using the data shown in Table 8-6. Beyond 18% of total individual risk, the results do not fit with the other percentages of risk and the values shown by the calculations are not realistic (Figure 8-14). This information simply means that the human health hazard individual risk due to flammability of natural gas does not go beyond 18% of individual risk.

Extensive pipeline networks for a natural gas supply system possess many risks. Appropriate risk management should be followed to ensure safety. Individual risk is one of the important elements for quantitative risk assessment. Considering the limitations in conventional risk assessment, a novel method is presented for measuring individual risk, combining all probable scenarios and parameters associated with practical situations, taking into account gas flammability. These parameters can be calculated directly by using the pipeline geographical and historical data. By using the proposed method, the risk management can be more appealing from a practical point of view. The proposed model is found to be innovative in using pipeline and incident statistical data. The method can be applied to pipeline management during the planning, design, and construction stages. It may also be employed for maintenance and modification of a pipeline network.

8.3   Natural Gas Supply and Processing

Natural gas is a widely used and abundant gas. It is attractive to consumers and institutions as it is clean, safe, and useful in a wide variety of applications. Natural gas has many individual hydrocarbon compounds,

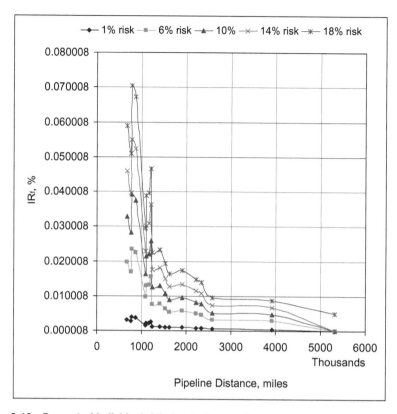

Figure 8-12   Percent of individual risk due to flammability with pipeline distance (1–18%).

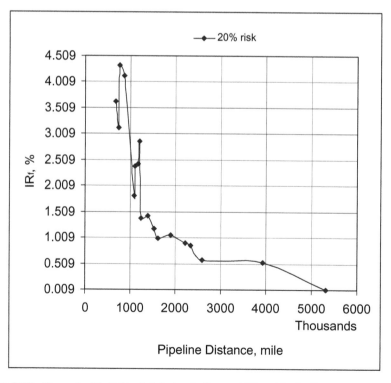

Figure 8-13   Percent of individual risk due to flammability with pipeline distance (20%).

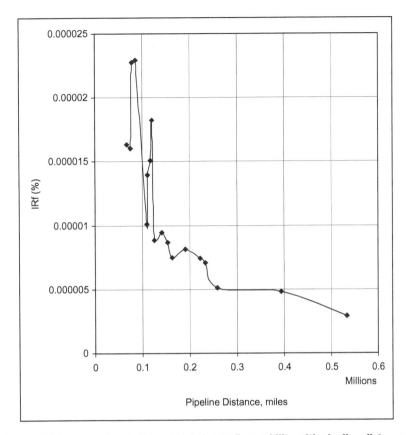

Figure 8-14 Percent of individual risk due to flammability with pipeline distance.

which together form a combustible mixture. Hydrocarbon compounds contain carbon atoms ranging in number from 1 to 4, and in the form C_nH_{2n+2}, namely methane, ethane, propane, and butane. Other impurities are also present, such as CO_2, O_2, N_2, H_2S, and other trace elements. Natural gas is largely composed of methane, ranging from approximately 70–90%, with the other hydrocarbons accounting for approximately 0–20%. When natural gas is largely methane, it is termed a "dry gas" and when it contains other hydrocarbons, it is termed a "wet gas".

Typically the impurities and other hydrocarbons are removed and either sold separately, such as butane and propane, or discarded as waste, such as H_2O and CO_2. Natural gas is typically measured in MCF (thousands of cubic feet), MMCF (millions of cubic feet), or TCF (trillions of cubic feet). The energy of natural gas is typically measured in BTUs, where 1 cubic foot of natural gas contains 1027 BTUs.

8.3.1 Life cycle of natural gas from well to wheel

Natural gas development and management from reservoir to the end user consists of several processes such as exploration, extraction, production, processing, transportation, storage, and distribution. Exploration is the preliminary processes of how the natural gas reserve is detected. Various geological exploration techniques are adopted in order to identify the reserves. Geological exploration is carried by surface and sub-surface geological examination of the structural rock properties through map study. With the advance of technologies, seismic exploration has become popular and is used to determine the natural gas reserves both onshore and offshore. Artificial vibrations are created on the surface and recorded to show how these vibrations are reflected back to the surface. Nowadays, a cross-sectional picture of underground rock formations

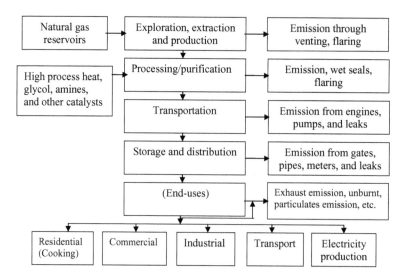

Figure 8-15   Natural gas production and end use pathways. After Chhetri et al. (2006).

are developed from 2D, 3D, and 4D seismic imaging. However, seismic exploration techniques use huge amounts of energy and create several environmental problems due to vibration and similar activities.

In order to extract the natural gas discovered in the deposits, onshore and offshore drilling activities are carried out after following the extraction process, which includes gaining permission to extract the gas, leasing, and taxes and royalties associated with the natural gas. Production and processing are carried out after the extraction, which includes natural gas production from the underground reservoir and removing impurities in order to meet certain regulatory standards before sending for burning. Various chemicals and catalysts are used during the processing of the natural gas. The purified natural gas is then transported in different forms, such as LPG, LNG, or gas hydrates and distributed to end users as per their demands. Figure 8-15 depicts the pathways of natural gas from well to end use.

8.3.1.1   Formation

Natural gas is typically formed from terrestrial sources of organic matter, but can also be formed from lacustrine and marine organic matter. Natural gas is generally cracked from kerosene at temperatures over 150°C (outside the oil window), thus generally found at depths greater than oil. However, natural gas can also be entrained in oil, and oil reservoirs can have a gas cap. Gas migration through reservoirs is typically fast, geologically speaking, so older gas can be found above younger oil.

Methane can also be formed biologically. Some tiny bacteria produce methane as a byproduct of their feeding process. A third way of producing methane is through reactions between hydrogen rich gases and carbon atoms deep within the Earth's crust.

As methane is a light gas it rises rapidly within the Earth through faults and permeable rocks. It is trapped by solid structures or impermeable fine-grained rocks, such as tight shales, mudstones, and limestones. Gas can also be trapped by impermeable or sealing faults.

8.3.1.2   History

Ancient stories state that natural gases seeping from the Earth's surface were ignited by lightning strikes, which puzzled past civilizations. The Chinese harnessed the natural gas from the seeps by using bamboos, using the gas to boil water. In 1785, Britain was the first country to use gas commercially to light houses and street lights. This natural gas was produced from coal. Natural gas from coal (now termed coal bed methane) was brought to the United States in 1816, to be used in street lighting. The gas from coal is believed to be not as efficient or environmentally-friendly as today's natural gas. This belief led the development of a series of "processing" schemes in order to remove undesired components from a natural gas stream. The assumption here is that natural gas processing improves the quality of the gas and increases efficiency. Recently, Islam et al. (2006) pointed out that this may not be the case, unless natural additives are used to process natural gas. With toxic chemical additives, even a small (usually below detection or regulatory limit) concentration will lead to numerous problems, particularly because when the gas is burned all chemicals are oxidized.

In 1821, the "father" of natural gas dug a 27-foot well in a stream in Fredonia, New York, where he had noticed bubbles rising to the surface. In 1859, Colonel Edwin Drake drilled the first well, and struck oil and natural gas at a mere 69 feet below the surface. The natural gas from this well was brought to the nearby town of Titusville, Pennsylvania via a 2-inch-diameter pipeline. This was the first indication that natural gas could be transported safely and relatively easily.

Throughout the 19th century, natural gas was used primarily for lighting. With the switch to electricity in the earlier 1900s, natural gas lights were converted to electricity. The first long pipeline was constructed in 1891, running from central Indiana to Chicago, a distance of some 120 miles. In the 1920s, larger effort was put into developing pipelines, but it was after World War II that significant advances were made in welding and pipe rolling, to construct reliable pipelines. Once reliable transportation methods of natural gas were established, more uses became prevalent, such as heating, household appliances, and electricity generation.

8.3.1.3 Natural gas reserves

In 2000, the *Oil and Gas Journal* and other world oil publications estimated world natural gas reserves to be in the order of 5210.8 TCF. Tables 8-6 and 8-7 show the world distribution of natural gas.

Table 8-8 shows the natural gas reserves in Canada. From the point of view of gas production, Canada is one of the most important countries in the world. It is the third largest producer, and second largest exporter of natural gas, producing 6.2 Tcf/yr, and the number one supplier to the United States, of some 3.6 Tcf /yr.

Comparing the 2004 Canadian reserves with the 2000 world reserves, shows that Canada has approximately 1% of the world's reserves. Canada also has large reserves of coal bed methane and tight gas. Most Canadians use natural gas in their homes for heating, and it is also widely used in industrial and commercial sectors.

Table 8-6 World natural gas distribution

Countries	% of World Reserves
Europe and Former USSR	42
Middle East	34
Africa and Far East	15
Central and South America	4
United States	3
Canada	1
Mexico	1

Data source: NGO (2005)

Table 8-7   The world distribution of natural gas

Country	Proved Reserves in 2003 (Tcf)	% World Reserves	Country	Proved Reserves in 2003 (Tcf)	% World Reserves
TOTAL WORLD	5501.4	100.00%	Norway	77.3	1.40%
Select Countries	5097.4	92.70%	Malaysia	75	1.40%
Russia	1680	30.50%	Turkmenistan	71	1.30%
Iran	812.3	14.80%	Uzbekistan	66.2	1.20%
Qatar	508.5	9.20%	Kazakhstan	65	1.20%
Saudi Arabia	224.7	4.10%	Netherlands	62	1.10%
United Arab Emirates	212.1	3.90%	Canada	60.1	1.10%
United States	183.5	3.30%	Egypt	58.5	1.10%
Algeria	159.7	2.90%	China	53.3	1.00%
Venezuela	148	2.70%	Libya	46.4	0.80%
Nigeria	124	2.30%	Oman	29.3	0.50%
Iraq	109.8	2.00%	Bolivia	24	0.40%
Indonesia	92.5	1.70%	Trinidad/Tobago	23.5	0.40%
Australia	90	1.60%	Rest of World	404.1	7.30%

Data from: *Oil and Gas Journal* (2003)

Table 8-8   Natural gas reserves, production, and export in 2004

Statistic	Volume
Reserves (end 2004)	56.5 Tcf
Production	17,000 MMcf/day (17 billion cu ft/day)
Export	9,700 MMcf/day (9.7 billion cu ft/day)

Source: www.capp.ca

Natural gas no longer comes second to crude oil in Canada, as gas is taking on a larger role as gas production is increasing faster than oil production. Royalties to Alberta are now larger for natural gas and its byproducts than crude oil and bitumen combined.

The main natural gas (and oil) producing area is the Western Canadian Sedimentary Basin, which is primarily in Alberta (83%), British Columbia (13%), and Saskatchewan (4%). Other natural gas producers include the Sable Offshore Energy Project in Nova Scotia, and minor amounts of natural gas in Ontario, Yukon, and the Northwest Territories.

There is great potential for natural gas discoveries in Canada, as shown in the following table from the National Energy Board. It should be noted that Table 8-8 shows proved reserves, while Table 8-9 shows proved and potential reserves.

Several areas in Canada have been identified as containing large deposits, but have yet to be developed. They have been classified by the National Energy Board as "Offshore" and "Frontier".

The offshore gas fields in production include Sable Island, which is producing from five different fields, as well as Hibernia, Terra Nova, and recently White Rose. These latter three re-inject the gas produced to stimulate the reservoir, as the infrastructure or local market does not exist for the gas. Sable Island pipes its gas to shore.

Frontier basins are those areas with large gas reserves, but no immediate plans for production or significant production due to the cost of development and distance from market. These include the Beaufort Sea, Mackenzie Delta, Arctic Islands, Labrador Basins, and the Grand Banks. Recent discoveries in the Mackenzie Delta (9 Tcf) and Prudhoe Bay in Alaska (28 Tcf) may go into production in the next few years

Table 8-9   Proved and potential gas reserves in Canada

Area	Remaining Established Reserves (Tcf)	Discovered Conventional Resources (Tcf)	Undiscovered Conventional Resources (Tcf)	Undiscovered Unconventional Resources (Tcf)	Total Remaining Resources (Tcf)
Western Canadian Sedimentary Basin	56.5		176	75	307.5
Alberta	45.3		138		183.3
British Columbia	8.1		30		38.1
Saskatchewan	2.8		2		4.8
Southern Territories	0.3		6		6.3
Eastern Canada	3.5	2	14		19.5
Ontario	0.5		1		1.5
Scotian Shelf	3	2	13		18
Frontier	0	33	270		303
Grand Banks/Labrador	0	9	36		45
Mackenzie/Beaufort	0	9	55		64
Artic Island	0	14	80		94
Other Yukon/NWT	0	1	10		11
Other Frontier	0	0	89		89
Total Canada	**59.9**	**35**	**460**	**75**	**630**

Source: CAPP (2005)

if a pipeline can link these fields to the existing pipelines in the area. It will be costly, but the reserves may make it worthwhile.

8.3.2   Gas processing

Processing of natural gas is required to remove impurities. A typical natural gas stream is a mixture of methane, ethane, propane, butane, other hydrocarbons, water vapor, oil and condensates, hydrogen sulfides, carbon dioxide, nitrogen, some other gases, and solid particles. The free water and water vapors are corrosive to transportation equipment. Hydrates can plug the gas accessories, creating various flow problems. Other gas mixtures, such as hydrogen sulfide and carbon dioxide, are considered to lower the heating value of natural gas, thus reducing its overall fuel efficiency. Gas processing is necessary to prevent corrosion, environmental hazards, and safety concerns associated with the transport of natural gas. This makes it mandatory that natural gas be purified before it is sent to transportation pipelines. There are certain restrictions imposed on the make-up of the natural gas that is allowed into the pipelines, called pipe "line quality" gas. Pipeline quality gas should not contain any other elements such as hydrogen sulfide, carbon dioxide, nitrogen, water vapor, oxygen, particulates, and liquid water that could have a detrimental effect on the pipeline and its operating equipment (EIA 2006). Even though hydrocarbons such as ethane, propane, butane, and pentanes have to be removed from natural gas, these products are also used for various other applications.

The presence of water in natural gas creates several problems. Liquid water and natural gas can form solid ice-like hydrates that can block valves and fittings in the pipeline (Nallinson 2004). Natural gas containing liquid water is corrosive, especially if it contains carbon dioxide and hydrogen sulfide. Water vapor in natural gas transport systems may condense causing a sluggish flow. It has further been argued that water vapor increases the volume of the natural gas, decreasing its heating value, which reduces the capacity of the transporting and storage system (Nallinson 2004). Hence, removal of free water, water vapors, and condensates are important steps during gas processing (Figure 8-15).

Other impurities of natural gas, such as carbon dioxide and hydrogen sulfide, generally called acid gases, must be removed prior to transportation of the gas (Chakma 1999). Carbon dioxide is a major greenhouse gas that contributes to global warming. Separating the carbon dioxide from the natural gas stream and its meaningful application, such as for Enhanced Oil Recovery (EOR), is of great importance. Even though hydrogen sulfide is not a greenhouse gas, it is a source of acid rain deposition, being toxic and corrosive and rapidly oxidizes to sulfur dioxide in the atmosphere (Basu et al. 2004). Also, oxides of nitrogen found in trace amounts in the natural gas can cause ozone layer depletion and global warming.

8.3.3   Current practices in gas processing

Natural gas obtained from reservoirs consists of a mixture of hydrocarbons such as methane, ethane, propane, butane, and pentanes. In addition to these hydrocarbons, raw natural gas also contains water vapor (H_2O), hydrogen sulfide (H_2S), carbon dioxide (CO_2), helium, nitrogen, and other compounds. CO_2 and H_2S are called acid gases. Hence, natural gas processing consists of separating various hydrocarbons and other fluids to produce "pipeline quality" dry natural gas. EIA (2006) illustrated a generalized natural gas processing schematic (Figure 8-16). A simplified version is shown in Figure 8-17. Note that water is removed, mainly to prevent corrosion problem. CO_2 is removed, mainly to improve the heating value and decrease corrosion effects. H_2S is removed, mainly for safety.

This generalized scheme includes all the necessary steps, depending on the types of ingredients available in a particular gas. The gas – oil separator unit removes any oil from the gas stream, the condensate separator removes free water and condensates, and the water dehydrator separates moisture from the gas stream. The other contaminants, such as CO_2, H_2S, nitrogen, and helium are also separated from the different units. The natural gas liquids, such as ethane, propane, butane, pentane, and gasoline are separated from methane by using cryogenic and absorption methods. The cryogenic process consists of lowering the temperature of the gas stream with a turbo-expander and external refrigerants. The sudden drop in temperature in the expander condenses the hydrocarbons in the gas stream, maintaining the methane in the gaseous form.

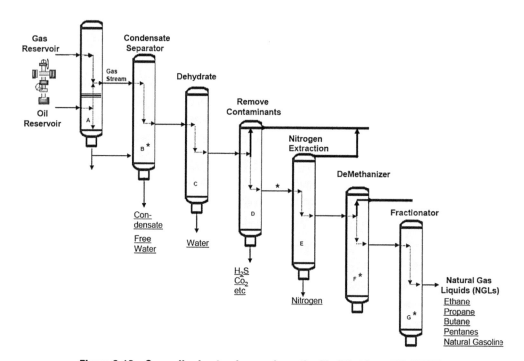

Figure 8-16   Generalized natural gas schematic. Modified from EIA (2006).

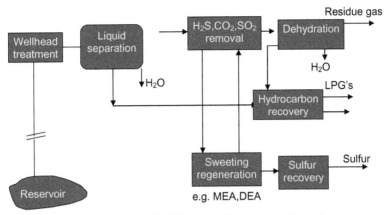

Figure 8-17 Schematic of the overall gas processing scheme.

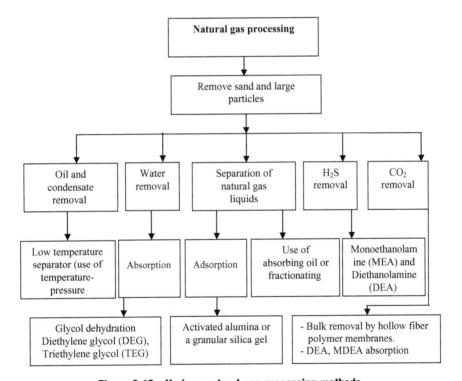

Figure 8-18 Various natural gas processing methods.

Figure 8-18 illustrates the details of removing each contaminant and chemical, as well as catalysts used during processing. Below is a detailed description of the removal of each pollutant.

8.3.3.1 Oil and condensate removal

Natural gas is dissolved in oil underground due to formation pressure. When natural gas and oil are produced, they generally separate, simply because of the decrease in pressure. The separator described here consists of a closed tank where gravity serves to separate out the heavier liquids, such as oil from lighter gases (EIA

2006). Sometimes, specialized equipment, such as the Low-Temperature Separator (LTS) are used to separate oil and natural gas (NGO 2005). When the wells are producing high-pressure gas along with light crude oil or condensate, a heat exchanger is used to cool the wet gas, and the cold gas then travels through a high-pressure liquid knockout that serves to remove any liquid into a LTS. The gas flows into the LTS through a choking mechanism, expanding the gas as it enters the separator. This rapid expansion of gas lowers the temperature in the separator. After liquid is removed, the dry gas is sent back through the heat exchanger, where it is warmed by the incoming wet gas. By changing the pressure at different sections of the separator, the temperature is varied, causing the oil and water to be condensed out of the wet gas stream. The gas stream enters the processing plant at high pressure (600 pounds per square inch gauge (psig) or greater) through an inlet slug catcher where free water is removed from the gas, after which it is directed to a condensate separator (EIA 2006).

8.3.3.2   Water removal

Water contained in a natural gas stream may cause the formation of hydrates when exposed to specific temperature and pressure conditions. Water in natural gas is removed by separation methods at or near the well head. However, this process cannot separate out all of the water. The water treatment process consists of dehydration of natural gas either by absorption or adsorption. In absorption, the water vapor is taken out by a dehydrating agent such as glycol (Nallinson 2004). In adsorption, the water vapor is condensed and collected on the surface. Adsorption dehydration can also be used, using dry-bed dehydrating towers, which contain desiccants such as silica gel and activated alumina, to perform the extraction.

Glycol dehydration: It is considered important to remove water vapor from the gas stream, which otherwise may cause hydrate formation if exposed to low temperatures that may plug the valves and fittings in gas pipelines (Twu 2005). Water vapor may further cause corrosion when it reacts with hydrogen sulfide or carbon dioxide present in the gas streams. Glycol is generally used for water dehydration or absorption (Nallinson 2004). Glycol has a chemical affinity for water. When it is in contact with a stream of natural gas containing water, it will absorb the water portion out of the gas stream (EIA 2006). Glycol dehydration involves using a glycol solution, either diethylene glycol (DEG) or triethylene glycol (TEG). After absorption, glycol particles become heavier and sink to the bottom of the contactor, from where they are removed. Glycol and water are then separated by boiling, as water boils at 212°F whereas glycol boils at 400°F (NGO 2005). The glycol is then reused in the dehydration process (Nallinson 2004).

Solid-desiccant dehydration: In this process, solid desiccants, such as activated alumina or granular silica gel materials, are used for adsorption in a two or more adsorption tower arrangements (Nallinson 2004). Natural gas is passed through these adsorption towers, with water being retained on the surface of these desiccants. The gas coming out from the bottom of the adsorption tower will be completely dry gas. This method is more effective than with glycol dehydrators and is best suited for large volumes of gas under very high pressure. Two or more towers are generally required as the desiccant in one tower becomes saturated with water and needs to regenerate the desiccant. Solid desiccant systems are more expensive than glycol dehydration processes.

The solid desiccant adsorption process consists of two beds, with each bed going through successive steps of adsorption and desorption (Rojey et al. 1997; Nallinson 2004). During the adsorption step, the gas to be processed is sent through the adsorbent bed and retains the water. When the bed is saturated, hot natural gas is sent to regenerate the adsorbent. After regeneration and before the adsorption step, the bed must be cooled. This is achieved by passing through cold natural gas. After heating, the same gas can be used for regeneration. Figure 8-19 shows the dehydration by adsorption in a fixed bed.

8.3.3.3   Separation of natural gas liquids

Natural gas liquids (NGLs) are saturated with propane, butane, and other hydrocarbons. NGLs have a higher value as separate products, so are separated from the natural gas stream. Moreover, reducing the

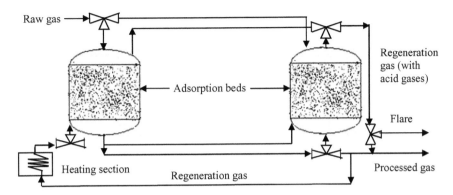

Raw gas

Regeneration gas (with acid gases)

Adsorption beds

Flare

Heating section

Regeneration gas

Processed gas

Figure 8-19   Dehydration by adsorption in fixed bed. Redrawn from Rojey et al. (1997).

concentration of higher hydrocarbons and water in the gas is necessary to prevent formation of hydrocarbon liquids and hydrates in the natural gas pipeline. The removal of NGLs is usually done in a centralized processing plant by a similar process to those used to dehydrate natural gas. There are two common techniques for removing NGLs from the natural gas stream, the absorption and cryogenic expander process.

8.3.3.4   The absorption method

This process is similar to adsorption by dehydration. The natural gas is passed through an absorption tower and brought into contact with the absorption oil, which soaks up a large amount of NGLs (EIA 2006). The oil containing NGLs exits the absorption tower through the bottom. The rich oil is fed into lean oil stills and the mixture is heated to a temperature above the boiling point of the NGLs but below that of the oil. The oil is recycled and the NGLs are cooled and directed to an absorption tower. This process allows recovery of up to 75% of butanes and 85–90% of pentanes and heavier molecules from the natural gas stream. If the refrigerated oil absorption method is used, propane recovery can reach up to of 90%. Extraction of the other, heavier NGLs can reach close to 100% using this process. Alternatively, a fractioning tower can be used where the varying boiling temperatures of the individual hydrocarbons in the natural gas stream can be used. The process occurs in stages as the gas stream rises through several towers where heating units raise the temperature of the stream, causing the various liquids to separate and exit into specific holding tanks (EIA 2006).

Pressure Swing Adsorption (PSA) is a gas separation technique in which the adsorption is regenerated by rapidly altering the partial pressure of the adsorbate in the gas phase. Motivated by its low energy requirements and low capital investment costs, tremendous progress has been made. In a sophisticated PSA process, only one component in the feed mixture, generally the weakly adsorbed one, is the desired product.

Feng et al. (1998) proposed a PSA process using a hollow fiber device. The invention constitutes of the use of a bundle of porous hollow fiber membranes to hold the adsorbent particles in place. The fiber wall, with pore openings in the hollow fibers, are arranged in a shell-and-tube configuration with proper potting of fiber ends. The adsorbent particles can be confined in either the tube or the shell side, while the gas stream is flowing through the other side and yet maintaining good contact with the adsorbent. Compared to the conventional fixed-bed adsorbed phase, the hollow-fiber device offers numerous advantages. For instance, pressure drop is reduced significantly due to unobstructed gas flow. Very small adsorbent particles can be used to reduce the intra-particle mass-transfer resistance, and the need for a highly porous adsorbent is minimized due to the very large specific external areas of the extremely fine particles. Also, a more uniform contact of gas to an adsorbent phase is achieved. The problems associated with gas flow channeling, because of low pressure drop and high mass transfer rates, can be solved by using fast PSA cycles to enhance the process capacity.

In almost all PSA processes, only one component in the feed mixture, generally the weakly adsorbed one, is the desired product. However, for improved economics it is important to recover as many of the components of the gas mixture as possible, by employing a three-stage pressure swing policy, in which the pressure is cycled through high, low, and vacuum pressures. A PSA process with intermediate feed inlet and side-stream outlet, called the Petluyk PSA process, has been proposed for the simultaneous separation of ternary gas mixtures (Dong et al. 1999).

Low concentrations of H_2S in natural gas can be handled by regenerable adsorbents such as activated carbon, activated alumina, silica gel, and synthetic zeolites. Non-regenerable adsorbents, i.e., zinc and iron oxides, have been used for natural gas sweetening. An experimental study of natural gas purification using a 5 A-zeolite was initiated in 1995 (Kikinides et al. 1995). In this work, it was suggested that the adsorbent could be used in an adsorption process to remove H_2S from natural gas. It also reported the successful removal of H_2S compounds from natural gas by Pressure Swing Adsorption using a zeolite of a pore size of percentage Amstrongs. A typical integrated processing scheme is to hydro-desulfurize sulfur compounds over a Co–Mo catalyst to H_2S and chemisorb the gas over ZnO. This scheme is effective for desulfurization on offshore platforms before pipeline transport of the gas. Mercaptans can be removed from natural gas by adsorption on NaX zeolites, pretreated with 5–10% aq. NaOH.

Absorption of regenerable solvents: Among the various acid gas removal processes, absorption by ethanolamine solution is the most commonly used. In this process the regeneration of the used solvent is carried out in a stripping column. This is the traditional method of removing H_2S from natural gas. In this process the sour natural gas is first treated with a solvent known as Diethanol amine (DEA) or Monoethanolamine (MEA), and sulfur is recovered by a Claus plant. The ethanolamine-based process for treatment of sour natural gas was discovered by Bottoms in 1930 (Chakma 1985).

Sufficient information on acid gas solubility in solvents, needed for the design of the gas plants at typical operating conditions, is available in the literature (Martin et al. 1978; Dingman et al. 1983; Mdax et al. 1987). Due to limitations of experimental techniques employed, it is difficult to obtain accurate experimental data under very low and very high partial pressure conditions (Astrita et al. 1983). Hu and Chakma (1990) developed a mathematical model for the prediction of equilibrium solubility of CO_2 and H_2S in aminoethanol aqueous solutions.

Non-regenerable chemical scavengers: In a number of situations, such as low-volume and low-pressure gases, the conventional ethanolamine process may not be economical. In such cases, the use of non-regenerable chemical scavengers is preferred (Chakma 1999). Leppin (1995) has evaluated three types of commercial scavenger, Nitite, no-nregenerable amine, and triazine. He has reported the successful application of the aforementioned agents, treating 16 MMSCFD of natural gas with 19 ppm H_2S and 0.5% CO_2 at 1000 psig. However, these scavengers are not usually environmentally-friendly, being toxic in nature and their safe disposal is costly. So with growing environmental concerns it is expected that this practice may no longer be allowed. If no economically viable technology is developed to handle these gases, the condensate producing wells may have to be shut down. Therefore, there is a need to develop other alternative technologies capable of handling low volume gases in an economical manner.

Newly developed techniques: The mixed amine process is more suitable at large-scale gas plants, because amine/Claus processes exhibit operating difficulties for small sulfur capacities (up to 5 tonnes/day). The process developed by VIIGAS (Russia), treats sour gas using an electric arc method, which is an alternative to the Claus unit coupled with alkanolamine solution treatment. The essence of the arc method whereby sour gas is delivered to an arc reactor through a plasmatron, where H_2S is dissociated at 2200–2370°F. Downstream, the mixture containing H_2, CO, CO_2, non-dissociated H_2S, and sulfur passes to the heat exchanger where it is water-cooled down to 1300°F. The mixture is then passed through a condenser where sulfur is condensed and cool gas is then subjected to burning in a furnace.

Liquid redox process: Low concentrations of H_2S in sour gases at high H_2S/CO_2 ratios are generally removed by liquid redox processes using transition metals, for example, Sulfolin and Stretford. VNIIGAS has developed a technique based on iron chelate solution, where H_2S reacts with an aqueous solution containing

Fe3$^+$-EDTA (FeY) and H_2S is oxidized to elemental sulfur. This technology was first applied to natural gas treatment in 1989. It can handle a stream of natural gas containing 6% CO_2 and 0.1% H_2S at 5.5 MPa.

Physical solvents: Physical solvents are commonly used for selective removal of H_2S over CO_2 and the process can be designed for either bulk acid gas removal or selective H_2S removal with specified degree of removal. Dimethyl ether of polyethylene glycol (Selexol), *N*-methyl pyrolidone (Purisol), Methanol (Recti-sol), and Propylene carbonate (Fluor) have been successfully used for natural gas treatment. Refrigerated methanol is well known as a selective physical solvent for the removal of H_2S from gases. Methanol, being a physical rather than a chemical solvent, absorbs by selective solubility the acid components present in the gases, not by chemical reactions as with amines (Minkkinen and Jonchere 1997). Claus-type recovery projects are complementary to this project if high H_2S loads are to be handled, which is already complicated in its operation.

Adsorption on solids: Low concentrations of H_2S in natural gas can also be handled well by regenerable adsorbents such as activated carbon, activated alumina, silica gel, and synthetic zeolites molecular sieve; MS). Non-regenerable adsorbents, i.e., zinc and iron oxides, have been used for natural gas sweetening.

Chi, et al. (1973) initiated an experimental study of natural gas purification using a 5 A zeolite. They suggested that this sorbent could be used in an adsorption process to remove H_2S from natural gas. The study on the successful application of these sorbents has been performed by different scientists (Kikkinides et al. 1995). They have also reported the removal of H_2S compounds from natural gas by pressure swing adsorption (PSA), using 5 A zeolites as sorbent. A typical integrated processing scheme is to hydrode-sul-furize sulfur compounds over Co–Mo catalyst to H_2S and chemisorb H_2S over ZnO. This scheme is good for desulfurization on offshore platforms before pipeline transport of the gas. Mercaptans can be removed from natural gas by adsorption on NaX zeolites, pretreated with 5–10% aq. NaOH.

8.3.3.5 The membrane separation

Various types of membranes can be used to remove water and higher hydrocarbons. The conventional membranes can lower the dew point of the gas. The raw natural gas is compressed and air cooled, which knocks out some water and NGLs. The gas from the compressor is passed through the membrane, which is permeable to water and higher hydrocarbons. The dry hydrocarbon depleted residual gas is then sent to the pipeline for use.

Recently, Basu et al. (2004), Akhtar (2003), and Bjorndalen et al. (2002) extensively studied membrane technology. Considering the importance and potentiality of membrane technique, the following sections elaborate different aspects of this technology.

A membrane can be defined as a selective barrier between two phases, the "selective" being inherent to a membrane or a membrane process (Mulder 1996). The membrane gas separation technology is over 15 years old and is proving to be one of the most significant unit operations. The technology inherits certain advantages over other methods that include compactness and light weight, low labor intensity, modular design permitting easy expansion or operation at partial capacity, low maintenance, low energy requirements, low cost, and environmental friendliness.

The progress in membrane science and technology was accelerated during the 1980s by the development and refinement of synthetic polymeric membranes. Membrane gas separation emerged as a commercial process on a large scale during this time. Also during this period, significant progress was made in virtually every aspect of membrane technology, including improvements in membrane formation process, chemical and physical structures, configurations, and applications. Chemical structures coupled with subtle physical properties of the membrane material influence the permeability and selectivity of a gas. The responses of a polymeric material to permeation are strongly influenced by the polarity and stearic characteristics of the polymer and permeate. The size and shape of bulky groups in both the polymer main chains and side chains determine certain fundamental properties, such as packing, density, and rigidity. These, in turn, influence the accessibility. An absence of such groups tends to increase the structural regularity, which favors increased density. The effect of lateral substituents on the backbone of aliphatic polyamides has a strong bearing on gas separation properties. Bulky hydrocarbon groups force parallel chain segments further apart, thereby increasing free volume and diminishing hydrogen bonding and rigidity, also increasing the permeability (Pandey and Chauhan 2001).

In the concept of membrane science, a synthetic membrane behaves as a thin barrier between two phases through which differential transport can occur. Driving forces that facilitate this transport are pressure, concentration, and electrical potential across the medium (Koros et al. 1996). The transport itself is a non-equilibrium process and the separation of chemical species results from differences in transport rates through the membrane.

Liquid membranes: Synthetic membranes can be divided into organic (polymeric or liquid) and inorganic membranes. Liquid membranes operate by immobilizing a liquid solvent in a microporous filter or between polymer layers. A high degree of solute removal can be obtained when using chemical solvents. When the gas or solute reacts with the liquid solvent in the membrane, the result is an increased liquid phase diffusivity. This leads to an increase in the overall flux of the solute. Furthermore, solvents can be chosen to selectively remove a single solute from a gas stream to improve selectivity (Astrita et al. 1983).

Polymeric membranes: Polymeric membranes have been developed for a variety of industrial applications, including gas separation. For gas separation, the selectivity and permeability of the membrane material determines the efficiency of the gas separation process. Based on flux density and selectivity, a membrane can be classified broadly into two classes, porous and nonporous. A porous membrane is a rigid, highly voided structure with randomly distributed interconnected pores. The separation of materials by a porous membrane is mainly a function of the permeate character and membrane properties, such as the molecular size of the membrane polymer, pore size, and pore-size distribution. A porous membrane is similar in its structure and function to the conventional filter. In general, only those molecules that differ considerably in size can be separated effectively by microporous membranes. Porous membranes for gas separation do exhibit high levels of flux but inherit low selectivity values. Microporous membranes are characterized by the average pore diameter, d, the membrane porosity, and tortuosity of the membrane.

There are several ways to prepare porous polymeric membranes, such as solution casting, sintering stretching, track etching, and phase separation. The final morphology of the membrane obtained will vary greatly, depending on the properties of the materials and the process conditions used. Nonporous or dense membranes have high selectivity properties but the rates of transport of gases through the medium are usually low. An important property of a nonporous dense membrane is that even permeant of similar sizes may be separated if their solubility in the membrane differs significantly. A dense membrane can be prepared by melt extrusion, where a melt is envisioned as a solution in which the polymer is both solute and solvent. In the solution-casting method, dense membranes are cast from polymer solutions prepared by dissolution of a polymer in a solvent vehicle to form a solvent. This is followed by complete evaporation of the solvent after casting.

Polymer membranes have gained popularity in isolating carbon dioxide from other gases (Gramain and Sanchez 2002). These membranes are elastomers formed from cross-linked copolymers of high molecular weights. They are prepared as thin films by extrusion or casting. They demonstrate unique permeability properties for carbon dioxide together with high selectivity toward H_2, O_2, N_2, and CH_4. These elastomers are very thin membranes (5–10 microns) and can be easily prepared by extrusion, solvent casting, or impregnation. The membranes are hydrophilic and the introduction of hydrophobic monomers provides regulation of water absorption. They are thermally stable up to 250°C and are biodegradable and nontoxic. Their applications include food packaging, natural or industrial gas treatment of confined atmospheres (airplane cabins, air-conditioned rooms), decarbonization of gases used in fuel cells, biomass treatment, and medical uses.

Different organic membranes: The effects of flexible hetro-atoms and cyclic structures, with nonlinear orientation in the polymer backbone, lead to a decreased structural regularity on permeation behavior (Zimmerman, 1984). The structural regularity favors close packing of molecules, which improve density and rigidity in both crystalline and amorphous phases. All these changes have some bearing on the effects of permeability and selectivity. Polyether sulfones (PES) have a relatively more regular structure and inherit a higher bulk density (1.37 g/cm^3) and an elevated glass transition T_g, (220°C). They also acquire a lower free volume when compared to polysulfones (PSF), which contain an aliphatic isopropyllidene group contributing less regularity, lower bulk density (1.23 g/cm^3), and lower T_g (190°C).

Isotropic Polyethersulfone (PESF) dense film have exhibited better selectivity for the commercially important gas pairs (CO_2/CH_4, He/CH_4, H_2/N_2, O_2/N_2), compared to bisphenol-A polysufone and cellulose acetate (Ellig et al. 1980).

Wang et al. (1990) reports that like polysulfone, PESF possesses properties of high mechanical and thermal stabilities. Moreover, it is generally easy to prepare the dried asymmetric membranes by the immersion phase inversion method using water as a coagulant. This polymer has only a moderate permeability. In order to use this polymer to prepare commercially attractive gas separation membranes, fabrication of ultra-thin skinned asymmetric hollow fiber membranes will be needed so as to increase permeation flux. Only a few studies on the preparation of PESF asymmetric hollow fiber membranes for gas separation have been reported in the literature. The PESF hollow fibers prepared at various spinning conditions exhibit reduced gas permeation behavior. The observed CO_2 pressure-normalized flux is less than 12 GPY at 24°C. In the later 1980s, PESF hollow fiber membranes with good permeance and selectivity were prepared from spinning solution containing 1 : 1 molar mixtures of propionic acid, NMP and high polymer concentration (more than 35 wt. %) (Kesting et al. 1989). The oxygen pressure-normalized flux was prepared to 13.1 GPU with O_2/N_2 selectivity of 5.1 at 50°C. Fritzsche et al. (1990) examined the skin layer structure of membranes.

Wang et al. (2000) made studies on the preparation and characterization of PESF hollow fiber membranes spun from moderate polymer concentration (25–30 wt%) and solvent systems containing various alcohols as non-solvent additives (NSA). NSA plays a dominant role in determining membrane structure and gas separation properties. They prepared the PESF hollow fiber membranes with the best combination of gas permeability and selectivity, by using ethanol as an additive in the various studies. Their studies also demonstrated that effective NSAs possess high affinities and diffusivities with the coagulant. It is also known that a certain amount of water can be added to the polymer solution before phase separation occurs. In addition, water and N, N-substituted amides (solvents) can form a complex through multiple hydrogen bonding. Therefore, the influence of water as an NSA on the structure and gas permeation properties of PESF hollow fiber membranes is of immediate interest. The authors demonstrated that silicon-coated PESF asymmetric hollow fiber membranes, with high pressure-normalized fluxes and ideal selectivities for gas pairs of He and N_2, were fabricated from an NMP/H_2O solvent system with mass ratio 8.4 : 1. In the presence of water as an additive, the viscosity of spinning solutions increased dramatically due to the formation of the $NMP : H_2O$ hydrogen-bonding complex. The observed pressure-normalized fluxes and ideal selectivities of the PESF membranes were higher than those of the hollow fiber membranes fabricated from the NMP/alcohol and NMP/propionic acid systems reported in literature.

Saha and Chakma (1992) suggested the attachment of a liquid membrane in a microporous polymeric membrane. They immobilized mixtures of various amines such as monoethanolamine (MEA), diethanolamine (DEA), amino-methyl-propanol (AMP), and polyethylene glycol (PEG) in a microporous polypropylene film and placed it in a permeator. They tested the mechanism for the separation of carbon dioxide from some hydrocarbon gases and obtained separation factors as high as 145.

Fuertes (2001) reported that when the size of micropores in the carbon membranes is in the 3–5 A range, gas molecules with effective diameters smaller than 4 A display significant differences in gas diffusivity, and mixtures of these gases can be effectively separated according to a molecular sieving mechanism. In this case, the gas transport rate through the membrane depends on the effective size of the gas molecules instead of the adsorption effects. Membranes with these characteristics are identified as Molecular Sieve Carbon Membranes (MSCM). Polymers (polyvinylidene chloride, polymides, poly furfuryl alcohol, phenolic regins) that are used to prepare MSCM, are initially crosslinked or become crosslinked during pyrolysis. This prevents the formation of large graphite-like crystals during carbonization and leads to the formation of disordered structures (non-graphiting carbons) with a very narrow porosity formed by micropores in the 3–5 A range (molecular sieve carbons). In this sense, an MSCM can be described as a thin molecular sieve carbon film. If the micropores in the carbon membrane are enlarged from 3 to 5 A, gas permeation through the membranes with enlarged micropores is governed by the adsorption strength instead of molecular size. This kind of membrane would be effective for separation of non-adsorbable or weakly adsorbable gases from adsorbable gases.

Inorganic membranes: The efficiency of polymeric membranes decreases with time due to fouling, compaction, chemical degradation, and thermal instability. Because of this limited thermal stability and susceptibility to abrasion and chemical attack, polymeric membranes have found application in separation processes where hot reactive gases are encountered. This has resulted in a shift of interest toward inorganic membranes.

Inorganic membranes are increasingly being explored to separate gas mixtures. Besides having appreciable thermal and chemical stability, inorganic membranes have much higher gas fluxes when compared to polymeric membranes. There are basically two types of inorganic membranes, dense (nonporous) and porous. Examples of commercial porous inorganic membranes are ceramic membranes, such as alumina, silica, titanium, glass, and porous metals, such as stainless steel and silver. These membranes are characterized by high permeabilities and low selectivities. Dense inorganic membranes are specific in their separation behaviours, for example, Pd-metal based membranes are hydrogen specific and metal oxide membranes are oxygen specific. Palladium and its alloys have been studied extensively as potential membrane materials. Air Products and Chemical Inc. developed the Selective Surface Flow (SSF) membrane. It consists of a thin layer (2–3 μm) of nano-porous carbon supported on a macroporous alumina tube (Rao et al. 1992). The effective pore diameter of the carbon matrix is 5–7 A (Rao and Sircar 1996). The membrane separates the components of a gas mixture by a selective adsorption-surface diffusion-desorption mechanism (Rao and Sircar 1993).

The larger polar molecules of a feed-gas mixture at the high-pressure zone of the membrane are selectively adsorbed on the carbon pore walls, followed by their selective diffusion along the pore walls to the low pressure side of the membrane, where they desorb into the low pressure gas phase. Thus, a gas stream enriched in the smaller and less polar components of the feed-gas stream mixture is produced as a low-pressure effluent gas from the SSF membrane system. However, a gas stream enriched in the larger and more polar components of the feed-gas mixture is produced as a low-pressure effluent gas. The performance of the SSF membrane for the separation of hydrocarbons from H_2 and CO_2 from H_2, CO_2 and CH_4 from H_2, H_2S from H_2, and H_2S from CH_4, are found in the literature (Parillo et al. 1997). Gas separation by means of microporous carbon membrane is based on interaction differences existing between components of a gas mixture with respect to the carbon membrane.

Materials that are thermally stable at temperatures above 400°C include ceramic membranes that are used to separate hydrogen from gassified coal (Fain et al. 2001). With ceramic materials, high separation factors have been achieved based on the ratios of individual gas permeances. This work focuses on the development of alumina membranes having nano-pores that can separate gases based on the molecular size. A layer of alumina having a mean pore diameter of about 7 nm is deposited in a stainless steel tube. Methods have been developed to reduce the pore diameter to values in the range between 0.5 and 3.0 nm. The membranes having the smallest estimated mean pore diameters have shown higher separation factors for helium and hydrogen than with other gases. These separation factors are determined from pure gas separations in these pore diameter ranges. For pure molecular flow, separation factors should increase with decreasing pore size, but in many cases, the separation factor decreases. This decrease occurs because of adsorption and surface flow of the heavier molecule. The separation factor continues to decrease until the pore size becomes small enough that the entrance into the membrane is restricted for the larger molecule. At this point, the separation factor increases rapidly as pore size is decreased. It is observed that the separation factors are directly proportional to temperature. This is because with increase of temperature, an unexpectedly large increase of permeance of hydrogen and helium is observed.

Hybrid membranes: Hybrid membranes occupy a special place in separation technology, due to the combination of thermodynamically based partitioning and kinetically based mobility discrimination in an integrated separation unit. The applicability of SSF membrane, which is a nano-porous carbon membrane, is being tested for separation of various gas mixtures. The separation mechanism of the SSF membrane consists of preferential adsorption of certain components of gas mixtures on the surface of the membrane and diffusion of the adsorbed molecules across the membrane pores. An experimental setup has been devized to study the permeability of a pure gas and to analyse the gas mixture separation runs (Paranjpe 2002). The future plan is to carry out the separation of hydrogen and hydrogen sulfide gases at a molar ratio 3 : 7. The experimental setup will also be used to study the permeability of a pure gas, as well as to carry out the gas mixture separation runs. Currently, the experiments are being conducted to detect the effects of feed composition, feed flow rate, feed pressure, and permeate pressure variations on the separation characteristics of various gas mixtures.

8.3.3.6 Mechanisms for gas separation

Various mechanisms for gas transport across membranes have been proposed, depending on the properties of both the permeant and the membrane. These include Knudsen diffusion, the molecular sieve effect and a solution diffusion mechanism. However, most of these models have been found to be applicable only to a limited number of gas/material systems. Different mechanisms (a–d) may be involved in the transport of gases across a porous membrane, giving a schematic representation of the mechanisms for the permeation of gases through porous as well as dense membranes (Baker 1995).

Mechanism (a): The permeation of gas through a porous membrane consists of Knudsen diffusion and Poiseuille flow, the properties of which are governed by the ratios of the pore radius (r) to the mean free path (λ) of the gas molecules. The mean free path is given by

$$\lambda = \frac{3\eta(\pi RT)^{0.5}}{2P(2M)} \qquad (8.13)$$

where

η = viscosity of the gas

R = universal gas constant

T = temperature

M = molecular weight

P = pressure.

If $\lambda/r \ll 1$ viscosity or Poiseuille flow predominates, the gas flux through the pore is described by

$$G_{vis} = \frac{r_2(P_1 - P_2)}{16LuRT} \qquad (8.14)$$

where

r = pore radius

P_1 = partial pressure of the gas on the feed side

P_2 = partial pressure of the permeate side

L = pore length

u = gas viscosity

G_{vis} = viscous flow

R = universal gas constant.

In Knudsen flow, r/λ is considerably less than unity and, therefore, there are more collisions with the pore walls than with other gas molecules. With every collision, the gas molecules are momentarily absorbed and then reflected in a random direction. As there are a less number of collisions among molecules than pore walls, each molecule will move independent of the others. Hence, the separation is achieved because different gaseous species are moving at different velocities.

The gas flux in such a case is described by

$$G_{mol} = \frac{8r(P_1 - P_2)}{3L(2\pi MRT)^{0.5}}$$
(8.15)

where

G_{mol} = molecular flow of the gas

r = pore radius

P_1 = partial pressure of the gas on the feedside

P_2 = partial pressure of the gas on the permeate side

L = pore length

M = molecular weight

R = gas constant

T = absolute temperature.

For Knudsen flow, the selectivity ratio or the separation factor for binary gas mixtures can be estimated from the square root of the ratio of the molecular weights. Knudsen separation can be achieved with membranes having pore sizes smaller than 50 nm. Based on Knudsen flow, separation factors for several gas pairs have been calculated and they represent ideal separation factors.

Mechanism (b): In order to function as a molecular sieve, membranes must have pore diameters that are in between those of the gas molecules to be separated. Separation factors greater than 10 should be achievable as the pores become smaller than approximately 0.5 nm. If the membrane has pore sizes between the diameter of the smaller and larger gas molecules, then only the smaller molecules can permeate and a very high separation would be achieved. In practical situations, there will be a distribution of pore size in the membranes, and thus the gas permeability is actually influenced by a combination of transport mechanisms. From a practical standpoint, as the pore size decreases, the membrane porosity is expected to decrease, resulting in a lower gas flow through the membrane. Therefore, the pore size and porosity must be balanced to produce an efficient membrane.

Mechanism (c): Gas separation can also be affected by partial condensation for some component of a gas mixture in the pores, with the exclusion of others, and the subsequent transport of the condensed molecules across the pores.

Mechanism (d): Selective adsorption of the more strongly adsorbed components of a gas mixture onto the pore surface, followed by the surface diffusion of the adsorbed molecule across the pore, can also facilitate the separation of gases.

A proper concentration gradient for the diffusing species must be imposed across the porous membrane to provide the driving force for transport by all four mechanisms. The selectivity of separation achieved by the Knudsen mechanism:

a) is generally a very low mechanism;

b) exhibits high selectivity and high permeability for the smaller components of a gas mixture;

c) requires the pore size of the membrane to be in the mesoporous size range (diameter > 30 A) so that condensation of the component of a gas mixture can take place. A very high selectivity of separation of the condensable component can be achieved by this mechanism. The condensation partial pressure of the component at the system pressure, the pore size, and the geometry of the membrane limit the extent of removal of that component from the gas mixture;

d) provides the most flexible and attractive choice for the practical separation of gaseous mixtures. The reason is that the separation selectivity is determined by the preferential adsorption of certain components of the gas mixture on the surface of the membrane pores, as well as by the selective diffusion of the absorbed molecules.

In meso- and micro-porous media, as in the case of inorganic membranes, if the relative pressure is increased, the adsorbed and capillary condensed materials permeate together. The porous medium behaves as a semi-permeable membrane through which the adsorbed permeate will flow freely while the weakly adsorbed component will be blocked. Thus, both the pore size and the physical-chemical nature of the pore surface play key roles in determining the separation efficiency of these membranes. Consequently, appropriate molecular engineering of their surface chemistries can alter the properties of the membranes.

The transport of gases through a dense polymeric membrane is usually described by the solution-diffusion mechanism. The important feature of dense membranes that are used in separation applications is the ability to control the permeation of different species. In the solution-diffusion mechanism, the permeants dissolve in the membrane material and then diffuse through the membrane down a concentration gradient. A separation is achieved between the different permeants because of differences in the amount of material diffusing through the membranes.

The solution-diffusion mechanism consists of three steps:

1. the absorption or adsorption at the upstream boundary;

2. activated diffusion (solubility) through the membrane; and

3. desorption or evaporation on the other side.

These solution-diffusion mechanisms are driven by a difference in thermodynamic activity at the upstream and downstream faces of the membrane as well as the interacting forces between the molecules of the membrane material and permeate molecules. The activity difference causes a concentration difference that leads to diffusion in the direction of decreasing activity. As in microporous membranes, differences in permeability of dense membranes also result from differences in the physical-chemical interactions of the gas species within the polymer. The solution-diffusion model assumes that the pressure within a membrane is uniform and the chemical gradient across the membrane is expressed only as a concentration gradient.

The schematic diagram shown in Figure 8-20 illustrates mass transfer across a regular membrane and a liquid membrane. In a regular membrane, the solute A passes through the membrane under pure diffusion mechanism, leaving the other gas behind. In the liquid membrane, the enhanced effect of having the liquid

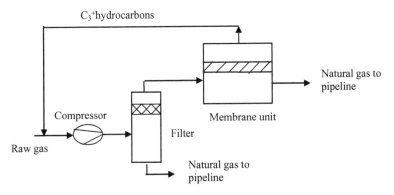

Figure 8-20   Membrane system for NGL recovery and dew point control. Redrawn from Mtrinc (2005).

solvent, B, can be seen by the additional flux of species AB. This system, with the additional flux, is known as facilitated transport. Without the effects of the liquid, mass transfer would be reduced to that of the regular diffusion membrane system, whereby the total flux only consists of the flux of species, A.

A synthetic membrane is a thin barrier between two phases through which differential transport can occur under a variety of driving forces, including pressure, concentration, and electrical potential across the membranes. Pressure difference across the membrane can facilitate reverse osmosis (RO), ultra-filtration, micro-filtration, gas separation, and pervaporation. Temperature difference across the membrane can facilitate distillation, whereas concentration differences can be used for dialysis and extraction. Electro-dialysis can also be performed with the help of membranes when an electric potential across the membrane is maintained. Differential transport occurs when the membrane restricts the transport of different species in some specific way. The transport is a non-equilibrium process and the separation of chemical species results from differences in transport rates through the membranes. Membrane separation is considered a non-equilibrium process because it is based on the selective permeation rate of the feed components. If a membrane system is allowed to move toward equilibrium, permeation would continue until the pressure and concentration of gases on both sides are equal (Koros et al. 1996).

The two main parameters defining the performance of a membrane are separation factor and permeance. In a binary system consisting of gases "a" and "b", with gas "a" as the fast permeating gas, the separation factor is defined as the concentration ratio of "a" to "b" in the permeate, and X_a and X_b are the concentration of the gases "a" and "b" in the feed.

$$\alpha_{a/b} = Y_a/Y_b \; / \; X_a/X_b \qquad (8.16)$$

where

α = separation factor

Y_a and Y_b = concentration of the gases "a" and "b" in the permeate

X_a and X_b = are the concentrations of gases "a" and "b" in the feed.

Permeance is defined as the volume of the feed passing through a unit area of membrane at unit time and under unit pressure gradient, expressed as (Tabe-Mohammadi 1999)

$$P = \frac{V}{At\Delta p} \qquad (8.17)$$

P = permeance

V = vol. of gas

A = area

t = time

Δp = pressure difference.

8.3.3.7   Recent developments related to separation of gases using membranes

Ward (1970) conducted pioneering work by demonstrating a general quantitative understanding of facilitated transport. He performed a detailed mathematical and experimental investigation of facilitated transport. The basic facilitated transport reaction scheme was modeled for a nitrous oxide/ferrous ion system. A mathematical model for a steady state system was developed.

Donaldson and Quinn (1975) also studied facilitated transport in liquid membranes, but they applied the theory to a carbon dioxide system. The model was developed as a means of determining diffusional and kinetic parameters for the reaction system. The derivation and solution of equations for the model were not explained clearly. The reaction system was a CO_2 hydration system for the formation of bicarbonate. This left the door open for future works in studying CO_2^- amine systems, which have evolved into a more common system for CO_2 capture.

Smith et al. (1977) wrote one of the first reviews on facilitated transport. They provided a comprehensive summary of the research into facilitated membranes at that time. They discussed facilitated oxygen transport, facilitated carbon dioxide transport, and a nitric oxide-ferrous chloride facilitated transport system. They provided a comprehensive document on the governing equations and the formulation of facilitation factor. One drawback of this paper is that the section of CO_2 systems was not for an amine system. At that time, CO_2-amine systems were becoming common and therefore we would expect to see researchers exploring that area more aggressively.

Two years later, Smith and Quinn (1979) developed a model for predicting the facilitation factors in facilitated transport systems. Using the standard reaction mechanism for facilitated transport, they solved 1D, steady state equations, considering both the physical diffusion and chemical reaction terms. They introduced the assumption of equal carrier and carrier complex diffusivities, such as $D_B = D_{AB}$, in order to simplify the analysis. The equations were solved using analytical techniques. Initially, they solved the system for nitric oxide-ferrous ion facilitated transport, obtaining comparable results to other studies. They also studied a CO_2–MEA system that had more complex chemistry. They did not clearly explain how the governing equations were incorporated into the model.

Since the reaction kinetics of CO_2-amine systems are not fully understood, researchers should be concise in describing which reaction equations they are using and what assumptions they are employing when simplifying their work. Although Smith and Quinn have calculated facilitation factors that compare with the literature, they did not present any figures that show the facilitation factor trends. Data analysis, be it experimental or mathematical, should include the presentation of charts that illustrates any trends in the data.

To counter the above problem, Folkner and Noble (1983) studied facilitated transport in membranes under transient conditions. Models were developed in 1D for the standard facilitated transport reaction mechanism under flat plate, cylindrical, and spherical geometries. The effects of the inverse Damkohler number, the equilibrium constant, the carrier concentration, and geometry on the facilitation factor were determined. The assumption that $D_B = D_{AB}$ was used. No real discussion was presented concerning how the equations were solved except to say a computer package was used. Very few authors seem to provide useful, easy-to-understand information in their articles about their specific modeling procedure. Researchers should be more willing to share verifiable information in the hope of aiding future projects and to add credibility to their findings.

Meldon et al. (1985) also modeled the standard reaction mechanism for facilitated transport under transient conditions. In their work, the permeation across the membrane was studied with emphasis on the permeation time lag. In this work they state that previous researchers may have made mistakes when analyzing transient systems with fast chemical reactions. They explain the error but few details of the numerical procedure is documented.

Basaran et al. (1989) studied facilitated transport using unequal carrier and complex diffusivities, such that D_B is not equal to D_{AB}. The standard reaction mechanism for facilitated transport was modeled, but for the first time the simplifying assumption that $D_B = D_{AB}$ was not used. The equations were solved by asymptotic analysis and numerically by the Galerkin finite element method. The model was solved for facilitation factors. The system seemed to be for a very generic case. A discussion that supported the reasons for modeling with unequal carrier and carrier complex diffusivities would add weight to their work. Mentioning scenarios for which the assumption $D_B = D_{AB}$ was inappropriate would have been beneficial.

Guha et al. (1990) investigated the steady state facilitated transport of CO_2 through an immobilized liquid membrane containing diethanolamine (DEA) solution. The effects of membrane thickness and downstream CO_2 partial pressure were investigated. The system was represented by a set of coupled diffusion reaction equations that were solved semi-analytically using matched asymptotic expansions and the regular perturbation method.

The system was also solved numerically, using a finite difference technique with non-uniform mesh size, and using orthogonal collocation on finite elements. There was little explanation of this procedure. The reaction mechanism for CO_2 DEA was presented, including the intermediate steps and overall reaction term. A good description of the development of the governing equations is given. Their work would have been more beneficial if the model was validated for other amines such as MEA, triethanolamine (TEA), and methyldiethanolamine (MDEA).

Davis (1991) modeled the unsteady state facilitated transport of oxygen across membranes containing haemoglobin and across membranes containing red blood cells. He used implicit finite difference methods to solve a set of nonlinear governing equations. The effect of various parameters including concentration, facilitation factor, upstream partial pressure, distance across the membrane, and time were considered. He suggested that his work could be improved by studying the effects of convection and by considering multi-dimensional diffusion.

Fan et al. (1992) studied the diffusion of gas through a membrane. In this work, a technique is proposed for estimating diffusion coefficients and solubilities. The governing equation used was Fick's second law of diffusion. Since their work operated solely on physical mass transfer, without chemical reactions, their system was simple mathematically. The methods of Laplace transformation and separation of variables were used to solve their equations. Results were compared to a three-point finite difference technique. Analytical solutions coincide with the numerical solutions. Although the paper is one of the more recent works being reviewed, it deals with a simplified system that did not include chemical reactions. Working models should offer the option of including a chemical reaction in the system.

Dindi et al. (1992) modeled a two-step reaction mechanism for facilitated transport. The so-called "parasitic binding" mechanism was solved for a 1D steady state flat plate geometry problem. The steady state equations were solved for the facilitation factor, investigating the effect of total carrier concentration, reaction equilibrium values, and membrane thickness. In this work, the previously used assumption of equal carrier and carrier complex diffusivities ($D_{AB} = D_B$) has been extended to $D_B = D_{AB} = D_{AB2}$. Even the first assumption is not always valid, care must be taken when extending such an assumption. Equations were solved using analytical techniques.

Jemma and Noble (1992) improved an earlier analytical solution for predicting facilitation factors in facilitated transport systems. They still used the standard reaction mechanism. The improvement is accomplished by assuming a small non-zero solute concentration of the exit of the membrane. This is probably more realistic since not all separation systems would be entirely efficient. Researchers need to begin investigating more complex reaction mechanisms for facilitated transport.

Teramoto et al. (1996) studied the facilitated transport of CO_2 through supported liquid membranes, based on the amines MEA and DEA. They conducted some interesting experimental work and found that the DEA system performed slightly better than the MEA system. The model developed was appropriate for a steady-state system, but was not explained clearly. He presented a set of algebraic equations that could be solved simultaneously to solve the facilitation factor. The results of the mathematical model matched reasonably well with the experimental data.

The first recorded experiment on the transport of gases and vapors in polymeric membranes was conducted in 1929. A few years later, separation of gases was achieved using natural rubber via Knudsen diffusion concept. Despite many experimental works, the progress of membrane separation techniques was slow in the early stages. The progress in membrane science and technology was accelerated during the 1980s by the development and refinement of polymeric membranes. Membrane gas separation emerged as a commercial process on a large scale during the 1980s (Kesting 1985). The commercial use of membranes for acid gas removal started in 1984 with the installation of the SACROC unit by Cynara (Parro 1999).

Alexander and Winnick (1995) developed an advanced process for the separation of H_2S from natural gas through an electrochemical membrane. In this process the electric field is employed as an alternative to the chemical potential driving force, developed due to pressure or concentration difference. H_2S is removed from natural gas by reduction to the sulfide ion and H_2 at the cathode. Sulfide ions from the anode are oxidized to elemental sulfur and are swept away by an inert gas stream. No absorbents are used and there is no need for subsequent treatment of a concentrated H_2S stream, as with gas sweetening technology.

While membranes are good for bulk acid gas removal, they must be combined with other processes when the acid gases are present in low concentrations. That is because at low levels of the partial pressure of acid gases, the driving force of the process is reduced.

8.3.3.8 Future scope of gas separation studies using membranes

The feasibility of using a novel carbon–multi-wall membrane for separating carbon dioxide from flue gas effluent from a power gas generation plant is being studied (Andrew et al. 2001). This is an innovative membrane system that provides an opportunity of high resistance to chemical and temperature constraints. This membrane consists of nano-sized tubes with pore sizes that can be controlled. This will enhance the kinetic and diffusion rates, which in turn will yield high fluxes. As the first step toward design of a working carbon–carbon nano-tube based membrane, specific goals have been included that require evaluation of separation mechanism, either diffusive or kinetic and a test to demonstrate proof of separation.

A fixed bed containing a specific amount of multi-walled nano-tubes may be tested for its ability to separate carbon dioxide from nitrogen at various proportions. The retention time of the carbon dioxide will be the barometer of efficiency of removal. Computational modeling can then be conducted to assess the interaction of various gas molecules and the graphite nano-tubes. Classical molecular dynamics simulations can then be used to determine the diffusive flow of pure molecules and molecular mixtures through the adsorbents acting as membranes. The specific molecules and binary systems that may be examined include methane, ethane, nitrogen, oxygen, nitrogen/carbon dioxide, and oxygen/carbon dioxide systems. The numerical simulation results provide an indication of how the structure and size of the molecules in the mixture influence the results. In addition, molecules with non-spherical aspect ratios exhibit different diffusive and separation behavior than spherical molecules.

Individual nano-tubes in bundles may be used to investigate the degree of separation for specific gases. Epoxy tube sheets on each end of the fiber bundle and O-rings are used to separate feed gas from permeate gas. Pressurized gas enters the membrane case, where it contacts the fiber bundle and flows radially inward. As the gas traverses the bundle, certain gases selectively permeate the fiber walls into the low-pressure bores of the fiber. The residual gas continues across the bundle and is collected in a central perforated tube. There are many inherent advantages of such a design. The hollow fiber technology maximizes surface area per unit volume relative to any competing membrane technology. Double-ended permeate flow minimizes pressure drop down the bore of the fiber and maximizes separation performance.

The mass transport in nano-materials have not been clearly understood (Wagner et al. 2002). Although previous transient models of separation and adsorption suggests that high selectivity is the origin of selective transport, recent analyses indicate that specific penetrant-matrix interactions actually dominate the effects at transition stage. The primary difficulties in modeling this transport are that the penetrants are in continuous contact with the membrane matrix material and the matrix has a common topology with multiple length scales. The objective is to understand and predict the transport of the smaller molecules. This will require combining quantum mechanics and molecular dynamics simulations to accurately predict thermodynamic solubilities and short-time tracer diffusivities in nano-structured membranes. These predictions can be validated by comparing with ongoing experiments on nano-porous carbons and polymer membranes.

The research can be focused on idealized models for nano-porous carbons and shifted toward real nano-porous carbons and polymers. In such cases, controlling the selective transport of small molecules is critical to membrane performance. In order to determine the effectiveness of a membrane for separation of gases, it is important to have an accurate estimate of the pore size distribution (Marzouki 1999). This is a key parameter in determining the separation factor of gases through nano-porous membranes. Active industrial collaboration may be required largely because of the potential uses of such membranes in gas separations, reverse osmosis, packaging, and fuel cells.

Unlike a multi-wall carbon membrane, SSP Silicon Products, Inc. has developed dimethyl silicone that possesses the unique ability to allow various gases to permeate rapidly through it. This phenomenon is due primarily to the flexible silicone–oxygen–silicone linking sites of the silicone chain and an absence of crystallinity in silicone rubber. Technically speaking, the process of permeation through a nonporous membrane is a three-stage activity. Whereas a porous material uses size exclusion as its method of separation, the process by which a nonporous membrane allows separation to occur is a much more complex means to an end. These steps are sorption in, diffusion through, and desorption from the membrane by the permeating gas. The rate of permeation is the product of diffusivity and solubility coefficients of the permeating gas. The solubility coefficients for gases into dimethyl silicone are comparable to those of most polymers but the diffusion rates through the silicone are nearly an order of magnitude greater than any other membrane polymers. Therefore, dimethyl silicone owes its rapid transport of gases to the high rate of diffusion and not solubility. Another key to a successful separation is the selectivity of the membrane.

Selectivity is the ratio at which one gas permeates *vs* the rate at which another gas permeates through the membrane. This factor is what ultimately determines the efficiency of a particular separation. Obviously, a membrane that permeates oxygen and nitrogen at the same rate would be worthless for the generation of oxygen rich or oxygen depleted air. Therefore, a high selectivity ratio of the component gases in the feed stream would result in effective separation. A comprehensive study should be conducted to study the effect of permeability of various gases and its effect in relation to other gaseous components. Although the driving force behind gas transfer through a membrane is pressure, it does not actually affect permeability. Furthermore, membrane thickness and temperature variables have little effect on permeability. Flux, on the other hand, is affected by pressure differentials and membrane thickness. A high-pressure differential and a very thin membrane will result in significant increases in flow rate of the permeate gas component.

8.3.3.9 The cryogenic expansion process

This process consists of dropping the temperature of the gas stream to lower levels. This can be done by the turbo-expander process. Essentially, cryogenic processing consists of lowering the temperature of the gas stream to around −120°F (EIA 2006). In this process, external refrigerants are used to cool the natural gas stream. The expansion turbine is used to rapidly expand the chilled gases, causing the temperature to drop significantly. This rapid temperature drop condenses ethane and other hydrocarbons in the gas stream, maintaining methane in gaseous form. This process recovers about 90 to 95% of the ethane (EIA 2006). The expansion turbine can be used to produce energy as the natural gas stream is expanded into recompressing the gaseous methane effluent. This helps to save energy costs in natural gas processing.

8.3.3.10 Sulfur and carbon dioxide removal

CO_2 and H_2S present in the natural gas are inert gases and have no heating value, thus reduce the heating value of natural gas (Nallinson 2004). Acid gases such as CO_2 and H_2S are separated from natural gas streams by contacting them with a chemical solvent such as alkanolamine (Chakma 1997). Acid gases are chemically absorbed by the solvent in an absorber and natural gas with much reduced acid gas content can be obtained. The chemical solvent containing the absorbed acid gases is regenerated to be used again in the absorption process. The hydrogen sulfide is converted to elemental sulfur and the CO_2 is released to the atmosphere. Since CO_2 is a greenhouse gas, releasing it into the atmosphere will have severe environmental impacts. With increasing awareness of its environmental impact and the ratification of the Kyoto protocol by most of the member countries, it is expected that the release of CO_2 into the atmosphere is to be limited.

Sulfur exists in natural gas as hydrogen sulfide (H_2S), which is corrosive. H_2S is called a sour gas in the natural gas industry. To remove H_2S and CO_2 from natural gas, amine solutions are generally used (EIA 2006). The natural gas is run through a tower, which contains the amine solution. This solution has an affinity for sulfur. There are two principal amine solutions used, monoethanolamine (MEA) and diethanolamine (DEA). Both DEA and MEA, in liquid form, will absorb sulfur compounds from natural gas as it passes through. The effluent gas is free of sulfur. The amine solution used can be regenerated (by removing the absorbed sulfur), allowing it to be reused to treat more natural gas. It is also possible to use solid desiccants like iron sponges to remove the sulfide and carbon dioxide. Amines solutions and different types of membrane technologies are used for CO_2 removal from natural gas streams (Wills et al. 2004). However, glycol and amines are highly toxic chemicals and have several health and environment impacts (Melnick 1992). Glycols become very corrosive in the presence of oxygen. CO_2 removal is also practised using molecular gate systems in which membranes of different sieve size, depending on the size of the molecule, are separated (Wills 2004).

The H_2S present in the natural gas stream is highly toxic, even if its concentration is below 5%. Moreover, the deposition of the elemental sulfur in tubular and flow lines can lead to decreased production, increased corrosion rates, and higher filtration costs. So it is necessary to remove the sulfur gas in order to keep the gas at acceptable specifications. The conventional employed treatment techniques are chemical based and expensive, along with serious environmental concerns. An environmentally-friendly, naturally occurring substance, such as limestone was investigated for its sulfur adsorption capability in downhole applications. It was found satisfactory for sulfur adsorption capability. The gas containing SO_2 was passed through the limestone packed column and a change in concentration at the outlet was noticed by bubbling the exit stream through a water beaker. The pH of water was monitored throughout the experiment.

The published literature on "separation of sour gas from natural gas stream" is reviewed in this section. Many people have performed studies on this issue and there are numerous reports on successful application of their techniques, published in different journals. Different processes developed by well-known organizations for natural gas treatments, which are proved successful on an industrial scale, have also been discussed in this section.

Direct oxidation to sulfur: Lower concentrations of H_2S in natural gas can be selectively oxidized over an activated carbon catalyst to elemental sulfur, water, and a small fraction of sulfur. Direct conversion or H_2S processes are basically used for H_2S removal in sulfur recovery and tail gas treating applications. Chaudhry and Tolefsson (1990) have reported their research to optimize the reaction conditions to develop a process for a commercial unit capable of processing sour natural gas containing 1.0% H_2S.

The LO-CAT II hydrogen sulfide oxidation process developed by ARI USA offers an environmentally sound, one step process for removing H_2S selectively from a gas stream and producing elemental sulfur. In the LO-CAT II auto-circulation unit, the acid gas is sparged through the LO-CAT solution in the absorption portion of the vessel. The H_2S is absorbed into the aqueous solution of chelated iron, and the resultant sulfide ions are oxidized to elemental sulfur (*Hydrocarbon Engineering Journal*, February, 1999). The other well-known process is Stretford (British Gas), which is based on catalytic liquid phase oxidation of H_2S using sodium metavenedate and anthraquinone disulfonic acid to obtain elemental sulfur.

Diaryl disulfide (DADS): Diaryl disulfide was first used successfully as a solvent to remove sulfur from tubular and flow lines at Exxon Co.'s US La Barge field in Wyoming, where there was the aforementioned problem of tubular plugging. DADS in its application is found preferable for sulfur deposition problems because it is cheaper and less volatile with no disposal needed when compared with the commonly used solvent dimethyl disulfide (DMDS). The sulfur uptake of DADS is observed to be far better than that of amine chemicals in the presence of a large amount of CO_2. The DADS treatment process applied at La Barge was batch wise in application (Voorhees and Kenlly 1991).

Scavengers and biocides: The use of scavengers and biocides is effective for H_2S control. These chemicals have been used both in topside equipment and downhole applications. Though typically effective, these chemicals are costly in application and may exhibit unintended toxicity in produced fluids (Sturman and Goeres 1999).

Treatment with NO_2/NO_3: The downhole injection of nitrite containing solutions into sour oil and gas wells under controlled conditions has been observed to effectively remove hydrogen sulfide (H_2S) from aqueous and gas phases. Souring control using nitrite has been successfully applied to a gas well in the San Juan Basin of New Mexico (Sturman and Coeres 1999). In a 36-hr down-hole squeeze in a gas well, injected

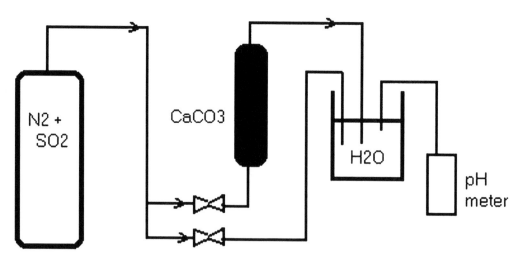

Figure 8-21 Experimental design of gas processing in laboratory conditions (Akhtar 2003).

NO_2— removed H_2S from produced gas for a period of 7 months. This treatment has proved an effective means of controlling H_2S in oil and gas wells and of removing iron sulfide scale for the near wellbore.

It can be concluded from the literature that there are a variety of processes for surface treatment of natural gas for sulfur removal, based on chemical treatment and mechanical separation. But studies performed on downhole sulfur removal from natural gas streams are limited and are mostly focused on chemical treatment. The information about the membrane application for downhole gas treatment is lacking in the literature. Moreover, the chemical treatment processes employed downhole are not continuous in their operation.

Natural separation agents: Al-Awadhy et al. (2005) conducted a series of experiments in order to separate sulfur from a crude oil stream using naturally occurring limestone. They reported that limestone has very high separation efficiency when it comes to removing sulfur. Akhtar (2003) investigated the possibility of removing sulfur dioxide using limestone impregnated with various concentrations of amines (MEA). His experimental setup is shown in Figure 8-21. By monitoring the pH of the water that captured effluent gases, he was able to demonstrate high removal efficiency using the amine-impregnated limestone. Some of these results are shown in Figure 8-22.

8.3.4 *Problems in natural gas processing*

Conventional natural gas processing consists of the application of various types of chemicals and polymeric membranes. The common chemicals used to remove water, CO_2, and H_2S are: diethylene glycol (DEG), triethylene glycol (TEG), monoethanolamines (MEA), diethanolamines (DEA), and triethanolamine (TEA). These are synthetic chemicals and have various health and environmental impacts. The detailed pathways of these chemicals are described below.

8.3.4.1 Pathways of glycols and their toxicity

Matsuoka et al. (2005) reported a study on electro-oxidation of methanol and glycol and found that electro-oxidation of ethylene glycol at 400 mV formed glycolate, oxalate, and formate (Figure 8-23). The study further reported that glycolate was obtained by three-electron oxidation of ethylene glycol, and was an electrochemically active product even at 400 mV, which led to further oxidation of glycolate. Oxalate was found stable and no further oxidation was seen and was termed as the non-poisoning path. The other product of glycol oxidation is called formate, which is termed as the poisoning path or CO poisoning path. A drastic

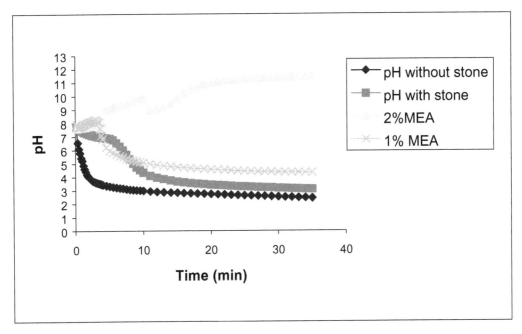

Figure 8-22 Concentration profile of SO₂ through and without limestone adsorption column (Akhtar 2003).

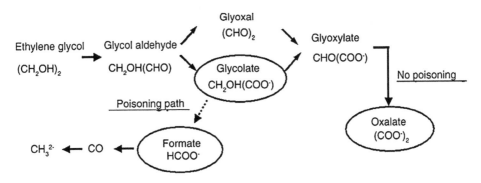

Figure 8-23 Ethylene glycol oxidation pathway in alkaline solution. After Matsuoka et al. (2005).

difference in ethylene glycol oxidation was noted between 400 and 500 mV. The glycolate formation decreased from 40–18% and formate increased from 15–20%. In the case of methanol oxidation, the formate was oxidized to CO_2 but ethylene glycol oxidation produces CO instead of CO_2 and follows the poisoning path over 500 mV.

The glycol oxidation produces glycol aldehyde as intermediate products. It is seen that as the heat is increased, the CO poisoning will increase. Wang et al. (2005) reported that oxidation of ethylene glycol, both on a bare catalyst surface as well as under steady-state conditions, completed oxidation to CO_2 is a minority reaction pathway (<6% at relevant currents). Formation of incompletely oxidized C_2 molecules prevail (contributions from partial oxidation C_1 products (formaldehyde, formic acid) are possible), indicating that C—C bond breaking is slow and rate limiting for CO_2 formation production occurs with CO poisoning.

Glycol ethers are known to produce toxic metabolites, such as the teratogenic methoxyacetic acid during biodegradation, so the biological treatment of glycol ethers can be hazardous (Fischer and Hahn 2005).

Abiotic degradation experiments with ethylene glycol showed that the byproducts are monoethylether (EGME) and toxic aldehydes, for example, methoxyacetaldehyde (MALD).

Glycol enters the body by inhalation or through the skin. Toxicity of ethylene glycol causes depression and kidney damage (MSDS 2005). Ethylene glycol in the form of dinitrate can be inhaled and passed through the skin. It can irritate the skin causing a rash or burning feeling on contact, can cause headache, dizziness, nausea, vomiting, abdominal pain, and a fall in blood pressure. High concentration levels can interfere with the ability of the blood to carry oxygen, causing headache, dizziness, and a blue color to the skin and lips (methemoglobinemia). It can cause breathing difficulties, collapse, and even death. This can damage the heart causing chest pain and/or increased or irregular heart rate (arrhythmia), which can be fatal. High exposure may affect the nervous system and may damage the red blood cells leading to anemia. The recommended airborne exposure limit is 0.31 mg/m^3 averaged over an 8-hr workshift. During a study of carcinogenetic and toxicity of propylene glycol on animals, the skin tumor incidence was observed (CERHR 2003).

8.3.4.2   Pathways of amines and their toxicity

Amines are considered to be toxic chemicals. It was reported that occupational asthma was found in a patient handling a cutting fluid containing diethanolamine (Piipari et al. 1998). DEA caused asthmatic airway obstruction at concentrations of 0.75 mg/m^3 and 1.0 mg/m^3. Toninello (2006) reported that the oxidation products of some biogenic amines also appear to be carcinogenic. DEA also reversibly inhibits phosphatidylcholine synthesis by blocking choline uptake (Lehman-McKeeman and Gamsky 1999). Systemic toxicity occurs in many tissue types including the nervous system, liver, kidney, and blood. Hartung et al. (1970) reported that inhalation by male rats of 6 ppm (25.8 mg/m^3) DEA vapor 8 hrs/day, 5 days/week for 13 weeks resulted in depressed growth rates, increased lung and kidney weights, and even some mortality. Rats exposed continuously for 216 hrs (9 days) to 25 ppm (108 mg/m^3) DEA showed increased liver and kidney weight and elevated blood urea nitrogen. Barbee and Hartung (1979) reported changes in liver mitochondrial activity in rats following exposure to DEA in drinking water. Mitochondrial changes were also observed.

Melnick (1992) reported that symptoms associated with diethanolamine intoxication include increased blood pressure, diuresis, salivation, and pupillary dilation (Beard and Noe 1981). Diethanolamine caused mild skin irritation in a rabbit at concentrations above 5%, and severe ocular irritation at concentrations above 50% (Beyer et al. 1983). Diethanolamine is a respiratory irritant and thus might exacerbate asthma, having a more severe impact on children than on adults (Chronic Toxicity Summary 2001). Diethanolamine is corrosive to eyes, mucous membranes, and skin. Liquid splashed in the eye causes intense pain and corneal damage, and permanent visual impairment may occur Prolonged or repeated exposure to vapors at concentrations slightly below the irritant level often results in corneal edema, foggy vision, and the appearance of halos around lights. Skin contact with liquid diethylamine causes blistering and necrosis Exposure to high vapor concentrations may cause severe coughing, chest pain, and pulmonary edema. Ingestion of diethylamine causes severe gastrointestinal pain, vomiting and diarrhea, and may result in perforation of the stomach.

8.3.4.3   Toxicity of polymer membranes

Synthetic polymers are made from the heavier fraction of petroleum derivatives. Petroleum products are refined by using highly toxic catalysts, chemicals, and excessive heat and pressures (Chhetri et al. 2006; Khan and Islam 2007; Khan 2006b). Hull et al. (2002) reported combustion toxicity of ethylene-vinyl acetate copolymer (EVA), a higher yield of CO, and several volatile compounds along with CO_2. Islam (2004) reported that oxidation of polymers produces more than 4000 toxic chemicals, 80 of which are known carcinogens. Hence, the natural gas that is separated through the polymer membranes might be contaminated with polymer and when the gas is burnt, various toxins are released that are harmful to humans and the environment.

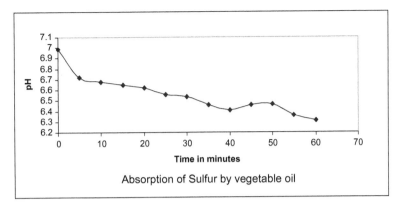

Figure 8-24   Decrease of pH with time, due to sulfur absorption in de-ionized water.

8.3.5   *Natural substitutes for natural gas processing chemicals*

8.3.5.1   Glycol

Glycol is one of the most important chemical used during the dehydration of natural gas. The impacts to human and environment due to the oxidation of glycol that generally produce carbon monoxide has been discussed. In search of the cheapest, most abundantly available material, clay has been considered as one of the better substitutes of toxic glycol. Clay is a porous material containing various minerals such as silica, alumina, and several others. Low et al. (2003) reported that the water absorption characteristics of sintered sawdust clay can be modified by the addition of saw dust particles to the clay. The dry clay as a plaster has a water absorption coefficient of 0.067–0.075 $kg/m^2S^{1/2}$ where weight of water absorbed is in kg, surface area in m^2, and time in seconds (EBW 2005). The preliminary experimental result was carried out in the Civil and Resources Engineering, Dalhousie University, indicating that clay can absorb considerable amounts of water vapor and can be efficiently used in dehydration of natural gas instead of synthetic glycol. Moreover, glycol can be obtained from some natural sources, and it is not toxic like synthetic glycol. Glycol can be extracted from the *Tricholoma matsutake* (mushroom), an edible fungus (Ahn and Lee 1986). Ethylene glycol is also found as a metabolite of ethylene, which regulates the natural growth of the plant (Blomstrom and Beyer 1980). Orange peel oils can also replace this synthetic glycol. Extracting these natural glycols would substitute for the synthetic toxic glycol.

8.3.5.2   Amines

Amines are used in natural gas processing to remove H_2S and CO_2. Monoethanolamine, diethanolamine, and triethanolamine are alkanolamine compounds. These are synthetic chemicals, the toxicity of which has been discussed above. Monoethanolamine is found in hemp oil that is extracted from the seeds of hemp (*Cannabis sativa*) plant. One hundred grams of hemp oil contain 0.55 mg of monoethanolamine. Moreover, an experimental study showed that olive oil and waste vegetable oil can absorb sulfur dioxide. Figure 8-24 indicates the decrease in pH of de-ionized water with time. This could be a good model to remove sulfur compounds from natural gas streams.

8.3.5.3   Membranes and absorbents

Bio-membranes have been developed recently. These membranes can be used instead of synthetic membranes (Basu et al. 2004). Similarly, natural absorbents such as silica gels can also be used for absorbing various contaminants from the natural gas stream.

Various natural gas processing techniques have been reviewed. Different chemicals and synthetic polymers such as glycols, amines, synthetic membranes, and some other adsorbents used for removing the impurities from the natural gas, are discussed. The pathways and toxicity of these chemicals are presented in this section. Some prospective natural material substitutes for these toxic chemicals are also presented. Glycol from natural sources such as citrus peel can be extracted to substitute for the synthetic glycol. Clay can be good as a substitute of synthetic glycol to dehydrate raw natural gas. Similarly, amines from natural sources, such as hemp seed oil are a good alternative for synthetic amines. Moreover, vegetable oils such as olive oil or waste cooking oil can absorb sulfur from the natural gas streams. Use of bio-membranes instead of synthetic membranes can also solve the environmental problem.

8.3.5.4 Natural gas storage

Natural gas is much harder to store than oil. Natural gas can be stored in tanks, or in underground reservoirs. Large storage facilities exist in Alberta and Ontario, at unregulated rates. Gas is usually consumed in greater abundance during winter than summer, and therefore gas in storage is depleted during the winter, but replenished in the summer. Also, this means that the pipeline can flow fully all year long. Typically, natural gas is stored in spherical tanks as the stresses on the steel from the pressure of the gas are more evenly distributed.

8.4   Oil Refining

Crude oil is a mixture of hydrocarbons. These hydrocarbon mixtures are separated into commercial products by numerous refining processes. Oil refineries are enormous complex processes. Figure 8-25 shows a complex oil refinery complex in Dartmouth, Nova Scotia. Refining involves a series of processes to separate and sometimes alter the hydrocarbons in crude oil. It relies on the basic difference between chemicals – in this case their boiling points. Figure 8-26 shows the major steps in the refining process. The first step is transportation and storage (Section 8.12).

In crude oil refining process, fractional distillation is the main process to separate oil and gas. For this process a distillation tower is used, which operates at atmospheric pressure, leaving a residue of hydrocarbons with boiling points above 400°C and more than 70 carbon atoms in their chains. Small molecules of hydrocarbons have low boiling points, while larger molecules have higher boiling points. The fractionating column is

Figure 8-25   An industrial refinery.

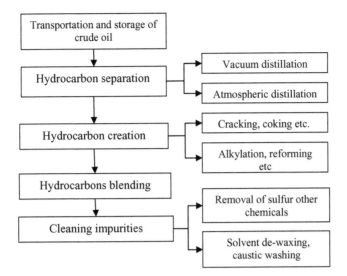

Figure 8-26   Major steps of process in oil refining.

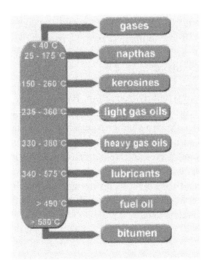

Figure 8-27   Pictorial view of fractional column.

cooler at the top than the bottom, so the vapors cool as they rise. Figure 8-27 shows the pictorial view of fractional column. It also shows the ranges of hydrocarbons in each fraction. Each fraction is a mix of hydrocarbons and each fraction has its own range of boiling points and is drawn off at a different level in the tower.

Petroleum refining has evolved continuously in response to changing consumer demand for better and different products from aviation gasoline and then for jet fuel, a sophisticated form of the original product, kerosene. A summary of a detailed process flow chart for the oil refining steps is presented in Table 8-10. The table also describes the different treatment methods for each of the refining phases.

8.4.1   Uses of catalysts or direct agents

A number of different solid catalysts are used in the oil refining process. They are in many different forms, from pellets to granular beads to dusts, made of various materials and having various compositions. Extruded

Table 8-10 Details of oil refining process and various types of catalyst used

Process	Description	Catalyst/Heat/ pressure used
Distillation process	It basically relies on the difference of boiling point of various fluids. Density also has an important role to pay in distillation. The lightest hydrocarbon at the top and heaviest residue at the bottom are separated.	Heat
Coking and thermal process	The coking unit converts heavy feedstocks into solid coke and lower boiling hydrocarbons products, which are suitable for converting to higher value transportation fuel. This is a severe thermal cracking process to form coke. Coke contains high boiling point hydrocarbons and some volatiles, which are removed by calcining at temperatures 1095–1260°C. Coke is allowed sufficient time to remain in high temperature heaters, insulated surge drums, hence called delayed cooking.	Heat
Thermal cracking	The crude oil is subjected to both pressure and large molecules are broken into small ones to produce additional gasoline. The naphtha fraction is useful for making many petrochemicals. Heating naphtha in the absence of air makes the molecules split into shorter ones.	Excessive heat and pressure
Catalytic cracking	Converts heavy oils into high gasoline, less heavy oils, and lighter gases. Paraffins are converted to C_3 and C_4 hydrocarbons. Benzene rings of aromatic hydrocarbons are broken. Rather than distilling more crude oil, an alternative is to crack crude oil fractions with longer hydrocarbons. Larger hydrocarbons split into shorter ones at low temperatures if a catalyst is used. This process is called catalytic cracking. The products include useful short chain hydrocarbons.	Nickels, zeolites, acid treated natural alumina silicates, amorphous and crystalline synthetic silica, alumina catalyst
Hydro-processing	Hydroprocessing (325°C and 50 atm) includes both hydro cracking (350°C and 200 atm) and hydrotreating. Hydrotreating involves the addition of hydrogen atoms to molecules without actually breaking the molecule into smaller pieces and improves the quality of various products (e.g., by removing sulfur, nitrogen, oxygen, metals, and waxes and by converting olefins into saturated compounds). Hydrocracking breaks longer molecules into smaller ones. This is a more severe operation using higher heat and longer contact time. Hydrocracking reactors contain fixed, multiple catalyst beds.	Platinum, tungsten, palladium, nickel, crystalline mixture of silica alumina. Cobalt and molybdenum oxide on alumina nickel oxide, nickel thiomolybdate tungsten and nickel sulfide and vanadium oxides, nickel thiomolybdate are in most common use for sulfur removal and nickel molybdenum catalyst for nitrogen removal.
Alkylation	Alkylation or "polymerization" – forming longer molecules from smaller ones. Another process is isomerization where straight chain molecules are made into higher octane branched molecules. The reaction requires an acid catalyst at low temperatures and low pressures. The acid composition is usually kept at about 50%, making the mixture very corrosive.	Sulfuric acid or hydrofluoric acid, HF (1–40°C, 1–10 atm). Platinum on $AlCl_3/Al_2O_3$ catalyst used as new alkylation catalyst
Catalytic reforming	This uses heat, moderate pressure, and fixed bed catalysts to turn naphtha, short carbon chain molecule fraction, into high-octane gasoline components – mainly aromatics.	Catalyst used is a platinum (Pt) metal on an alumina (Al_2O_3)
Treating non-hydrocarbons	Treating can involve chemical reaction and/or physical separation. Typical examples of treating are chemical sweetening, acid treating, clay contacting, caustic washing, hydro-treating, drying, solvent extraction, and solvent de-waxing. Sweetening compounds and acids de-sulfurize crude oil before processing and treat products during and after processing.	

pellet catalysts are used in moving and fixed bed units, while fluid bed processes use fine, spherical, particulate catalysts. Catalysts used in processes that remove sulfur are impregnated with cobalt, nickel, or molybdenum. Cracking units use acid-function catalysts, such as natural clay, silica alumina, and synthetic zeolites. Acid-function catalysts impregnated with platinum or other noble metals are used in isomerization and reforming. Used catalysts require special handling and protection from exposure, as they may contain metals, aromatic oils, carcinogenic polycyclic aromatic compounds, or other hazardous materials, and may also be pyrophoric. Table 8-10 presents the different catalysts used in the different process. Table 8-11 shows the various processes and products during the refining process.

8.4.2 Waste generation

Boilers, process heaters, and other process equipment are responsible for the emission of particulates, carbon monoxide, nitrogen oxides (NO_x), sulfur oxides (SO_x), and carbon dioxide. Catalyst changeovers and cokers release particulates. Volatile organic compounds (VOCs) such as benzene, toluene, and xylene are released from storage, product loading and handling facilities, and oil–water separation systems, and as fugitive emissions from flanges, valves, seals, and drains (World Bank 1998). It is reported by World Bank (1998) that for each ton of crude processed, emissions from refineries may be approximately as:

- *Particulate matter*: 0.8 kg, ranging from less than 0.1 to 3 kg.
- *Sulfur oxides*: 1.3 kg, ranging 0.2–06 kg; 0.1 kg with the Claus sulfur recovery process.
- *Nitrogen oxides*: 0.3 kg, ranging 0.06–0.5 kg.
- *Benzene, toluene, and xylene (BTX)*: 2.5 g, ranging 0.75 to 6 g; 1 g with the Claus sulfur recovery process. Of this, about 0.14 g benzene, 0.55 g toluene, and 1.8 g xylene may be released per ton of crude processed.
- VOC emissions depend on the production techniques, emissions control techniques, equipment maintenance, and climate conditions and may be 1 kg/t of crude processed (ranging from 0.5 to 6 kg/t of crude).

Table 8-11   Various processes and products in oil refining

Process name	Action	Method	Purpose	Feedstock(s)	Product(s)
FRACTIONATION PROCESSES					
atmospheric distillation	separation	thermal	separate fractions	desalted crude oil	gas, gas oil, distillate, residual
vacuum distillation	separation	thermal	separate without cracking	atmospheric tower residual	gas, gas oil, lube, residual
CONVERSION PROCESSED – DECOMPOSITION					
catalytic cracking	alteration	catalytic	upgrade gasoline	gas oil coke, distillate	gasoline, petrochemical feedstock
coking	polymerize	thermal	convert vacuum residuals	gas oil coke, distillate	gasoline, petrochemical feedstock
hydro-cracking	hydrogenate	catalytic	convert to lighter hydrocarbons	gas oil, cracked oil residual	lighter higher-quality products
hydrogen steam reforming	decompose	catalytic/ thermal	produce hydrogen	de-sulfurized gas, O_2, steam	hydrogen, CO, CO_2
steam cracking	decompose	thermal	crack large molecules	atm tower, heavy fuel/distillate	cracked naphtha, coke, residual
viscosity breaking	decompose	thermal	reduce viscosity	atm tower residual	distillate tar

Table 8-11   Continued

Process name	Action	Method	Purpose	Feedstock(s)	Product(s)
CONVERSION PROCEESES – UNIFICATION					
alkylation	combining	catalytic	unit olefins and iso-paraffins	tower iso-butane/ cracker olefin	iso-octane (alkylate)
grease compounding	combining	thermal	combine soap and oils	lube oil, fatty acid, alky metal	lubricating grease
polymerizing	polymerize	catalytic	unite 2 or more olefins	cracker olefins	high-octane naphtha, petrochemical stocks
CONVERSION PROCESSES – ALTERATION OR REARRANGEMENT					
catalytic reforming	alteration/ dehydration	catalytic	upgrade low octane naphtha	coker/hydro- cracker naphtha	high octaner, reformate/ aromatic
isomerization	rearrange	catalytic	straight chain to branch	butane, pentane, hexane	iso-butane/pentane/ hexane
TREATMENT PROCESSES					
amine treating	treatment	absorption	remove acidic contaminants	sour gas, hydrocarbons w/CO_2 and H_2S	acid free gases and liquid hydrocarbons
desalting	dehydration	absorption	remove contaminants	crude oil	Desalted crude oil
drying	treatment	absorption/ therm	remove H_2O and sulfur compounds	liquid hydrocarbon, LPG, alky feedstock	sweet and dry hydrocarbons
furfural extraction	solvent extr.	absorption	upgrade mid distillate and lubes	cycle oils and lube feedstocks	high-quality diesel and lube oil
hydro-de- sulfarization	treatment	catalytic	remove sulfur, contaminants	high-sulfur residual/ gas oil	de-sulfurized olefins
Hydro-treating	hydrogenation	catalytic	remove impurities, saturate hydrocarbons	residuals, cracked Hydrocarbons	cracker feed, distillate, lube
phenol extraction	solvent extr.	absorption/ therm	improve viscosity index, color	lube oil base stocks	high-quality lube oils
solvent de-asphalting	treatment	absorption	remove asphalt	vacuum tower residual, propane	heavy lube oil, asphalt
solvent de-waxing	treatment	cool/filter	remove wax from lube stocks	vacuum tower lube oils	de-waxed lube basestock
solvent extraction	solvent extr.	absorption/ precipitation	separate unsaturated oils	gas oil, reformate, distillate	high-octane gasoline
sweetening	treatment	catalytic	remove H_2S, convert mercaptan	untreated distillate/ gasoline	high-quality distillate/gasoline

Source: OSHA (2005)

8.4.3 Pollution prevention and control

Petroleum refineries are complex plants, and the combination and sequence of processes is usually very specific to the characteristics of the raw materials (crude oil) and the products. Specific pollution prevention or source reduction measures can often be determined only by the technical staff. However, there are a number of broad areas where improvements are often possible, and site-specific waste reduction measures in these areas should be designed into the plant and targeted by management of operating plants. World Bank (1998) reported different pollution prevention control measures. Following are efforts that should be considered (World Bank 1998):

Reduction of air emissions:

- Minimize losses from storage tanks and product transfer areas by methods such as vapor recovery systems and double seals.
- Minimize SO_x emissions either through desulfurization of fuels, to the extent feasible, or by directing the use of high-sulfur fuels to units equipped with SO_x emissions controls.
- Recover sulfur from tail gases in high-efficiency sulfur recovery units.
- Recover non-silica-based (i.e., metallic) catalysts and reduce particulate emissions.
- Use low-NO_x burners to reduce nitrogen oxide emissions.
- Avoid and limit fugitive emissions by proper process design and maintenance.
- Keep fuel usage to a minimum.

Elimination or reduction of pollutants:

- Consider reformate and other octane boosters instead of tetraethyl lead and other organic lead compounds for octane boosting.
- Use non-chrome-based inhibitors in cooling water, where inhibitors are needed.
- Use long-life catalysts and regenerate to extend the catalysts' life cycle.

Recycling and reuse:

- Recycle cooling water and, where cost-effective, treated wastewater.
- Maximize recovery of oil from oily wastewaters and sludges. Minimize losses of oil to the effluent system.
- Recover and reuse phenols, caustics, and solvents from their spent solutions.
- Return oily sludges to coking units or crude distillation units.

Operating procedures:

- Segregate oily wastewaters from stormwater systems.
- Reduce oil losses during tank drainage carried out to remove water before product dispatch.
- Optimize frequency of tank and equipment cleaning to avoid accumulating residue at the bottom of the tanks.
- Prevent solids and oily wastes from entering the drainage system.
- Institute dry sweeping instead of washdown to reduce wastewater volumes.
- Establish and maintain an emergency preparedness and response plan and carry out frequent training.
- Practice corrosion monitoring, prevention, and control in underground piping and tank bottoms.
- Establish leak detection and repair programs.

8.4.4   Suggested guidelines

Currently refineries use various toxic and expensive catalyst leaving a number of environmental and health problems. The following suggestions are recommended as alternative for the refineries.

Use of solid acid catalyst for alkylation: Refiners typically use either hydrofluoric acid (HF), which can be deadly if spilled, or sulfuric acid, which is also toxic and increasingly costly to recycle. Refineries can use a solid acid catalyst, unsupported and supported forms of heteropolyacids and their cation exchanged salts, which has recently proved effective in refinery alkylation. A solid acid catalyst for alkylation is less widely dispersed into the environment compared to HF. Changing to a solid acid catalyst for alkylation would also promote more safety at a refinery. Solid acid catalysts are an environmentally-friendly replacement for liquid acids, used in many significant reactions, including alkylation of light hydrocarbon gases to form iso-octane (alkylate) used in reformulated gasoline. Use of organic acids and enzymes for various reactions is to be promoted.

Use of nature based less or non toxic catalyst: The catalysts that are in use today are very toxic and wasted after a series of use. This will create pollution to the environment, so using catalysts with fewer toxic materials significantly reduces pollution. Use of Nature-based less toxic catalysts such as zeolites, alumina, and silica are to be promoted. Various biocatalyst and enzymes, which are nontoxic and from renewable origin, are to be considered for future use.

Use of bacteria to breakdown heavier hydrocarbons to lighter ones: Since the formation of crude oil is the decomposition of biomass by bacteria at high temperature and pressure, there must be some bacteria that can effectively break down the crude oil into lighter products. A series of investigations are necessary to observe the effect of bacteria on the crude oil.

Zero waste approach: Any chemical or industrial process should be designed in such a way that a close-loop system is maintained so that all wastes are absorbed within the assimilative capacity of the Earth. Recycling the resources and using them for alternative use will lead to a zero waste approach.

Use of cleaner crude oil: Crude oil itself is comparatively cleaner than distillates as it contains less sulfur and toxic metals. The use of crude oil for various applications is to be promoted. This will not only help to maintain the environment because of its less toxic nature but also be less costly as it avoids expensive catalytic refining processes.

Use of gravity separation systems: Heavier fractions can be settled out through the density difference method. Various settling tanks in different stages can be designed and allow sufficient time to settle the fractions based on their density. Though it will not solve all the problems, some of the environmental problems can be reduced by this method. This will be less costly compared to other processes but more time consuming.

8.4.5   No-flaring technique

Flaring is a common technique in oil refinery, to burn off low-quality gas (Figure 8-28). With increasing awareness of the environmental impact and the ratification of the Kyoto protocol by most of the member countries, it is expected that gas flaring will no longer be allowed in the near future. This will require significant changes in the current practices of oil and gas production and processing. The low-quality gas that is flared contains many impurities and during the flaring process, toxic particles are released into the atmosphere. Acid rain, caused by sulfur oxides in the atmosphere, is one of the main environmental hazards resulting from this process. Moreover, flaring of natural gas accounts for approximately a quarter of the petroleum industries emissions (UKOO 2002).

Recently, Bjorndalen et al. (2005) developed a novel approach to avoid flaring during petroleum operations. Petroleum products contain materials in various phases. Solids in the form of fines, liquid hydrocarbon, carbon dioxide, and hydrogen sulfide are among the many substances found in the products. According to

Figure 8-28 Oil flaring.

Bjorndalen et al. (2005), by separating these components through the following steps, no-flare oil production can be established:

- Effective separation of solid from liquid
- Effective separation of liquid from liquid
- Effective separation of gas from gas

Many separation techniques have been proposed in the past (Basu et al. 2004; Akhtar 2003). However, few are economically attractive and environmentally appealing. This option requires an innovation approach that is the central theme of this chapter. Once the components for no-flaring have been fulfilled, value added end products can be developed. For example, the solids can be used for minerals, the brine can be purified, and the low-quality gas can be re-injected into the reservoir for EOR. Figure 8-29 outlines the components and value added end products that will be discussed below.

8.4.5.1 Separation of solid–liquid

Even though numerous techniques have been proposed in the past, little improvement has been done in energy efficiency of solid–liquid separation. Most techniques are expensive, especially if a small unit is operated. Recently, a patent has been issued in the United States. This new technique of removing solids from oil is an EVTN system. This system is based on the creation of a strong vortex in the flow to separate sand from oil. The principle is that by allowing the flow to rotate rapidly in a vortex, centrifugal force can be generated. This force makes use of the density differences between the substances. The conventional filtration technique requires a large filter surface area (in the case of high flow rate), replacement of filter material, and back flush. The voraxial technique eliminates this problem. Moreover, it is capable of maintaining high gravity or "g" forces, as well as high flow rates, which will be effective in oil-fines separation (EVTN 2003). This product shows great potential for the separation of liquid and fines.

Use of surfactants from waste can be another possible technique to separate solids from liquid. The application of the combination of waste materials with water to separate fines from oil is attractive due to the relatively low cost and environmentally sound nature. Waste products such as cattle manure, slaughter house waste, okra, orange peels, pine cones, wood ash, paper mill waste (lignosulfate), and waste from the forest

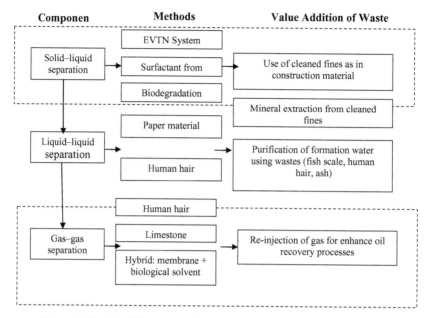

Figure 8-29   Breakdown of no-flaring method (Bjorndalen et al. 2005).

industry (ferrous chloride) are all viable options for the separation of fines and oil. Cattle manure and slaughter house waste is plentiful in Alberta, where flaring is common. Researchers from the UAE University have determined that okra is known to act like a soap (Chaalal 2003). Okra extract can be created through pulverization methods. Orange peel extract should also be examined as it is a good source of acid. A study conducted at Dalhousie University determined that wood ash can separate arsenic from water and therefore may be an excellent oil/fines separator (Rahman et al. 2004). Pine cones, wood ash, and other plant materials may also be viable. Industrial wastes, such as lignosulfate and ferrous chloride, which have been beneficial in the industrial areas of cellulose production and sewage treatment, respectively, can be potential separators.

Finally, a biodegradation method for stripping solid waste from oily contaminants is currently under study as a collaborative effort between Dalhousie and UAE universities. Thermophilic bacteria are found to be particularly suitable for removing low-concentration crude oils from solid surface (Tango and Islam 2002). Also, it has been shown that bioremediation of flare pits, which contain many of the same substances that need to be removed for an appealing no-flare design, has been successful (Amatya et al. 2002).

Once the fines are free from liquids, they can be used for other applications, for example, as a substitution of components in construction materials. Drilling wastes have been found to be beneficial in highway construction (Wasiuddin 2002) and an extension of this work can lead to the usage of fines. Studies have shown that the tailings from oil sands are high in titanium content. With this in mind, the evaluation of valuable minerals in fines will be conducted. To extract the minerals, chemical treatment is used to modify the surface of minerals. Treatment with a solution derived from natural material has great potential. Microwave heating has the potential of assisting this process (Haque 1999; Hua et al. 2002), enhancing selective floatability of different particles. Temperature can be a major factor in the reaction kinetics of a biological solvent with mineral surfaces. Various metals respond in a different manner under microwave conditions, which can make significant changes in floatability. The recovery process will be completed by transferring the microwave-treated fines to a flotation chamber. Finally, atomic absorption spectrometer can characterize the flotation products.

8.4.5.2 Separation of liquid–liquid

Once the oil has been separated from fines via water and waste material, it must be separated from the solution itself as well as from formation water. Oil–water separation is one of the oldest practices in the oil industry, as it is almost impossible to find an oil reservoir absolutely free of connate water. In fact, the common belief is that all reservoirs were previously saturated with water and after oil migration only a part of the water was expelled from the reservoir and replaced by oil. There are two sources of water that cause water to flow to the wellbore. The first source is the connate water that usually exists with oil in the reservoir, saturating the formation below the oil–water contact (OWC) or in the form of emulsion, even above the OWC. The other source is associated with water flooding that is mainly considered in secondary oil recovery.

In almost all oil reservoirs, connate water coexists with oil as a percentage filling pore spaces. In some reservoirs, where water pressure is the main driving mechanism, water saturation may even exceed oil saturation as the production continues. Therefore, when the wellbore is operational, oil production mixed with water is inevitable. As production continues, more water invades the oil zone and watercut in production stream increases consequently.

Before taking an oil well into production, the oil zone is perforated above the OWC to allow oil flow into the wellbore. The fact that part of the formation above the OWC is still saturated with water consequently causes water production. The problem becomes more severe with time as the continuous drainage of the reservoir causes the OWC to move upward, resulting in excessive water production. Because of this typical phenomenon occurring in oil reservoirs, it is inevitable to avoid water production.

Moreover, water flooding is an obvious practice for enhancing oil recovery after oil production decline. Encountering problems of high water production and early breakthroughs are common obstacles of water flooding practice, which in turn cause high production costs. Water production is not only associated with high cost considerations (installation and operation) but also is a major contributing factor to the corrosion of production facilities and reservoir energy loss. Moreover, the contaminated water can be an environmental contamination source, if it is not properly disposed of.

Water production can be tolerated to a certain extent, depending on the economic health of a given reservoir. The traditional practice of the separation of oil from water is applied after simultaneous production of both and then separating them in surface facilities. Single stage or multistage separators are installed where both oil and water can be separated by gravity segregation.

Downhole water oil separation has been investigated since the early days of producing oil from petroleum reservoirs. Hydrocyclone separation has been used (Bowers et al. 2000) to separate oil from water at the surface but its small size made its application downhole even more attractive. This technique is most efficient at 25 to 50% water content in the outlet stream, to give a water stream as clean as possible with a few exceptions under ideal conditions. This makes the consideration of this technique limited to a number of circumstances where high water cut is expected and costs involved are justified.

Stuebinger et al. (2000) compared the hydrocyclone to gas–oil–water segregation and concluded that all downhole separations are still premature and suggested more research and investigation for optimizing and enhancing these technologies.

Recently, the use of membrane and ceramic materials has been proposed (Fernandez et al. 2001; Chen et al. 1991). While some of them show promise, these technologies do not fall under the category of economically appealing. The future of liquid-liquid separation lies within the development of inexpensive techniques and preferably downhole separation technology.

The first stage is the material screening, searching for potential material that can pass oil but not water. The key is to consider the fundamental differences between oil and water in terms of physical properties, molecular size and structure, and composition. These materials must in essence be able to adsorb oil and at the same time prevent water from passing through. Having discovered at least one material with this feature, the next stage should be to study the mechanism of separation. Understanding this mechanism would assist in identifying more suitable materials and better selection criteria. The third stage is material improvement.

Testing different downhole conditions using selected material for separation and the possibility of improving material by mixing, coating, or coupling with others should be investigated. The outcome of this stage would be a membrane sheet material that gives best results and a suitable technique that optimizes the procedure. Investigating the effect of time on the separation process mainly, any fouling problem and its remedy should be carried out in this stage. Eventually, a new completion method should be designed. The material with the ability to separate oil and water can also be used in aboveground separation units. The main advantage of this technique will be to reduce the size of the separation units and to increase their efficiency. This is critical, especially in offshore rigs where minimizing the size of different units is of essence and any save in space would substantially reduce production cost.

Preliminary studies and initial laboratory-scale experiments show that the use of a special type of long fiber paper as the membrane could be a good start (Khan and Islam 2006e). These preliminary experiments have been performed during the material selection stage with encouraging results. A laboratory-scale experimental setup consists of a prototype-pressurized reservoir containing oil-water emulsion and tubing on which the perforated section is wrapped with the membrane. The oil-water emulsion has been produced and it is found that the selected material gives 98–99% recovery of oil without producing any water. These findings are subject to the laboratory conditions in which ΔP is kept about 20 psi. Continuous shaking was employed to maintain the emulsion intact during the whole process of separation. An emulsion made up of varying ratios of oil and water was used in different sets of experiments. Almost all the oil–water ratios gave the same separation efficiency with the material used.

The above-mentioned paper material is made of long fibrous wood pulp treated with water proofing material as a filtering medium. This treatment prevents water from flowing through and at the same time allows oil to pass easily. The waterproofing agent for paper used in these experiments is "rosin soap" (rosin solution treated with caustic soda). This soap is then treated with alum to keep the pH of the solution within the range of 4~5. Cellulose present in the paper reacts reversibly with rosin soap in the presence of alum and forms a chemical coating around the fibrous structure, thus acting as a coating to prevent water to seep through it. This coating allows long chain oil molecules to pass, making the paper a good conductor for the oil stream. Because of the reversible reaction, it was also observed that the performance of the filter medium increases with the increased acidity of the emulsion and vice versa. It was also observed that the filter medium is durable to make its continuous use over a long time, keeping the cost of replacement and production down. Various experiments are being done to further strengthen the findings for longer time periods and for high production pressures. It must be noted that the material used as a filtering medium is environmentally-friendly and can easily be modified to suit downhole conditions. Based on the inherent property of the material used, some other materials can also be selected to give the equivalent good results, considering the effect of surrounding temperature, pressure, and other parameters present in downhole conditions.

Human hair has great potential in removing oil from the solution. Hair is a natural barrier against water but it easily absorbs oil. This feature was highlighted during a US DOE-funded project in 1996 (reported by CNN in February, 1997). A doughnut-shaped fabric container filled with human hair was used to separate oil from a low-concentration oil-in-water emulsion. The Dalhousie petroleum research team has later adapted this technique with remarkable success in both the separation of oil and water as well as heavy metals from aqueous streams.

Purification of the formation water after the oil separation will ensure an all-round clean system. Wastes such as fish scales have been known to absorb lead, strontium, zinc, chromium, cobalt (Mustafiz 2002; Mustafiz et al. 2002), and arsenic (Rahaman 2003). Also, wood ash can adsorb arsenic (Rahman 2002; Rahman et al. 2004). Both of these waste materials have great prospects to be implemented in the no-flare process.

8.4.5.3 Separation of gas–gas

The separation of gas is by far the most important phase of no-flare design. Current technologies indicate that separation may not be needed and the waste gas as a whole can be used as a valuable energy income stream. Capstone Turbine Operation has developed a micro-turbine, which can generate up to 30 kW of power and consume 9000 ft³/day of gas (Capstone 2003). Micro-turbines may be especially useful for off-shore applications where space is limited. Another possible use of the waste gas is to re-inject it into the reservoir for pressure maintenance during EOR processes. The low-quality gas can be re-pressurized via a compressor and injected into the oil reservoir. This system has been tested at two oil fields in Abu Dhabi, with much success (Cosmo Oil 2003). Also, simple incineration instead of flaring to dispose of solution gas has been proposed (Motyka and Mascarenhas 2002). This process only results in a reduction of emissions over flaring. The removal of impurities in solution gas via separation can be achieved both downhole and at the surface, thus eliminating the need for flaring (Bijorndalen et al. 2005).

Many studies have been conducted on the separation of gases using membranes. In general, a membrane can be defined as a semi-permeable barrier that allows the passage of select components. An effective membrane system will have high permeability to promote large fluxes. It will also have a high degree of selectivity to ensure that only a mass transfer of the correct component occurs. For all practical purposes, the pore shape, pore size distribution, external void, and surface area will influence the separation efficiencies (Abdel-Ghani and Davies 1983). Al Marzouqi (1999) determined the pore size distribution of membranes, by comparing two models and validating them with experimental data. Low concentrations of H_2S in natural gas can be handled well by regenerable adsorbents such as activated carbon, activated alumina, silica gel, and synthetic zeolite. Non-regenerable adsorbents, i.e., zinc and iron oxides, have been used for natural gas sweetening. Many membrane systems have been developed including polymer, ceramic, hybrid, liquid, and synthetic.

Polymer membranes have gained popularity in isolating carbon dioxide from other gases (Gramain and Sanchez 2002). These membranes are elastomers formed from crosslinked copolymers of high molecular weights. They are prepared as thin films by extrusion or casting. They demonstrate unique permeability properties for carbon dioxide together with high selectivity toward H_2, O_2, N_2, and CH_4. Ceramic membranes are used to separate hydrogen from gasified coal (Fain et al. 2001). With ceramic materials, high separation factors have been achieved based on the ratios of individual gas permeances. Hybrid membranes combine thermodynamically based partitioning and kinetically based mobility discrimination in an integrated separation unit. Unfortunately, the permeability of common membranes is inversely proportional to selectivity (Kulkarni et al. 1983). Thus the development of liquid membranes has lead to systems that have both a high permeability and high selectivity. Liquid membranes operate by immobilizing a liquid solvent in a micro-porous filter or between polymer layers. A synthetic membrane is a thin barrier between two phases through which differential transport can occur under a variety of driving forces including pressure, concentration, and electrical potential across the membranes. Pressure difference across the membrane can facilitate reverse osmosis, ultra-filtration, micro-filtration, gas separation, and pervaporation. Temperature difference across the membrane can facilitate distillation, whereas concentration difference can be used for dialysis and extraction.

The feasibility of using a novel carbon – multi-wall membrane for separating carbon dioxide from flue gas effluent from a power gas generation plant is being studied (Andrew et al. 2001). This membrane consists of nano-sized tubes with pore sizes that can be controlled. This will enhance the kinetic and diffusion rates, which in turn will yield high fluxes.

The mass transport in nano-materials has not been clearly understood (Wagner et al. 2002). Although previous transient models of separation and adsorption suggest that high selectivity is the origin of selective transport, recent analyses indicate that specific penetrant-matrix interactions dominate the effects at the transition stage. The primary difficulties in modeling this transport are that the penetrants are in continuous contact with the membrane matrix material and the matrix has a common topology with multiple length scales. In order to determine the effectiveness of a membrane for separation of gases, it is important to have an accurate estimate of the pore size distribution (Marzouki 1999). This is a key parameter in determining the separation factor of gases through nano-porous membranes.

Membrane systems are highly applicable for separation of gases from a mixture. Continuing enhancement in technological development of membrane systems is a natural choice for the future. Proven techniques

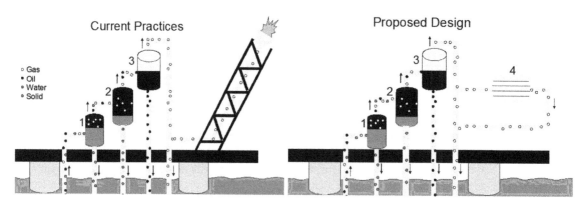

Figure 8-30   Pictorial representation of flaring techniques and proposed no-flaring methods. After Bjorndalen et al. (2005).

include materials such as zeolite (Izumi et al. 2002; Romanos et al. 2001; Robertson 2001; Jeong et al. 2002) when combined with pressure swing adsorption (PSA) techniques. PSA works on the principle that gases tend to be attracted to solids and adsorb under pressure. Zeolite is expensive and therefore fly ash along with caustic soda has been used to develop a less expensive zeolite (Indian Energy Sector 2002).

Waste materials can also be an attractive separation media due to the relatively low cost and environmentally sound nature. Since caustic soda is a chemical, other materials such as okra can be a good alternative. Rahaman et al. (2003) have shown that charcoal and wood ash is comparable to zeolite for the removal of arsenic from wastewater. Many of the waste materials discussed in the solid-liquid separation section also have the potential as effective separation materials. Carbon fibers (Fuertes 2001; Park and Lee 2003; Gu et al. 2002) as well as palladium (Lin and Rei 2001; Chang et al. 2002; Karnik et al. 2002) fibers have been studied extensively for separation. A low-cost alternative to this technology is human hair since it is a natural hollow fiber. Wasiuddin et al. (2002) outlined some of the potential uses of human hair by confirming that it is an effective medium for the removal of arsenic from water.

As well as improving new-age techniques (e.g., polymeric membranes), research will be performed to test new materials, hybrid systems (e.g., solvent/membrane combination), and other techniques that are more suitable for the Atlantic region. For gas processing research, focus will be given to improve existing lique-faction techniques, addressing corrosion problems (including bio-corrosion), and treatment of regenerated solvents.

8.4.5.4   Overall plan

Figure 8-30 compares flare design to the overall plan for no-flare design. Point 1 represents solid-liquid separation, Point 2 liquid-liquid separation, and Point 3 liquid-gas separation. Since current techniques to separate liquids from gas are relatively adequate, solid–liquid–gas separation is not considered in this chapter. Point 4 represents the new portion of this design. It shows the addition of the gas-gas separator. For the no-flare system to be effective, each one of these points must perform with efficiency.

8.5   Sustainable Oil Refining Model

To achieve sustainability in transportation and refining operations, the green supply chain is proposed. The framework proposed supply chain model is developed based on the work of Lakhal and H'Mida (2003) and Lakhal et al. (2007). It analyses the structure of the supply chain from production, transportation, and distribution to end users (Figure 8-2). The specific aspects of the model include:

- The actual contaminants through the supply chain;
- Analysis of operations, process, materials design, and selection, according to environmental policy.

The research asserts that environmental practices would accrue competitive benefits to petroleum companies and enhance corporate performance (Sharma 2001). This section defines attributes of the green supply chain for an oil refinery, using the framework developed by Lakhal and H'Mida (2003; 2004) to assess greenness efforts of an oil refinery through its supply chain. The section proceeds to develop the concept of the green supply chain.

The ideal green supply chain is an "Olympic" green supply chain, similar to the Olympic logo. It has five zeros, each zero representing the zero waste objective. According to this model, the ideal "Olympic" oil refinery green supply chain should be characterized by:

(i) *Five zeros of waste or emissions (corresponding to the five circles in the Olympic flag)*:
- Zero emissions (air, soil, water, solid waste, hazardous waste)
- Zero waste of resources (energy, materials, human)
- Zero waste in activities (administration, production)
- Zero use of toxics (processes and products)
- Zero waste in product life cycle (transportation, use, end-of-life)

(ii) *Green inputs and outputs*: The zero waste approach is defended by the Zero-waste Organization (Zero Waste 2005) using a visionary goal of zero waste to represent the endpoint of "closing-the-loop", so that all materials are returned at the end of their life as industrial nutrients, thereby avoiding any degradation of nature. A 100% efficiency of use of all resources – energy, material, and human – is promoted by Zerowaste, working toward a goal of reducing costs, easing demands on scarce resources, and providing greater availability for all. These principles of Zerowaste applies to products reduce impact during manufacture, transportation, during use, and at end of life. For petroleum as a unity of analysis, the concept of the green supply chain is illustrated in Figure 8-31. Such an approach, which is always the norm in Nature, is only beginning to be proposed in the petroleum sector (Bjorndalen et al. 2005) or even in the renewable energy sector (Khan et al. 2005b).

We now proceed with an analysis of the five zeros, outputs, and inputs, in order to elaborate the parameters of the "Olympic" green supply chain. Parameters of an "Olympic" refinery:

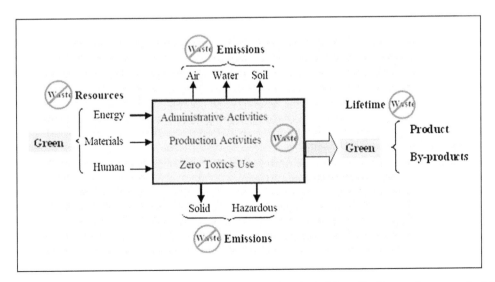

Figure 8-31   The concept of the "Olympic" green supply chain adapted from the Zerowaste approach. Adapted from Lakhal et al. (2006).

Table 8-12   Emissions from a refinery

Materials transfer and storage

Air releases: VOCs

Hazardous/solid wastes: anthracene, benzene, 1,3-butadiene, cumene, cyclohexane, ethylbenzene, ethylene, methanol, naphthalene, phenol, PAHs, propylene, toluene, 1,2,4-trimethylbenzene, xylene

Separating hydrocarbons

Air releases: carbon monoxide, nitrogen oxides, particulate matter, sulfur dioxide, VOCs

Hazardous/solid wastes: ammonia, anthracene, benzene, 1,3-butadiene, cumene, cyclohexane, ethylbenzene, ethylene, mercury, methanol, naphthalene, phenol, PAHs, propylene, toluene, 1,2,4-trimethylbenzene, xylene

Source: Environmental Defense (2005)

Table 8-13   Primary wastes from oil refinery

Cracking/coking	Alkylation and reforming	Sulfur removal
Air releases: carbon monoxide, nitrogen oxides, particulate matter, sulfur dioxide, VOCs	Air releases: carbon monoxide, nitrogen oxides, particulate matter, sulfur dioxide, VOCs	Air releases: carbon monoxide, nitrogen oxides, particulate matter, sulfur dioxide, VOCs
Hazardous/solid wastes, wastewater: ammonia, anthracene, benzene, 1,3-butadiene, copper, cumene, cyclohexane, ethylbenzene, ethylene, methanol, naphthalene, nickel, phenol, PAHs, propylene, toluene, 1,2,4-trimethylbenzene, vanadium (fumes and dust), xylene	Hazardous/solid wastes: ammonia, benzene, phenol, propylene, sulfuric acid aerosols or hydrofluoric acid, toluene, xylene Wastewater	Hazardous/solid wastes: ammonia, diethanolamine, phenol, metals Wastewater

Source: Environmental Defense (2005)

8.5.1   Zero emissions (air, soil, water, solid waste, hazardous waste)

The five primary activities in refinery processes are materials transfer and storage, separating hydrocarbons (e.g., distillation), creating hydrocarbons (e.g., cracking/coking, alkylation, and reforming), blending hydrocarbons, and removing impurities (e.g., sulfur removal) and cooling. Tables 8-12 to 8-14 (Environmental Defense 2005) enumerates the primary emissions at each activity level. We count 6 primary air release emissions and 23 primary hazardous/solid wastes.

- **Primary Air release**: carbon monoxide, nitrogen oxides, particulate matter, particulate matter, sulfur dioxide, and Volatile Organic Compounds VOCs.

- **Primary Hazardous/solid wastes**: 1,2,4-trimethylbenzene, 1,3-butadiene, ammonia, anthracene, benzene, copper, cumene, cyclohexane, diethanolamine, ethylbenzene, ethylene, hydrofluoric acid, mercury, metals, methanol, naphthalene, nickel, PAHs, phenol, propylene, sulfuric acid aerosols or toluene, vanadium (fumes and dust), and xylene.

Among these, sulfur dioxide, sulfuric acid aerosols, and VOCs are closely controlled. The sulfur components cause acid rain and VOCs are performance indicators that measure the quantity of smog-inducing VOCs released into the air by a refinery.

Table 8-14   Pollution prevention options for different activities in material transfer and storages

Cracking/coking	Alkylation and reforming	Sulfur removal	Cooling
Using catalysts with fewer toxic materials reduces the pollution from "spent" catalysts and catalyst manufacturing.	Using catalysts with fewer toxic materials reduces the pollution from "spent" catalysts and catalyst manufacturing.	Use "cleaner" crude oil, containing less sulfur and fewer metals. Use oxygen rather than air in the Claus plant reduces the amount of hydrogen sulfide and nitrogen compounds produced.	Ozone or bleach should replace chlorine to control biological growth in cooling systems. Switching from water cooling to air cooling could reduce the use of cooling water by 85%.

Source: Environmental Defense (2005)

8.5.2   *Zero waste of resources (energy, material, and human)*

The most important resource in the refinery process is energy. Unlike the manufacturing industry, labor costs do not constitute a high percentage of expenses in a refinery. In this continuous process, there is no waste of materials in general. The waste of human resources could be measured by ratios of accident (number of work accidents/number of employers) and absenteeism due to illness (number of days lost for illness/ number of work days × number of employees). In the "Olympic" refinery, ratios of accident and absenteeism would be near to zero.

The refining process uses a lot of energy: Typically, approximately 2% of the energy contained in crude oil is used for distillation. The "advanced" or "progressive" distillation, involving the use of modern chemical engineering techniques, cuts down the energy required to distil crude oil by 30 to 65%. This technique requires large-scale rebuilding of distillation units to enable separation of crude oil components. This would avoid the ineptness of conventional distillation processes, like heating oil at high temperatures and separating products as they cool. Advanced or progressive distillation may also include more effective heat exchange, further reducing the energy required for distillation. Considering this potential reduction of energy consumption, it can be estimated that an "Olympic" refinery would use only 1% contained in the crude.

8.5.3   *Zero waste in administration activities*

Considerable cost savings could be achieved by using resource-efficient products and good environmental practices. Good practices should target energy-efficiency, waste reduction, water conservation, and other resource-efficient practices for the environment. By taking advantage of these practices, refineries can avoid resource waste and save money. An "Olympic" refinery should have a list of good practices and encourage employers to respect them. For example, one workstation (computer and monitor) left running after business hours, causes power plants to emit nearly one ton of CO_2 per year. That emission could be cut by 80% if the workstation is switched off at night and set to "sleep mode" during idle periods in the day. If every computer and monitor in the United States was turned off at night, the nation could shut down 8 large power stations and avoid emitting 7 million tons of CO_2 every year (Nichols et al. 2001).

8.5.4   *Zero use of toxics (processes and products)*

A number of procedures are used to turn heavier components of crude oil into lighter and more useful hydrocarbons. These processes use catalysts or materials that help chemical reactions without being used up themselves. Table 8-15 shows different toxic catalysts and base metals. Refinery catalysts are generally toxic and must be replaced or regenerated after repeated use, turning used catalysts into a waste source. The refining process uses either sulfuric acid or hydrofluoric acid as catalysts to transform propylene, butylenes, and/or isobutane into alkylation products or alkylate. Vast quantities of sulfuric acid are required for this

Table 8-15   Catalysts and materials used to produce catalysts base metals and compounds

Name of Catalysts	Name of metals base
Activated alumina, Amine, Ammonia, Anhydrous hydrofluoric acid Anti-foam agents – e.g., oleyl alcohol or Vanol, Bauxite, Calcium chloride, Catalytic cracking catalyst, Catalytic reforming catalyst, Caustic soda, Cobalt molybdenum, Concentrated sulfuric acid, Demulsifiers – e.g., Vishem 1688, de-waxing compounds (catalytic) – e.g., P4 Red, wax solvents	Aluminum (Al), Aluminum Alkyls, Bismuth (Bi), Chromium (Cr), Cobalt (Co), Copper (Cu), Hafnium (Hf), Iron (Fe), Lithium (Li),
Diethylene glycol, Glycol – Corrosion inhibitors), Hydrogen gas, Litharge, Na MBT (sodium 2-mercaptobenzothiazole) – Glycol corrosion inhibitor (also see the taxable list for Oil Refining – Corrosion inhibitors), Na Cap – glycol corrosion inhibitor (also see the taxable list for Oil Refining – Corrosion inhibitors), Nalcolyte 8103, Natural catalysts – being compounds of aluminum, silicon, nickel, manganese, iron, and other metals, Oleyl alcohol – anti-foam agent, Triethylene glycol, wax solvents – de-waxing compounds	Magnesium (Mg), Manganese (Mn), Mercury (Hg), Molybdenum (Mo), Nickel (Ni), Raney Nickel, Phosphorus (P), Potassium (K), Rhenium (Re), Tin (Sn), Titanium (Ti), Tungsten (W), Vanadium (V), Zinc (Zn), Zirconium (Zr), and more to write.

Source: CTB (2006)

process. Hydrofluoric acid (HF), also known as hydrogen fluoride, is extremely toxic and can be lethal. Using catalysts with fewer toxic materials significantly reduces pollution.

Eventually, organic acids and enzymes, instead of catalysts must be considered. Thermal degradation and slow reaction rates are often considered to be the greatest problems of using organic acid and catalysts. However, recent discoveries have shown that this perception is not justified. There are numerous organic products and enzymes that can withstand high temperatures and many of them induce fast reactions. More importantly, recent developments in biodiesel indicate that the process itself can be modified in order to eliminate the use of toxic substances (Table 8-16). The same principle applies to other materials, for example, corrosion inhibitors, bactericides, etc. Often, toxic chemicals lead to high corrosion vulnerability and even more toxic corrosion inhibitors are required. The whole process spirals down to a very unstable process, which can be eliminated with the new approach (Al-Darbi et al. 2002).

8.5.5   *Zero waste in product life cycle (transportation, use, and end-of-life)*

The complex array of pipes, valves, pumps, compressors, and storage tanks at refineries are potential sources of leaks into air, land, and water. If they are not contained, liquids can leak from transfer and storage equipment and contaminate soil, surface water, and groundwater. This explains why, according to industrial data, approximately 85% of monitored refineries have confirmed groundwater contamination as a result of leaks and transfer spills (EDF 2005).

To prevent the risks associated with transportation of sulfuric acid and on-site accidents associated with the use of hydrofluoric acid, refineries can use a solid acid catalyst that has recently proven effective for refinery alkylation. A solid acid catalyst for alkylation is much less able than HF to disperse into the environment in a short time frame. Changing to this method would promote inherent safety at a refinery, rather than merely improving accident mitigation and response.

An "Olympic" green refinery supply chain should have storage tanks and pipes above ground to prevent groundwater contamination. There is room for improving the efficiency of these tanks with natural additives. Frequently, the addition of synthetic materials make an otherwise sustainable process unstable. Advances in using natural materials for improving material quality have been made by Saeed et al. (2003).

Table 8-16   Chemicals used in refining

Chemicals used in refining	Purpose
Ammonia	control corrosion by HCL
Tetraethyl lead (TEL) and tetramethyl lead (TML)	additives to increase the octane rating to
Ethyl tertiary butyl ether (ETBE), methyl tertiary butyl ether (MTBE), tertiary amyl methyl ether (TAME),	increase gasoline octane rating and reduce carbon monoxide
Sulfuric Acid and Hydrofluoric Acid	alkylation processes, some treatment processes.
Ethylene glycol	dewatering
Toluene, methyl ethyl ketone (MEK), methyl isobutyl ketone, methylene chloride, ethylene dichloride, sulfur dioxide	dewaxing
zeolite, aluminum hydrosilicate, treated bentonite clay, Fuller's earth, bauxite, and silica-alumina	catalytic cracking
nickel	catalytic cracking
Granular phosphoric acid	polymerization
Aluminum chloride, hydrogen chloride	isomerization
Imidazolines and surfactants	oil soluble corrosion inhibitors
Amino ethyl imidazoline	
Hydroxy-ethyl imidazoline	
Imidazoline/amides	
Amine/amide/DTA	
Complex amines	water soluble corrosion inhibitors
Benzyl pyridine	
Diamine amine morpholine	neutralizers
Imidazolines	emulsifiers
Sulfonates	
Alkylphenolformaldehyde, polypropeline glycol	desalting and emulsifier
Cobalt molybdate, platinum, chromium alumina	
$AlCl_3$-HCl, copper pyrophosphate	

8.5.6   Green outputs

Sulfur is the most dangerous contaminant among refinery output products. When fuel oils are combusted, the sulfur in them is emitted into the air as sulfur dioxide (SO_2) and sulfate particles (SO_4). Emissions of SO_2, along with emissions of nitrogen oxides, are a primary cause of acidic deposition (i.e., acid rain), which has a significant effect on the environment, particularly in central and eastern Canada (2002). Fine particulate matter ($PM_{2.5}$), of which sulfate particles are a significant fraction (30–50%), may adversely affect human health.

In this chapter, the greenness effort has been defined as the difference between regulated sulfur levels and average real sulfur levels. This difference should be positive or equal to zero. The idea behind it is that a petroleum company must comply with any regulations that may be in place. It is clear that each greenness effort would have to be specific to its particular region. Even in one country, it could vary from one region to another. In Canada, for example, there is currently no regulated national standard for sulfur in either HFO or LFO. Provinces regulate the sulfur content in HFO at various levels, ranging from 1.1% wt. up to 3.0% wt. For LFO, the sulfur content is regulated at 0.5% wt (Canada 2001). Table 8-2 indicates the regulation level of petroleum products in Canada. If one company offers better products then required, the difference is the greenness effort made by that company.

Table 8-17   Level regulation for petroleum products

Fuel Type	Sulfur	Benzene
Heavy fuel oil, **HFO**	1% by weight[1]*	
Motor gasoline	30 mg/kg[2]	1% by volume[5]
Light fuel oil, **LFO**	0.1% by weight[3]*	
Diesel fuel	15 mg/kg[4]	
Aviation gasoline	N/A	
Lubricants	N/A	

[1] Starting January 1, 2002 in Europe
[2] Starting January 2005 in Canada
[3] Starting January 1, 2008 in Europe
[4] Starting June 1, 2006 in Canada
[5] Effect in July 1999 in Canada
N/A: Non yet available
* There is no Canadian standard for this product

8.5.7   Green inputs

In the oil refining process, the quality of the crude will affect the quality of the refined products. Using "cleaner crude oil" containing less sulfur and fewer metals will prevent pollution. Less effort and treatment will be needed to remove impurities, and smaller amounts of pollutants will enter the environment through refining and the use or disposal of products. Also, practically all steps of crude oil refining and subsequent synthesis (e.g., making of plastics) must be overhauled. It is possible that by using proper environmental science, the process can be reversed to create "good" final products that are environmentally-friendly (Rahbar and Islam 2005).

It is important to note that the "greening" of a process cannot be accomplished unless every item is checked for sustainability. Sustainability can be only assured by using natural materials (Chhetri and Islam 2006a). In this process, the role of catalysts and other additives cannot be ignored. Proper analysis of such roles requires the knowledge of the science of intangibles, something that cannot be detected using a conventional detector. For instance, catalysts have been known for the longest time to be environmentally benign because they would not release any appreciable amount. "Appreciable" here relates to what can be detected. It turns out that any catalyst would release some amount, which would end up with the end user.

Because most of the gas or oil is actually burned while the rest of them are exposed to slow oxidation, any molecule of a toxic catalyst will end up being oxidized, releasing harmful products. Similarly, heat sources should also be checked for sustainability. Whenever there are alternate and more natural forms of energy available, they should be used. For instance, there is no reason why a sunny site should not use solar heating. Such a process can generate sufficient heat to provide us with a cheap and environmentally-friendly alternative to electrical or combustion heating. Note, however, that the solar energy has to be direct. Recently, Khan and Islam (2007) have introduced a series of such processes that are highly efficient, inexpensive, and most importantly, environmentally-friendly.

Similarly, during the polymerization process, if no toxic chemical is added and synthesis of materials takes place under natural environmental conditions, the plastic that will be produced will be environmentally-friendly. It is important in this process not to introduce any toxic chemical throughout the process, as tested by the Khan and Islam (2006e) criterion. Polymers produced with this process will remain environmentally-friendly and can be used in numerous applications, such as packages, paintings, hard composites, etc.

8.6   Corrosion in Petroleum Structures

In this section, focus is on marine and thermo-chemically active environments. This should cover harsh environments such as bottomhole conditions, etc. Because most of the corrosion activities are attributed to microbial actions, focus is given to microbially enhanced corrosion.

8.6.1   Corrosion in the marine and chemical environment

Marine and chemical environments are severely corrosive. However, the degree of severity depends on several variables such as humidity, temperature, chloride and other chemical contents, wind, and sunlight. Saline particles in the marine atmosphere accelerate metallic corrosion processes as chloride increases the solubility of the corrosion products. Marine chlorides dissolved in the layer of moisture also considerably raise the conductivity of the electrolyte layer on metal and tend to destroy the passive film existing on metallic surfaces. The overall reaction that occurs when steel is immersed in sea water or a corrosive liquid environment can be written as:

$$2Fe + 2H_2O + O_2 \rightarrow 2Fe + 4OH— \rightarrow 2Fe(OH) \tag{8.18}$$

However, ferrous hydroxide (corrosion products) reacts with salt to form ferrous chloride:

$$Fe(OH)_2 + 2NaCl \rightarrow FeCl_2 + 2NaOH \tag{8.19}$$

In later stages of corrosion, ferrous chloride reacts with water to form hydrochloric acid:

$$FeCl_2 + 2H_2O \rightarrow Fe(OH)_2 + 2HCl \tag{8.20}$$

Ingress of chloride, abundant in the marine environment at discontinuities due to rust, is the greatest cause of pitting. However, it has been observed in many studies that the pH inside the pit could reach as low as 2 due to the effect of HCl.

Offshore structures are often exposed simultaneously to five different zones of corrosion: the first zone is the subsoil zone, the structure then passes up to submerged zone area, tidal, splash zone, ending with the marine atmosphere zone. The most severe attack occurs in the splash zone because of alternate wetting and drying and also due to aeration.

In recent years, the construction of offshore structures has significantly expanded and the techniques of coating methods and anti-corrosives have reached a high stage of development. Since most of these structures are made of steel and designed to operate in an aggressive environment over a long period of time, the coating systems must give good performance with minimum maintenance costs (Equation 8.18). In addition, the cathodic protection, which is the normal privation method that combines with coating especially in the splash zone, is not reliable. The problem in this zone is that the surfaces are intermittently wetted so the cathodic protection system used for the rest of the structure will not work. Resistance to water, ease of application, impact resistance, long-term stability, and resistance to bacteria and other microorganisms are the most important factors to be considered in choosing a coating in marine environments.

Many researchers have undertaken studies of marine corrosion behavior involving different coatings. Some have developed indoor techniques (in the laboratory) to reduce the time exposure frame, while others have carried out a long outdoor exposure test, which takes a minimum of five years to get reliable data. Appleman (1992) has compared several laboratory-accelerating tests of coating within the exterior marine environment. One of the remarkable conclusions addressed was that a reliable accelerated laboratory test method for predicting field performance and durability of coating systems is still missing. In study performed by Chengde (1995), the characteristic, fabrication, construction technique, and quality control requirement for external pipelines coatings in seawater was discussed. He has also reported that some coating systems have been used successfully in this harsh environment. Szokolik (1992) carried out experimental work to evaluate the use of a single inorganic zinc coat for oil and gas facilities in marine environments. It was found that the performance of a single coat of zinc performed significantly better than most multi-coat systems. However, the sensitivity of zinc to acid and alkalyzed make the use of top coating very important. Munger (1992) described the use of various coating systems in the five different zones of corrosion in the marine environment.

Salt spray test according to ASTM B117 is the most widely used coating evaluation test. Salt solution is pumped into a nozzle where it meets a jet of humidified and compressed air, forming a fine droplet spray under high humidity and temperature.

All the existing methods of corrosion tests are time consuming and prone to error. The proposed research is aimed at developing a test that is short, yet accurate.

8.6.2   Microbially-influenced corrosion (MIC)

Most types of microbes in water or soils have the potential to participate in corrosion. However, the most important group of bacteria associated with corrosion is sulfate reducing bacteria (SRB). SRB are indigenous to oilfields and are known to cause severe operational problems, which significantly increase the cost of various operations in the entire petroleum industry. These operational problems include corrosion, hydrocarbon souring, increased formation of emulsions and suspended solids, and reservoir plugging. In addition, SRB produces hydrogen sulfide that increases safety and environmental concerns. Despite efforts made by numerous scientists to determine the mechanism of MIC, the exact mechanism in which SRB influence the corrosion of iron is still subject of controversy. These bacteria reduce inorganic sulfate to sulfide, which may remain as dissolved sulfide, are liberated as hydrogen sulfide gas, or may react with ferrous ions to produce the characteristic black precipitate of iron sulfides. There is no doubt that the overall reaction can be described by

$$4Fe + SO_4^{2-} + 4H_2O \rightarrow FeS + 3Fe(OH)_2 \tag{8.21}$$

However, Costello (1974) studied the problem of anaerobic corrosion of ferrous metals in detail, using modern electrochemical techniques and concluded that the Miller and King theory (1971) of an iron-iron sulfide galvanic cell was substantially correct. The specific cathodic reaction in the presence of H_2S is most likely to be

$$2H_2S + 2e^- \rightarrow 2HS^- + H_2 \tag{8.22}$$

According to Costello, the bacterial hydrogenous system may play a secondary role by removing molecular hydrogen, H_2, favoring the production of H_2S. Many other researchers have put forward another approach that anaerobic corrosion by SRB is caused by a highly active volatile phosphor compound that reacts with bulk iron to form iron phosphate. The mechanism proposed by Cord-Ruwish et al. (1998) integrates electrochemical phenomena with microbial physiology. More recent studies have conclusively shown that sulfate reduction can occur with cathodically formed hydrogen. Pope et al. (1992) proposed a general model for the development of MIC activities due to SRB. Recent reviews by Stott (1993) and Javaherdahti (1999) summarized different corrosion mechanisms. The most commonly encountered SRB is known as Desulfovibro. SRB will only flourish and cause damage if they obtain sufficient sulfate. The usual limitation for the bacteria to be grown is the carbon source. However, in many oilfield water systems, mixed populations of organisms were not predominant to carbon source and the limitation was due to sulfates.

The development of appropriate methods for mitigation and prevention of MIC in the industry have been the ultimate goal of researchers since its inception. Uncontrolled growth and activity of these microorganisms in oilfields can create environmental and operational problems. Mitigation of MIC can be achieved only after careful monitoring and consideration of the situation. Otherwise, considerable time and money may be spent performing unnecessary treatments. Many methods have been used in the detection of MIC bacteria. The common way in monitoring of oil and gas systems is to monitor the bacterial contamination by collecting fluid samples and culturing these samples for bacteria growth. Another method, used extensively, is monitoring the growth of biofilms. Biofilm probes have been developed for these purposes. A rigorous program of monitoring should be instituted early in the life of an operating system.

Coatings are the primary means of combating external corrosion. Although coatings are routinely evaluated for resistance to variety of environments (e.g., offshore), few coatings have been developed with consideration to their resistance to MIC. However, for protective coatings to be effective, they must be resistant to microbial degradation. Almost all cases of MIC on external surfaces are associated with desponded coatings or other areas shielded by cathodic protection (CP). When a pipeline is buried in clay, cathodic protection is highly recommended to minimize or prevent corrosion. In many cases the CP current is not evenly distributed on the pipeline surface. This may allow microbes to colonize and initiate MIC. An example illustrated by Pope et al. (1992) concluded that coatings are the only method that can be used to prevent external MIC on pipelines. Another effect that SRB can cause on CP is that the bacteria may colonize on sacrificial anodes and reduce the output by 75% (Jones et al. 1992).

Jack et al. (1996), have studied many field cases of MIC caused by SRB and found that it was associated with desponded of plastic tape coatings. They have reported an increase in bacteria population at these sites, in spite of the fact that overall CP levels were above -1.0 V, which is normally required on the pipeline. In addition, large amounts of white siderite (ferrous carbonate) and iron sulfides (usually amorphous) were found in the cross-section of the corrosion nodules. Several other studies noticed that a high percentage of external MIC occurred in connection with desponded coating and followed the same general pattern as quoted by Jack et al. (1996). Jones et al. (1992) showed that bacteria rapidly attack many types of coatings with subsequent corrosion occurring under the coating. Ray et al. (1997) studied blistering and delimitation of marine coatings used to protect carbon steel. Laboratory tests were conducted to evaluate alkyd coating, epoxy polymide, and polyurethane. Jones-Meehan et al. (1992) investigated microbial attack on protective coatings using mixed communities consisting of strict and facultative anaerobes. (SRB and non-sulfate reducers) coating deterioration was observed using SEM/EDS, ESEM/EDS and EIS.

However, many workers anticipated that blistering or unbonding occurs on the surface of the coating as a result of microbial attack in the areas between the substrate surface and the coating. Nevertheless, Ray et al. (1997) reported that the mechanism for microbial degradation of organic coating includes direct attack by enzymes or acids, blistering due to gas evolution, and polymer destabilization by concentrated sulfides. Moreover, Stranger-Johanessen (Oslo Research Institute) found that blisters did not exert deteriogene activity on new coating while the microorganisms found on the outer surface do. Therefore, it was concluded that the microorganisms attack the coatings from outside by means of excreted metabolites, which may change the coatings chemical and physical properties in such a way that blistering and debonding can occur.

Recently, our research group published a series of papers on the feasibility of using biological additives for preventing microbial corrosion. Also, advances were made for developing accelerated methods for determining both onset and growth patterns of corrosion (Al-Darbi et al. 2002).

8.6.3   A new accelerated test

Recently, Muntasser and Islam (2002) have developed a new accelerated test that can screen out additives within a week. Previously such test would take many weeks. The principle of this new test is that corrosion in metals occurs due to oxidation-reduction reactions. These reactions depend on the type of metal and its surrounding environment. The key point is that there is an exchange of electrons and production of ions. In the case where there are two metals, the more anodic (this property has been determined and well documented in the galvanic scale) metal will lose an electron, and give up an ion to solution, thus decreasing its solid mass. The cathode will accept the electron, and will attract positive ions in an attempt to balance this new charge. It will then react with the ion, and depending on the ion, it may increase its solid mass, but usually it will react with the ions to produce a by-product such as hydrogen gas. If there is only one metal present, this process will still take place, except that the cathode and anode will be a small region of centralized charge, and these areas of charge can change locations many times during the corrosion process.

Corrosion is governed by several key parameters that need to be identified in order to devise a test that is fast and accurate. The first parameter is the transfer of electrons. Second, this is driven by a potential difference due to an area of centralized charge, or a greater tendency of one metal to hold electrons compared to another metal. Third, a liquid (electrolyte) is required in order to allow for the transfer of ions. This does not necessarily require submersion since liquid can form in the tiny pores of a metal that is close to the dew point temperature at a given pressure.

There are some other factors that need to be considered. First, the galvanic scale is solution-dependant and is usually used for salt water. It is assumed that salt water has a standard composition. However, this assumption must be reconsidered. In offshore rigs, it is not possible that they could be built in an area that has a geology leading to a high concentration of some chemical, which can cause excessive corrosion to certain metals. There are many metals and alloys, each having different properties and it is possible that reactions could occur at an extremely high rate if the conditions are conducive. In the case of ships, they enter many waters with varying amounts of pollutants. This means that the galvanic scale could really be tested, as could the corrosion resistance of a ship, if it is anchored somewhere, such as near a sewage outlet. These types of factors need to be assessed in detail for specific situations, since it would be nearly impossible to predict every possible situation. For example, if a rig is to be built in a particular area, corrosion testing should be conducted using electrolyte samples taken from that particular area. This includes varying depths, as concentrations are sure to vary with distance from a contamination source.

This brings up another interesting point, as to whether the composition of the solution varies with seasons, or with time in general. This could be due to increased traffic in an area, or even due to the building of new plants upstream. Other factors include biological factors, as there must be organisms that could affect metals. This could include microorganisms that feed on metals or produce corrosion conducive environments. There may also be factors such as a region of high plant growth, large concentrations of fish carcasses, waste, or even a mating ground. These conditions could cause reactions, or encourage the growth of organisms that could affect corrosion. Finally, the stress levels that metal encounters is another factor that needs to be considered. The stress level could be due to mechanical or thermal changes. Both changes should be considered in order to develop an accelerated test.

The development of a single technique that could model all possible scenarios is practically impossible. Instead, we should resort to simulating a site-specific case. A study of an area needs to be done to investigate all the conditions in that area. This would include all geological, biological, meteorological, and whatever else could be deemed interesting. This approach would be effective for a stationary rig. However, in the case of a ship, we should consider the locations where the vessel travels through (e.g., a tanker which travels from Canada to Britain), and then establish a specific critical path. This could be, say, either the Arctic, or some tropical areas. Such a vessel would be exposed to various conditions during its operation, and so we need to determine where it will spend most of its stationary time. This would indicate that we should consider, say, the main ports and determine whether there are any conditions specific to that area. Therefore, a statistical analysis would also have to be conducted. We cannot possibly, by current techniques, protect an object from every possible corrosive environment. Therefore, we should consider the areas that would have maximum exposure to corrosive environment in a given vessel. Then it would be only in special situations where the vessel is exposed to a highly polluted environment that we should be concerned with brief exposures.

Once we has determined which environments are going to play the most critical role in the service life of an object, we could then start to identify the factors requiring simulation. The area of greatest concern is the splash zone, as in the case of a ship or an offshore rig. This is the area that would see the most friction due to contact with the waves and debris. It is also the area that is exposed to agitation, which would in turn be associated with a high gas concentration in the fluid and thus accelerated corrosion. It is also the area that would be exposed to both light, and the electrolyte as opposed to the other extremes such as the very top, or bottom of an object. This means that if we could develop a protection method that would meet the needs of this harsh environment, it would then be more than adequate for assessing any other sections of the ship. Of course, there are other areas that would never see this type of harsh treatment (e.g., the radar tower), therefore we should determine the worst-case scenario for a given specific component.

In developing this test procedure, all possibilities must be considered. It is also known that a ship's hull becomes charged as it travels through the Earth's magnetic field, and also due to its internal machineries (e.g., generators and engines). It is also well established that electricity is a direct cause of corrosion – so how would a ship's hull that is being magnetically charged affect its rate of corrosion? Could we take advantage of this effect to hinder corrosion rate or would it accelerate the simulation of corrosion?

Other factors that should be considered, and which may accelerate the corrosion test, could include the enhancement of the high-energy area (perhaps by polishing with increasingly finer sand papers). Also, the use of different solutions of higher reaction rates could be considered to accelerate the occurrence of corrosion. Among others, fatigue stress could be modeled with imposed stress in a tri-axial machine. Note that, in theory, a tri-axial state should not have a pronounced effect on the fatigue rate, but this needs to be quantified. We could also model coatings' failure due to various combination of stresses caused by the change in thermal and mechanical loading conditions.

8.6.4 Background of chemical additives and new biopolymers

In recent years, the construction of petroleum structures (particularly offshare ones) has significantly expanded, and the techniques for coating methods and anti-corrosive and anti-fouling protection have reached a high stage of technological development. Since many offshore structures are designed to operate in aggressive environments for long periods, the coating systems must provide good protection, with low and economical maintenance.

Since of the evolution of applications with demanding performance criteria, attention has shifted considerably from conventional to newly emerging engineering materials. Material development is now focused on high performance polymers and composites for a wide verity of applications. While polymer-based coatings have been used successfully to prevent corrosion in other parts of the world, few such coatings appear to have success in the marine environment.

8.7 Hydrate Problems and Some Suggestions

One of the most critical challenges facing oil and gas producers today is assuring multiphase flow through long subsea pipelines. As production systems continue to go deeper, flow assurance becomes a major issue for offshore production teams, where the operator is faced with several challenges associated with multiphase flow (e.g., gas hydrate formation, wax deposition on walls, corrosion, and scale build-up on walls). The avoidance or remediation of these problems is the key aspect of flow assurance that enables the design engineer to optimize the production system and to develop safe and cost-effective operating strategies for the range of expected conditions including startup, shutdown, and turndown scenarios.

Prevention of hydrate plugs is a key component of a flow assurance strategy and often the most expensive to implement. To date, the phenomenon of gas pipeline plugging due to gas hydrates is still an important industrial problem, which can led to safely hazards to production/transportation systems and to substantial economic risks. Therefore, an understanding of how, when, and where hydrates form is necessary to overcome hydrate problems. This chapter answers these crucial questions as well as providing significant information on the best methods to prevent hydrates in deepwater production operations.

8.7.1 Gas-hydrates

Gas hydrate is an ice-like crystalline solid, called a clathrate, which occurs when water molecules form a cage-like structure around smaller guest molecules. The most common guest molecules are methane, ethane, propane, isobutane, normal butane, nitrogen, carbon dioxide, and hydrogen sulfide, of which methane occurs most abundantly in natural hydrates. While many factors influence hydrate formation, the two major conditions that promote hydrate formations are:

(1) the gas being at the appropriate temperature and pressure; and

(2) the gas being at or below its water dew point.

Other factors that determine hydrate formation include mixing, kinetics, type of physical site, surface for crystal formation, agglomeration, and salinity of the system.

Hydrates are known to occur when natural gas and water coexist at elevated pressure and reduced temperature. Hence, when the multiphase fluid produced at the wellhead flows through the subsea pipelines, it becomes colder, which means that most subsea pipelines could experience hydrates at some point in their operating envelope. Moreover, shut-in and startup are also primary times when hydrates form. On shut-in, the line temperature cools rapidly to that of the ocean floor so that the system is almost always in the hydrate region if the line is not depressurized. At that condition, multiple hydrate plugs can form.

8.7.2   Possible problems

Although gas hydrates may be of potential benefit both as a most important source of hydrocarbon energy and as a means of storing and transmitting natural gas, they represent a severe operational problem as the hydrate crystals may deposit on the pipe wall and accumulate to large plugs, which can block pipelines and cause huge industrial loses or shutdown of the production. Acceleration of these plugs when driven by a pressure gradient can also cause considerable damage to production facilities, and thereby creating a severe safety and environmental hazard. For this reason, the hydrate forming in offshore gas transmission pipelines should be prevented effectively and economically to guarantee the pipelines running normally.

8.7.3   Hydrate preventing techniques

There are few methods of preventing hydrate formation in offshore production systems. The permanent solution is removal of water prior to pipeline transportation, using an offshore dehydration plant, which is not often cost effective. Another means of combating hydrate formation is to avoid the regions of pressure and temperature where hydrates form. It may be possible to keep the fluid warmer than hydrate formation temperature (with the inclusion of a suitable margin for safety) or operate at a pressure less than the hydrate formation pressure. However, the line-depressurization approach is not practical in long and high-pressure gas transmission pipelines. In addition, gas decompression at the wellhead results in lowering the temperature below freezing and thus creating favorable conditions for hydrate formation. In general, at the well site two methods are applicable, namely thermal and chemical.

8.7.3.1   Thermal methods

These methods use either the conservation or introduction of heat in order to maintain the flowing mixture outside the hydrate formation range. Heat conservation has been the normal technique and is simply insulation. This method can be feasible for some subsea applications depending upon the fluid being transported, the tie back distance, and topsides capabilities of the host platform. However, the design of such systems typically seeks a balance between the high cost of the insulation, the intended operability of the system, and the acceptable risk level. A number of different concepts are available for introducing additional heat to a pipeline. The simplest is an external hot-water jacket, either for a pipe-in-pipe system or for a bundle.

Other methods use either conductive or inductive heat tracing. There is concern over the reliability of conductive systems. The direct electrical heating system consists of feeder cable installed piggy-back to the subject to be heated. Once power is supplied, the magnetic field generated around each cable induces electrical currents in the pipe wall, thus heating it. The electrical rating of the system depends on the heat requirement, pipe material, and the pipe length. This system provides environmentally-friendly fluid-temperature control without flaring for pipeline depressurization. The effect is to increase production as there is no time lost by unnecessary depressurization, pigging, heating-medium circulation, or removal of hydrate blockage.

Recently, another method consisting of using an exothermic chemical reaction has been developed by Petrobras (Khalil 1995). This technique is not applicable for continuous heat input but has been used in a batch process to melt hydrate deposits.

8.7.3.2 Chemical inhibition

An alternative to the thermal processes previously described is chemical inhibition. Chemical inhibitors are injected at the wellhead and prevent hydrate formation by depressing the hydrate temperature below that of the pipeline operating temperature. This method is expensive, although less than the thermal stimulation method, due mainly to the cost of high porous chemicals. However, hydrate inhibition using chemical inhibitors is still the most widely used method, and the development of alternative, cost-effective and environmentally acceptable hydrate inhibitors is a technological challenge for the oil and gas production industry.

8.7.3.3 Thermodynamic inhibitors

Traditionally, the most common chemical additives used to control hydrates in gas production systems have been methanol, ethylene glycol, or triethylene glycol at a high enough concentration. These chemicals are called "thermodynamic inhibitors" and have the effect of shifting the hydrate formation loci to the left, causing the hydrate formation point be displaced to a lower temperature and/or a high pressure.

The thermodynamic inhibitors selection process often involves comparison of many factors including capital/operating cost, physical properties, safety, corrosion inhibition, gas dehydration capacity, etc. However, a primary factor in the selection process is whether or not the spent chemical will be recovered, regenerated, and re-injected. Typically, methanol is used in a non-regenerable system because it is a relatively inexpensive inhibitor and, therefore, the economics of methanol recovery will not be favorable in most cases. Often when applying this inhibitor, there is a significant expense associated with the cost of "lost" methanol.

However, since methanol has a lower viscosity and lower surface tension, this makes effective separation easy at cryogenic conditions (below −13°F), so is usually preferred. In many cases hydrate plug formation is prevented by the addition of glycols (usually ethylene glycol because of its lower cost, lower viscosity, and lower solubility in liquid hydrocarbons) to depress the hydrate formation temperature. But in order to be effective, glycols must be added at rates of up to 100% of the weight of water. Since glycols are expensive inhibitors, there is a definite need for extra, costly, and space consuming, onshore or offshore plants for their regeneration. Therefore, new, cost-effective, and environmentally acceptable hydrate inhibitors that allow multiphase fluids to be transported untreated over long distances have been under increasing investigation by the oil and gas industry. These new hydrate inhibitors can lead to substantial cost savings, not only for the reduced cost of the new inhibitor but also in the size of the injection, pumping, and storage facilities. Thus, it is possible to redesign production facilities on a smaller scale. Such a changeover in hydrate inhibitor technology may also be an environmental requirement in the near future.

8.7.3.4 Low dosage hydrate inhibitors

Recently, two new types of low-dosage inhibitors have been developed, which will enable the subsea gas transmission pipelines to handle increased gas volumes without additional glycol injection, or extra glycol recovery units. Replacing traditional hydrate inhibitors with these new generation of hydrate inhibitors can also lead to substantial cost savings, not only in the reduced cost of the new inhibitor but also in transportation costs and size of the injection, pumping, and storage facilities. Thus, it is possible to redesign production facilities on a smaller scale. These chemicals are called "kinetic inhibitors" and "anti-agglomerant inhibitors". Most commercial kinetic inhibitors are high molecular weight polymeric chemicals, which may prevent crystals nucleation or growth during a sufficient delay compared to the residence time in the pipeline. Kinetic inhibitors should be considered for residence times of less than a few minutes. Kinetic inhibitors are relatively insensitive to the hydrocarbon phase and may therefore turn out to be applicable to a wide range of hydrocarbon systems. However, the industrial application of kinetic inhibitors depends on the repeatability of multiphase pipeline testing results among laboratory, pilot plant, and field, and the transferability among different plants.

In contrast to other types of inhibitors, anti-agglomerators (i.e., growth and slurry additives) inhibit hydrate plugging, rather than hydrate formation. They allow hydrates to form but keep the particles small and well dispersed so that fluid viscosity remains low, allowing the hydrates to be transported along with the produced fluids. However, these additives are not as well advanced at present and are currently undergoing flow loop tests.

As stated earlier, the choice between inhibitor alternatives must be based on physical limitations as well as economics. However, operating conditions may also limit the number of available choices. For example, in a recent project carried out by Baker Petrolite, it was shown that under severe conditions, the required dosage of an anti-agglomerator, unlike thermodynamic and kinetic inhibitors, does not increase as the degree of subcooling increases. Therefore, this method of treatment is a cost-effective solution for the control of gas hydrates.

8.7.4   Most promising area of research

All the technologies discussed above deal with chemicals, some more toxic than others. Chemicals hold no promise for the long-term applications. In fact, some North Sea operators have forbidden the use of anti-agglomerates because of their toxic nature. Our research group led the research that would eventually develop technologies that would be environmentally benign. A strain of bacteria was found in Nova Scotia that would diminish the formation of hydrate while thriving under high pressure (tested at 300 psi) and low-temperature conditions (4°C), in presence of hydrocarbon. This research has been reported in a thesis by Jorge Paez (MS thesis, Chemical Engineering, Dalhousie University 2001). This technology can be considered to be a breakthrough technology in this area, as this will develop the first biological treatment for hydrate prevention. The implication of this research is tremendous.

Chapter 9

Decommissioning of Drilling and Production Facilities

9.1   Introduction

The decommissioning of oil and gas structures is an issue that has gained a great deal of attention. It is the final phase of oil and gas operations that starts by unplugging and abandoning the well, removing the infrastructure, doing remediation work, and clearing debris from the project site. The most difficult ones among all decommissioning activities are those that pertain to offshore activities. All indications point to the peak years of offshore platform decommissioning occurring before 2010. It is hard to reach an agreement on how many platform offshore oil and gas installations exist. In 2001, Ferreira and Suslick (2001) suggested a number of 7270 oil offshore installations around the world, distributed over more than 53 countries. Over 4000 are situated in the Gulf of Mexico, some 900 in Asia, some 700 in the Middle East, and around 1000 in the North Sea and northeast Atlantic (UKOOA 2005). A 20-year plan for a platform is usual, but it is common to have a life cycle between 30 to 40 years. At the end of this period, excepting relocation, platforms are decommissioned. Figure 9-1 illustrates a typical life cycle for an oil offshore platform. This figure highlights the environmental damage caused during its life cycle.

In the oil and gas industry, operational phases, such as exploration, development, and production are well studied and reported (Khan and Islam 2006a–d; Khan et al. 2006c), but little emphasis is placed on the decommissioning phase (Wood 2005). Figure 9-2 shows the different phases and related cash flow. The decommissioning of offshore oil and gas operations is a complex process, going through stages of planning, gaining government approval, and implementing the removal, disposal, or reuse of a structure when it is no longer needed for its current purpose.

This chapter examines the decommissioning phase of petroleum operations. It provides a brief overview of the historical perspective of petroleum platform decommissioning, environmental issues, and types of platforms. This chapter also investigates the current practices of drill rig decommissioning, including the regulations and standards, techniques, and legal framework. A case study of a decommissioning project on the Scotian Shelf in Canada is described. It also suggests importing novel materials to permeate the petroleum industry as this industry ventures into more challenging environments. Finally, this chapter analyzes the sustainability of current practices and suggests alternative techniques to achieve sustainability in oil and gas decommissioning.

9.2   Historical Analysis

According to our research, it seems that the first recorded offshore platform decommissioning was from the Gulf of Mexico's Outer Continental Shelf (OCS) in 1973. Ever since, platforms have been removed from the OCS at a rate of about 100 a year, according to data from the US Department of the Interior's Minerals Management Service (Poruban 2001). Perhaps the most publicized decommissioning case is that of Brent Spar. Its owner, Shell UK Exploration and Production spent $36 million on research to find a widely acceptable way to dispose of Brent Spar and it took two years (between 1995 and 1997) to come up with the principle "don't dump, reuse or recycle" (Knott 1998).

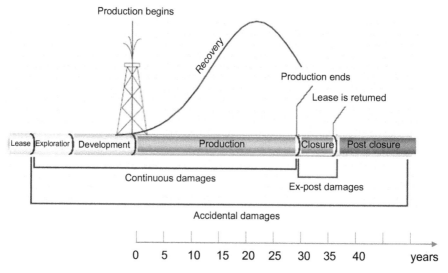

Figure 9-1   Oil platform life cycle and environmental damage. Adapted from Ferreira et al. (2004).

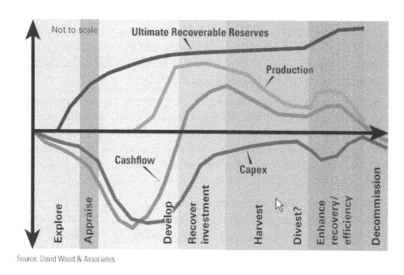

Figure 9-2   Petroleum operation profit center life cycle.

9.3   Type of Oil Platforms/Platform Structures

In offshore oil and gas operations, many different types of platforms are used. They are steel jacketed platform (shallow water), concrete gravity structure, steel gravity structure, floating production system, steel jacketed platform (deepwater), compliant tower, and tension leg platform (Figures 9-3 and 9-4). They range from small shallow structures to heavy structures for deepwater. Submersible rigs are used in shallow water, generally ranging up to 25 m depth (UKOOA 2005). In the fully blasted position, the rig hull on the seafloor serves as foundation support for drilling operations and resists environmental loads caused by waves, winds, and currents.

Jack-up rigs operate in water depths of 20–125 m. The maximum depths depend on the expected waves and winds at the site. In a moderate site, like the Scotian Shelf, jack-up rigs are limited to approximately 75–90 m

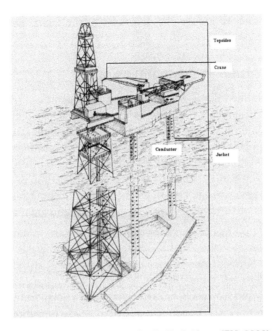

Figure 9-3   Sketch of a typical oil platform (EIA 2002).

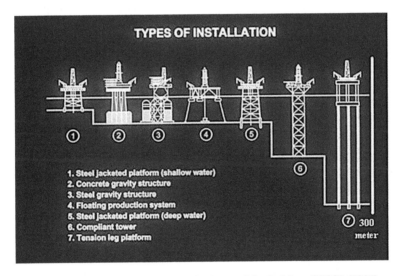

Figure 9-4   Different types of oil platforms. Adapted from UKOOA (2005).

of water depth. They consist of self-contained legs lowered to contact the seabed (Figures 9-3 and 9-4). A mat-supported rig is used in soft seafloor sediments. The main challenges facing the industry are the many different types and designs of structures, meaning that there is not a single tried-and-tested method for removing and deconstructing all of the installations.

A larger type of oil and gas platform includes oil and gas processing equipment, living spaces for crews, and working deck. Oil and gas supply pipes are also attached vertically to the rigs that are anchored at the base and protrude through the water surface to support the topside structural components. These components include the crossbeams, legs, and the piles inside the legs. Near the surface, horizontal crossbeams occur about every 15–20 m (50–66 ft) and, at greater depths, every 30–40 m (99–132 ft). Horizontal, diagonal, and oblique crossbeams extend both around the perimeter of the jacket and reach inside and across the platform (Schroeder and Love 2004).

9.4   Environmental Issues in Decommissioning

The main source of pollution in the oil and gas exploration is drill cuttings (the collective name for drilling mud, specialty chemicals, and fragments of reservoir rock) being deposited on the seafloor. For example, in 1996, in the North Sea, drill cuttings were estimated to be 7 million m^3 (De Groot 1996), but this amount was updated to 12 million m^3 in 2000 (OLF 2000). It is clear that drill cuttings are a complex mixture having an adverse impact on the environment (Breuer et al. 2004). Each pile of drill cuttings represents a unique combination of sediment characteristics, contaminants, and benthic community and each is affected by the local hydrodynamic regime. However, the constituents found in drill cuttings piles include heavy metals, barite, bentonite, specialty chemicals, hydrocarbons, organic contaminants, and radioactive materials.

Available information on drill cuttings accumulations mainly focuses on three chemical groups, hydrocarbons, heavy metals, and to a lesser degree, radionuclides. Although there is uncertainty as to the extent to which oil concentrations reduce over time, it appears that most oil is dispersed during initial deposition. Thereafter, further degradation of oil in the pile will be very slow. In the case of drill cuttings piles, which contain a number of potentially hazardous chemicals, synergistic effects from multiple contaminants should be considered. Grants and Briggs (2002) conducted a toxicity study of sediments from around a North Sea oil platform, concluding that the sediment around the platform was very toxic. Dichloromethane extracted from sediments close to the platform are very toxic to Microtox, even after removal of sulfur.

The toxic pollutants released from the platform extend as far as 600 m from the platform. Around 100 to 300 m from the platform sediments caused 100% mortality of *Corophium*. Marsh (2003) studied the archived drilling records of the drill cuttings oil at the North West Hutton Platform and concluded that a least 130 chemical products were used. He adds that safety and toxicity information is generally sparse, even for the better-reported products. Furthermore, the obstacles facing researchers wishing to obtain data on the safety and (eco-)toxicity is that many data are easily available and that many websites charge registration fees. However, information may also be commercially confidential, even for products no longer in use. There is no standardized international legacy database on all the chemical products which are, or have been, used on oil platforms.

For the purposes of decommissioning, total amounts of wastes generation are required to be estimated. It helps to plan and develop appropriate measures during the decommissioning phase. Wastes generation information, the life cycle of petroleum operations, and its different activities, were collected through a review of government documents and reports, and published papers, such as CNSOPB activities reports (CNSOPB 2005; Khan and Islam 2003a, 2006b; EPA 2000). Different types of wastes generation were estimated following the methodology of EPA (2000). Using Equations 9.1–9.8, wastes generation has been estimated (EPA 2000).

The dry drill cuttings volume is estimated based on Equation 9.1. In this estimation, the dry drill cuttings are equivalent to gauge hole volume plus washout:

$$\text{Drilling hole volume (ft}^3) = \{\text{length (ft)} \times \pi \ [\text{diameter (ft)/2}]2\} \times (1 + \text{washout fraction of } 0.075) \quad (9.1)$$

$$\text{Drill cuttings (bbls)} = \text{hole volume (ft3)} \times 0.1781 \text{ bbls/ft3} \quad (9.2)$$

$$\text{Drill cuttings (lbs)} = \text{drill cuttings (bbls)} \times 910 \text{ lbs/bbl} \quad (9.3)$$

$$\text{Total wastes (TW)} = (\text{base fluid}) + (\text{water}) + (\text{barite}) + (\text{drill cuttings}) \quad (9.4)$$

$$\text{TW} = (\text{RF} \times \text{TW}) + \{[\text{RF} \times (\text{WF/SF})] \times \text{TW}\} + \{[\text{RF} \times (\text{BF/SF})] \times \text{TW}\} + (\text{DF} \times \text{TW}) \quad (9.5)$$

where

TW = total waste (whole drilling fluid + dry cuttings), in lbs

RF = retort weight fraction of synthetic base fluid

Table 9-1   Characteristics of SBF drilling mud (EPA 2000)

Waste Characteristics	Value
SBF formulation	**(by weight)**
Synthetic base fluid	47%
Barite	33%
Water	20%
Percent (vol.)	
formation oil	0.2%
SBF density	**280 lbs/bbl**
SBF drilling fluid	
density	9.65 lbs/ga
Barite density	1,506 lbs/bbl
Water	350 lbs/bbl
Drilling cuttings	910 lbs/bbl
Priority pollutant organics	**lbs/bbl of SBF**
Naphthalene	0.0010024
Fluorene	0.0005468
Phenanthrene	0.0012968
Phenol	0.000003528
Priority pollutant metals	**mg/kg Barite**
Cadmium	1.1
Mercury	0.1
Antimony	5.7
Arsenic	7.1
Beryllium	0.7
Chromium	240.0
Copper	19.7
Lead	35.1
Nickel	13.5
Selenium	1.1
Silver	0.7
Thallium	1.2
Zinc	200.5

WF = water weight fraction from drilling fluid formulation

SF = synthetic base fluid weight fraction from drilling fluid formulation

BF = barite weight fraction from drilling fluid formulation

DF = drill cuttings weight fraction, calculated as follows

$$DF = 1 - \{RF \times [1 + (WF/SF) + (BF/SF)]\} \qquad (9.6)$$

$$TW = \text{drill cuttings (lbs)} \qquad (9.7)$$

Waste components are estimated following Equations 9.4 and 9.5. In order to calculate TW, Equations 9.4 and 9.5 are first used to calculate DF (Equation 9.6). Then TW is calculated following Equation 9.7. Input data to estimate the emissions of petroleum operations are shown in Tables 9-1 and 9-2. Theses data have been gathered from different sources such as EPA (2000) and Wenger et al. (2004). In this estimation, 10.2% (wt./wt.) standard (baseline) solids control have been taken into account (EPA 2000).The whole drilling fluid volume is estimated following Equation 9.8.

$$\text{Whole SBF volume (bbls)} = \text{synthetic base fluid (bbls)} + \text{water (bbls)} + \text{barite (bbls)} \qquad (9.8)$$

The formation oil in whole mud discharged is 0.2% (vol.), is calculated based by Equation 9.8.

Table 9-2   Characteristics of SBF drilling mud (Wenger et al. 2004).

Waste Characteristics	Value
Priority pollutant organics	**lbs/bbl of SBF**
Naphthalene	0.0010024
Fluorene	0.0005468
Phenanthrene	0.0012968
Phenol	0.000003528
Non-conventional Organics	**lbs/bbl of SBF**
Alkylated benzenes	0.0056429
Alkylated naphthalenes	0.0530502
Alkylated fluorenes	0.0063859
Alkylated phenanthrenes	0.0080683
Alkylated phenols	0.0000311
Total biphenyls	0.0104867
Total Dibenzothiophenes	0.0004469
Non-Conventional Metals	**mg/kg Barite**
Aluminum	9,069.9
Barium	588,000
Iron	15,344.3
Tin	14.6
Titanium	87.5
Conventional	**lbs/bbl of SBF**
Total oil as SBF	190.5
Total oil as formation oil	0.588
TSS as barite	133.7

9.5   Toxicity and Degradation of Waste Generation

9.5.1   Bio-geo transformation of oil in marine environment

Immediately following the release of oil into the marine environment, its chemical composition begins to change, a process referred to as weathering. The rate of weathering is dependent on both existing environmental conditions, as well as the initial composition of the oil. The weathering process may take over one year to complete. There are several steps in the weathering process, such as spreading, evaporation, dissolution, dispersion, emulsification, sedimentation, photolysis, and bioaccumulation (Figure 9-5). Discharged oil spreads over the surface of water as a slick, with tides, waves, currents, and winds greatly accelerating the spread and influencing direction. Within the first few days, the volatile components of crude oils evaporate. Depending on specific gravity, evaporation can account for the loss of 30–70% of the spilled oil. For example, lighter oils have a greater evaporation rate (Patin 1999; NEC 2003). Some of the components of evaporated oils are oxidized with the UV radiation of sunlight, producing toxic acidic and phenolic compounds that are highly toxic. However, it is reported that the concentration of these oxidized toxic compounds is relatively low and their impact is not ecologically significant (Kingston 2002).

Waves, currents, and upwelling help to form water-in-oil emulsion, in which water droplets are incorporated into the oil. This process generates "chocolate mousse" or "mousse". Emulsification greatly alters the behavior of the oil in the marine environment, as it turns the oil into a denser, semi-solid. Generally, 3–15% of crude oil dissolves in seawater, however, the solubility rate of fluoranthene is low. According to Prager (1998), it is only 0.2–0.26 mg/L. In rough sea conditions, oil undergoes a mixing process known as dispersion. Dispersion takes place vertically as well as horizontally. This process removes the oil by natural means.

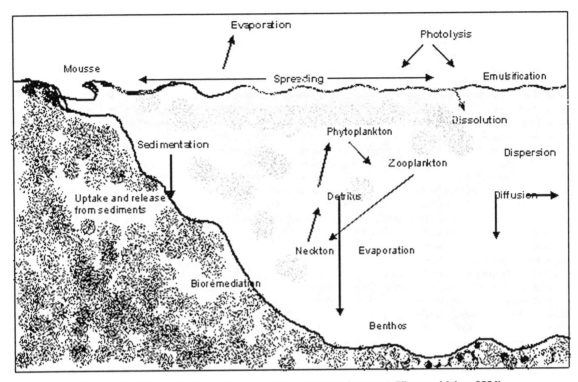

Figure 9-5   Oil weathering pathways in the marine environment (Khan and Islam 2004).

For example, the dispersed oil, in the form of droplets, is bio-accumulated by different microorganisms. Oil droplets attached to suspended solids sink to the sea bottom and are then broken down through microbial degradation.

9.5.2   Transformation of oil compounds

Several simulation tests were run for fluoranthene in the hypothetical environment. The study area was designed as a marine region. In the simulation tests, fluoranthene discharges were used in a similar fashion to those in real oil and gas operations. Sources of fluoranthene from offshore installations include produced water, drilling fluids, seepages from storage, transportation activities, working decks, and other activities. For this simulation, it was assumed that fluoranthene was continuously released directly into the marine environment, at a rate of 27 kg/h over a study area of 55,600 m^2. The emission rates in water, soil, air, and sediment were decided as 20, 4, 2, and 1 kg/hr, respectively. The result of the fluoranthene simulation is illustrated in Figure 9-6.

The rectangular and cloud shapes of Figure 9-6 represent the main ecosystem components of the study area, of soil, water, sediments, and air. The arrows in the figure represent the emission, reaction, advection, and intermediate exchanges of the chemicals among the compartments. The value of residing amounts of oil, percentage, fugacity value, and concentrations are given within each component. These pictorial presentations give us the comparative scenario of all compartments, which may be helpful in deciding which compartment will require the most attention and most response effort, in the event of a real accident. Whereas the fugacity model incorporates the concept of a whole picture, the total environment approach, non-fugacity models only focus on the effects within a particular ecosystem component (Mackay 1985). For example, non-fugacity models assess emission status in water, soil, and air separately (Mackay and Paterson 1991). Fugacity environmental modeling implements the concept of "unit world" and examines impacts in air, water, soil, and sediment compartments together (Mackay 2004).

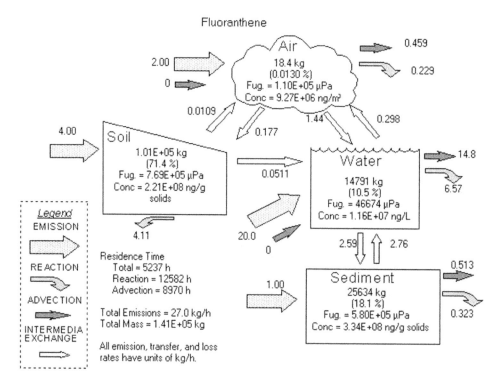

Figure 9-6   Diagram of fugacity model output for fluoranthene.

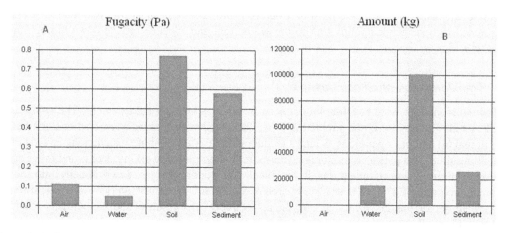

Figure 9-7   Concentration (B) and fugacity (A) of fluoranthene are shown in air, water, soil, and sediment.

Based on the emission rate, inflow concentration, and other physiochemical parameters, the total amount of fluoranthene found in air, water, soil, and sediment were 18.4, 14,791, 101,000, and 25,634 kg, respectively (Figure 9-7). The highest volume of fluoranthene was found in the soil compartment, while the lowest amount was found in the air compartment (Figure 9.7B). Comparatively lower amounts of fluoranthene were found in the water, because this oil compound has low water solubility. According to Prager (1998), the high molecular weight and relative non-polarity of fluoranthene makes this compound virtually insoluble (0.20–0.26 mg/L) in water. However, results of the model indicated that there were 73,115.91 kg of fluoranthene found in the water component, including exclusively water, suspended solids, and fish.

Table 9-3   Phase properties of the fluoranthene in the different compartments

Compartments	Z mol/m³.pa	Amount kg	Amount %	Conc. g/m³	Conc. μg/g	Conc. mol/g
Air: Bulk	4.15E-04	90.75047	1.30E-02	9.27E-03	7.688335	9.27E-03
Air vapor	4.10E-04	89.68726	0.012833	9.16E-03	7.59826	9.16E-03
Aerosol	243,196.5	1.063208	1.52E-04	5,431,488	2,263,120	5,431,488
Water: bulk	1.223846	73,115.91	10.4616	11.55574	11.55566	11.55574
Water	1.044894	62,424.46	8.93184	9.866049	9.866049	9.866049
Particles	34,126.93	10,194.18	1.458607	322,231.6	134,263.2	322,231.6
Fish	8,323.642	497.277	7.12E-02	78,593.08	78,593.08	78,593.08
Soil: Bulk	1,706.66	498,977	71.39482	265,639.6	177,064.6	265,639.6
Air	4.10E-04	2.40E-02	3.43E-06	6.39E-02	52.95398	6.39E-02
Water	1.044894	91.64887	1.31E-02	162.6365	162.6365	162.6365
Solid	3412.693	498,885.3	71.3817	531,181.6	221,325.7	531,181.6
Sediment: Bulk	1365.913	12,6714.4	18.1306	160,214.5	125,167.6	160,214.5
Water	1.044894	77.54703	1.11E-02	122.5607	122.5607	122.5607
Solid	6825.386	126,636.8	18.1195	800,582.2	333,575.9	800,582.2

Fugacity (Pa) of fluoranthene for the bulk compartments are shown in Figure 9-7(A) and with detailed results presented in Table 9-3. The Z value in soil, sediment, air, and water are 1,706.66, 1,365.91, 4.15E-04, and 1.22 mol/m³. Pa, respectively (Table 9-3). The highest fugacity value was found in soil, followed by sediment, air, and water.

Very little fluoranthene was emitted in the air. However, in the air compartment, 1044 kg of fluoranthene was recorded, derived mainly from intermediate exchange between water and soil. Water and soil media released fluoranthene at rates of 60.3 and 8.51 kg/hr to air. The total overall residence time of fluoranthene in the system is 263 hrs, whereas its reaction and advection times were found to be 343 and 1119 hrs, respectively.

Fluoranthene is a six-ringed, highly stable PAH compound that has a long residence time in soil, and is therefore considered as a highly toxic compound (Prager 1998; NRC 2003). Higher residence time of this compound will cause more damage and risk to inhabited flora and fauna in the area. It is reported that following 7 applications of polynuclear aromatic hydrocarbons (containing oily sludge) to soil over a 2-yr period, that was then monitored for an additional 1½ yrs, fluoranthene residue in the soil at the end of the 2-year remediation period was decreased by 39% in the following 1½ yrs (Prager 1998; NIOSH 1977). In the sludge remediation experiment, 4.7% of the applied fluoranthene remained after 3½ years (NIOSH 1977).

The water component has three sub-compartments; suspended particles, fish, and exclusively water. The fluoranthene concentrations among these three compartments are 10,194.18, 497.27, and 73,115.91 kg, respectively (Table 9-3). In this simulation, a significant portion of fluoranthene accumulated in the fish sub-compartment. This finding requires further study to examine which areas of the body are responsible for the accumulation of fluoranthene and how readily this compound is broken down by fish. Additional questions that should be explored are as to what is the bio-concentration and residence time of fluoranthene in fish and what is the uptake and loss rate in fish body and its different organs?

The second highest amount of fluoranthene was found in the suspended particles of water (Table 9-3). These particles are mainly comprised of phytoplankton, zooplankton, and silt contents. In this study area, the silt concentration is comparatively low due to insignificant discharge of silt from the river networks. Therefore, the suspended particles are mainly composed of the plankton communities in the water, which are primary components of the food chain. While it is reported that marine bacteria are capable of degrading fluoranthene, this model output cannot clearly identify qualitative and quantitative data regarding effects on the plankton

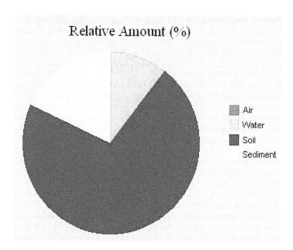

Figure 9-8 Relative amount of fluoranthene into air, water, soil, and sediment compartments.

Table 9-4 Calculated intermediate transport of fluoranthene

Intermediate Transport	Half-time hr	Equiv. Flow (m^3/h)	Rate Trans (Kg/h)
Air to water	0.367174	155,709.6	1.44376
Air to soil	2.996098	19,082.32	0.176934
Water to air	1432.174	25.80691	0.298218
Water to sediments	164.6722	224.4459	2.593639
Soil to air	266,567.4	4.12E-05	0.010934
Soil to water	57,055.54	1.92E-04	5.11E-02
Sediment to water	268.3835	1.72E-02	2.75796

community and the ability and rate of biodegradation. Prager (1998) reported that in crude oil enriched with fluoranthene, incubated with coastal sediment in a flowing-seawater system, 1.9–2.4% of the chemical was removed per week, translating to a half-life of 143–182 days or a degradation rate of 2 ng/g-soil/hr at a concentration of 10 µg/g. However, in a pilot wastewater treatment plant, no fluoranthene was lost due to biodegradation (KOECT 1984). This simulation study can provide a biodegradation rate value of this compound.

The transport rates, half-time in intermediate transport and equivalent flow for fluoranthene, are calculated and shown in Table 9-4. The highest transport rate for fluoranthene has been found to be between air and water. This transport pattern of oil compounds in different compartments and sub-compartments contributes to the understanding of what components may be most adversely affected in the case of a real spill. Results such as these will help to develop management and response plans accordingly. If it is known in which compartment the spill has occurred, it will be easier to anticipate the effect level by observing the transport rates. For example, if a spill occurred in the soil, it is less dangerous than a spill in water, because the transport rate between oil to other media is less than the water to media. In considering the transport rate, managers can take precautionary measure to determine vulnerability of marine zones.

9.5.3 Uptake of oil compounds by fish

The compartmental model found 497.27 kg of fluoranthene accumulated in the fish sub-compartment. Considering the toxicity level of this compound (Table 9-5), this finding requires further study to know what might happen once it enters the fish body. To find out the answer to this question, a simulations test was

Table 9-5 Toxicity level of different marine organisms

Name of Species	LC$_{50}$ /EC50	Source
Mysid shrimp	40 mg/L/96 hr (static conditions unmeasured)	USEPA 1980
Bluegill	3,980 mg/L/96 hr (static conditions unmeasured)	USEPA 1980
Polychaete	500 mg/L/96 hr (static conditions unmeasured)	USEPA 1980
Selenastrum capricornutum (alga)	54,400 mg/L/96 hr, toxic effect: cell numbers; 54,600 mg/L/96 hr, toxic effect: chlorophyll a (static conditions unmeasured)	Verschueren 1983
Skeletonema costatum (alga)	45,000 mg/L/96 hr, toxic effect: chlorophylla; 45,600 mg/L/96 hr, toxic effect: cell numbers (static conditions unmeasured)	USEPA 1980
Cyprinodon variegates (sheepshead minnow)	560,000 mg/L/96 hr (static conditions unmeasured)	USEPA 1980

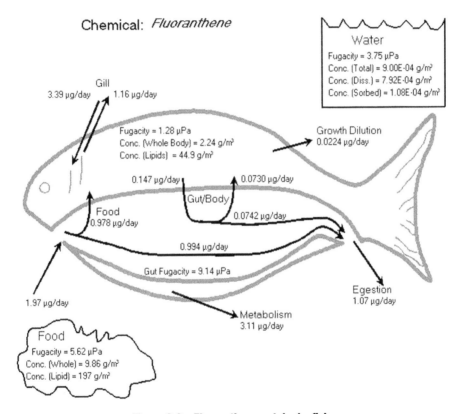

Figure 9-9 Fluoranthene uptake by fish.

run using the "Fish model". Results can be found in Figure 9-9. In this figure, the daily uptake and loss rates are shown using different arrow signs. Generally, fish uptake of any toxic compound is through the mouth or gills. Total fluoranthene uptake was estimated to be 5.357899 µg/day, where 3.38 µg enters via the mouth, and the remaining 1.97 µg crosses the gills.

Following uptake, compounds are either released through various excretion processes or metabolized by the fish. In this study, the loss of the fluoranthene was identified as release through gills, metabolism, growth dilution, and egestion (gross). Total losses were estimated at 5.35 µg/day, and these losses through the gills, metabolism, growth dilution, and egestion were 1.15, 3.11, 2.24, and 1.06 µg/day respectively.

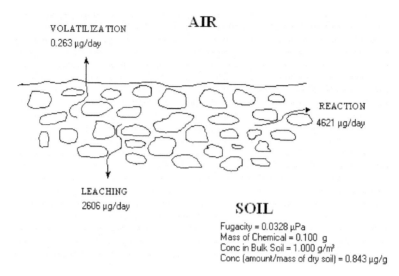

Figure 9-10   Pictorial view of fluoranthene transmission to soil.

9.5.4   Soil contamination

Figure 9-10 shows the three different transformation process oil compounds in the soil, volatilization, leaching, and reaction. This simulation test shows that when 1.0 gm/m³ of oil are released into the soil from the oil platform decommissioning activities, the rate of leaching, volatilization to air, and reaction rate are 2606 ug/day, 0.263 ug/day, and 4621 ug/day, respectively. Among all the processes, the reaction process transformed the highest amount of fluoranthene. The leaching process is the second largest and volatilization is lowest amount.

9.6   Decommissioning Regulations

Until recently, the term "decommissioning" was not even used with regard to the termination of offshore oil activities. International regulation on such activities was first seen in the 1958 Geneva Convention on the Continental Shelf. Further regulations on the "abandonment" of offshore oil structures were developed, such as the 1982 United Nations Convention on Law of the Sea (UNCLOS), the 1989 International Maritime Organization (IMO), and the 1992 Convention of the Protection of the Marine Environment of the North-East Atlantic (OSPAR). The regulations set by these agencies are outlined below (Hamzah 2003).

9.6.1   International laws

9.6.1.1   1958 Geneva Convention on the Continental Shelf

This convention called for the complete removal of any installation that is abandoned or disused. It also outlines the responsibilities and duties of the states conducting oil and gas activities with regard to the continental shelf. Specifically, these activities are to be done in a way that does not interfere with the rights of other states. However, this convention does not contain specific regulations for the methods of disposing of abandoned or disused structures.

9.6.1.2   1982 United Nations Convention on Law of the Sea (UNCLOS)

This international treaty on ocean governance came into force in 1994. Article 60(3) of this convention states:

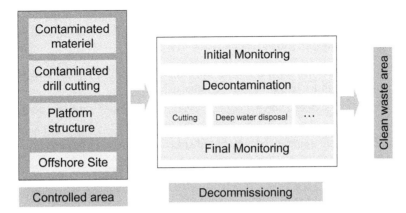

Figure 9-11   **Decommissioning process of offshore platforms that shows how a contaminated site is cleaned up.**

Any installations or structures which are abandoned or disused shall be removed to ensure safety of navigation, taking into account any generally accepted international standards established in this regard by the competent international organization. Such removal shall also have due regard to fishing, the protection of the marine environment and the rights and duties of other States. Appropriate publicity shall be given to the depth, position and dimensions of any installations or structures not entirely removed.

(UNCLOS)

This is to say that partial removal of marine structures may be permitted so long as the remaining structure does not have adverse effects on the natural environment, rights and duties of states, and navigational safety. This convention was the first to address the issue of sustainability with regard to disposal of offshore structures.

9.6.1.3   1989 International Maritime Organization (IMO)

These guidelines and standards specify that the decision to allow an offshore installation, structure, or parts thereof, to remain on the seabed should be based on a case-by-case evaluation (CNSOPB 2004). Matters to evaluate include, but are not limited to potential effects on the marine environment, on navigation or other users of the sea, and potential future effects relating to deterioration of materials. The factors have been considered as part of the EA of the Cohasset Phase II Decommissioning Project. Other factors, such as costs and risk of injury to personnel, have not received detailed consideration as part of the EA but will receive additional consideration by the CNSOPB as part of its regulatory responsibility.

The IMO guidelines and standards require monitoring of the accumulation and deterioration of material left on the seabed, to ensure there is no subsequent adverse impact on navigation, other uses of the sea, or the marine environment. This requirement is reflected in the EA by the requirement for a monitoring program at the site. Also among the standards in the IMO document is the requirement for the entire removal of abandoned or disused installations or structures standing in less than 75 m of water and weighing less than 4000 tonnes.

9.6.1.4   1992 Convention of the Protection of the Marine Environment of the North-East Atlantic (OSPAR)

OSPAR is a regional treaty, only applicable in Europe, specifically to activities in the North Sea. The regulations for the decommissioning of offshore structures set by this convention are more stringent than the

IMO guidelines (Hamzah 2003) and currently assumes the complete removal of the structure and transport to shore for decommissioning, with limited allowances for alternatives (Ekins et al. 2006). Indeed, the OSPAR Decision 98/3 calls for:

(i) the complete removal and onshore reuse, recycling, or disposal of all topside structures and sub-structures or jackets weighing less than 10,000 tonnes;

(ii) the complete removal of sub-structures weighing over 10,000 tonnes, with the possibility to leave footings in place; and

(iii) a consideration to derogate concrete gravity based structures, floating concrete installations, and any concrete anchor-base (*ibid*).

9.6.2   Specific national standards

Given that this is the proposed project in the first decommissioning of an offshore oil facility in Canada, international experience with decommissioning, primarily from the North Sea and the United States, was reviewed by RAs. Key international decommissioning practices pertinent to the proposed project are summarized below.

Norway: All platforms are required to be removed, except those over 10,000 tonnes may be considered for *in-situ* disposal. A preferred option is to abandon pipelines and associated materials *in situ* provided they do not impede other users of the sea (Norwegian Petroleum Directorate 2003).

United Kingdom: In the United Kingdom, complete decommissioning processes are set by the Department of Trade and Industry (DTI). However, a derogation under OSPAR must occur with international consultation (Ekins et al. 2006). Specific to the disposal methods, all platforms are required to be removed, except those over 10,000 tonnes may be considered for *in-situ* disposal. Pipelines are addressed case by case. Major pipelines are candidates for abandonment *in situ*. Small diameter flowlines that are neither trenched nor buried should normally be removed (DTI 2000).

United States: The Minerals Management Service (MMS) of the United States requires that oil platform structures be removed to a depth of 15 feet below the mud-line within 1 year of an OCS lease termination (Schroeder and Love 2004). However, this requirement can be altered to allow for artificial reef structures. Also, pipelines and associated materials can be abandoned *in situ* provided they do not constitute a hazard to navigation or commercial fishing (MMS 2002).

9.6.2.1   Application of Canadian rules and regulations

Department of Fisheries and Ocean (DFO) and Environment Canada (EC), primarily have mandates relating to the maintenance and health of the natural environment and the decommissioning of offshore oil structures (CNSOPB 2004). The EC is the leading federal department in promoting a variety of policies and programs concerning the environment, including the *Toxic Substances Management Policy* and *Pollution Prevention – A Federal Strategy for Action* (CNSOPB 2004).

The Department of Fisheries and Ocean Canada (DFO) administers a number of statutes, including the *Fisheries Act* and *Oceans Act*. Section 35 of the *Fisheries Act* protects the fish habitat from harmful alteration, disruption, or destruction from works or undertakings (CNSOPB 2004). However, the Act does allow for harmful alteration, disruption, or destruction if it has been authorized by the Minister of Fisheries. Marine mammals are protected under the *Marine Mammal Regulations of the Fisheries Act*.

9.6.3   Canada-Nova Scotian Offshore Petroleum Board (CNSOPB) Law

The CNSOPB was established in 1990 following proclamation of the *Canada-Nova Scotia Offshore Petroleum Resources Accord Implementation Act, S.C.* 1988, c.28 by the federal government and the

Table 9-6 Involvement of federal agencies with decommissioning processes

Table 9-6 Involvement of federal agencies with decommissioning processes

Agency or department	Responsible authority	No role
CNSOPB	X	
Environment Canada	X	
DFO	X	
Health Canada		X
National Energy Board		X
Industry Canada		X
Department of National Defense		X
Transport Canada		X

Canada-Nova Scotia Offshore Petroleum Resources Accord Implementation (Nova Scotia) Act, SNS, 1987, c.3 by the provincial government (CNSOPB 2004).

The CNSOPB is responsible for:

- the enhancement of safe working conditions for offshore operations;

- protection of the environment during offshore petroleum activities;

- management and conservation of offshore petroleum resources;

- ensuring compliance with the provisions of the Accord Implementation Acts, that deal with Canada-Nova Scotia employment and industrial benefits;

- rights issuance and management, Environmental Screening Report Cohasset Phase II Decommissioning 13; and

- resource evaluation, data collection, duration, and distribution.

Table 9-6 illustrates how various federal departments are involved with the decommissioning processes of offshore oil and gas platforms.

9.7 Current Practices of Decommissioning

Major steps of offshore oil and gas operations are seismic, drilling, production, transportation, and decommissioning (Figure 9-12). The main focus of this research is the decommission phase of the offshore supply chain. The input and output of every phase are also shown in this figure. Three major input and six types of wastes outputs are identified. Current practices of offshore platform decommissioning are discussed in the following subsection.

The present decommissioning practice is considered as the only "environmentally appropriate" way to remove offshore oil and gas platforms. However, decommissioning operations take several months (Figure 9-12), but the whole process can take as long as three years. To complete the process, the operator of an offshore oil and gas installation has to plan, gain government approval, and implement the removal, disposal, or reuse of a structure after completion of production. There are three stages in the decommissioning process – planning, permitting, and implementation. The major stages of offshore platform decommission can be summarized as (UKOOA 2005):

- Different decommissioning options are developed, assessed, and selected by balancing environmental factors, cost, technical feasibility, health and safety, and public acceptability factors;

- The operator applies to the government to cease production, having proved the reservoir is no longer viable. The government grants a Cessation of Production permit or "COP". The wells are then securely plugged deep below the surface;

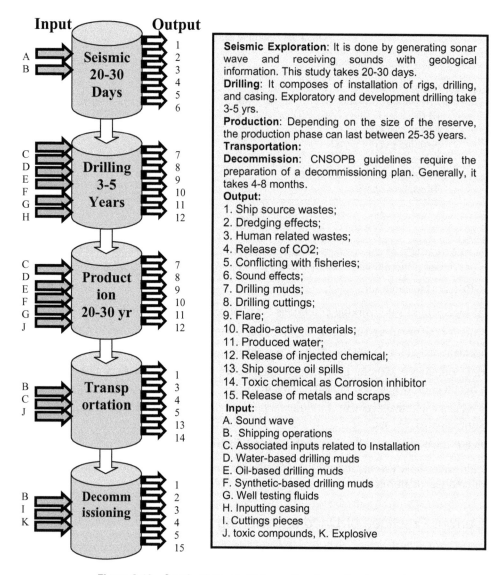

Figure 9-12 Supply chain of offshore oil and gas development.

- The operator gains government approval to proceed with its recommended decommissioning option and offshore operations begin to remove all or parts of the structure to shore;
- The parts of the structure removed to shore are then reused, recycled, or disposed of.

Detailed phases of offshore operations are presented in Figure 9-13. This flow chart shows different components of decommissioning. In the figure it is shown that oil and gas processing equipments and piping are completely removed. The decommissioning process considers the total pipelines run from all platforms either to shore or to other platforms that collect the oil or gas. They are generally shipped to shore for disposal.

The deck and jacket of a rig are the most concerning parts for disposal. There is a strict legal framework of national, regional, and international regulations governing how operators decommission disused offshore

Table 9-7   Decommissioning scenarios

Entities	Disposal Alternatives				
I Topsides	**A** Jacket Topsides: Lift and transport to shore for recycling Tank Topsides: Lift and transport to shore for recycling			**B** Jacket Topsides: Lift and transport to shore for recycling Tank Topsides Remove at shore and recycle after tow shore on substructure	
II Substructures	**A** Jackets: Reef in-place Tank: Reef in place	**B** Jackets: Reef at Tank. Tank: Reef in place	**C** Jackets: Recycle onshore Tank: Leave in-place	**D** Jackets: Recycle onshore Tank: Refloat and deposit in deep water	**E** Jackets: Recycle onshore Tank: Refloat and recycle
III Pipelines	**A** Remove to shore for recycling of materials		**B** Leave buried in-place		
IV Cuttings	**A** Slurrification and re-injection	**B** Remove to shore for disposal	**C** Leave in-place	**D** Cover with gravel	
V Seabed	**A** Remove debris		**B** Leave debris in-place		

facilities (UKOOA 2005). Under current regulatory requirements, more that 90% of the structures are needed to be completely removed. The removed portions are reused as platforms or disposed of onshore. At present, a more flexible and phased approach is used. It suggested immediate and total removal of offshore structures (mainly platforms) weighing up to 4000 tons in the areas with depths less than 75 m, and after 1998 – at depths less than 100 m. The rest, 10%, which comprise very large and heavy steel or concrete installations, are allowed to be partially removed, which is known as toppling.

In deeper waters, removing only the upper parts from above the sea surface to 55 m depth and leaving the remaining structure in place is permitted (Patin 1999). The removed fragments can be either transported to shore or buried in the seabed. This approach considers the possibility of secondary use of abandoned offshore platforms for other purposes.

At present, platform decommissioning alternatives fall into four general categories – complete removal, partial removal, toppling, and leave-in-place (Figure 9-13). Some studies propose the decommissioning alternatives are to leave partial or platform on site. These develop artificial reefs, providing a substrate for marine organisms.

Table 9-7 details a decommissioning scenario of different components of an offshore platform. Considering some different alternatives, the decommissioning of a given platform will follow up to 160 scenarios: $2 \times 5 \times 2 \times 4 \times 2$. An example of a scenario that could be recommended is the shaded area A,C,B,C,A. This is the privileged Ekofisk Group in the Ekofisk Area (the Norwegian Offshore area). The selection criteria may be technical, safety, environmental, social, and economic aspects of each Disposal Alternative, as well of the needs of other users of the sea and the physical and operational limitations.

Knowing how much material to handle in a decommissioning operation will help estimate the operations cost. Some cost estimation models are based on the weight of the structure. Each platform is unique, but we may define a common offshore platform in the middle of the size range, which may contain materials as set out in Table 9-8.

Offshore oil platforms are composed of several portions, jacket, leg, deck, conductors, topside equipment, and bottom-side equipment. The jacket is the main supporting structure of the rig and holds the deck and topside equipment. The legs or jacket legs provide a foundation for the platform, acting much like the roots of a tree. Decks are placed on top of the jacket to house and provide space for drilling and processing equip-

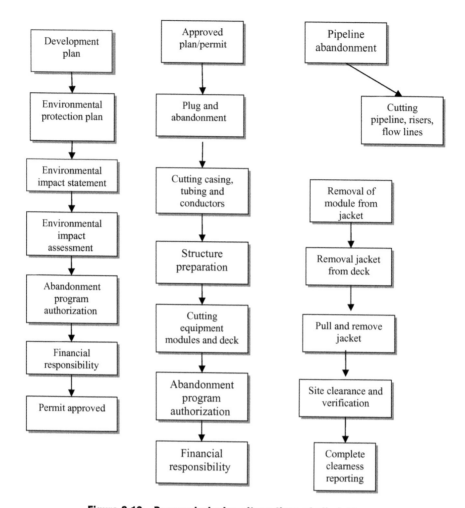

Figure 9-13   Decommissioning alternatives of oil platforms.

ment, a heliport, living quarters, and other infrastructure (Kaisera and Byrd 2005). Topside equipment such as compressors, cranes, drills, heat exchangers, meters, power generation units, pumps, separators, scrubbers, and tanks are also kept within the decks. Conductors are used to carry the oil and gas to the surface and finally, the bottomside houses equipment such as cables, manifolds, pipelines, flowlines, risers, umbilicals, and wellheads (Kaisera and Byrd 2005).

Figure 9-13 shows the different steps of offshore structure decommission. Operational activities can take just a few months, but the administrative process from planning to site clean-up takes approximately three years. The process includes planning, environmental screening, environmental impact assessment, abandonment program authorization, financial responsibility, and authorization approval. After receiving approval from the proper authority, the decommissioning activities are carried out. This generally starts with unplugging the well and the parts of the structure can then be removed. There are many different ways to reuse, recycle, or dispose of the pieces (Abraham 2001).

At the end of production, a well is abandoned by injecting cement plugs downhole to seal the wellbore to prevent future leakage. Cement is placed across the open perforations and squeezed into the formation to seal off all production intervals and to protect aquifers (Kaisera and Byrd 2005). After cementing, the production casing is cut off and removed from the top of the cement. A cement plug is placed over the casing stub. The rest of the casing strings over the plug, which usually extend to the surface, are cut off. After plugging the well and removing the casing tube, the conductors are cut off. Generally, a sand cutter, a mechanical cutting tool, or explosives, are used to dismantle the conductors of the oil and gas platform. In a typical operation, the tubing and production casings are cut using a jet cutter and a small, less than 5 lb, explosive blast, and the strings are cut using a mechanical cutter (Kaisera and Byrd 2005). Finally, other parts of the platform are dismantled using different mechanical processes, where particularly explosives are used.

Currently, the different parts of the jacket and decks are either reused or disposed of. Figure 9-14 shows the decommissioning alternatives of oil and gas platforms. For example, the jacket and deck can be left-in, eliminating three complete processes, complete removal, partial removal, and toppling. Decommissioning process alternatives are discussed in the following sections.

9.7.1   Reuse

The oil and gas platforms are large structures and are expensive to build. After oil and gas extraction activities are completed, the structure, or part of it, is sometimes reused. The reuse of the structure as a future oil and gas platform is the best option, but it is necessary to consider whether it has sufficient structural integrity. Abraham (2001) suggests using the remaining structures as prisons, military outposts, fish farms, casinos, luxury hotels, waste disposal facilities, marine research facilities, wind and wave action electricity genera-

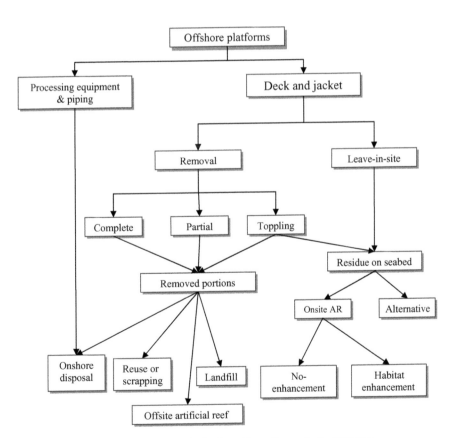

Figure 9-14   Decommissioning alternatives of oil platforms.

tion plants, chemical plants, nuclear power plants, communications facilities, navigation sites, and meteorological centers.

9.7.2   Removal

According to current legal requirements, offshore oil and gas facilities must be removed. The removal of the structure can be partial, complete, or toppled down. Figure 9-15 shows three different methods of removal. These rigorous mechanical processes can take a few weeks to a few months. The removal process is done mechanically, cutting the structure into manageable pieces. The use of explosives is also common in the dismantling process, but can cause many environmental problems (Kaiser and Byrd- 2005). Explosives have been known to kill and scatter turtles and fish (Abraham 2001). These negative effects on fish and the corresponding effects on fisheries has been the focus of attention for both government and fishing groups. Complete offshore structure removal activities create other environmental issues such as the alteration of seabottom ecology as a result of dragging rigs to shore. The removal process is also highly energy consuming as four different types of vessels are required to be used at the same time. Finally, the removal processes poses human health hazards, including exposure to toxic and dangerous materials such as asbestos, methane, and radioactive materials as well as the risk of fatalities and injuries (Abraham 2001).

9.7.3   Residue on the seabed

When partial removal takes place, portions of the platforms, such as jackets of bases, remain on the seafloor. Even after the complete removal of major structures, many small parts are left, including unused pipes and stabilizer mattresses, which are typically made of concrete. It is reported that these materials left on the seafloor affect the normal benthic habitat (Abraham 2001). The presence of larger pieces left behind may cause mechanical damage to fishing gear and can create obstacles in shipping navigation. Abandoned offshore oil and gas facilities on the seafloor also create environmental issues as the pieces may contain hazardous wastes such as heavy metal sludge, PCB fluids, halon gases, and hydrocarbon compounds. Some of these compounds have been found to remain for a period as long as ten years in the marine environment (Wells et al. 1997; Khan and Islam 2005d,f). The effects of these compounds on marine organisms can be chronic or lethal, depending on their concentrations (Wells et al. 2005).

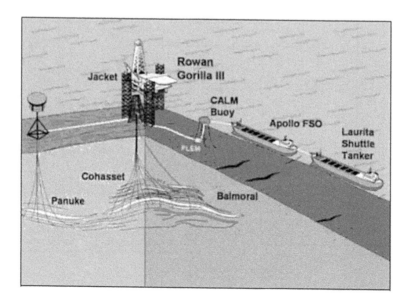

Figure 9-15   Schematic diagram of Cohasset Project on the Scotian Shelf, Canada (CNSOPB 2004).

9.7.4 Artificial reefs

Artificial reefs have become a popular coastal management practice in many countries as they benefit sessile species, benthic fish, and algal growth (Baine 2001). To establish a reef, different shapes and structures are made with concrete, rocks, vessels, plastics, wood, and steel. It is reported that using decommissioned offshore facilities to create artificial reefs in the Gulf of Mexico have benefited recreational fishermen, commercial fishermen, and recreational divers (Abraham 2001). However, unplanned and unjustified development of artificial reefs might create wastes on the seafloor, so the materials used in the composition of the rig must be considered in order in ensure environmental health.

9.8 Case Studies

In this case study, the Cohasset Decommission Project on the Scotian Shelf, Canada is selected. The project is located approximately 256 km southeast of Halifax and approximately 41 km from Sable Island. Initially, the project was approved in 1990 and was Canada's first offshore oil development when it began production in 1992 (CNSOPB 2004). Production continued until 1999, by which time 7.1 million m^3 (44.5 million barrels) of oil had been produced (CNSOPB 2004). The Cohasset Decommissioning Project is separated into two phases (Table 9-8), activities of phase I and phase II.

9.8.1 Description of equipment

The amount of materials required to construct the offshore oil and gas production facilities is immense. Material includes steel, pipe, tubing, and concrete. The Cohasset is the part of Scotian Offshore Energy Project (SOEP), located 200 km off Nova Scotia, 10–40 km north of the Scotian Shelf. For example, the SOEP project used more than 900,000 m of pipe, 109 million kg of steel, 53,000 m of chrome tubing, and 160 million kg of concrete (Khan and Islam 2006c). Table 9-9, shows the amount of materials used in the SOEP and Cohasset Project. The Cohasset structure consists of two platforms and subsea equipment, which are described in the following sub-section.

9.8.1.1 Platforms

According to CNSOPB (2004), the Panuke wellhead jacket weighs approximately 1600 tonnes and is a steel tripod type structure with a 4-m-diameter central column, which houses 5 24-inch conductors, 2 risers, and

Table 9-8   Detailed description of first and second phases of Cohasset Decommissioning Project

Phase I	Phase II
Approved by CNSOPB 2000	Began in the Summer 2003
– Well abandonment	– The Partial Removal Option and the total removal of platform
– Flushing and cleaning of process and utility systems	
– Ensuring all platforms, installations, vessels, and pipeworks were gas and oil free	– both options would involve the removal and reuse, recycling, or disposal of platforms on-land, but differ in their treatment of subsea equipment
– Preserving fixed components for future use or removal	– The Partial Removal Option, which is preferred by the proponent, would treat, and abandon *in situ*, offshore subsea equipment not posing a snagging hazard
– Removing the CALM Buoy	
– Demobilizing and removing mobile components	
– Depressurizing and de-energizing platforms and subsea facilities	– The Total Removal Option would involve the removal and on-land management of all subsea equipment

Adapted from Ekins et al. (2006)

**Table 9-9   The amount of materials required to construct
the production facilities**

Sable Islam Offshore Project		Cohasset Project	
Materials	Weight (kgs)	Materials	Weight (Tonnes)
Steel	109 million	Panuke jacket	1,600
Pipe	900,000	Panuke deck	750
Tubing	53,000 m	Cohasset jacket	1,900
Concrete	160 million	Cohasset deck	350

2 J-tubes. This platform consists of 3 skirt piles, which are 60 inches in diameter, and 2 docking piles. They also reported that the deck structure weighs approximately 750 tonnes and has 4 levels, namely lower, mezzanine, top deck, and helideck.

The Cohasset wellhead jacket weighs approximately 1900 tonnes and is a steel 4-legged structure that housed 10 24-inch conductors (now removed), 4 risers, and 2 J-tubes (CNSOPB 2004). There are a total of 4 skirt piles that are 72 inches in diameter. The deck of Cohasset structure weighs approximately 350 tonnes and has 3 levels, namely lower, mezzanine, and top decks (CNSOPB 2004).

All the platform-associated material of Panuke and Cohasset will be removed and taken to onshore as part of both decommissioning options under consideration (CNSOPB 2004). According to the decommissioning plan, the total platform will be removed, piles will be cut at least two meters below the seafloor, well conductor stubs will be removed at least four/tow meters below the seafloor, and the two docking piles will be removed (CNSOPB 2004). Explosives are commonly used in other parts of world, including the United States, but in this decommissioning project CNSOPB strictly advised not to use explosives in the removal of infrastructure. According to CNSOPB (2004), piles will be cut using abrasive cutting techniques.

9.8.1.2   Subsea equipment

According to CNSOP (2004), the Subsea materials includes 2 approximately 10-km-long subsea interfield flowlines that once connected the Panuke and Cohasset platforms, 2 pipeline end manifolds (PLEMs), a power cable approximately 10 km long, 2 approximately 2.5 km-long export flowlines from each of the jackets to each of the PLEMs; and 510 stabilizing concrete mattresses (Table 9-10).

In the case of the partial removal option, either the PLEMs or PLEM topsides would be removed to prevent snagging hazards to fishing equipment. Flowlines and power cables would be disconnected, the ends secured, and abandoned on the seafloor at the site. Stabilization mattresses would also be abandoned. It is noted that flexible flowlines were employed in project activities to facilitate future removal. According to CNSOPB (2004), all subsea equipment would be removed for the Total Removal Option (Figure 9-15).

9.8.1.3   Management of equipment

Approximately 4700 tonnes of platform materials will be handled, recycled, and/or disposed of as part of the Partial Removal Option, while an additional 3200 tonnes of subsea equipment will require management as part of the Total Removal Option. Platform material is anticipated to consist primarily of steel and marine growth. Subsea equipment consists of a variety of materials. For example, mattresses are constructed primarily of concrete, PLEMs of steel, and flowlines of a selection of materials. Recovered materials will be delivered to a third-party contractor for reuse, recycling, or disposal onshore in compliance with applicable legislation and standards.

Table 9-10 Subsea equipment type, description, status, and management options

Equipment type	Descriptions
Interfield production flowlines	8″ inner diameter flexible flowline used for transferring unprocessed production fluids between the two platforms. Approximately 10 km long. Flange-connected to rigid risers at each of the platforms, held in place on the seafloor by stabilization mattresses.
	Contain 31 anodes (84 kg each or total of 2604 kg). Anodes are composed primarily (>99.995%) of zinc, but contain small amounts of other metals, such as aluminum, cadmium, copper, iron, lead, and silicon.
Interfield water injection flowline	Approximately 10 km long. Contains 34 anodes (63 kg each or total of 1242 kg). See above Interfield production flowlines re: anodes
Export flowlines	The Cohasset export flowline is a 6″ ID, approximately 2.3 km flexible line used to transfer crude oil from the production facilities on the rig to the tanker. It is flange-connected to the rigid pipe riser at the Cohasset platform and the PLEM. A similar arrangement is installed at the Panuke platform.
	Contain four anodes (63 kg each or a total of 252 kg). See Interfield production flowlines re: anodes.
Interfield Pirelli Power Communication cable	3″ outer diameter, 110 km, 3.8 KV power cable, laid alongside the two interfield flowlines. Provided electrical power and communications to either platform. Includes fiber-optics control umbilical that allows remote control of the wellheads and manifold.
PLEMs	The PLEMs are located approximately 2.5 km from the respective platforms, and consist of steel structures that house the manifold end of the export flowline. The Panuke PLEM is the largest of the two, measuring 6.2 m × 6.2 m × 1.3 m.
	Two anodes on each PLEM (322 kg each or a total of 1288 kg). Anodes are composed primarily (>99.5%) of zinc, but contain small amounts of other metals, such as aluminum, cadmium, copper, iron, lead, and silicon.
Stabilization mattresses	Vary in design, size, and weight and are composed of concrete sections connected by polypropylene rope. The purpose is to ensure the subsea equipment is not moved or shifted from its location by waves or currents.
	There are 512 mattresses giving a total weight of approximately 1735 tonnes.

9.8.2 *Operating schedule*

CNSOPB (2004) reported that the whole decommission work and the removal of materials have been estimated to take approximately 4 weeks for the Partial Removal Option and between 8 and 15 weeks for the Total Removal Option. The project has been assessed for the period of 2004–2009 inclusive. This timeframe includes a subsea survey of the field to verify that the work program has met its stated objectives.

9.9 Sustainability of Offshore Platform Decommissioning

At present, it is considered that the best solution of offshore platforms is to cut the structures into small manageable pieces, lift them onto barges, and bring them back to shore for reuse, recycling, or disposal (Khan and Islam 2003b; UKOOA 2005). This is also considered as the "only environmental way" to handle the abandoned structures. This process is often a dangerous, lengthy, costly, and weather sensitive procedure (Patin 1999; UKOOA 2005). However, there are better alternatives to obtain ecological economic benefit. The enormous structures can be used as fish shelters/habitats and artificial reefs. However, the industry is currently not considering this alternative, sponsoring instead a number of different projects to develop new technologies that would cut down the time spent offshore by lifting larger pieces of the structure in one go and floating them back to shore. The industry is keen on speeding up the time taken to remove a structure rather than developing a sustainable technique.

The ecological, safety and risks, and economic issues of present practices are considered. In the decommissioning phase an offshore operation has to submit the environmental impact assessment of the decommissioning process. Mostly it is reported that there is no significance environmental impact due to this process. However, if the structures are used in a sustainable way, such as development of artificial rigs or fish shelters, then the ecological consequences are much more acceptable than present practices.

The structure dismantling is more labor-intensive, and involves complex and potentially hazardous operations. The risk analysis suggests that the probabilities of fatal injuries are higher. If the same structure is used as artificial rigs than the risk factor is six times less than present practices. Onshore dismantling also involves potential exposure to the LSA scales, asbestos, etc. and therefore requires strict health and safety controls throughout. The dismantle process also involves much greater engineering complexity of rotating the rig and dismantling. It is reflected in the different initial cost estimates – £12 million for deep water disposal and £46 million for horizontal dismantling (UKOOA 2005).

9.9.1 Sustainability of current practices

The main objectives of current decommissioning practices are to fulfill the regulatory requirements and the issue of sustainability is not taken into consideration. Recently, sustainability issues in energy development have been introduced by Khan and Islam (2005c) and Khan et al. (2005b, 2006a, 2006b). They proposed that a technology is sustainable if it assimilates the long-term functional efficiency of natural systems. As Nature is infinitely functional, technology must be functional for an infinite duration. This approach to technology has been applied effectively in renewable (Khan et al. 2006b) and non-renewable energy development and technological evaluations (Khan and Islam, 2005c). To achieve overall sustainability in oil and gas operations, from technology development to company operation, the "Olympic" green supply chain model has been proposed by Khan et al. (2006c). Figure 9-16 shows that a sustainable technology will be beneficial in terms of total environmental costs, with and unsustainable technology going in the opposite direction.

Table 9-12 shows the daily and total project air emissions of PM, NOx, SO_2, VO_2, VOC, and CO_2 from the Cohasset Decommission Project on the Scotian Shelf.

These are direct emissions from the decommissioning activities. In addition, total hazardous wastes, such as LSA scale, heavy metal sludge, PCB fluids, halon gases, and oil compounds, have been reported in the project location (Abraham 2001). In terms of partial removal of structures, facilities are painted and con-

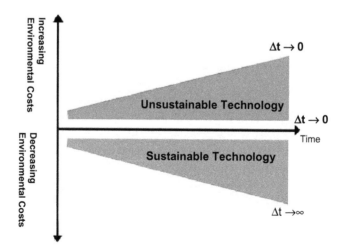

Figure 9-16   Direction of sustainable and unsustainable technology.

Table 9-11   Materials contained in a mid-size large steel structure

	Steel Tons	Aluminum Tons	Copper Tons	Non-metals Tons	Total Tons
Topside	20,000	20	200	300	20,520
Jacket	9,000	500		700	10,200
Footings	10,000	300		1,000	11,300
Total structure	39,000	820	200	2,000	42,020

Table 9-12   Emissions associated with maximum level of activity (CNSOPB 2004)

Emission	Project emissions (tonnes)	Daily emissions (tonnes/day)
PM	14	0.1
NOx	499	5
SO_2	58	0.6
VOC	31	0.3
CO_2	37,000	35

structed using many different components, all of which decompose at different rates with varying effects on marine species and ecosystems. The complete removal of offshore oil structures is more acceptable when considering the cumulative environmental effects on the marine ecosystem (Abraham 2001).

However, Abraham (2001) argues that the total removal of offshore structures has adverse effects on the health and safety of workers. It is argued that exposure to toxic and dangerous materials such as asbestos, methane, and radioactive materials, and the risk of fatalities and injuries that may arise from dangerous work, such as diving and the use of explosives in the cleaning and dismantling of an offshore structure, are clear hazards to human welfare. There are also the environmental risks of bringing sections of offshore facilities to land for disposal, including the possibility of sinking material during transport and air pollution resulting from transportation. Such risks may be greater than the certain environmental consequences of leaving all or a portion of a facility on the seafloor (Abraham 2001). However, in order to determine the most sustainable decommissioning practice, both short- and long-term impacts need to be assessed.

Table 9-13 shows the environmental, ecological, and technological factors that might affect the valued ecosystem components in the case of the Cohasset Decommission Project on the Scotian Shelf. These assumptions are based on the ecological and technological status proposed by the current decommissioning project framework (CNSOPB 2004).

9.10   Alternative Approaches

To achieve sustainability in the decommissioning process, new and innovative methods should be developed. For example, the large oil rigs can be converted into marine life habitat spaces. Furthermore, the use of explosives in the dismantling of oil rigs should be avoided and less environmentally destructive cutting techniques should be used. Considering the materials used to make the platforms, the rigs could be converted into reefs wherever suitable. It has been reported that the deployment of an artificial reef structure on the seabed has an immediate positive impact on habitat restoration (Wilding and Sayer 2002). According to

Table 9-13   Potential impacts of Cohasset Decommissioning Project

VEC	Descriptions
Air quality	The principal sources of direct emissions will be power requirements of vessels: a mid-sized construction vessel – 7,000 hp (assumed to be mostly in maneuvering mode), heavy lift barge – 15,000 hp (assumed operating capacity 10%) anchor handling tugboat – 1,500 hp (assumed to be mostly in maneuvering mode), and transportation vessel – 12,000 hp (assumed to be mostly in maneuvering mode).
Water quality	Project-related discharges to the marine environment are considered in the proponent's EA in terms of predicted quantities and general characteristics, as well as applicable standards. The oil in water concentration was tested for flowlines and risers and determined to be less than 5 mg/L of oil in water in all cases. This concentration is within the standard for permitted oil in water content of 15 mg/L5.
Sediment quality	For the Total Removal Option, based on available data, EC has concluded that most of the sediment to be disposed of at sea is unlikely to be contaminated and therefore not likely to cause significant adverse environmental effects, for the sediment lying in close proximity to the two platforms.
Marine Fish	There are 256 different species of marine fish that might be present within the study area. Although the release of residual water from export flowlines could potentially affect marine fish, the proponent's EA states that any chemicals that may be discharged from the flowlines were approved for discharge and ranked within the less hazardous CNSOPB Offshore Chemical Selection Guidelines (OCSG) to pose little or no risk. Potential impacts on marine fish from project activities can result from a change in water quality, sediment disturbance and loss of habitat. Noise levels associated with project activities are not expected to result in significant harm to fish. Sediment in the water can abrade fish membranes and decrease visibility affecting fish that feed by sight. As it settles it can smother benthic eggs and larvae. The Sable Island Bank is characterized by sandy sediments with high levels of natural disturbance, and because effects are predicted to be short-term and localized, marine fish populations would be expected to recover quickly from any disturbance. The RAs agree that sediment disturbance would be greater for the Total Removal Option, but it is still not expected to have significant effects for marine fish.
Marine benthos	Level of natural disturbance. The benthic community is expected to recover from the disturbance associated with the full removal of subsea structures. The size and dimensions of structures to be removed is such that they are expected to become alternately covered and uncovered over time. The long-term habitat value of these structures is likely to be low under these conditions. The abandonment of project components on the seafloor for the Partial Removal Option could result in leaching of metals or other substances into the water. The materials to be left *in situ* are considered to be fairly inert, and are not expected to lead to significant levels of contamination during decomposition.
Marine birds	The proponent lists: (1) the effects of noise from short-term, localized sources; (2) collisions with vessels; and (3) pollution from potential accidental spills as the main issues related to marine birds. Potential impacts from project activities include oiling from spills, disturbance or displacement from noise, and by vessels or helicopters, and attraction to the project area by lights or discharge of food waste. The impacts of oiling can be considered with respect to murres because this species is common in the project area, susceptible to oiling, and faced by existing cumulative stresses. The attraction of birds to project activities may result from lighting or the discharge of food waste from vessels. Night-flying seabirds, such as storm petrels, are most at risk of attraction to light, particularly during the summer months. In the winter months, dovekies may be attracted to lights under certain environmental conditions. Many species of land birds undertaking an oversea migration are known to be attracted to lights. This attraction could

Table 9-13   Continued

VEC	Descriptions
	lead to collisions with project vessels, predation by other species (e.g., gulls), and use of energy reserves.
Special areas	The proponent commits to following its Code of Practice for Sable Island and thus not coming within 2 km of Sable Island under normal operating conditions. The proponent's EA states that potential impacts of the project to Sable Island include disturbance from an emergency helicopter landing on the island and effects to intertidal habitat and vegetation as a result of an accidental spill. However, it also maintains that the potential for an emergency landing on the island is unlikely given routine maintenance and preventative measures. In the unlikely event of a spill, the proponent indicates that the affected area would be localized and spills are not predicted to reach Sable Island even if winds are blowing toward the island.
	The potential for impacts to the island is minimized given the distance between the project site and Sable Island (i.e., 41 km). In minimizing the potential for interactions with the transiting vessels/helicopters, it is recommended that these remain as far as possible from the island
Commercial fisheries	Cutting pilings to more than 2 m below the seafloor and well conductors to more than 4 m below the seafloor: – removing PLEMs or PLEM topsides to prevent snagging hazards – submitting locations of infrastructure left on the seafloor and/or removed to update hydrographic charts Commercial fisheries gear, interference with fishing vessels and interruption of fishing activity during project activities, and possible loss of fishing grounds. Potential impacts from the Total Removal Option would include interference with fishing vessels and interruption of fishing activity during project activities. The proponent bases much of its analysis for the Partial Removal Option on claims that the flowlines will be self-burying and that although they may become unburied during punctual events such as severe storms, they are expected to bury again after such events. The results of the September 2004 subsea survey inspection indicate that the self-burial process predicted in their earlier submissions is now essentially complete. Since uncertainty remains as to the rate at which covering and uncovering of the subsea equipment occurs and the effect of severe storms on the process, a monitoring and reporting program will be necessary. Compensation for any damage to fishing gear will also be required.
Species at risk	The potential impacts from project activities could be the same for COSEWIC-listed species as for those that are not listed. However, the threshold at which an effect is considered significant could be much lower. For example, an impact on one individual of a species at risk, especially if that species is listed as endangered, may be considered a significant effect if it jeopardizes the species' survival or recovery
Malfunctions and accidental events	The proponent's EA indicates that there are several unplanned situations that might be encountered during decommissioning, including small spills, collisions, and extreme weather conditions. The potential for accidental spills is primarily related to the operation of the heavy-lift barge and support vessels. On-board the vessels and barge are small amounts of marine diesel fuel and lube oil that could be accidentally released.
Marine mammals	Sable Bank is an important foraging area for seals and other marine mammals. While the project area does not overlap with any marine mammal conservation areas (e.g., right whale sanctuaries or the Gully Marine Protected Area) marine mammals are expected to be present. Potential impacts from project activities on marine mammals include oiling from spills, vessel collisions, and disturbance from noise and other waste emissions.

Source: CNSOPB (2004)

Table 9-14   Materials remaining at Brent Spar

Material	Weight (Tonnes)
Steel + other metals	7,570
Sand/scale	90
Aluminum	29
Copper	14
PCBs	0.002
Haematite/concrete	6,800
Hydrocarbons	51
Zinc	14
Other heavy metals	>1
total	14,568

Adapted from Post Note (1995)

several studies, algae begins to grow immediately, sessile organisms will settle, drifting plankton acquire substances that provide shelter, and reef associated fish numbers increase with the growing of their food components (Wilding and Sayer 2002). By establishing armors, a completely new reef-based community can be developed that enhances fisheries as well as ecosystem productivity.

Different materials are used to build the rigs and some of them remain even after decommission of the platforms and other project facilities (Table 9-14). Considering these problems, alternatives materials can be used for different petroleum operations materials. For example, fiber-reinforced plastic (FRP) materials are a good alternative.

Recently, Taheri et al. (2005) proposed a most significant import of materials technology into petroleum operations. According to them, the FRP materials for fabrication of various structural components have increased at an exponential rate in the past few decades, and the technology has not permeated into the petroleum industry to any appreciable degree. Attributes such as lightweight, high specific strength and stiffness, excellent fatigue properties, better corrosion resistance, thermal insulation, vibration damping, and tailorability are some of the properties associated with FRP. They are generally more expensive than traditional materials such as steel and concrete. However, as their use increases, their cost decreases, as has been the case in the past decade. Even at the present time, if the long-term durability of these materials is considered, and compared against the maintenance cost associated with the use of steel liners, it can be appreciated that the higher initial cost of FEP can be easily offset and their use, even today, can be justified. Moreover, by appropriate design and hybridization, using less expensive fibers such as E-glass, along with more expensive fibers, such as carbon, the production cost of composite structures can be further lowered significantly. For instance, the cost of carbon fibers and prepregs has been reduced by more than 50% in the past two decades.

One of the most interesting applications of FRP in recent years, in relation to the oil and gas industry, has been the recent certification of an FRP drilling riser for use in aggressive, high-risk applications that will enable oil production in water depths approaching 3000 m. The riser, which is the pipeline that connects the drilling platform or drill ship on the water surface to the wellbore at the seabed, is made up of individual 15–25 m-long FRP pipes and connected together by metallic joints. The challenges for such pipelines are that they must have adequate strength and must be durable to withstand abrasion and corrosive chemicals while assuring flow. Mooring lines are another application that has been recently identified as the one that can benefit greatly from FRP materials. This is highlighted by a recent call for proposals on the topic by the relevant industry. This follows the continuation of the efforts that have been devoted in the past 20 years in using FRP components to replace traditional materials in topside piping, gratings, ladders, handrails, wellhead enclosures, and buoyancy elements.

Finally, our research group has recently investigated a series of natural fibers that can bring down the cost significantly (Saeed et al. 2003). It should be noted that the price of FRP materials is rapidly decreasing. Indeed, the average cost of FRP dropped from $25/lb in 1988 to around $6/lb in 1998. Today, the average cost of what used to be "specialty fiber" has further dropped to $0.50/lb. Salama of Conoco (see OilOnline), attributes this reduction in cost to improvements in technology and process as well as lower raw materials cost. Salama also believes that FRP applications become more justified for use in tendons in deepwater (6000–12,000 ft), since steel tendons start to be challenged by hydrostatic pressure on the outside, which can cause collapse. This approach sets the stage for converting waste into value-added materials, considered to be the wave of the future.

In a design study presented at the Third International Conference on Composite Materials for Offshore Operations (CM003, sponsored by the Composites Engineering and Applications Center at the University of Houston), Chevron and CSO Aker Maritime (Houston, Texas) demonstrated that an FRP riser system (with composite air cans for buoyancy) can decrease the baseline cost of a truss spar hull design by more than 50%. In the area of composite application for downhole applications, Leighton et al. (1998) and Saltel et al. (1998) suggested an inflatable composite sleeve to repair damaged tubing as well as to shut off undesired perforations. The sleeve is run-in on an electric wireline, inflated to push the composite against the inside of the casing, and then heated to polymerize the resins.

The question may arise why this technology has not penetrated the petroleum industry? There appears to be some widespread myths and misconceptions about FRP materials. They are discussed below.

The oldest misconception is: "The fibers tend to wick liquid, leading to blistering." This used to be the case in the early to mid-1970s, when resin chemistry had not fully matured. However, this myth cannot be taken seriously, when ships, minesweepers, and yachts made of FRP have been in service for more than three decades now. This myth cannot be even remotely justified when considering the existence of the $84 million Visby Corvette, a 73 m-long stealth ship made of carbon-vinyl-ester with a sandwich core, by a Swedish manufacturer. Moreover, Mata-bromobiphenol blister-resistance epoxy resins were developed in the early 1990s (Wang et al. 1991). Polyproplene glycol meleate resins are also proved to resist blistering. Norwood and Hotton (1993) also stated that the use of even general purpose polyester (isophthalic acid-neopontyl glycal-based) are more than adequate for resisting blistering.

The next misconception among engineers unfamiliar with FRP materials is that they are not as strong as steel. FRP materials exhibit much stronger specific strength than steel in its unidirectional form. Typical unidirectional graphite-epoxy has a tensile strength of 1400 MPa (AS-H3501 carbon epoxy, this is a commonly used FRP), as opposed to 250 MPa yield strength and 400 MPa ultimate strength, for A36 structural steel. However, this would be an unfair comparison, because most structures are not made of unidirectional composites, and when fibers are oriented in multi-directions, the overall strength is lower (so that loads from other directions could be resisted effectively). Nevertheless, even that strength, especially in the sense of relative strength, would be much higher than that of steel. The following graph, extracted from the course notes offered by Dick Hadcock of Grammon Aerospace in 1988, illustrate the fact. Figure 9-17 shows the specific properties of composites.

Even in the oil and gas industry, the fiberglass wetwells that are being marketed by L.F. Manufacturing, Inc. of East Giddings, Texas, at a competitive price, form an excellent example of how FRP materials can be used in the petroleum industry. The wetwells are 36 inch diameter – 15-feet-long large pipes, with flexural strength of 56 Ksi, with a wall thickness of merely 0.3 inches.

Because the petroleum industry deals with wells that have rather narrow annuli, another common misconception is that FRP materials will take up more space, and so cannot be used in well tubulars. Here is an example comparing two pipes, one made of a common grade of structural steel (A36), and the other one made of AS-H3501 carbon epoxy, with the following properties:

$$E_{11} = 138 \text{ GPa}$$

$$E_{22} = 8.9 \text{ GPa}; \ \Lambda_{12} = 0.30$$

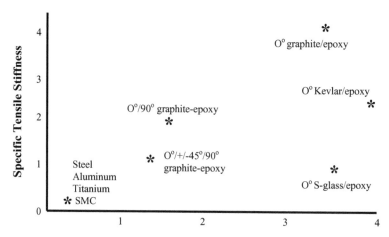

Figure 9-17   Specific properties of composites. From course notes by Dick Hadcock of Grammon Aerospace (1988).

$$G_{12} = 7.1 \text{ GPa}; \ S_L^+ = 1440 \text{ MPa}$$

$$S_L^- = 1440 \text{ MPa}; \ S_T^+ = 51 \text{ MPa}$$

$$S_T^- = 206 \text{ MPa}; \ S_{LT} = 93 \text{ MPa and Specific Gravity of } 1.8$$

where

E_{11} and E_{22} = elastic moduli measured along the fiber and perpendicular to it, respectively

G_{12} = the shear modulus

S = the strength with the subscripts referring to the measured quantity along (L) and transverse (T) to the fiber direction, and the superscripts referring to the direction of the applied load during the measurement of the strength (tension or compression). The pipes were assumed to have a mean diameter of 300 mm (12″) and a length of 12 m.

The FRP pipe was made of 12 layers of 0.125 mm-thick carbon-epoxy (thus having a total thickness of 1.5 mm), with a lay-up sequence of $(\pm 55.3)_{3,S}$, which is considered to be an optimal fiber-orientation for pipes and pressure vessels. This pipe could safely endure 3.2 MPa of pressure, according to first-ply failure based on the Tsai-Wu failure criteria. An equivalent steel pipe, made of structural steel with yield strength of 250 MPa, would have to have a wall-thickness of 1.85 mm to safely endure 3.2 MPa internal pressures, without yielding.

The next misconception arises from the lack of understanding of mechanics of such systems. People assume that connections are a particular problem with FRP materials. Connections are problems in virtually any mechanical system, ranging from biomedical applications to space shuttle technologies. In this respect, more progress has been made in FRP materials than any other. Today, nano-elastomers reinforced adhesive can be purchased with 400% higher toughness than the adhesive produced only two years ago. The properties of adhesives that were produced two years ago simply could not be compared with those produced in the 1980s, because their properties are significantly improved. Today, researchers are working toward improving the adhesives strength and toughness by enclosing whiskers and nano-tubes.

Another misconception attached with FRP arises from the lack of understanding of the creeping behavior of FRP materials. The creep in FRP can be effectively controlled. The following comments should be considered:

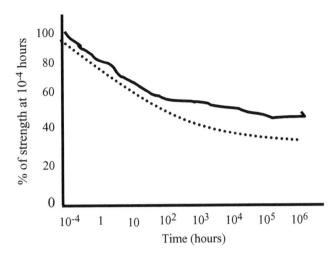

Figure 9-18 Strength reduction with time for FRP materials. From Holmes and Just (1983).

1. Unidirectional carbon exhibit excellent creep resistance (Martine 1993).

2. Such fiber loses only 9% of their modulus (rigidity) in a period of 20 years (Hallaway 1990; Martine 1993).

3. Limitation of initial strain on matrix, when multidirectional FRP is used, should be the design constraint for creep. By limiting the axial strain to 0.2%, the transfer cracking of resin is minimized, in turn improving the creep performance. When selecting resins, the resin should have glass transition temperature 50°C above the service temperature to further reduce the creep.

4. Figure 9-18 for glass GRP (which creeps more than carbon FRP), illustrates that the reduction in strength is not particularly high (Holmes and Just 1983). They report that the level of creep noted in the graph should be what most GRP would experience.

5. FRP materials are tailorable, and economically viable composites by a proper combination of fibers (hybridization) can be designed.

Figure 9-18 shows the strength reduction with time for FRP material. With the proposed technology, opportunities will open up for other novel materials to permeate the petroleum industry it expands into more challenging environments, such as deepwater offshore operation, harsh climate terrain, etc. (Taheri et al. 2005).

9.11 Guidelines for Sustainable Management

The "Olympic" framework for green decommissioning of an offshore platform is introduced to achieve sustainability (Lakhal et al. 2005; Khan et al. 2006c,e). A detailed description of the "Olympic" green supply chain is described in Chapter 2. This concept could be relevant to the offshore platform life cycle. However, in this chapter we will adapt it and apply it for the decommissioning phase. The term "Olympic" is used to characterize:

(*i*) *Five zeros of waste or emissions* (*corresponding to the five circles in the Olympic flag*):

- Zero emissions (air, soil, water, solid waste, hazardous wastes)
- Zero waste of resources (energy, materials, human)
- Zero waste in activities (administration, production, transportation)
- Zero use of toxics
- Zero waste left on the site

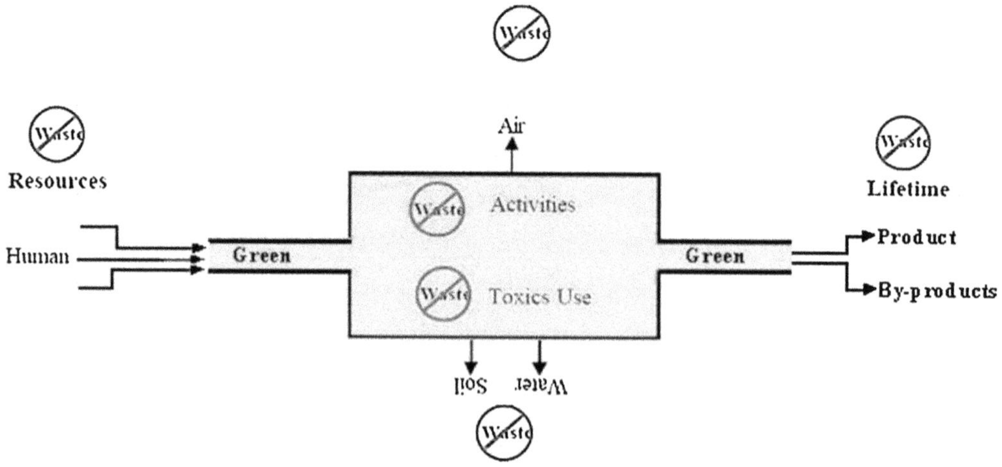

Figure 9-19   The proposed "Olympic" green supply chain for offshore oil and gas operations.

(ii) Green inputs and outputs

Major parameters of an "Olympic" green supply chain is elaborated in respect to offshore platform decommissioning (Figure 9-19). It shows the green input and green output and five zero wastes objective.

9.11.1 (i) Zero emissions (air, soil, water, solid waste, hazardous wastes)

In general, there are many wastes that are released during offshore structure dismantling processes. For example, it is reported that there is a residual oily sludge left in the many used tanks (Post Note 1995). If this sludge is not be pumped out, the seawater will become contaminated by additives (glyoxal for removing hydrogen sulfide) and other contaminants that dissolve in seawater (e.g., oil from the sludge, zinc from sacrificial anodes used to control corrosion of the tank). Some key waste elements of the offshore structure, Brent Spar are shown in Table 9-14 (Post Note 1995). In addition to large amounts of solid structural wastes, there are many other wastes, such as hydrocarbons and PCBs, which can be emitted into seawater, air, and sediments.

The proposed "Olympic" green supply chain makes sure none of these compounds are released in the marine environment. This includes all kind of wastes, which include air, soil, water, solid waste, and hazardous waste. However, the present practices do not consider zero emission. In the planning process, the emission prevention methods should be developed.

9.11.2 (ii) Zero waste of resources (energy, materials, and human)

In this study, it is revealed that the present decommissioning process of offshore oil and gas platform is not efficient with respect to energy, materials, and humans. The common decommissioning technique is to cut down the thousands of tonnes of platform into manageable pieces and bring it onshore for disposal. This process is highly energy consuming. In addition, an enormous oil platform consists of many valuable materials. Table 9-14 shows that a single rig has 7570, 29, 14, and 14 tonnes of steel, aluminum, cupper, and zinc, respectively. These components are not wastes, but valuable resources. There are also huge amounts of hydrocarbons left in unused oil tanks. In the earlier section, it is also stated that the dismantling process is highly risky. Most of the time dynamite or other explosive are used to blow-out or break up the structure. Use of explosives is risky to humans and marine life. There is no clear study on how much marine species can be damaged due to their use. Virtually all fish and many invertebrates associated with platform structure next to the seafloor will die from this first zone of injury (Schroeder and Love 2004). Alternative uses of offshore platforms (AROR), discussed above, will not only save these marine organism, but will also increase fisheries production.

9.11.3 (iii) Zero waste in administration activities, production, and transportation

The present dismantling of offshore platform involves a long administrative process. The operations might take a short time such as couple of months, but the total administrative process takes up to three years. The less time spent by the workers offshore, the less their exposure to risk of injury.

Earlier, it is also discussed that the present decommissioning process is highly expensive. Therefore, it would be more cost-effective to ensure that parts of the structures could be reused, either whole or in part, for other projects rather than being scrapped. In addition, the use of offshore structures as artificial reefs not only reduces dismantling and transportation costs, but also brings other benefits, such as in fisheries productions. According to Dybas (2005), oil companies can save between $400 and $600 million per rig by converting them into artificial reefs instead of offshore decommissioning. Therefore, the conversion of AROR as well as other proposed models is economically profitable. For example, it has been recently reported that existing oil rigs have spawned lush marine habitats that are home to a profusion of rare corals, with 10,000 to 30,000 fish per rig (Dybas 2005).

9.11.4 (iv) Zero use of toxics

There is always the risk of toxic release in the marine ecosystem. For example, in the decommissioning process, the stressful rotation of the buoy into a horizontal position can release contaminated ballast water, which would enter the more sensitive coastal environment. Also, if the structure sank this would create hazards to navigation and fishing. The environmental effects would be physical and localized, such as burial of organisms. Releases of toxic compounds around the rigs are also evident (Section 9.3). Application of the "Olympic" green supply chain considers any toxic releases, which is not considered in the present practices.

9.11.5 (v) Zero waste left on the site

Application of the "Olympic" green supply chain ensures zero wastes in the decommissioned site. The ideal situation is that all the platform parts, topsides, jackets, and footings, should be removed to shore. Figure 9-20 shows the decommissioning process of offshore platforms. The process considers all contaminated areas, including contaminated material and drill cuttings, to go through decommissioning process. After

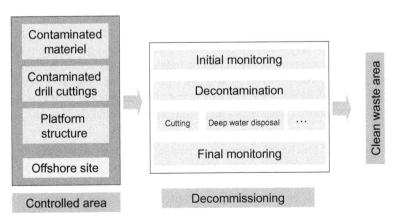

Figure 9-20 **Decommissioning process of offshore platforms that shows how a contaminated site is cleaned up.**

Table 9-15   The amount of materials required to construct the production facilities. Data source: Khan et al. (2006c)

Materials	Quantity
Steel	109 million
Pipe	900,000 meters
Tubing	53,000 meters
Concrete	160 million

decommissioning, the contaminated site will be a clean waste area. This operation could involve around US $27 million (Ekins et al. 2006).

In current dismantling practices, removal of main structure, such as the rig, is the end of the decommissioning process. However, even if the main structure is removed, there often remains associated infrastructure, such as pipelines and cables (UKOOA 2005). The average production period of offshore production is more than 15 years. During this period, an enormous amount of debris falls overboard, including scaffolding poles, cabling, and lumps of concrete. These products are not removed as part of decommission project. However, the principle of the "Olympic" green supply chain is to ensure zero waste left on the site. The main objective is to avoid adversely affecting the marine environment or other users of the sea, such as shipping and fishermen.

Major components left behind after the decommissioning process are pipelines, debris, and drill cuttings. It is reported that production facilities, such as decommissioning of pipelines and cables, is regulated separately from platforms under the Pipeline Safety Regulations 1996 (UKOOA 2005). For decommissioning disused pipelines, there are no international guidelines. However, there are huge amounts of pipes used in production facilities. For instance, the Sable Island Offshore project (SIOP) used 900,000 meters of pipes (Table 9-15). The application of the "Olympic" green supply chain makes sure that these wastes are removed.

Once a structure has been removed, a comprehensive survey of the site is carried out and any material left behind is then picked up and disposed of onshore. Another major amount of waste that should be removed are drill cuttings. It is reported that after removing platforms, large amounts of drilling cutting remain in the North Sea and other parts of the world (Patin 1999; UKOOA 2005). A £6 million industry-wide research project was undertaken to solve this problem. The research suggested that options range from natural degradation for low environmental impact accumulations to covering or retrieval of those with more significant impacts (UKOOA 2005).

9.12 Ecological and Economic Benefit of Artificial Reefs from Oil Rigs

In offshore operations, different types of oilrigs are installed according to water depth. Oil rigs are huge structures weighing thousands of tonnes. By law, abandoned rigs must be decommissioned and removed from marine waters. Generally, these rigs are cut down to manageable size and brought back to shore. The AROR model is proposing to use these rigs as artificial rigs. The main objectives of the AROR are to use the abandoned structures for fisheries yield and production, for recreational activities, to prevent trawling, to repair degraded marine habitats, and overall economic social benefits (Cripps and Ababel 2002; Baine 2001).

Artificial reefs have been used for coastal management (Baine 2001). For establishing a reef, different shapes of structures are made with concrete, rocks, used tiers, vessels, plastics, wood, and steels. Popular prefabricated concrete structures are often used for making an artificial reef. Targeting specific desired organisms to inhabit particular materials can be included. For example, to develop an oyster bed, the natural shells are collected and attached to the rigs. Sessile organisms such as algae, sponges, gorgonians, and other benthic organisms, will then become attached to the deployed structure, in this case on the rigs (Figure 9-21). In

Figure 9-21   Abandoned oil rigs (left) can be converted into an artificial reef (right). Modified from API (2005).

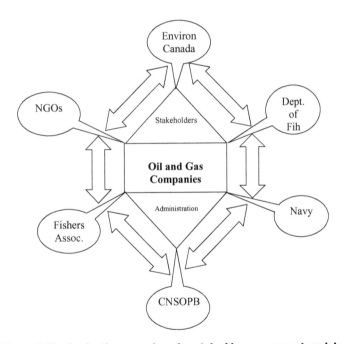

Figure 9-22   Implanting agencies of sustainable management models.

the benthic environment, space and shelter are limited. Therefore, these structures provide shelter for many marine organisms and protect them from predators. Immediately after deploying the artificial rigs, a few organisms will begin to grow in association with other fauna and flora, forming symbiotic and predatory relationships. As a result, a reef-based food chain will develop, with the AROR providing food sources for comparatively large organisms, as well as recreational species, such as crabs, larger fish species, and lobsters.

In the proposed AROR models, the abandoned rigs can be kept in their original sites to establish a reef community or can be transported to another planned site giving a choice of type of seafloor conditions, for example sand or muddy bottom or shallow or deep water areas. It is reported that wherever artificial reefs are established benthic communities become productive. For example, the sand and mud habitat areas may not be as diverse as an original reef community, but they can still produce the productive benthic flora and fauna in establishing AROR.

In summary, it has been reported that deployment of an artificial reef structure on the seabed has an immediate positive impact on habitat restoration (Wilding and Sayer 2002). According to different studies it has been observed that algae begins to grow immediately, sessile organisms start to settle, drifting plankton acquires substances that provide shelter, and reef-associated fish start to increase with the growing of their food components (Wilding and Sayer 2002). By establishing armors, a totally new reef-based community will develop that enhances the fisheries as well as ecosystem productivity. The AROR is one of the most effective ways of managing abandoned rigs in the context of ecosystem improvement.

It is recently reported that existing oil rigs have spawned lush marine habitats that are home to a profusion of rare corals and 10,000 to 30,000 fish per rig (Dybas 2005). Figure 9-22 shows details of the coordination process of the proposed models. According to Dybas (2005), oil companies can save between $400 and $600 million per rig by converting them into artificial reefs instead of offshore decommissioning. Therefore, the conversion of AROR as well as other proposed models are economically profitable and can be implemented with the coordination of other governmental agencies and NGOs.

Chapter 10

Summary and Conclusions

10.1 Summary

Modern management style is inspired by short-term gain. This short-term approach is inherently anti-Nature. It creates conflict with the natural ability of human beings to serve Humanity, and prevents them from achieving their originally intended objectives. If Nature is perfect (this is the first premise of this book), an anti-Nature approach cannot be sustainable. This perfection, or balance, of Nature is continuous, both in space and time. To understand this perfect nature of Nature, we must use the science of intangibles. This is because the tangible aspects of anything do not go beyond a tiny element in space, Δs, i.e., some arbitrary change in space approaching zero; and an even smaller element in time, Δt, going to zero – meaning that "time" is literally right now.

Since the Industrial Revolution, models have been based only on what can be characterized as the science of tangibles, founded on these assumptions about space and time. Short-term models are only successful for a finite time. Inherently, they must implode in the long term. It is only a matter of time before the harmful effects of such models become apparent and are ubiquitous in contemporary society. In medical science, this *modus operandi* amounts to treating the symptom. In economic development, it amounts to increasing wasteful habits to increase GDP. In business, it amounts to maximizing quarterly income, even if it means resorting to corruption. In psychology, it means maximizing pleasure and minimizing pain (both in the short term). In social science, it amounts to creating a culture of fear. In mathematics, it means obsessions with numbers and exact (and unique) solutions. In technology, it means promoting comfort despite the expense of long-term damage. In religion, it means obsession with ritual and short-term gain. In philosophy, it has meant positivism, behaviorism . . . and – most importantly – mechanical materialism.*

This approach was summarized by George Orwell, who wrote in his book *1984*, "He who controls the past, controls the future; he who controls the present, controls the past." This matter of controlling the present to control the past, and therefore controlling the future, has become painfully clear in the Information Age.

---

* By *mechanical* materialism, we mean the view that asserts matter as everything, but then goes on to define it without recognizing that the mode of its very existence is motion. In other words, it is the view that "the only matter that matters" (so to speak) is that which exists in the shortest possible time span and smallest possible extent of space. Therefore, energy's presence in such a space-time frame is accidental and not inherent. Right away, out leaps a serious difficulty: if the law of conservation of matter and energy is true, and matter and energy only change form, one into the other, where is the energy coming from? That is just the starting point of absurdities. Lord Kelvin's second law of thermodynamics takes this same energy, whose origins are assumed, without being anywhere defined or connected to the existence of matter in the first place, and declares that it will dissipate in an accumulating state of entropy and then everything – matter as well as energy – will end in the so-called "heat death of the universe".

Is the notion, that motion is the mode of existence of matter, separable from the notion that the world has a material existence, regardless of our cognition of it? For mechanical materialists, such a separation is essential. Any notion derived from the reference to "external force" in Newton's first law, that matter in infinite space and time (as distinct from some finite matter, in the form of discrete objects) could be "at rest", is aphenomenal and anti-Nature. Yet, this is the assumption that underlies and justifies the idea that matter in the environment is just there for the taking. Matter's mode of existence – how it comes to be, and what is likely to happen to it in the environment, which is a living system regardless of whether anything within it is "organic".

This book unravels the mysteries of the short-term approach and proposes long-term solutions for every problem encountered by the petroleum industry. Because the petroleum industry is the driver of all energy solutions today, and it is energy solutions that dictate our civilization, any positive change in the petroleum industry can set off a technological revolution. This book is precisely one of those rare efforts into exploring why technologies and management styles have failed Humanity and why science has betrayed our conscience. Answers to these questions are needed to change the spiraling downward pathway that we have traveled so far, throughout the modern age. This revolution is needed to reverse the current trends of the world that have been characterized as a "technological disaster" by many, including some Nobel Laureates in Science. By introducing a system that is based on knowledge, rather than perception, the book offers hope for a civilization cured of Enron, global warming, and diabetes, all of which have emerged as a symptom of the short-term model that we termed "aphenomenal". This book has a single conclusion, that by taking the long-term approach, both the short term and long term will be prosperous. Who will invoke this change and how? Five years ago, a vice president of a major oil company in Canada asked the principal author, "But, Dr. Islam, how do you plan to introduce long-term thinking?" This book answers that question and shows how anyone can invoke these changes, be it a manager in large corporation or an engineer in a small firm.

10.2   Fundamental Misconceptions in Technology Development

Long-term thinking is promoted through elimination of misconceptions. It is misconceptions and false perceptions that have led to eliminating long-term considerations to make any scheme look appealing, depending on how much information has been hidden. Sustainability can only come through knowledge. For knowledge, misconceptions and false perceptions must be eliminated from the root of the thought process. By tracking the first assumptions of all engineering and natural science "laws", the following misconceptions were identified (Zatzman et al. 2006).

10.2.1   Chemicals are chemicals

This misconception allowed Paul Hermann Müller, credited with inventing dichlorodiphenyltrichloroethane (DDT) and awarded a Nobel Prize in medicine and physiology, to glamorize the making of synthetic products. This old misconception got a new life, inspired by the work of Linus Pauling (the two-time Nobel prize winner in chemistry and peace), who justified using artificial products by saying that the artificial and natural both have similar tangible (at $\Delta t = 0$) features. Synthetic chemicals were acceptable because no matter which individual chemical constituent you chose, it existed in Nature anyway. The fact that almost none of the synthetic combinations of these otherwise natural-occurring individual elements had ever existed in Nature was ignored. A top scientist with DuPont claimed polyvinyl chloride (PVC) gives out dioxin (through oxidation) which also occurs in Nature; therefore, synthetic PVC should not be harmful – even if the "naturally-occurring" version of the resulting dioxin existed only briefly (so briefly that the natural and artificial sources couldn't be discerned), in transition to something else. The mere fact that something "exists" (often it is a matter of definition) in Nature tells us nothing about the mode of its existence. This is crucial, when it is remembered that synthetic products are used up and then dumped as waste into the environment. No consideration is given – at the time when their introduction is being planned – to the consequences of their possible persistence or accumulation in the environment.

All synthetic products "exist" (as long as the focus is on the most tangible aspect) in Nature, in a timeframe in which $\Delta t = 0$. That is the mode of their existence and that is precisely where the problem lies. With this mode, we can justify the use of formaldehyde in beauty products, anti-oxidants in health products, dioxins in baby bottles, bleaches in toothpaste, all the way up to every pharmaceutical product promoted today. Of course, the same focus, based on the science of the tangible, is applied to processes as well as products. Accordingly, nuclear fission in an atomic weapon is the same as what is going on inside the sun (the common saying is that "there are trillions of nuclear bombs going off every second inside the sun"). In other words: "energy is energy"! This misconception can make nuclear energy appear clean and efficient in contrast to "dirty", "toxic", and even "expensive" fossil fuel. All nuclear physicists seem to have this misconception, including those who won Nobel prizes in physics (e.g., Enrico Fermi, Ernest Rutherford).

10.2.2 If you cannot see it, it does not exist

Even the most militant environmental activists will admit that the use of toxic chemicals is more hazardous if the concentration is high and the reaction rate is accelerated (through combustion, for example). The entire chemical industry became engaged in developing catalysts that are inherently toxic and anti-Nature (through purposeful denaturing). The use of catalysts (always extremely toxic because they are truly denatured – i.e., concentrated) was justified by saying that catalysts by definition cannot do any harm because they only help the reaction but do not participate. The excuse is: "we are replacing enzyme with catalyst." Enzymes allegedly do not participate in the reaction either. The case in point is the 2005 Nobel Prize in Chemistry, which was awarded to three scientists. Yves Chauvin explained how metal compounds can act as catalysts in "organic synthesis". Richard Schrock was the first to produce an efficient metal-compound catalyst and Robert Grubbs developed an "even better catalyst" that is "stable" in the atmosphere. This invention is claimed to create processes and products that are "more efficient", "simpler" to use, "environmentally friendlier", "smarter", and less hazardous. Most importantly, the entire invention is called a great step forward for "green chemistry".

Another example relates to PVC. The defenders of PVC often state, just because there is chlorine in PVC, it does not mean that PVC is bad. After all, they argue, chlorine is bad as an element, but PVC is not an element, it is a compound. Here two implicit spurious assumptions are invoked:

1. chlorine can and does exist as an element;
2. the toxicity of chlorine arises from it being able to exist as an element.

This misconception, combined with "chemicals are chemicals", makes up an entire aphenomenal process of "purifying" through concentration. This "purification" scheme would go on to be used from oil refining to uranium enrichment. The truth, however, is that if the natural state of a matter, or the characteristic time of a process is violated, the process becomes anti-Nature. For instance, chemical fertilizers and synthetic pesticides can increase the yield of a crop, but that crop would not provide nutrition similar to the one produced through organic fertilizer and natural pesticides. H_2S is essential for human brain activities, yet concentrated H_2S can kill. Water is essential to life, yet "purified" water can be toxic and leach out minerals rather than nourishing living bodies.

Many chemical reactions are valid or operative only in some particular range of temperature and pressure. Temperature and pressure are themselves neither matter nor energy, and are therefore not deemed to be participating in the reaction. The reaction itself cannot even take place without these "conditions of state" being present within definite thresholds. However, because temperature and pressure are themselves neither matter nor energy undergoing a change, they are deemed not part of the reaction. It is just as logical to separate the act of shooting someone from the intention of the individual who aimed the gun and pressed the trigger. Temperature and pressure are but calibrated (i.e., tangible) measures of indirect/oblique (i.e., intangible) indicators of energy (heat, in the case of temperature) or mass (per unit area, in the case of pressure). Instead of acknowledging that, logically, these things are obviously involved in the reaction in some way that is different from the ways in which the tangible components of the reaction are involved, we simply exclude them on the grounds of this perceived intangibility.

It is the same story in mathematics: data output from some process is mapped to some function-like curve that tends to include discontinuities where data could not be obtained. Calculus teaches that we cannot differentiate across the discontinuous bits because "the function does not exist there." So instead of figuring out how to treat – as part of one and the same phenomenon – both the continuous areas and the discontinuities that have *been* detected, the derivative of the function obtained from actual observation according only *to* the intervals in which the function "exists", is treated as definitive – clearly a fundamentally dishonest procedure.

10.2.3 Simulation = emulation

We proclaim that we emulate Nature. It is stated that engineering is based on Natural Science. Invention of toxic chemicals and anti-Nature processes are called Chemical Engineering. People seriously thought of sugar as concentrated sweet, just like honey, so the process must be the same: "we are emulating Nature." This is the first thought, but then – because honey cannot be mass-produced and sugar can – we proceeded to "improve" Nature. This, in itself, will take on a life of its own. The entire chemical and pharmaceutical industry engage in denaturing first, then reconstructing to maximize productivity and tangible features of the product, intensifying the process that converts real to artificial.

There are countless simulators and simulations "out there". Routinely, they disregard, or try to linearize/smooth over, the non-linear changes-of-state. The interest and focus is entirely on what can be taken from Nature. This immediately renders such simulators and simulations useless for long-term planning. Simulators and simulations start by focusing on some structure or function of interest, which they then abstract. This process of abstraction negates any consideration of the natural-whole in which the structure or function of interest resides, or operates, or thrives; anything constructed or implemented according to the norms of such an abstraction must inevitably leave a mess behind.

Emulations and emulators, on the other hand, start with what is available in Nature, and how to sustain that, rather than what they would like to take from Nature, regardless of the resulting mess. A great confusion about this distinction is widespread. For example, people speak and write about how the computer "emulates" the computational process of an educated human – yet another abstraction, in which principles of the computer's design mimic an abstraction of the principles of how computation is supposed to work. But from the computer side of this scenario, the physical real-world computer circuitry that is implementing this abstraction of principles operates – in reality – according to the carefully-engineered limitations of semiconducting materials, which have been selected and refined to meet certain output criteria of the power supply and gate logic on circuit boards inside the computer. Second, from the computation side, the human brain as a biochemical complex incorporating matter-that-thinks does not compute according to the abstract principles of this process. It might most generously be called an abstraction being analogized to another abstraction, but it cannot be properly described as emulation of any kind. Emulation requires establishing a $1:1$ correspondence between operation "X_c" of the computer and operation "X_b" of the human brain. Obviously such a correspondence does not yet exist. Brain research is very, very, very far from being capable of establishing such a thing. We have already found many examples where the two processes are exactly the reverse of one another; from the chip overheating whereas the brain cools, to the division operator in the brain compared to the addition operator in the computer. This is repeated too often not to be reflecting some deeper underlying meaning.

10.3   Understanding Nature and Sustainability

In order to reverse the current "technological disaster", we must identify an alternative model to replace the currently used implosive one. The only sustainable model is that of Nature. It is important to understand what Nature is and how to emulate it. Even though many claims have been made about such emulation, no modern technology truly emulates the science of Nature. It has been the opposite: observations of Nature have rarely been translated into pro-Nature technological development. Today, some of the most important technological breakthroughs have been mere manifestations of the linearization of Nature Science; Nature linearized by focusing only on its external features. For example, computers process information exactly opposite to how the human brain does. Turbines produce electrical energy while polluting the environment beyond repair, even as electric eels produce much higher-intensity electricity while cleaning the environment. Batteries store little electricity while producing extremely toxic spent materials. Synthetic plastic materials look like natural plastic, yet their syntheses follow an exactly opposite path. Furthermore, synthetic plastics do not have a single positive impact on the environment, whereas natural plastic materials do not have a single negative impact. In medical science, every promise made at the onset of commercialization has proved to be the opposite of what really happened: witness Prozac, Vioxx, Viagra, etc. Nature, on the other hand, does not allow a single product to impact on the long-term negatively. Even the deadliest venom (e.g., cobra, poisoned-arrow tree frog) has numerous beneficial effects in the long-term. This catalogue carries on in all directions: microwave cooking, fluorescent lighting, nuclear energy, cellular phones, refrigeration cycles, and combustion cycles. In essence, Nature continues to improve matters in its quality, as modern technologies continue to degrade the same into baser qualities.

Nature thrives on diversity and flexibility, gaining strength from heterogeneity, whereas the quest for homogeneity seems to motivate much of modern engineering. Nature is non-linear and inherently promotes multiplicity of solutions. Modern Applied Science, however, continues to define problems as linearly as possible, promoting "single"-ness of solution, while particularly avoiding nonlinear problems. Nature is inherently sustainable and promotes zero-waste, both in mass and energy. Engineering solutions today start with a "safety factor" while promoting an obsession with excess (hence, waste). Nature is truly transient, never showing any exact repeatability or steady state. Engineering, however, is obsessed with standards and replication, always seeking "steady-state" solutions.

Table 10-1 shows the major differences between natural products and artificial products. Note that the features of artificial products are only valid for a time, t = "right now" ($\Delta t = 0$). This implies that they are only claims and are not true. For instance, artificial products are created on the basis that they are identical, the first premise of mass production and the economics of volume. No components can be identical, let alone products. Similarly, there is no such state as steady state. There is not a single object that is truly homogeneous, symmetrical, or isotropic. This applies to every claim on the right-hand side of Table 10-1. Note that

Table 10-1 Typical features of natural processes as compared to the claims of artificial processes. Modified from Khan and Islam (2007)

Nature ($\Delta t \rightarrow \infty$) (Real)	Artificial ($\Delta t \rightarrow 0$) (Aphenomenal)
Complex	Simple
Chaotic	Steady, periodic, or quasi-periodic
Unpredictable	Predictable
Unique (every component is different)	Non-unique, self similar
Productive	Reproductive
Non-symmetric	Symmetric
Non-uniform	Uniform
Heterogeneous, diverse	Homogeneous
Internal	External
Anisotropic	Isotropic
Bottom-up	Top-down
Multifunctional	Single-functional
Dynamic	Static
Irreversible	Reversible
Open system	Closed system
True	False
Self healing	Self destructive
Nonlinear	Linear
Multi-dimensional	Unidimensional
Infinite degree of freedom	Finite degree of freedom
Non-trainable	Trainable
Infinite	Finite
Intangible	Tangible
Open	Closed

the moment an artificial product comes into existence, it becomes part of Nature, therefore is subject to natural behavior. This is equivalent to saying, "You cannot create anything, there is already everything." The case in point can be derived from any theories or "laws" advanced by Newton, Kelvin, Planck, Lavoisier, Bernoulli, Gibbs, Helmholz, Dalton, Boyles, Charles, and a number of others who serve as the pioneers of modern science. Each of their theories and laws had the first assumption that would not exist in Nature, either in content (tangible) or in process (intangible).

10.4   Solutions that would Turn an Implosive Process Around

Nature is perfect, hence inherently sustainable. Sustainability can come only through emulating Nature. Table 10-2 shows the contrast between a sustainable process and an unsustainable process. Any time the anti-Nature scheme is implemented, the overall process is bound to be unsustainable, in which the original intention of the process will never be achieved (beyond $\Delta t = 0$). This process starts with this intention. If the intention is natural (real) and the process followed is natural (real), only then will the stated objective will be attained. By emulating nature (from source, i.e., intention to the process, i.e., the pathway), this approach of obliquity will indeed make sure all objectives, short-term or long-term, will be achieved.

Sustainable development starts with the source, i.e., a natural intention. The intended objective is achievable only if the source itself is natural. We cannot achieve an objective that is anti-Nature. That is the translation of "You cannot fight Nature." Here it is important to be honest, even with oneself. Nature never lies. For sustainability, honesty is not the best policy, it is the only policy. Having the correct intention is a necessary condition, but it is not sufficient for achieving sustainability. The process has to be natural. It means the features of natural processes are not violated (Tables 10-1 and 10-2). This is the only management style and technology development mode that can assure sustainability. It is important to screen current practices *vis-à-vis* natural processes and determine sustainability. It is equally important to check new technologies and ensure natural process is being followed. In predicting behavior and functionality of these techniques, we must be equipped with innovative tools and nonlinear mathematical models. By using nonlinear mathematics, even the most simplified reservoir prediction can be made closer to reality. By using innovative tools, the true picture of the reservoir as well as the production streams will emerge and the guesswork from engineering practices will be eliminated.

In order to make the management process sustainable, misconceptions highlighted above (Section 10.2) must be eliminated. In the past, a holistic approach to solving petroleum problems was missing. The current model is based on conforming to regulations and reacting to events. It is reactionary because it is only reactive and not fundamentally pro-active. Conforming to regulations and rules that may themselves not be based on any sustainable foundation can only increase long-term instability. Martin Luther King, Jr. famously pointed out, "We should never forget that everything Adolf Hitler did in Germany was 'legal'." Environmental regulations and technology standards are such that fundamental misconceptions are embedded in them: they follow no natural laws. A regulation that violates natural law has no chance to establish a sustainable environment. What was "good" law and "bad" law for Martin Luther King, Jr., is true law and false law, respectively.

Table 10-2   True difference between sustainable and unsustainable processes

Sustainable (Natural)	Unsustainable (Artificial)
Progressive/youth measured by the rate of change	Non-progressive/resists change
Unlimited adaptability and flexibility	Zero-adaptability and inflexible
Increasingly self evident with time	Increasingly difficult to cover up aphenomenal source
100% efficient	Efficiency approaches zero as processing is increased
Can never be proven to be unsustainable	Unsustainability unravels itself with time

Under today's regulations, crude oil is considered to be toxic and undesirable in a water stream, whereas the most toxic additives are not. For instance, a popular slogan in the environmental industry has been, "Dilution is the solution to pollution." This is based on all three misconceptions in Section 10.2., yet all environmental regulations are based on this principle. The tangible aspect, such as the concentration, is considered – but not the intangible aspect, such as the nature of the chemical, or its source. Hence, "safe" practices initiated on this basis are inevitably unsafe in the long run. Environmental impacts are not a matter of minimizing waste or increasing remedial activities, but of humanizing the environment. This requires the elimination of toxic waste altogether. Even nontoxic waste should be recycled 100%. This involves not adding any anti-Nature chemicals to begin with and then making sure each produced material is recycled, often with value addition. A zero-waste process has 100% global efficiency attached to it. If a process emulates Nature, such high efficiency is inevitable. This process is the equivalent of the greening of petroleum technologies. With this model, no one will attempt to clean water with toxic glycols, remove CO_2 with toxic amides, or use toxic plastic paints to "green". No one will inject synthetic and expensive chemicals to increase EOR production. Instead, we will settle for waste materials or naturally available materials that are abundantly available and pose no threat to the ecosystem.

In economic analysis, intangible costs and benefits must be included. Starting from understanding the true nature of energy pricing, the scheme that converts waste into valuable materials can turn the economic onus of the petroleum industry into an economic boon. The current economic development models lack the integration of intangibles, such as environmental impact, the role of human factors, and long-term implications (Zatzman and Islam 2006). These models are based on ancient accounting principles that the Information Age has rendered obsolete. This points to the need for a thorough renovation of economic theory that takes into account the roles of intangible factors such as intention, the actual passage of historical time, information, etc. so that a new approach to energy pricing is provided with a basis more solid and lasting than that of some short-term policy objective. This economic development model will make unsustainable practices exuberantly costly while making sustainable practices economically appealing.

10.5 Conclusions and Recommendations

Based on the theories advanced in this book, the following conclusions and recommendations can be made.

10.5.1 Management

Talks of sustainable and holistic management practices abound. However, a scientifically sustainable management style must be based on a time-tested criterion. Recent criteria advanced by Khan and Islam (2007) demonstrate the value of a bottom-up management and technology development criterion. This approach includes intangibles, which are essential components of knowledge and are opposite to perception. This management style is truly sustainable and should form the basis of petroleum resource management. Applying this model will require changes of attitude and proper education in science of intangibles. This is the only model that can change the management model once glamorized by Enron (based on short-term gain).

10.5.2 Exploration

Exploration for oil and gas is sustainable as long as various scientific criteria are met. With the guidelines proposed in this book (Chapter 3), exploration activities can be carried out in any part of the world, including the ecologically sensitive regions. It is recognized that there need to be advances in imaging techniques that are based on truly natural processes, which would make the exploration process environmentally acceptable.

10.5.3 Drilling and production operations

It has already been recognized through the advent of under-balanced drilling, that the use of numerous thickening agents, emulsifiers, stabilizers, etc. are overkill and drilling can be performed with water or even lighter mud. It is also true that every chemical additive to drilling mud has a substitute in Nature or in natural waste products (that are 100% recyclable). This applies to emulsifiers, foaming agents, thickeners, thinners, and everything else that is added to drilling mud. Even during water drilling, the source water should not be obtained through an anti-Nature process that is both expensive (both in the short-term and long-term) and inefficient. Waterjet drilling has tremendous potential, as does foam drilling, not just for drilling the formation but also for the perforations. The latter is a new invention and can replace the use of explosives in completion operations. A drilling process can be rendered more effective by using the sonic-while-drilling process that can generate data, readily usable for more accurate drilling, as well as for reservoir character-ization and realtime reservoir monitoring.

Production processes should be rendered zero-waste. In addition, by using only natural products for various applications, ranging from corrosion inhibition to hydrate control, from emulsion breakers to wax cleaner, we can eliminate the vulnerability due to the use of toxic chemicals that invariably get into the production stream of oil, gas, and water.

The future of production operations and drilling lies in using environment-friendly tubulars. More research in sustainable technology can open up opportunities for truly sustainable oil and gas operations.

10.5.4 Waste management

The current waste management is based on scientifically spurious assumptions. It is recognized that sustain-able waste management cannot be based on waste minimization, as no amount of waste is acceptable if long-term is included in the equation. Today's total production of oil and gas and associated water is so huge, we cannot benefit from just reducing waste, but must implement waste conversion. Long-term is the essence of the management model that relies heavily on intangibles. Waste management must be tackled from the source, by eliminating the source of wastes that are not readily accepted by Nature. By using proper science, every artificial component added to the natural stream of crude oil should be replaced by natural materials, much of which are either available in abundance or are produced as a fully-recyclable waste of a natural process.

10.5.5 Reservoir engineering and well logging

Current tools use toxic or radioactive elements to image downhole. This practice does not have to continue. At present there are many alternatives that could use natural fluids or energy that can be used to image the reservoir. Among mathematical tools, the most common has been the use of linear models, the only model that cannot model reality (Nature is nonlinear). Nonlinear models are presented and they should be used for analysis, instead of settling for software packages that are high on graphics and "ease of operations". The framework of an *in situ* relative permeameter is presented. This can revolutionize well logging and core analysis that have suffered from sampling errors and fundamental contradictions in terms of representative elemental volume (REV).

10.5.6 Enhanced Oil Recovery (EOR)

EOR operations have been slated in the past for being too expensive, ineffective, and environmentally haz-ardous. Sustainable practices should involve the use of waste materials from natural sources that will bring double dividends in terms of economic benefit and environmental consequences. In order to increase efficiency, the reasons behind low sweep efficiency are identified (e.g., viscous fingering) and remedies provided. A useful guideline for designing EOR is also provided.

10.5.7 Transportation, processing, and refining

Numerous practices that are anti-Nature (both in content and in process) are identified and alternate solutions provided. By using renewable materials and energy for any process, we can increase efficiency, economic benefits, and environmental acceptability. For every application, from natural gas processing to oil refining, from natural gas liquefaction to plastic manufacturing, examples of sustainable technologies are provided. By using these technologies, we can indeed even produce plastic materials that are environmentally acceptable.

10.5.8 Decommissioning

The current decommissioning regulations and practices are not conducive to good environment or sustainability. Pro-active decommissioning techniques are proposed. They can render current liabilities into assets of the future, without jeopardizing the environment. With the scheme outlined in the book, the petroleum industry does not have to be told by the legislators how to care for the environment.

10.5.9 Research

Sustainable means natural. The essence of Nature is dynamism, another word for change. No positive changes can be invoked unless thorough research is performed in order to discover the truth. This research will continually unravel the mysteries of Nature and create new knowledge, which will be the basis of the proposed sustainable management practice management. This closes the loop from Management → Research → Knowledge → Management. This is the essence of Knowledge-based decision making, which must be the driver of any sustainable technology.

References

Abdel-Ghani, M.S. and Davies, G.A. (1983) Simulation of non–woven fibre mats and the application to coalescers. *Chemical Engineering Science*, **40**, 117–29.

Abernethy, E.R. (1976) Production increase of heavy oils by electromagnetic heating. *Journal of Canadian Petroleum Technology*, July–September.

Abou-Kassem, J.H., Farouq Ali, S.M., and Islam, M.R. (2006) *Petroleum Reservoir Simulations*. Gulf Publishing Company, Texas.

Abou-Kassem, J.H., Zatzman, G.M., and Islam, M.R. (2007) Newtonian Mechanism and the Non-Linear Chaos of Nature: Preliminary Investigations in the Mathematics of Intangibles, *J. Nature Sci. and Sust. Tech.*, in press.

Abraham, P.D. (2001) Offshore oil and gas facility decommissioning in nova Scotia and Newfoundland. *Dalhousie Law Journal*, **24**, 333.

Ackman, R.G., Heras, H., and Zhou, S. (1996) Salmon lipid storage sites and their role in contamination with water-soluble petroleum materials. *Journal of Food Lipids*, **3**(3), 161–70.

Adkins, J.P. (1992) Microbial enhanced oil recovery from unconsolidated limestone cores. *Journal of Geomicrobiology*, **10**, 77–86.

Adrianto, L., Matsuda, Y., and Sakuma, Y. (2005) Assessing local sustainability of fisheries system: a multi-criteria participatory approach with the case of Yoron Island, Kagoshima prefecture, Japan. *Marine Policy*, **29**, 9–23.

Agarwal, P.N. and Puvathingal, J.M. (1969) Microbiological deterioration of woollen materials. *Textile Research Journal*, **39**, 38.

Agbon, I.S. and Intevep, S.A. (2000) *Social Responsibility and the Sustainable Economic Development of Oil and Gas Producing Communities in Nigeria*. SPE International Conference on Health, Safety, and the Environment in Oil and Gas Exploration and Production, SPE Paper No. 61102, June 26–28, Stavanger, Norway.

Agha, K.R., Belhaj, H., and Islam, M.R. (2003) *Transient Thermal Deformations of Rocks During Laser Drilling*. Proceedings of Oil and Gas Symposium, CSCE Annual Conference, refereed proceeding, Moncton, June.

Agha, K.R., Belhaj, H.A., Mustafiz, S., Bjorndalen, N., and Islam, M.R. (2004) Numerical investigation of the prospects of high energy laser in drilling oil and gas wells. *Petroleum Science and Technology*, **22**(9–10), 1173–86.

Ahamed, M.K. (1984) Behundi net fishery in the brackishwater area of Patuakhali with emphasis on bionomics and mortality of shrimps. *Research Report Freshwater*, Fisheries Research Station, Chandpur, Bangladesh, **8**.

Ahn, J.S. and Lee, K.H. (1986) Studies on the Volatile Aroma Components of Edible Mushroom (*Tricholoma matsutake*) of Korea. *Journal of the Korean Society of Food and Nutrition*, **15**, 253–7 (cited in BUA, 1994).

Ahnell, A. and O'Leary, H. (1997) Drilling and production discharges in the marine environment, in *Environmental technology in the oil industry* (ed. S.T. Orszulik), Blackie Academic & Professional, London, UK, pp. 181–208.

Akhtar, J. (2002) *Numerical and Experimental Modelling of Contaminant Transport in Ground Water and of Sour Gas Removal from Natural Gas Streams*. Chalmers University of Technology, Goteborg, Sweden.

Al Marzouqi, M.H.H. (1999) Determining pore size distribution of gas separation membranes from adsorption isotherm data. *Energy Sources*, **21**(1–2), 31–8.

Al-Darbi, M., Muntasser, Z., Tango, M., and Islam, M.R. (2002) Control of microbial corrosion using coatings and natural additives. *Energy Sources*, **24**(11), 1009–18.

Al-Darbi, M.M., Saeed, N.O., Ackman, R.G., Lee, K., and Islam, M.R. (2003) *Vegetable and Animal Oils Degradation in Marine Environments.* Proceedings of Oil and Gas Symposium, CSCE Annual Conference, refereed proceeding, Moncton, June.

Al-Darbi, M.M., Saeed, N.O., Ajijolaiya, L.O., and Islam, M.R. (2006) A Novel Oil Well Cementing Technology Using Natural Fibers, *J. Pet. Sci. Tech.*, **24**(12), 1267–82.

Al-Maghrabi, I., Bin-Aqil, A.O., Chaalal, O., and Islam, M.R. (1999) Use of thermophilic bacteria for bioremediation of petroleum contaminants. *Energy Sources*, **21**(1/2), 17–30.

Al-Maghrabi, I., Chaalal, O., and Islam, M.R. (1998) *Thermophilic Bacteria in UAE Environment Can Enhance Biodegradation and Mitigate Wellbore Problems.* 8th ADIPEC, SPE Paper No. 49545, October.

Alexender, S.R. and Winnick, J. (1995) Removal of hydrogen sulfide from natural gas through an electrochemical membrane separator. *American Institute of Chemical Engineers Journal*, **41**, 523.

Alguacil, D.M., Fischer, P., and Windhab, E.J. (2006) Determination of the interfacial tension of low density difference liquid – liquid systems containing surfactants by droplet deformation methods. *Chemical Engineering Science*, **61**, 1386–1394.

Ali, M. and Islam, M.R. (1998) The effect of asphaltene precipitation on carbonate rock permeability: an experimental and numerical approach. *SPE Production and Facilities*, **August**, 78–83.

Almalik, M.S., Attia, A.M., and Jang, L.K. (1997) Effects of alkaline flooding on the recovery of Safaniya crude oil of Saudi Arabia, *Journal of Petroleum Science and Engineering*, **17**, 367–72.

Amanullah, M. and Tan, C.P. (2001) A field applicable laser-based apparatus for mudcake thickness measurement, SPE-68673, in *SPE Asia Pacific Oil and Gas Conference and Exhibition Proceedings*, Jakarta, April 17–19: Society of Petroleum Engineers, CD-ROM, 11 p.

Amatya, P.L., Hettiaratchi, J.P.A., and Joshi, R.C. (2002) Biotreatment of flare pit waste. *Journal of Canadian Petroleum Technology*, **41**(9), 30–6.

Amin, R. and Waterson, A. (2002) *The Challenges of Multiphase Flow.* Trident Consultants, Kuala Lumpur, Malaysia.

AMITECH (Academic of Technology) (1997) Salt production. A daily internet edition of Bangladesh news. April 12, (1997)

Andrew, R., Hilding, J., and Pardue, J. (2001) Separation of CO2 from flue gases by Carbon-Multi wall nano tube membranes. August. Grant No. DE-FG2600-NT40825.

API (American Petroleum Institue) (2005) Don't Scrap the Platform; Create New Worlds! American Petroleum Institute, Washington, USA <www.api.org> [Accessed on January 15, 2006].

Appleman, B.R. (1992) *Bridge Paint: Removal, Containment, and Disposal.* Transportation Research Board, Washington, DC. Report No. TRB-NCHRP-SYN/176.

Appleton, A.F. (2006) Sustainability: a practitioner's reflection. *Technology in Society*, **28**(1–2), 3–18.

Arangath, R., Ogoke, V.C., and Onwusiri, H.N. (2002) *A Cost-Effective Approach to Improve Performance of Horizontal Wells Drilled in High Permeability Formations.* SPE International Symposium and Exhibition on Formation Damage Control, SPE Paper No. 73786, February 20–21, Lafayette, LA.

Arrington, J.R. (1960) Size of crude reserves is key to evaluating exploration programs. *Oil and Gas Journal*, **58**(9), 30–134.

Arsic, B., Oka, S., and Radovanovic, M. (1991) Characterization of limestone for SO_2 absorption in fluidised bed combustion. *FBC Technology and the Environmental Challenge*, Hilger, London.

Aslam, J. and Alsalat, T. (2000) *High-Pressure Water jetting: An Effective Method to Remove Drilling Damage.* SPE International Symposium on Formation Damage Control, SPE Paper No. 58780, February 23–24, Lafayette, LA.

Astrita, G., Savage, D., and Bisio, A. (1983) *Gas Treating with Chemical Solvents.* John Wiley and Sons, New York.

ASRD (Alberta Sustainable Resource Development) (2002) *Policy and Procedures Document for Submitting the Geophysical Field Report Form.* Alberta Sustainable Resource Development, Edmonton, AB.

Atkinson, G. (2000) Technology and sustainable development, in *Frameworks to Measure Sustainable Development.* OECD, Paris.

ATSDR (Agency for Toxic Substances and Disease Registry) (2006) *Medical Management Guide Line for Sodium Hydroxide*, http://www.atsdr.cdc.gov/MHMI/mmg178.html <Accessed on June 07, 2006>.

Attanasi, E.D. and Root, D.H. (1994) The enigma of oil and gas field growth. *American Association of Petroleum Geologists Bulletin*, **78**(33), 321–32.

Attanasi, E.D., Mast, R.F., and Root, D.H. (1999) Oil, gas field growth projections – wishful thinking or reality. *Oil and Gas Journal*, **97**(14), 79–81.

Au (1993) *Cited from, Communication and Echolocation.* <http://www.seaworld.org/infobooks/Bottlenose/echodol.html>

Auramo, J., Aminiff, A., and Punakivi, M. (2002) Research agenda for e-business logistics based on professional opinions. *International Journal of Physical Distribution & Logistics*, **32**, 513–23.

Austria, G., Savage, D., and Bisio, A. (1983) *Gas Treating with Chemical Solvents*. John Wiley and Sons, New York.

Aycaguer, A.C., Lev-On, M., and Winer, A.M. (2001) Reducing carbon dioxide emissions with enhanced oil recovery projects: a life cycle assessment approach. *Energy & Fuels*, **15**(2), 303–8.

Aylor, W.K.Jr. (1999) Measuring the impact of 3D seismic on business performance. *Journal of Petroleum Technology*, **June**, 52–6.

Babckock, R.E., Spillman, R.W., Goddin, C.S., and Cooley, T.E. (1988) Natural gas cleanup: a comparison of membrane and amine treatment processes. *Energy Progress*, **8**(3), 135–42.

Baine, M. (2001) Artificial reefs: a review of their design, application, management and performance, *Ocean & Coastal Management*, **44**, 241.

Baker Hughes Inc. (2001). *Environmental Affairs: Regulatory Requirements*. Baker Hughes Inc., Texas.

Baker Petrolite (2001) *Deepwater Flow Assurance, Oil and Gas Production*, Baker Hughes Inc., Texas.

Baker, R.W. (1995) Membrane technology, in *Kirk Othmer Encyclopedia of Chemical Technology*, Vol. 16 (ed. K. Othmer). Wiley, Singapore.

Banat, I.M. (1995) Biosurfactants production and possible uses in microbial enhanced oil recovery and oil pollution remediation: A review. *Resource Technology*, **51**, 1–12.

Bansal, A. and Islam, M.R. (1991) *State-of-the-art Review of Non-Destructive Testing with Computer Assisted Tomography*. SPE-22127, Proceedings of the International Arctic Technology Conference, Anchorage, (1991).

Bansal, K.M. and Sugiarto (1999) *Exploration and Production Operations – Waste Management A Comparative Overview: US and Indonesia Cases*. SPE Asia Pacific Oil and Gas Conference, SPE54345, April 20–22, Jakarta, Indonesia.

Barakat, H.Z. and Clark, J.A. (1966) On the solution of the diffusion equations by numerical methods. ASME Trans. *Heat Transfer*, **88**, 83.

Barasan, O.A., Burban, P., and Auvil, S.R. (1989) Gases separation using Membrane Systems. *Industrial and Chemical Engineering Research*, **28**, 108.

Barbee, S.J. and Hartung, R. (1979) Diethanolamine-induced alteration of hepatic mitochondrial function and structure. *Toxicology and Applied Pharmacology*, **47**, 431–40.

Barla, P., Doucet, J.A., and Green, J.D.M.S. (2000) Protecting habitats of endangered species on private lands: analysis of the instruments and Canadian policy. *Canadian Public Policy*, **26**(1), 95–110.

Basaran, O.A., Burban, P.M., and Auvil, S.R. (1989) *Facilitated Transport with Unequal Carrier and Complex Diffusivities. Industrial and Chemical Engineering Research*, **28**, 108–19.

Bassiouni, Z. (1994) *Theory, Measurement, and Interpretation of Well Logs*, First printing, Henry L. Doherty Memorial Fund of AIME, SPE, Richardson, TX.

Basu, A. (2005) Experimental and Numerical Studies of a Novel Technique for Abatement of Toxic Metals from Aqueous Steams, Faculty of Engineering, Dalhousie University, PhD Dissertation.

Basu, A., Akhtar, J., Rahman, M.H., and Islam, M.R. (2004) A review of separation of gases using membrane systems. *Petroleum Science and Technology*, **22**(9–10), 1343–68.

Basu, A., Akhter, J., Rahman, M.H., and Islam, M.R. (2004) A review of separation of gases using membranes. *Journal of Petroleum Science and Technology*, **22**(9–10), 1343–68.

Batarseh, S. (2001) Application of Laser Technology in the Oil and Gas Industry: An Analysis of High Power Laser-Rock Interaction and its Effect of Altering Rock Properties and Behavior. Colorado School of Mines, PhD Dissertation.

Bauerle, D. (2000) Laser processing and chemistry, in *Targeted Literature Review: Determining the Benefits of StarWars Laser Technology for Drilling and Completing Natural Gas Wells*, 2nd edn (eds R.M. Graves and D.G. O'Brien). GRI-98/0163, July 1998, Springer.

BBC (2005) Oil worker deaths inquiry begins. *BBC Online News.* <http://news.bbc.co.uk/> [Accessed on October 30, 2005].

Beard, R.R. and Noe, J.T. (1981) Aliphatic and alicyclic amines, in *Patty's Industrial Hygiene and Toxicology*, Vol. 2B, 3rd edn (eds G.D. Clayton and F.E. Clayton). John Wiley & Sons, Inc., New York, pp. 3135–73.

Beck, J.V. (1947) Penn grade progress on bacteria for releasing oil from sands. *Producers Monthly*, **11**, 13–19.

Belhaj, H., Mousavizadegan, H., and Islam, M.R. (2006a) *Reservoir Permeability: Innovative Measurements Based on a Novel Concept*. SPE Western Regional/AAPG Pacific Section/GSA Cordilleran Section Joint Meeting, SPE Paper No. 100427, May 8–10, Anchorage, AK.

Belhaj, H., Mousavizadegan, H., Ma, F., and Islam, M.R. (2006b) *Three-Dimensional Permeability Utilizing a New Gas-Spot Permeameter*. SPE Gas Technology Symposium, SPE Paper No. 100428, May 15–17, Calgary, AB.

Bentsen, R.G. (1985) A new approach to instability theory in porous media. *Journal of Society of Petroleum Engineering*, **October**, 765–79.

Bergman, P.D., Drummond, C.J., Winter, E.M., and Chen, Z.-Y. (1996) Disposal of Power Plant CO2 in Depleted Oil and Gas Reservoirs in Texas, Proceedings of the third International Conference on Carbon Dioxide Removal, Massachusetts Institute of Technology, Cambridge, MA, USA, September 9–11.

Berry, J.A. and Wells, G. (2004) Integrated fate modeling for exposure assessment of produced water on the sable island bank (Scotian Shelf, Canada). *Environmental Toxicology and Chemistry*, **23**(10), 2483–93.

Beyer, K.H.Jr., Bergfeld, W.F., Berndt, W.O., Boutwell, R.K., Carlton, W.W., Hoffmann, D.K., and Schroeter, A.L. (1983) Final report on the safety assessment of triethanolamine, diethanolamine, and monoethanolamine. *Journal of the American College Toxicology*, **2**, 183–235.

Bhide, B.D. and Stern, S.A. (1993) Membrane processes for the removal of acid gases from natural gas. I. Process configurations and optimization of operating conditions, *Journal of Membrane Science*, **81**, 209–37.

Biazar, J. and Islam, R. (2004) Solution of wave equation by adomian decomposition method and the restrictions of the method. *Applied Mathematics and Computation*, **149**, 807–14.

Biazar, J., Babolian, E., and Islam, R. (2004) Solution of the system of ordinary differential equations by adomian decomposition method. *Applied Mathematics and Computation*, **147**, 713–19.

BioBasices (2006) Microbial Enhanced Oil Recovery, The Science and the issues, BioPortal, Government of Canada, <http://www.biobasics.gc.ca/english/View.asp?x=793> [Accessed on August 12, 2006].

Biron, M., Campbell, R., and Moryasu, M. (2000) Historical review (1994–1998) and assessment of the 1999 exploratory snow crab (Chionoecetes opilo). Fishery off southwestern Nova Scotia (NAFO Division 4X). DFO Can. Stock Assess. Sec. Res. Doc. 2000/18.

Bjorndalen, N., Belhaj, H.A., Agha, K.R., and Islam, M.R. (2003) *Numerical Investigation of Laser Drilling*. SPE Eastern Regional/AAPG Eastern Section Joint Meeting, SPE Paper No. 84844, September 6–10, Pittsburgh, PA.

Bjorndalen, N., Mustafiz, S., and Islam, M.R. (2004) *A Zero-Waste Oil and Gas Production Scheme*. 8th Mediterranean Petroleum Conference and Exhibition, Tripoli, Libya. **January** 20–22.

Bjorndalen, N., Mustafiz, S., and Islam, M.R. (2003) Numerical modeling of petroleum fluids under microwave irradiation for improved horizontal wells performance. *International Communications in Heat and Mass Transfer*, **30**(6), 765–74.

Bjorndalen, N., Mustafiz, S., and Islam, M.R. (2005a) The effect of irradiation on immiscible fluids for increased oil production with horizontal wells. Proceedings of IMECE05 2005 ASME International Mechanical Engineering Congress and Exposition November 5–11, 2005, Orlando, Florida USA. IMECE2005-81752.

Bjorndalen, N., Mustafiz, S., and Islam, M.R. (2005b) No-flare design: converting waste to value addition. *Energy Sources*, **27**(4), 371–80.

Blasco, F., Saenger, P., and Janodet, E. (1996) Mangroves as indicators of coastal change. *CATENA*, **27**(3–4), 167–78.

Blomstrom, D.C. and Beyer, E.M.Jr. (1980) Plants metabolise ethylene to ethylene glycol. *Nature*, **283**, 66–8.

Boesch, D.F., James, J.N., Cacchione, D.A., Geraci, J.R., Neff, J.M., Ray, J.P., and Teal, J.M. (1987) An assessment of the long-term environmental effects of US offshore oil and gas development activities: future research needs, in *Long-Term Environmental Effects of Offshore Oil and Gas Development* (eds D.F. Boesch and N.N. Rabalais). Elsevier Applied Science, London and New York.

Bolli, A. and Emtairah, T. (2001) *Environmental Benchmarking for Local Authorities: From Concept to Practice*, Environmental Issues. International Institute for Industrial Environmental Economics, Sweden.

Bolze, D.A. and Lee, M.B. (1989) Offshore oil and gas development: Implications for wildlife in Alaska. *Marine Policy*, **13**(3), 231–48.

Borwankar, R.P. and Wasan, D.T. (1986) The kinetics of adsorption of ionic surfactants at gas-liquid surfaces. *Chem Eng Sci*, **41**, 199.

Bounicore, A.J. and Davis, W.T. (eds) (1992) *Air Pollution Engineering Manual.* Van Nostrand Reinhold, New York.

Bowers, B.E., Brownlee, R.F., and Schrenkel, P.J. (2000) Development of a downhole oil/water separation and reinjection system for offshore application. SPE 63014. *SPE Production & Facilities*, **15**(2), 115–22.

Brandt/EPI (1996) *The Handbook on Solids Control & Waste Management*, 4th edn. Brandt/EPI™ <http://www.brandtvarco.com/html/pdf/Solids.pdf> [Accessed on June 23, 2005].

Breuer, E., Stevenson, A.G., Howe, J.A., Carroll, J., and Shimmield, G.B. (2004) Drill cutting accumulations in the Northern and Central North Sea: a review of environmental interactions and chemical fate. *Marine Pollution Bulletin*, **48**, 12–25.

Brown, L., Postel, S., and Flavin, C. (1991) From growth to sustainable development, in *Environmentally Sustainable Economic Development* (ed. R. Goodland). UNESCO, Paris, pp. 93–8.

Browning, J.V., Miller, K.G., and Pak, D.K. (1996) Global implications of Eocene Greenhouse and Doubthouse sequences on the New Jersey coastal plain – The Icehouse cometh. *Geology*, **24**, 639–42.

Broz, J.S., French, T.R., and Caroll, H.B. (1985) *Blocking of High Permeability Zones in Steamflooding by Emulsions.* Proceedings of the 3rd UNITAR Conference on HeavyCrude and Tar Sands.

Bubela, B. (1985) Effect of biological activity on the movement of fluids through porous rocks and sediments and its application to enhanced oil recovery. *Journal of Geomicrobial*, **4**, 313–27.

Burk, J.H. (1987) Comparison of sodium carbonate, sodium hydroxide, and sodium orthosilicate for EOR. *SPE Reservoir Engineering*, **2**(1), 9–16.

Buset, P., Riiber, M., and Eek, A. (2001) *Jet Drilling Tool: Cost-Effective Lateral Drilling Technology for Enhanced Oil Recovery.* Coiled Tubing Roundtable, SPE Paper No. 68504, SPE/IcoTA, March 7–8, Houston, Texas.

Butler, R.M. and Mokrys, I.J. (1991) A new process (VAPEX) for recovering heavy oils using hot water and hydrocarbon vapor. *Journal of Canadian Petroleum Technology*, **30**(1), 97–106.

Buzek, J., Podkanski, J., and Warmuzinski, K. (1997) The enhancement of the rate of adsorption of CO_2 in amine solutions due to the maragoni effect. *Energy Conversion and Management*, **38**(Supp 1), S69–74.

CAAP (2005) <www.capp.ca> [Accessed on June 2, 2006].

Campbell, T.C. (1977) A comparison of sodium orthosilicate and sodium hydroxide for alkaline waterflooding. *Journal for Petroleum Technology*, SPE Paper No. 6514, 1–8.

Campbell, T.C. (1982) The role of alkaline chemicals in the recovery of low-gravity crude oils. *Journal of Petroleum Technology*, **SPE Paper no 8814-PA**, 2510–16.

Campbell, T.C. and Krumrine, P.H. (1979) Laboratory Studies on Alkaline Water Flooding, SPE 8328, presented at the 54th Annual Fall Technical Conference and Exhibition, Las Vegas, Nevada.

Canada Nova Scotia Offshore Petroleum Board (CNSOPB) (2001) Strategic Environmental Assessment, Parcels No. 1–9, Call for Bids NS01-1.

Canada Nova Scotia Offshore Petroleum Board (CNSOPB) (2002) *Environmental Protection Board. White Page.* <http://www.cnsopb.ns.ca/Environment/evironment.html> [Accessed on April 21, 2002].

Canada, E. (2001) *A Review of International Initiatives to Accelerate the Reduction of Sulphur in Light and Heavy Fuel Oil.* Environment Canada, Ottawa.

Canada, E. (2002) *Setting Canadian Standards for Sulphur in Heavy and Light Fuel Oils.* Environment Canada, Ottawa: <http://www.ec.gc.ca/energ/fuels/reports/cnslt_rpts/disc_paper/toc_summ_f.htm> [Accessed on August 12, 2006].

Canadian Environmental Modeling Centre (CEMC) (2003) Level III Version 2.70 Released March 2002, University, Peterborough, Ontario, Canada.

Canesis (2004) *Wool the Natural Fiber.* Canesis Network Ltd, Private Bag 4749, Christchurch, New Zealand.

CAPP (Canadian Association of Petroleum Producers) (2001) *Technical Report.* Produced Water Waste Management. August (2001)

CAPP (2002) *Cited in National Inventory Report, 1990–2004* – Greenhouse Gas Sources and Sinks in Canada, Environment Canada, <http://dsp-psd.pwgsc.gc.ca/collection/En49-5-5-10-2-2002E.pdf> [Accessed on October 11, 2006].

Capstone (2003) Microturbine: Application Overview – Resource Recovery. Available <http://www.globalmicroturbine.com/pdf/gas_oil_flaring/caps_oilandgas.PDF> March 11.

Carcoana, A. (1992) *Applied Enhanced Oil Recovery.* Prentice Hall Inc., Englewood Cliffs, NJ.

Cariou, R., Jean-Philippe, A., Marchand, P., Berrebi, A., Zalko, D., Andre, F., and Bizec, B. (2005) New multiresidue analytical method dedicated to trace level measurement of brominated flame retardants in human biological matrices. *Journal of Chromatography,* **1100**(2), 144–52.

Carroll, J.J. (2003) *Natural Gas Hydrates: A Guide for Engineers.* Gulf Professional Publishing, Houston, TX.

Carstens, J.P. and Brown, C.O. (1971) *Rock Cutting By Laser.* 46th Annual Fall Meeting of SPE, SPE Paper No. 3529, October 3–6. New Orleans, LA.

Cashdollar, K.L., Zlochower, I.A., Green, G.M., Thomas, R.A., and Hertzberg, M. (2000) Flammability of methane, propane, and hydrogen gases. *J. Loss Prev. Process. Ind.,* **13**, 327–40.

Castanier, L.M. (1989) Steam additives: fieldprojectsofthe eighties. *Journal of Petroleum Science andEngineering,* **2**, 193.

Caurant, F., Bustamante, P., and Bordes, M. (1999) Bioaccumulation of cadmium, copper and zinc in some tissues of three species of marine turtles stranded along the French Atlantic coasts. *Marine Pollution Bulletin,* **38**(12), 1085–91.

CDC (Centers for Disease Control) Report (2001) *National Report on Human Exposure to Environmental Chemicals.* Centers for Disease Control and Prevention, National Center for Environmental Health, Division of Laboratory Sciences, Mail Stop F-20, 4770 Buford Highway, NE, Atlanta, Georgia 30341-3724, NCEH Pub No: 05-0725.

CEF Consultants Ltd (1998) Exploring for Offshore Oil and Gas (Nov). No. 2 of Paper Series on *Energy and the Offshore.* Halifax, NS, <http://www.cefconsultants.ns.ca/2explore.pdf> [Accessed on May 18, 2005].

CEF (2005) *Strategic Environmental Assessments of the Misaine Bank Area,* prepared for Canada–Nova Scotia Offshore Petroleum Board, Halifax, Canada.

CEIA (Centre d'Estudis d'Informaci Ambiental) (2001) *A New Model of Environmental Communication from Consumption to Use of Information.* European Environment Agency, Copenhagen.

CERHR (2003) NTP-CERHR Expert Panel Report on Reproductive and Developmental Toxicity of Propylene Glycol. *National Toxicology Program,* US Department of Health and Human Services, NTP-CERHR-PG-03.

Center for Energy (2006) *Green House Gases: What Industries Are Doing?* <http://www.centreforenergy.com> [Accessed on March 7, 2007].

Chaalal, O. (2003) Personal communication with R. Islam, February 27.

Chaalal, O., Tango, M., and Islam, M.R. (2005) A new technique of solar bioremediation. *Energy Sources,* **27**(3), 361–70.

Chaffey, D.R. and J.H. Sandom (1985). Sundarban Forest Inventory Project, Bangladesh: A glossary of vernacular plant names and a field key to the trees. Overseas Development Authority (UK), Land Resources Development Center, Surrey.

Chakma, A. (1976). Acid gas re-injection – a practical way to eliminate Co2 emissions from gas processing plants. *Energy Conversion Management*, **38**, 205–9.

Chakma, A. (1994). Separation of acid gases from power plant flue gas streams by formulated amines, in *Gas Separation*, (ed. E.F. Vansant), Elsevier Science Publishers, pp. 727–37.

Chakma, A. (1999) Formulated solvents: new opportunities for energy efficient separation of acid gases. *Energy Sources*, **21**, 1–2.

Chakma, A. and Jha, K.N. (1992) *Heavy Oil Recovery from Thin Pay Zones by Electromagnetic Heating.* 67th Annual Technical Conference and Exhibition, SPE Paper No. 24817, October 4–7, DC, USA.

Chang, K.-Y., Lin, W.-H., and Chang, H.-F. (2002) Simulation of hydrogen production from dehydrogenation of ethanol in a palladium membrane reactor. *Journal of the Chinese Institute of Chemical Engineers*, **33**, 225.

Chang, Y. (1987) Preliminary studies assessing sodium pyrophosphate effects on microbial mediated oil recovery. *Annals of the New York Academy of Sciences*, **506**, 296–701.

Chaudhry, A.I. and Tolelfeson, E.L. (1990) Catalyst modification and process design considerations for the oxidation of low concentrations of hydrogen sulfide in natural gas. *Canadian Journal of Chemical Engineering*, **68**, 449.

Cheema, T.J. and Islam, M.R. (1995) A new modeling approach for predicting flow in fractured formations, in *Groundwater Models for Resources Analysis and Management* (ed. El-Kady). Lewis Publishers, Boca Raton, FL, pp. 327–38.

Cheevaporn, V. and Menasveta, P. (2003) Water pollution and habitat degradation in the Gulf of Thailand. *Marine Pollution Bulletin*, **47**(1–6), 43–51.

Chemistry Store (2005) <http://www.chemistrystore.com/index.html> [Accessed on June 07, 2006].

Chen, K., Hu, G.Q., and Lenz, F. (2001) Effects of doubled atmospheric CO2 concentration on apple trees: I. Growth analysis. *Gartenbauwissenschaft*, **66**(6), 282–8.

Chenard, P.G., Englehardt, F.R., Blare, J., and Hardie, D. (1989) Patterns of oil-based drilling fluid utilization and disposal of associated wastes on the Canadian offshore frontier lands, in *Drilling Wastes* (eds F.R. Engehardt, J.P. Ray and A.H. Gillam). Elsevier Applied Science, New York, pp. 119–36.

Cheng, K.H. (1986) Chemical Consumption during Alkaline Flooding: A Comparative Evaluation, SPE/DOE Fifth Symposium on Enhanced 011Recowy of the Society of Petroleum Engineers and the department of Energy held in Tulsa, OK, April 20–23, 1986, SPEIDOE 14944.

Chengde, Z. (1995) Corrosion Protection Of Oil Gas Seawater Submarine Pipeline For Bohai Offshore Oil Field, Petroleum Engineering held in Saijing, PR China, November 14–17, 1995, SPE 29972.

Chhetri, A.B and Islam, M.R. (2006a) Towards producing a true green biodiesel. *Energy Sources* (submitted).

Chettri, A.B. and Islam, M.R. (2006b) A critical review of electromagnetic heating for enhanced oil recovery. *Journal of Petroleum Science and Technology* (submitted).

Chhetri, A.B. and Islam, M.R. (2007) *Inherently Sustainable Technology Development.* Nova Science Publishers, New York.

Chhetri, A.B., Khan, M.I., and Islam, M.R. (2006) A novel sustainably developed cooking stove. *Journal Nature Science and Sustainable Technology*, **1**(4), 1–28.

Chhetri, A.B., Zaman, M.S., and Islam, M.R. (2006) *Effects of Microwave Heating and Refrigeration on Food Quality.* 2nd International Conference on Appropriate Technology, July 12–14. Zimbabwe.

Chi, C., Its Des, W., and Lee, H. (1973) Natural Gas Purification by 5A Molecular Sieves and ign Method. *AIChE Symposium Series*, **134**(69), 95–101.

Chilingar, G.V. and Islam, M.R. (1995) Mathematical modeling of three-dimensional microbial transport in porous media. *International Journal of Science and Technology*, **64**, 2, 55.

Choi, E.S., Cheema, T.J., and Islam, M.R. (1997) A New dual porosity/dual permeability model with non-Darcian flow through fractures. *Journal of Petroleum Science and Engineering*, **17**, 331–44.

Chronic Toxicity Summary (2001) *Chronic Toxicity Summary – Diethanolamine. Determination of Noncancer Chronic Reference Exposure Levels*, Batch 2B, December.

ClearTech (2006) Industrial Chemicals, North Corman Industrial Park, Saskatoon S7L 5Z3, Canada, <http://www.cleartech.ca/products.html> [Accessed on May 08, 2006].

CMAI (2005) *Chemical Market Associates Incorporated.* <www.kasteelchemical.com/slide.cfm> [Accessed on May 20, 2006].

CNSOPB (1996) *Offshore Waste Treatment Guidelines.* Canada-Nova Scotia Offshore Petroleum Board, Ottawa.

CNSOPB (2002) *Offshore Waste Treatment Guidelines.* Canada-Nova Scotia Offshore Petroleum Board, Ottawa.

CNSOPB (2004) *Environmental Screening Report, Cohasset Phase II Decommissioning, Encana Corporation.* Canada-Nova Scotia Offshore Petroleum Board, Fisheries and Oceans Canada, Environment Canada.

CNSOPB (2005) *Activity Report (2005)* Canada-Nova Scotia Offshore Petroleum Board, Halifax, Nova Scotia. November 5, (2004).

Cobscook Bay Research Center (2001) *Phytoplankton and Harmful Algal Blooms. Maine Phytoplankton Monitoring Program.* Eastport, Maine. <http://www.cobscook.org/> [Accessed on December 12, 2002].

Colella, W.G., Jacobson, M.Z., and Golden, D.M. (2005) Switching to U.S. hydrogen fuel cell vehicle fleet: the resultant changes in emissions, energy use, and greenhouse gases. *Journal of Power Sources*, **150**, 150–82.

Collins, N., Cook, J., Reece, M., Martin, S., Pitt, R., Canning, S., Stewart, P., and MacNeil, M. (2001a) *Environmental Impact Assessment of a 2D seismic Survey in Sydney Bight.* CEF Consultants Ltd., Communications, Environmental and Fisheries Consultants. Halifax, NS.

Collins, N., Cook, J., Jensen, M., Martin, S., and Burns, R. (2001b) *Preliminary Environmental Impact Assessment of Exploratory Drilling in Sydney Bight.* Hunt Oil Company of Canada and TotalFinaElf E&P, Calgary, Alberta.

Commission of the European Communities (1993) Technoeconomic Study on the Reduction Measures, based on Best Available Technology, of Water Discharges and Waste Generation from Refineries, DG XI A3, Brussels.

Commission of the European Communities (1991) Technical Note on the Best Available Technologies to Reduce Emissions of Pollutants into the Air from the Refining Industry. DG XI A3, Brussels.

Continental Shelf Associates Inc. (1997) *Metals and Organic Chemicals Associated with Oil and Gas Well Produced Water: Bioaccumulation, Fates and Effects in the Marine Environment.* Gulf of Mexico Produced Water Bioaccumulation Study.

Cooke, C.E.Jr., Williams, R.E., and Kolodzie, P.H. (1974) Oil recovery by alkaline water flooding, Paper Number 4739-PA. *Journal of Petroleum Technology*, 1356–74.

Cord-Ruwisch, R., Lovley, D.R., and Schink, B. (1998) Growth of Geobacter sulfurreducens with acetate in syntrophic cooperation with hydrogen-oxidizing anaerobic partners. *Applied and environmental microbiology*, **64**, 2232–6.

Cosmo Oil (2003) *Environmental Performance – Oil Field Development, Crude Oil Transport.* <http://www.cosmo-oil.co.jp/eng/envi/2001/pdf/11_25.pdf> [Accessed on March 11].

Costanza, R., Cumberland, J., Daly, H., Goodland, R., and Norgaard, R. (1997) *An Introduction to Ecological Economics, International Society for Ecological Economics.* St. Lucie Press, Florida.

Couper, A. (1993) *Physical Methods of Chemistry*, 2nd edn, Vol. IXA, Part A, p. 1. Wiley-Interscience, New York.

Couturier, M.F., Karidio, I., and Steward, F.R. (1993) Study on the rate of breakage of various Canadian limestones in a circulating transport reactor, in *Circulating Fluidised Bed Technology IV* (ed. A.A. Avidan). American Institute of Chemical Engineering, New York.

Cox, A. (1999) A research agenda for supply chain and business management thinking. *Supply Chain Management*, **4**(4), 209–14.

CTB (Consumer Taxation Branch) (2006) *Non-taxable/Taxable List: Catalyst or Direct Agents for the Oil Refining Process.* <http://www.sbr.gov.bc.ca/ctb/Tax_Lists/Oil_and_Gas_Catalyst(non-taxable_taxable> [Accessed on June 2, 2006].

CWRT (Center for Waste Reduction Technologies, American Institute of Chemical Engineers) (2002) *Collaborative Projects: Focus area Sustainable Development: Development of Baseline Metrics.* <http://www.aiche.org/cwrt/pdf/BaselineMetrics.pdf> [Accessed on May 28, 2005].

Cranford, P.J., Gordon, D.C., Jr, Hannah, C.G., Loder, J.W., Miligan, T.G., Muschenheim, D.K., and Shen, Y. (2003) Modeling potential effects of petroleum exploration drilling on northeastern Georges Bank scallop stocks. *Ecological Modeling*, **166**, 19–39.

Cripps, S.P. and Ababel, J.P. (2002) Environmental and socio-economic impact assessment of reef, a multiple platform rigs-to-reefs development. *ICES Journal of Marine Science*, **59**, S300.

Croce, B., Stagg, R., and Everall, N. (1997) Ecotoxicological determination of Pigmented Salmon Syndrome – a pathological condition of Atlantic salmon associated with river pollution. *Ambio*, **26**(8), 505–10.

Cronin, M.A. and Bickham, J.W. (1998) A population genetic analysis of the potential for a crude oil spill to induce heritable mutations and impact natural populations. *Ecotoxicology*, **7**(5), 259–78.

Currie, D.R. and Isaacs, L.R. (2005) Impact of exploratory offshore drilling on benthic communities in the Minerva gas field, Port Campbell, Australia. *Marine Environmental Research*, **59**, 217–33.

Currie, D.R. and Isaacs, L.R. (2005) Impact of exploratory offshore drilling on benthic communities in the Minerva gas field, Port Campbell, Australia. *Marine Environmental Research*, **59**, 217–33.

Daly, H.E. (1992) Allocation distribution and scale: towards an economics that is efficient just and sustainable. *Ecological Economics*, **6**, 185–93.

Daly, H.E. (1999) *Ecological Economics and the Ecology of Economics*, essay in criticism, Edward Elgar, UK.

Darcy, H.P.G. (1856) *Les Fontaines Publiques de ville de Dijon, Exposition et Application des Pricipes a Suivre et des Furmules a Emplyer dans les Questions de Distribution d'Eau.* Victor Dalmont, Paris.

Darton, R. (2002) *Sustainable Development and Energy: Predicting the Future.* Proceedings of the 15th International Conference of Chemical and Process Engineering, Prague, August.

Darwish, N.A., Al-Mehaideb, R.A., Braek, A.M., and Hughes, R. (2004) Computer simulation of BTEX emission in natural gas dehydration using PR and RKS equations of state with different predictive mixing rules. *Environmental Modelling & Software*, **19**, 957–65.

Davidson, L.B. (1969) The effect of temperature on the relative permeability ratio of different fluid pairs in two-phase systems. *Journal of Petroleum Technology*, **August**, 1037–46.

Davis, R.A., Thomson, D.H., Malme, C.I., and Malme, C.I. (1998) *Environmental Assessment of Seismic Explorations.* Canada/Nova Offshore Petroleum Board, Halifax, NS, Canada.

Davis, S. (1991) Unsteady facilitated transport of oxygen in hemoglobin-containing membranes and red cells. *Journal of Membrane Science*, **56**, 341–58.

Dayton, P.K. (1986) Cumulative impacts in the marine realm, in *Cumulative Environnemental Affects: A Binational Perspective.* Canadian Environmental Assessment Research Council, pp. 79–84. The Canadian Environmental Assessment Research Council, Ottawa, Canada.

De Groot, S.J.D. (1996) Quantitative assessment of the development of the offshore oil and gas industry in the North Sea. *ICES Journal of Marine Science*, **53**, 1045–50.

Deakin, S. and Konzelmann, S.J. (2004) Learning from Enron. *Corporate Governance*, **12**(2), 134–42.

DeMontigny, D., Tontiwachwuthikul, P., and Chakma, A. (2006) Using polypropylene and polytetrafluoroethylene membranes in a membrane contactor for CO_2 absorption. *Journal of Membrane Science*, **277**, 99–107l.

Dennis, L. (1995) *Natural Gas Production: Performance of Commercial Technology.* SPE/EPA Exploration and Production Environmental Conference, SPE Paper No. 29743, March 27–29, Houston, TX.

Department of Fisheries and Oceans (DFO) (1999) *Canadian Atlantic Integrated Fisheries Management Plan Bigeye, Yellowfin (Thunnus obesus). (Thunnus albacares). Albacore Tunas (Thunnus alalunga), 1998–1999.* Department of Fisheries and Oceans. <http://www.mardfo-mpo.gc.ca/fisheries/res/imp/98othon.html> [Accessed on April 22, 2002].

Department of Fisheries and Oceans (DFO) (2002) *GroundFish. Integrated Fisheries Management Plan. Scotia-Fundy Fisheries, Maritime Region. April 1, 2000–March 31, 2002.* Department of Fisheries and Oceans, Moncton, New Brunswick.

Department of Trade and Industry (2000) *Guidance Notes for Industry. Decommissioning of Offshore Installations and Pipelines under the Petroleum Act.* Offshore Decommissioning Unit, Atholl House, AB.

DeVries, A.J. (1958) Foam stability. *RubberChemistry and Technology*, **5**, 1142–205.

Dewalt, B.R., Vergne, P., and Hardin, M. (1996) Shrimp aquaculture development and the environment: people, mangroves and fisheries on the Gulf of Fonseca, Honduras. *World Development*, **24**(7), 1193–208.

Dewulf, J., Van Langenhove, H., Mulder, J., van den Berg, M.M.D., van der Kooi, H.J., and de Swaan Arons, J. (2002) Illustrations towards quantifying the sustainability of technology. *Green Chemistry*, **2**, 108–14.

Dewulf, J. and Langenhove, H.V. (2004) Integrated industrial ecology principles into a set of environmental sustainability indicators for technology assessment. *Resource Conservation & Recycling*, **43**(4), 419–32.

deZabala, E.F. and Radke, C.J. (1982) *The Role of Interfacial Resistances in Alkaline Water Flooding of Acid Oils*. 1982 SPE Annual Conference and Exhibition, SPE Paper No. 11213, New Orleans, pp. 26–9.

deZabala, E.F., Vislocky, J.M., Rubin, E., and Radke, C.J. (1980) *A Chemical Theory for Linear Alkaline Flooding*. The 5th SPE International Symposium on Oilfield and Geothermal Chemistry, SPE Paper No. 8997, Stanford, California, pp. 199–211.

DFO (2003) *State of the Eastern Scotian Shelf Ecosystem*. DFO Can. Sci. Advis. Sec. Ecosystem Status Rep. 2003/04.

Dilgren, R.E., Deemer, A.R., and Owens, K.B. (1982) *Laboratory Development and Field Testing of Steam/Non-condensable Gasfoams for Mobility Control in Heavy Oil Recovery*. California Regional Meeting, SPE Paper No. 10774, March 24–26.

Dincer, I. and Rosen, M.A. (2005) Thermodynamic aspects of renewable and sustainable development. *Renewable & Sustainable Energy Reviews*, **9**, 169–89.

Dincer, I. and Rosen, M. (2004) Exergy as a drive for achieving sustainability. *International Journal of Green Engineering*, **1**(1), 1–19.

Dindi, A., Noble, R.D., Yu, J., and Koval, C.A. (1992) Experimental and modeling studies of a parasitic binding mechanism in facilitated membrane transport. *Journal of Membrane Science*, **66**, 55–68.

Diviacco, P. (2005) An open source, web based, simple solution for seismic data dissemination and collaborative research. *Commuters & Geosciences*, **31**, 599–605.

Djabbarah, N.F., Weber, S.L., Freeman, D.C., Muscatello, J.A., Ashbaugh, J.P., and Covington, T.E. (1990) *Laboratory Design and Field Demonstration of Steam Diversion Withfoam*. 60th California Regional Meeting, SPE paper No. 20067, April 4–6, Ventura.

DOE (U.S. Department of Energy) (2002) <http://www.fe.doe.gov/oil_gas/drilling/laserdrilling.shtml> [Accessed on October 7].

DOE (1999) *Environmental Benefits of Advanced Oil and Gas Production Technology*, DOE-FE-0385, U.S. Department of Energy, Office of Fossil Energy, Washington, DC. <http://www.osti.gov/bridge/product.biblio.jsp?osti_id=771125>.

DOE (2004) *Enhanced Oil Recovery/CO2 Injection, US Department of Energy*, U.S. Department of Energy, 1000 Independence Ave., SW, Washington, DC, 20585.

DOE, (2005) *The Drilling Waste Management Information System, Fact Sheet – The First Step: Separation of Mud from Cuttings, Argonne National Laboratory, Natural Gas & Oil Technology Partnership Program*. <http://web.ead.anl.gov/dwm/index.cfm> [Accessed on October 22, 2006].

Donaldson, T.L. and J.A Quinn, (1975) Carbon dioxide transport through enzymatically active synthetic membranes. *Chemical engineering Science*, **30**, 103–15.

Dong, F., Hongmei, L., Kodama, A., Goto, M., and Hirose, T. (1999) The Petlyuk PSA process for the separation of ternary gas mixtures: exemplification by separating a mixture of CO_2-CH_4-N_2. *Separation and Purification Technology*, **16**, 159.

Donnelly, K., Beckett-Furnell, Z., Traeger, S., Okrasinski, T., and Holman, S. (2006) Eco-design implemented through a product-based environmental management system. *Journal of Cleaner Production*, **14**(15–16), 1357–67.

Doscher, T.M. (1967) Technical problems in in situ methods for recovery of bitumen from tar sands. *Proceedings 7th World Petroleum Congress*, **3**, 628.

Doscher, T.M. and Hammershaimb, E.C. (1981) *Field Demonstration of Steam Drive with Ancillary Materials.* SPE r DOE 2nd Symposium on Enhanced OilRecovery, SPE r DOE paper No. 9777, April 5–8, Tulsa, Okla.

Dosouky, S.M. (1996) Modelling and laboratory investigation of microbial enhanced oil recovery. *Journal of Petroleum Science and Engineering*, **15**, 309–20.

Draft Technical Background Document. Environment Department, Washington, DC.

Du, R., Feng, X., and Chakma, A. (2006) Poly (N,N-dimethylaminoethyl methacrylate)/polysulfone composite membranes for gas separations. *Journal of Membrane Science*, **279**, 76–85.

Dunn, K. (2003) *Caveman Chemistry*, chapter 8. Universal Publishers, USA.

Dunn-Norman, S., Gupta, A., Summers, D.A., Koederitz, L.F., and Numbere, D.T. (2002) *Recovery Methods for Heavy Oil in Ultra-Shallow Reservoirs.* The SPE Western Regional/AAPG Pacific Section Joint Meeting, SPE Paper No. 76710, May 20–22, Anchorage, AK, USA.

Dutta, N.N., Baruah, A.P., and Phukan, P. (1990) Natural gas sweetening: an overview. *Chemical Engineering World*, **25**.

DWMIS (Drilling Waste Management Information System) (2005) *The Drilling Waste Management Information System, Fact Sheet – The First Step: Separation of Mud from Cuttings, Argonne National Laboratory, Natural Gas & Oil Technology Partnership program.* <http://web.ead.anl.gov/dwm/index.cfm> [Accessed on October 22, 2006].

Dybas, C.L. (2005) *Marine Life Complicates Removal of Old Oil Rigs, the Washington Post Monday*, April 11, 2005; Page A10.

Dyer, S.B., Huang, S.S., Farouq Ali, S.M., and Jha, K.N. (1994) Phase behaviour and scaled model studies of prototype Saskatchewan heavy oils with Carbon dioxide. *JCPT* **October**, 42–8.

Dykstra, H. and Parsons, R.L. (1950) The Prediction of Oil Recovery by Waterflood, *Secondary Recovery of Oil in the United States*, American Petroleum Institute, New York, pp. 160–74; presented as paper #801-24k at the Spring Meeting of the Pacific Coast District, Division of Production, American Petroleum Institute, Los Angeles, May 6–7, 1948.

Earlougher, R.C. Jr. (1977) *Advances in Well Test Analysis*, 2nd edn, Monograph Series, SPE of AIME, New York, Dallas, USA.

EBW (2005) <www.ecobuildnetwork.org/pdfs/Straube_Moisture_Tests.pdf> [Accessed on June 2, 2006].

Editorial (1999a) Suncor starts construction of $C2B oil sands expansion. *Engineering and Mining Journal*, **200**(5), 10.

Editorial (1999b) Syncrude sets record for first quarter production. *Canadian Chemical News*, **51**(6), 8.

Edmonds, B., Moorwood, R.A.S., and Szczepanski, R. (1998) *Hydrate Update*, GPA Spring Meeting, Darlington, May.

Edmondson, T.A. (1965) Effect of temperature on water flooding. *Journal of Canadian Petroleum Technology*, **October–December**, 236–42.

Edmunds, N. (1999) On the difficult birth of SAGD. Distinguished Author Series. *Journal of Canadian Petroleum Technology*, **38**(1), 14.

Ehrlich, R. (1970) *The Effect of Temperature on water-oil Imbibition Relative Permeability.* Eastern Regional Meeting of SPE, SPE Paper No. 3214, November 5–6, 1970, Pittsburgh, PA.

Ehrlich, R. and Wygal, R.J. Jr. (1977) Interrelation of crude oil and rock properties with the recovery of oil by caustic water flooding. *Society of Petroleum Engineering Journal*, Paper No. 5830-PA, August, 1977 issue.

EIA (2002) *Energy Information Administration International Energy Annual 2002*, 1000 Independence Avenue, SW, Washington, DC 20585.

EIA (Energy Information Administration) (2003) *Annual Energy Outlook 2003 with Projections to 2025*, Report No.: DOE/EIA-0383(2003), Energy Information Administration, Washington, DC 20585.

EIA (2004a) *International Energy Outlook 2004*, Greenhouse Gases general information. Energy Information Administration, Environmental Issues and World Energy Use. EI 30, 1000 Independence Avenue, SW, Washington, DC 20585.

EIA (2004b) *World Proved1 Reserves of Oil and Natural Gas, Most Recent Estimates.* 1000 Independence Avenue, SW, Washington, DC, USA.

EIA (2006) *Energy Information Administration, Office of Oil and Gas*, Natural Gas Division, January. <www.eia.doe.gov/pub/oil_gas/natural_gas/feature_articles/2006/ngprocess/ngprocess.pdf> [Accessed on July 22, 2006].

Eissen, M. And Metzger, J.O. (2002) Environmental performance metrics for daily use in synthetic chemistry. *Chemistry A European Journal*, **8**, 3581–5.

Ekins, P., Vanner, R., and Firebrace, J. (2006) Decommissioning of offshore oil and gas facilities: a comparative assessment of different scenarios. *Journal of Environmental Management*, **79**, 420–38.

Elkamel, A., Al-Sahhaf, T., and Ahmed, A.S. (2002) Studying the interactions between an Arabian heavy crude oil and alkaline solutions. *Journal Petroleum Science and Technology*, **20**(7&8), 789–807.

Ellig, D.L., Althouse, F.P., and McCandless, C. (1980) Separation of gases using membrane systems. *Journal of Membrane Science*, **6**, 250.

Elliott, G. (2005) *Personal Communication*, Executive Chairman, Titanium Corporation, Toronto, Ontario, Canada.

Emery, L.W., Mungan, N., and Nicholson, R.W. (1970) Caustic slug injection in the singleton field. *Journal of Petroleum Technology*, **22**, 1569–76.

Energy Information Administration (1990) The domestic oil and gas recoverable resource base – supporting analysis for the national energy strategy: U.S. Department of Energy, Energy Information Administration SR/NES/90-05.

Energy Information Administration (1999) Annual Energy Review 1998: U.S. Department of Energy, Energy Information Administration DOE/EIA-0384 (98), **July**, pp. 119.

Engas (1993) Effects of Seismic Shooting on catch and catch-availability of cod and haddock. *Fisken og Havet*, **9**, 99–117.

Enviro Voraxial Technology Inc. (EVTN) (2003) *Technological Overview.* <http://www.evtn.com/techoverview.htm> [Accessed on March 3].

Environment Canada (1973) *Guidelines on the Use and Acceptability of Oil Spill Dispersants.* Environment Canada, Environmental emergency Branch, Ottawa, Ontario.

Environment Canada (1999) *Compliance and Enforcement Policy for Wildlife Legislation.* Environment Canada, Ottawa, Ontario.

Environment of Canada (2005) *Petroleum Refining, Environmental Protection Branch*, Atlantic Region, 45 Alderney Drive, Dartmouth, NS, Canada.

Environmental Defense (2004) <http://www.environmentaldefense.org/> [Accessed on June 2, 2006].

EOR Survey (2004) 2004 worldwide EOR survey. *Oil & Gas Journal's 2004*, **102**(14), 53.

EPA (2000) *Development Document for Final Effluent Limitations Guidelines and Standards for Synthetic-Based Drilling Fluids and other Non-Aqueous Drilling Fluids in the Oil and Gas Extraction Point Source Category.* EPA-821-B-00-013, U.S. Environmental Protection Agency, Office of Water, Washington, DC 20460, December, <http://www.epa.gov/waterscience/guide/sbf/final/eng.html>.

EPA (2002) *Drilling Wastes Characterization, Control, and Treatment Technologies*, chapter VII. United States Environmental Protection Agency. Office of Water, Washington, DC.

EPA/DOE (1999) *Environmental Benefits of Advanced Oil and Gas Production Technology.* DOE-FE-0385, U.S. Department of Energy, Office of Fossil Energy, Washington, DC <http://www.osti.gov/bridge/product.biblio.jsp?osti_id=771125> [Accessed on October 12, 2006].

Ertekin, T., Abou-Kassem, J.H., and King, G.R. (2001) *Basic Applied Reservoir Simulation.* SPE textbook series, 7, Richardson, Texas: SPE.

ESCAP (Economic and Social Commission for Asia and the Pacific) (1987) *Coastal Environmental Management Plan for Bangladesh*, II-Main Report, Bangkok.

Esteban, A., Hernandez, V., and Lunsford, K. (2000) *Exploit the Benefits of Methanol*, Proceedings of 79th GPA Annual Convention, Atlanta, GA.

Fabbrocino, G., Iervolino, I., Orlando, F., and Salzano, E. (2005) Quantitative risk analysis of oil storage facilities in seismic areas. *Journal of Hazardous Materials*, **A123**, 61–9.

Fain, D.E., Marshall, B.B., Raj, T., Addock, K.D., and Powell, L.E. (2001) *Ceramic Membranes for High Temperature Hydrogen Separation.* 15th Annual Conference on Fossil Energy Materials, April 30–May 2, 2001, Oak Ridge, Tennessee.

Fall, M.L. and Shulman, H.S. (2003) *Comparison of Energy Consumption for Microwave Heating of Alumina, Zirconia, and Mixtures.* 105th Annual American Ceramic Society National Convention in Nashville Tennessee, April 28, 2003.

Fall, P.L. (2005) Vegetation change in the coastal-lowland rainforest at Avai'o'vuna Swamp, Vava'u, Kingdom of Tonga. *Quaternary Research*, **64**(3), 451–9.

Falls, A.H., Hirasaki, G.J., Patzek, T.W., Gaughtz, D.A., Miller, D.D., and Ratulowski, T. (1988) Development of a mechanistic foam simulator: the population s. balance and generation by snap-off. *SPE Reservoir Engineering*, **August**, 884–92.

Falls, A.H., Musters, J.J., and Ratulowski, J. (1989) The apparent viscosity of foams in homogeneousbeadpacks. *SPE Reservoir Engineering*, **May**, 155–64.

Fan, D., White, R.E., and Gruberger, N. (1992) Diffusion of gas through a membrane. *Journal of Applied Electrochemistry*, **22**, 770–2.

Fanchi, J.R. (1990) *Feasibility of Reservoir Heating by Electromagnetic Irradiation.* 69th Annual Technical Conference and Exhibition, SPE Paper No. 28619, September 23–26, New Orleans, USA.

FAO (1984) *Integrated Development of Sundarbans, Bangladesh.* Fisheries integrated development in the Sundarbans, based on the work of H.R. Rabanal. FAO: TCP/BGD/2309.

FAO (1994) Mangrove forest guidelines. *FAO Forestry Paper*, **117**, 319pp.

Farouq Ali, S.M. (1997) Heavy Oil Research Laboratories, Edmonton, Alberta, Canada, Personal communication.

Farouq Ali, S.M. (2004) *The Unfulfilled Promise of Enhanced Oil Recovery – What More Lies Ahead?* SPE Distinguished Lecture Series, Halifax, 2003–2004, May 14, 2006.

Farouq Ali, S.M., Redford, D.A., and Islam, M.R. (1987) *Scaling Laws for Enhanced Oil Recovery Experiments.* Proceedings of the China-Canada Heavy Oil Symposium, Zhu Zhu City, China.

FBB (Friends of Bristol Bay) (2005) The Impacts of Seismic Surveys on marine mammals and fish, Friends of Bristol Bay, PO Box 334, Naknek AK 99633, Phone/Fax: (907) 246-FISH, <www.akmarine.org> [Accessed on June 24, 2006].

Feng, X., Pan, C.Y., McMinis, C.W., Ivory, J., and Ghosh, D. (1998) ••. *American Institute of Chemical Engineers Journal*, **44**(7), 2210.

Fenn, O. (1989) The use of water jets to assist free-rolling cutters in the excavation of hard rock. *Tunnelling and Underground Space Technology*, **4**(3), 409–17.

Fernandez, L.G., Soria, C.O., Garcia Tourn, C.A., and Izquierdo, M.S. (2001) *The Study of Oil/Water Separation in Emulsion by Membrane Technology.* SPE Latin American and Caribbean Petroleum Engineering Conference, SPE Paper No. 69554, March 25–28, Buenos Aires, Argentina.

Ferreira, D.F. and Suslick, S.B. (2001) Identifying potential impacts of bonding instruments on offshore oil projects. *Resources Policy*, **27**, 43–52.

Ferreira, D., Suslicka, S., Farleyc, J., Costanzac, R., and Krivovc, S. (2004). A decision model for financial assurance instruments in the upstream petroleum sector. *Energy Policy*, **32**, 1173–84.

Fischer, A. and Hahn, C. (2005) Biotic and abiotic degradation behaviour of ethylene glycol monomethyl ether (EGME). *Water Research*, **39**, 2002–7.

Fisheries and Oceans Canada (FOC) (2001) *Atlantic salmon Maritime Province overview for 2000 Fisheries and Oceans Canada, Gulf Fisheries Management and Maritimes Regions.* Stock Status Report D3-14(2001)(revised). Fisheries and Oceans Canada, Maritime Region, Moncton, New Brunswick.

Folkner, C.A., and Noble, R.D. (1983) Transient response of facilitated transport membranes. *Journal of Membrane Science*, **12**, 289–301.

Forest, J. and Oetti, M. (2003) Rigorous simulation supports accurated refinery decisions, in *Proceedings of Fortune International Conference on Foundations of Computers-Aided Process Operations* (eds I.E. Grossman and C.M. McDonald). Coral Springs, CAChE, pp. 273–80.

Frauenfeld, D., Lillico, C., Jossg, G., and Vilcsak, S. (1998) Evaluation of partly miscible processes for Alberta heavy oil reservoirs. *Journal of Canadian Petroleum Technology*, **37**(4), 17–24.

Freeborn,W.K., Skoreko, F.A., and Luhning, R.A. (1989) *AWACT: Anti-water Coning Technology.* Proceedings of the AOSTRA Conference, Edmonton, June.

Frempong, P. (2005) *Ultrasonic of pore fluid pressure and scaling to field measurement.* PhD dissertation, Department of Civil and Resource Engineering, Dalhousie University.

Frempong, P.K., Butt, S., and Islam, M.R. (2003) *Linking Acoustic and Mechanical Properties of Materials with Particular Reference to Reservoir Rocks: A Critical Literature Review.* Proceedings of Oil and Gas Symposium, CSCE Annual Conference, refereed proceeding, Moncton, June.

French, D.P. (1998) Evolution of oil trajectory, Fate and Impact Assessment Models, in *Oil Spill 98 Conference, 1998.* Wessex Institute of Technology, Ashurst Lodge, Ashurst, Southampton, UK.

French, D. (2001) Modeling oil spills impacts, in *State of the Art in Oil Modelling and Processess* (eds R. Garcia-Martinez and C. Brebbia). WIT Press, Ashurst Lodge, Ashurst, Southampton, UK.

Friends of Bristol Bay (2005) The Impacts of Seismic Surveys, on marine mammals and fish. PO Box 334, Naknek AK 99633, Phone/Fax: (907) 246-FISH, <www.akmarine.org>, [Accessed on July 19, 2006].

Fritzsche, A.K., Cruse, C.A., Murphy, M.K., and Kesting, R.E. (1990) Polyethersulfone and polyphenylsulfone hollow fiber trilayer membranes spun from Lewis acid: base complexes – Structure determination by SEM, DSC, and oxygen plasma ablation. *Journal of Membrane Science*, **54**(1), 29–50.

Frostman, L.M. (2000) *Anti-aggolomerant Hydrate Inhibitors for Prevention of Hydrate Plugs in Deepwater Systems.* Proceedings of SPE Annual Technical Conference and Exhibition, October 1–4, 2000, pp. 573–9, Dallas, TX.

Fuertes, A.B. (2001) Effect of air oxidation on gas separation properties of adsorption-selective carbon membranes. *Carbon*, **39**, 697.

Funston, R., Ganesh, R., and Leong, L.Y.C. (2002) *Evaluation of Technical and Economic Feasibility of Treating Oilfield Produced Water to Create a 'New' Water Resource.* <www.gwpc.org/Meetings/PW2002/Papers/Roger_Funston PWC2002.pdf> [Accessed on June 15, 2005].

Gabos, S. (2000) *Solution Gas Flaring and Public Health*, March 16, 2000, Cited in Canadian Public Health Association. Canadian Public Health Association 2000 Resolutions & Motions, Resolution No. 3.

Gaddy, D.E. (1998) Next-generation drilling equipment pushes back water depth barrier. *Oil & Gas Journal*, **96**(9), 74 (3 pages).

Gagnon, M.M. and D.A. Holdway. (1999) Metabolic enzyme activities in fish gills as biomarkers of exposure to petroleum hydrocarbons. *Ecotoxicology and Environmental Safety* **44**(1), 92–9.

Gahan, B.C., Parker, R.A., Batarseh, S., Figueroa, H., Reed, C.B., and Xu, Z. (2001) *SPE Annual Technical Meeting and Dinner.* September 30–October 3, New Orleans, LA.

Gal, T., Stewart, T.J., and Hanne, T. (eds) (1999) *Multicriteria Decision Making Advances in MCDM Models*, Algorithms, Theory and Applications, Kluwer Academic Publishers.

Gatlin, C. (1960) *Petroleum Engineering – Drilling and Well Completions.* Prentice Hall, Inc., Englewood Cliffs, NJ.

Geankopolis, C.J. (1993) *Transport Processes and Unit Operations*, 3rd edn. Prentice Hall Inc., Englewood Cliffs, NJ.

GEMI (2004) *Enhancing Supply Chain Value Through Environmental Excellence.* Global Environmental Management Initiative, Washington, DC 20005.

George, D.S. and Islam, M.R. (1998) *A Micromodel Study of Acidizationfoam.* SPE Paper No. 49489, ADIPEC 98, October.

Georgiadis, P. and Vlachos, D. (2004) The effect of environmental parameters on product recovery. *European Journal of Operational Research*, **157**, 449.

GESAMP (I MO/FAO/UNESCO/WMO/WHO/IAEA/UNEP) Joint Group of Experts on the Scientific Aspects of marine Pollution (1993) *Impacts of Oil and Related Chemicals and Wastes on the Marine Environment.* GESAMP Reports and Studies No. 50. International Maritime Organization, London.

Gessinger, G. (1997) Lower CO_2 emissions through better technology. *Energy Conversion and Management*, **38**, 25–30.

Gibson, B. (1991) Should environmentalists pursue sustainable development? *Probe Post*, **Winter**, 22–5.

Gleick, J. (1987) *Chaos – Making a New Science.* Penguin Books, New York.

Goggin, D. (1998) *Geologically-Sensible Modeling of the Spatial Distribution of Permeability in Eolian Deposits: Page Sandstone, Northern Arizona.* PhD Thesis, Petroleum Engineering Department, University of Texas at Austin.

Goggin, D.J., Thrasher, R., and Lake, L.W. (1988) A theoretical and experimental analysis of minipermeameter response including gas slippage and high velocity flow effects. *In Situ*, **12**(1&2), 79–116.

Goodwin, S. and Hunt, A.P. (1995) *Prediction, Modeling and Management of Hydrates Using Low Dosage Additives*, Advances in Multiphase Operations Offshore Conference, November 29–30, 1995, London, UK.

Government of Nova Scotia 2001, Nova Scotia Energy Policy (2001) Government of Nova Scotia, Halifax, Canada.

Government of the People's Republic of Bangladesh (2006) <http://www.bangladeshgov.org> [Accessed on August 22, 2006].

GPSA (1998) *Gas Processors Supplies Association Engineering Data Book*, 11th edn. Gas Processors Suppliers Association, Tulsa, OK.

Gramain, P. and Sanchez, J. (2002) Membrane pour la separation selective Gazeuse. French Patent. WO 022245, March.

Grant, A. and Briggs, A.D. (2002) Toxicity of sediments around a North Sea oil platform: are metals or hydrocarbons responsible for ecological impacts? *Marine Environmental Research*, **53**, 95–116.

Graves, R.M., Gahan, B.C., and Parker, R.A. (2002) *Comparison of Specific Energy between Drilling with High Power Lasers and other Drilling Methods.* SPE Annual Technical Conference and Exhibition, Paper No. 77627, September 29–October 2, 2002, San Antonio, TX, USA.

Graves, R.M., and O'Brien, D.G. (1998a) *StarWars Laser Technology Applied to Drilling and Completing Gas Wells.* 1998 SPE Annual Technical Conference and Exhibition, SPE Paper No. 49259, September 27–30, 1998, New Orleans.

Graves, R.M. and O'Brien, D.G. (1998b) Targeted Literature Review: Determining the Benefits of StarWars Laser Technology for Drilling and Completing Natural Gas, GRI-98/0163, July.

Gray, L.C. (1914) Rent under the assumption of exhaustibility. *Quart. J. Econ.*, **28**, 466–89.

Green, D.W. and Willhite, G.P. (1998) *Enhanced Oil Recovery, Richardson.* Henry L. Doherty Memorial Fund of AIME, Society of Petroleum Engineers, TX, pp. 240–92.

GRI (Global Reporting Initiative) (2002) *Sustainability Reporting Guidelines.* GRI, Boston, MA.

Grigg, R.B. and Schechter, D.S. (1997) *State of the Industry in CO2 Floods.* Annual Technical Conference and Exhibition, SPE Paper No. 38849, October 5–8, San Antonio, TX.

Grogan, A.T. and Pinczewski, W.V. (1987) The role of molecular diffusion processes in tertiary CO2 flooding. *Journal of Petroleum Technology*, **May**, SPE-12706.

Gu, C., Gao, G.-H., Yu, Y.-X., and Nitta, T. (2002) Simulation for separation of hydrogen and carbon monoxide by adsorption on single-walled carbon nanotubes. *Fluid Phase Equilibria*, **194–197**, 297.

Guha, A.K., Majumdar, S., and Sirkar, K.K. (1990) Facilitated transport of CO_2 through and immobilized liquid membrane of aqueous diethanolamine. *Industrial and Chemical Engineering Research*, **29**, 2093–100.

Gunal, O.G. and Islam, M.R. (2000) Alteration of ahphaltic crude rheology with electromagnetic and ultrasonic irradiation. *Journal of Petroleum Science and Engineering*, **26**, 263–72.

Guzzella, L., Roscioli, C., Viganò, L., Saha, M., Sarkar, S.K., and Bhattacharya, A. (2005) Evaluation of the concentration of HCH, DDT, HCB, PCB and PAH in the sediments along the lower stretch of Hugli estuary, West Bengal, northeast India. *Environment International*, **31**(4), 523–34.

HAL and GMA (Hatch Associates Limited and Griffiths Muecke Associates) (2000) Workshop on cumulative environmental affects assessment and monitoring on the Grand Banks and Scotian Shelf. Environmental Studies Research Funds Report ESRF 137, Ottawa, Ontario.

Hallada, M.R., Walter, R.F., and Seiffert, S.L. (2000) *Proceeding of SPIE*. 4065, pp. 614.

Hamzah, B.A. (2003) International rules on decommissioning of offshore installations: some observations. *Marine Policy*, **27**, 339–48.

Handy, L.L., Amaefule, J.O., Ziegler, V.N., and Ershaghi, I. (1982) Thermal stability of surfactants for reservoir applications. *Society of Petroleum Engineering Journal*, **October**, 72230.

Haque, K.E. (1999) Microwave energy for mineral treatment processes – a brief review. *International Journal of Mineral Processing*, **57**(1), 1–24.

Harding, G. (1986) Cultural carrying capacity: a biological approach to human problems. *Bio Science*, **36**, 599–606.

Harding, G. (1993) *Living within Limits*. Oxford University Press, New York.

Hartung, R., Rigas, L.K., and Cornish, H.H. (1970) Acute and chronic toxicity of diethanolamine. *Toxicol. Appl. Pharmacol*, **17**, 308 (abstract).

Hawken, P. (1992) *The Ecology of Commerce*. Plenum, New York.

Haynes, H.J., Thrasher, L.W., Katz, M.L., and Eck, T.R. (1976) Enhanced Oil Recovery, An Analysis of the Potential for Enhanced Oil Recovery from Known Fields in the United States, National Petroleum Council, 1625K Street, NW, Suite 600, Washington, DC. 20006.

Henda, R., Herman, A., Gedye, R., and Islam, M.R. (2005) Microwave enhanced recovery of nickel-copper ore: communication and floatability aspects. *Journal of Microwave Power & Electromagnetic Energy*, **40**(1), 7–16.

Hendrichs, H. (1975) The status of the tiger (Panthera tigiris) (Linne) in the Sundarbans mangrove forest (Bay of Bengal). *Saugetier-kundliche Mitteilugen*, **23**, 161–99.

Hibernia Environmental Assessment Panel (HEAP) (1986) *Hibernia Development Project: Report of the Environmental Assessment Panel*. Ministry of Supply and Services Canada, Ottawa, Ontario.

Hirasaki, G.J. (1989) The steam-foam process. *Journal of Petroleum Technology*, **May**, 449.

Hirasaki, G.J. and Lawson, J.B. (1985) Mechanism of foam flow through porous media: apparent viscosity in smooth capillaries. *Society of Petroleum Engineering Journal*, **April**, 176–90.

Hirst, A.G. and Rodhouse, P.G. (2000) Impacts of geophysical seismic surveying on fishing success. *Reviews of Fish Biology and Fisheries*, **10**, 113–18.

Hisham, M., Ghazi Al-Enezi, E., and Hughes, R. (1995) Modelling of enrichment of natural gas wells by membranes. *Gas Sep. Purif.* **9**(1), 3–11.

Hitzman, D.O. (1988) Review of microbial enhanced oil recovery field tests. In symposium on Applying Microorganisms to Petroleum Technology, pp. VI-1–VI-41.

Hobbs, B.F. and Meier, P. (2000) Energy decisions and the environment. A Guide to the Use of Multicriteria Methods (International Series in Operations Research & Management Science) Kluwer Academic Publishers, Massachusetts.

Ho-Cheng, H. (1990) A failure analysis of water jet drilling in composite laminates. *International Journal of Machine Tools and Manufacture*, **30**(3), 423–9.

Holdway, D.A. (2002) The acute and chronic effects of wastes associated with offshore oil and gas production on temperate and tropical marine ecological processes. *Marine Pollution Bulletin*, **44**, 185–203.

Hong, N.P. (1996) Mangrove destruction for shrimp rearing in Minh Hai, Vietnam: its damage to natural resources and environment. *SEAFDEC Asian Aquaculture*, **18**, 7–11.

Hood, M., Nordlund, R., and Thimons, E. (1990) A study of rock erosion using high-pressure water jets. *International Journal of Rock Mechanics and Mining Science & Geomechanics*, **27**(2), 77–86.

Horne, R.N. (1997) *Modern Well Test Analysis*; *A Computer – Aided Approach*, 2nd edn, Petroway, Inc., 926 Bautistu, Palo Alto, California, USA.

Hossain, M.E., Ketata, C., Khan, M.I., and Islam, M.R. (2006c) Waterjet – a sustainable and effective drilling technique. *Journal of Nature Science and Sustainable Technology*, in press.

Hossain, M.E., Khan, M.I., Ketata, C., and Islam, M.R. (2005) Flammability and individual risk assessment for natural gas pipelines. *International Journal of Risk Assessment and Management*, in press.

Hossain, M.E., Khan, M.I., Ketata, C., and Islam, M.R. (2006a) Computation of flammability and a measure of individual risk assessment for natural gas pipeline. *International Journal of Risk Assessment and Management*, in press.

Hossain, M.E., Khan, M.I., Ketata, C. and Islam, M.R. (2006b) SEM-based structural and demrial analysis of paraffin wax and beeswax for petroleum appliations. *International Journal of Material Properties and Technology*, Submitted.

Hossain, M.E., Khan, M.I., Ketata, C., and Islam, M.R. (2007) Human Health Risks Assessment Due to Natural Gas Pipeline Explosions, *Int. J. Risk Assessment and Management*, in press.

Hossain, M.S. (2001) Biological aspects of the coastal and marine environment of Bangladesh. *Ocean & Coastal Management*, **44**(3–4), 261–82.

Hotelling, H. (1931) The Economics of Exhaustible Resources. *Journal of Political Economy*, **39**, 137–75.

Hu, W. and Chakma, A. (1998) Modelling of equilibrium solubility of CO_2 and H_2S in aqueous diglycolamine (DGA) solutions. *Canadian Journal of Chemical Engineering*, **68**, 409–12.

Hua, Y., Lin, Z., and Yan, Z. (2002) Application of microwave irradiation to quick leach of zinc silicate ore. *Minerals Engineering*, **15**(6), 451–6.

Huang, T. and Rudnicki, J.W. (2006) A mathematical model for seepage of deeply buried groundwater under higher pressure and temperature. *Journal of Hydrology*, **327**(1–2), 42–54.

Huizing, A. and Dekker, H.C. (1992), Helping to pull our last planet out of the red: an environmental report of BSO/ORIGIN. *Accounting, Organisations and Society*, **17**, 449–58.

Hull, T.R., Quinn, R.E., Areri, I.G., and Purser, D.A. (2002) Combustion toxicity of fire retarded EVA. *Polymer Degradation and Stability*, **77**, 235–42.

Human Resources Development Canada (HDRC) (2001) Industry Profile: Human Resources Profile: Fish Products Industries and Services Incidental to Fishing.

Hunt, A. (1996) Fluid properties determine flow line blockage potential. *Oil & Gas Journal*, **94**(29), 62–6.

Hydrocarbon Engineering Journal (1999) *Offshore Sour Gas Treatment*, February.

IChemE (Institute of Chemical Engineers) (2002) The sustainability metrics, sustainable development progress metrics recommended for use in the process industries, 2002 <http://www.getf.org/file/toolmanager/O16F26202.pdf> [Accessed on June 10, 2005].

IEA (International Energy Agency GHG) (1995) *R&D Program Report*, Carbon Dioxide Utilization, IEA Greenhouse Gas R&D Program, Cheltenham, UK.

Imai, S. (2001) Case study of Japanese companies': environmental accounting in Asia, In *International Forum on Business and the Environment*, pp. 13–25. Ministry of the Environment, Japan, September 27–29.

Indian Energy Sector (2002) *Managing Fly Ash*, <http://www.teriin.org/energy/flyash.htm> [Accessed on March 11, 2005].

Ingram, D. (1986) Is RF energy hazardous to your health? *CQ radio amateur Journal*, **42**, 102–4

Innovation (2002) Achieving excellence: investing in people, knowledge and opportunity Canada's innovation strategy Canada, Canada's Innovation Strategy. Ottawa, Ontario.

IPCC (2001) *Climate Change 2001: The Scientific Basis*. Contribution of the working group I to the third Assessment report of the Intergovernmental Panel on Climate Change, Cambridge University Press.

IRIN (2006) *NIGERIA: More than 150 Killed in Pipeline Blast*, UN Office for the Coordination of humanitarian affairs, <www.IRINnews.org>.

Islam, M.R., Erno, B.P., and Davis, D. (1992) Hot water and gas flood equivalence of in Situ combustion. *Journal of Canadian Petroleum Technology*, **31**(8), 44–52.

Islam, H. (2001) Sundarbans Forest Facing Serious Threats, *Wnwire*. United Nations Foundation, April 26.

Islam, K.L. and Hossain, M.M. (1986) Effects of ship scraping activities on the soil and sea environment in the coastal area of Chittagong, Bangladesh. *Marine Pollution Bulletin*, **17**(10), 462–3.

Islam, M.R. (2004) Unraveling the mysteries of chaos and change: the inherently sustainable technology development. *EEC Innovation*, **2**, 108.

Islam, M.R. (1993) Oil recovery from bottom water reservoirs. *Journal of Petroleum Technology*, **June**, 514–16.

Islam, M.R. (1996) Emerging technologies in enhanced oil recovery. *Energy Sources*, **21**, 97–111.

Islam, M.R. (2000) State-of-the-art of Enhanced Oil Recovery, in *Energy State of the Art 2000*, (ed. C.Q. Zhou), International Science Services, Charlottesville, VA, pp. 325–44.

Islam, M.R. (2002) Mathematical Modelling of Abnormally High Formation Pressures, in *Origin and Prediction of Abnormal Pressures* (ed. G.V. Chilingar, Serebryakov, V.A., and Robertson, J.O., Jr.), Elsevier, Amsterdam, pp. 311–51.

Islam, M.R. (2003) Revolution in education, in *EECRG*, Nova Scotia, Canada [ISBN 0-9733656-1-7].

Islam, M.R. (2005) Beyond synthetic plastic, in *Executive Times*, March, Dhaka, Bangladesh.

Islam, M.R. and Chakma, A. (1992) A new recovery technique for heavy-oil reservoirs with bottom water. *SPE Reservoir Engineering*, **May**, 180–6.

Islam, M.R. and Farouq Ali, S.M. (1989a) Water flooding oil reservoirs with bottomwater. *Journal of Canadian Petroleum Technology*, **28**, 59–66.

Islam, M.R. and Farouq Ali, M.S. (1989b) *Numerical Simulation of Alkaline/Cosurfactant/Polymer Flooding*; Proceedings of the UNI-TAR/UNDP Fourth Int. Conf., Heavy Crude and Tar Sand.

Islam, M.R. and Farouq Ali, S.M. (1990) New scaling criteria for chemical flooding experiments. *Journal of Canadian Petroleum Technology*, **29**(1), 30–6.

Islam, M.R. (2001a) *Emerging Technologies in Subsurface Monitoring of Petroleum Reservoirs*. SPE Latin America and Caribbean Petroleum Engineering Conference, SPE Paper No. 69440. March 25–28, Buenos Aries.

Islam, M.R. (2001b) Foreword, *Petroleum Geology of the South Caspian Basin: Description of Selected Oil and Gas Reservoirs* by L.A. Buryakovsky, G.V. Chilingar and F. Aminzadeh. Gulf Publishing, Boston, MA.

Islam, M.R. (2002) Emerging technologies in subsurface monitoring of petroleum reservoirs. *Petroleum Res. Journal*, **13**, 33–46.

Islam, M.R. and Abou-Kassem, J.H. (1998) *Experimental and Numerical Studies of Electromagnetic Heating of a Carbonate Formation*. Proceedings of the SPE Annual Technical Conference and Exhibition, SPE Paper No. 49187, New Orleans, USA.

Islam, M.R. and Bentsen, R.G. (1986) A dynamic method for measuring relative permeability. *J. Can. Pet. Tech.*, **25**(1), 39–50.

Islam, M.R. and Bentsen, R.G. (1987) Effect of different parameters on two-phase relative permeability. *AOSTRA Journal Research*, **3**, 69–90.

Islam, M.R. and Chakma, A. (1990) Mathematical modelling of visbreaking in a jet reactor. *Chem. Eng. Sci.*, **45**(8), 2769–75.

Islam, M.R. and Chakma, A. (1993) Storage and utilization of CO_2 in petroleum reservoirs – a simulation study. *Energy Conversion and Management*, **34**(9), 1205–12.

Islam, M.R. and Farouq Ali, S.M. (1991) Scaling of in-situ combustion experiments. *Journal of Petroleum Science and Engineering*, **6**, 367–79.

Islam, M.R. and George, A.E. (1991) Sand control in horizontal wells in heavy oil reservoirs. *Journal of Petroleum Technology*, **43**(7), 844–53.

Islam, M.R. and Wellington, S.L. (2001) *Past, Present, and Future Trends in Petroleum Research*. SPE Western Regional Meeting Held in Bakersfield, SPE Paper No. 68799, March 26–30, 2001, California.

Islam, M.R. and Zatzman, G.M. (2005) *Unravelling the Mysteries of Chaos and Change: the Knowledge-based Technology Development*, Proceedings of the First International Conference on Modeling, Simulation and Applied Optimization, Sharjah, U.A.E. February 1–3, 2005.

Islam, M.R. (1985) *Impact of flow regimes on oil-water relative permeability*, MSc Thesis, Department of Mining, Metallurgical and Petroleum Engineering, University of Alberta.

Islam, M.R. (1990) *Mathematical Modelling of Multiphase in a Horizontal Well Incorporating Flow Through Perforations.* Invited speaker, SPE Forum Series in Europe, Grindelwald, Switzerland, September 16–21.

Islam, M.R. (1994) A new recovery technique for gas production from Alaskan gas hydrates. *Journal of Petroleum Science and Engineering*, **11**(4), 267–81.

Islam, M.R. (1998) *Simultaneous CO2 Sequestration and Enhanced Oil Recovery in Heavy Oil Formations*, PTRC Internal report.

Islam, M.R. (1999) Emerging technologies in EOR. *Energy Sources*, **21**, 27–111.

Islam, M.R. (2000) *Emerging Technologies in Subsurface monitoring of Oil Reservoirs.* Paper No. CIM 2000-89, Canadian International Petroleum Conference, Calgary, June.

Islam, M.R. (2004) Inherently-sustainable energy production schemes. *EEC Innovation*, **2**(3), 38–47.

Islam, M.R. (2005a) Knowledge-Based Technologies For The Information Age, JICEC05-Keynote speech, Jordan International Chemical Engineering Conference V, September 12–14, 2005, Amman, Jordan.

Islam, M.R. (2005b) Unraveling the mysteries of chaos and change: knowledge-based technology development. *EEC Innovation*, **2**(2–3), 45–87.

Islam, M.R. (2005c) Unravelling the mysteries of chaos and change: The Knowledge-based technology development, Invited speech, *Joint Session of Fifth International Conference on Composite Science and Technology and First International Conference on Modeling, Simulation and Applied Optimization*, February, 2005, Sharjah, UAE.

Islam, M.R. (2005d) Developing knowledge-based technologies for the information age through virtual universities, E-transformation Conference, April 21, (2005) Turkey.

Islam, M.R. (2006) *Computing for the Information Age*, Keynote speech, Proceeding of the 36th International Conference on Computers and Industrial Engineering, June, Taiwan.

Islam, M.R. and Chakma, A. (1988) *Role of Dynamic Interfacial Tensions in Numerical Simulation of Cosurfactant/ Alkaline/Polymer Floods.* AIChE Ann. Meet., Paper No. 98-e, November, Washington, DC.

Islam, M.R. and Chakma, A. (1991) Mathematical modelling of enhanced oil recovery by alkali solutions in the presence of cosurfactant and Polymer. *Journal of Petroleum Science and Engineering*, **5**, 105–26.

Islam, M.R. and Farouq Ali, S.M. (1994) Numerical simulation of emulsion flow through porous media. *Journal of Canadian Petroleum Technology*, **33**(3), 59–63.

Islam, M.R., Chakma, A., and Jha, K. (1994) Heavy oil recovery by inert gas injection with horizontal wells. *Journal of Petroleum Science and Engineering*, **11**(3), 213–26.

Islam, M.R., Chakma, A., and Nandakumar, K. (1990) Flow transition in mixed convection in a porous medium saturated with water near 4 C. *Can. J. Chem. Eng.*, **68**, 777–85.

Islam, M.R. and Gianetto, A. (1993) Mathematical modelling and scaling up of microbial enhanced oil recovery. *Journal of Canadian Petroleum Technology*, **32**(4), 30–6.

Islam, M.R., Harding, T., Stewart, B., and Moberg, R. (1999) *Heavy Oil Research in Canada*, IEA Workshop on EOR, September, Paris, France.

Islam, M.R., Mustafiz, S., Belhaj, H., and Mousavizadegan, S. (2008) *A Handbook of Knowledge-Based Reservoir Simulation.* Gulf Publishing Company, Austin, TX.

Islam, M.R., Verma, A., and Ali, S.M.F. (1991) In situ combustion – the essential reaction kinetics, Heavy Crude and Tar Sands – Hydrocarbons for the 21st Century, 4, UNITAR/UNDP.

Islam, M.R., Zatzman, G., Mustafiz, S., and Khan, M.I. (2006) Greening of petroleum operations. *Journal of Nature Science and Sustainable Technology*, in press.

Islam, M.S. (2003) Perspectives of the coastal and marine fisheries of the Bay of Bengal, Bangladesh. *Ocean & Coastal Management*, **46**(8), 763–96.

Islam, M.R. and Chilingar, G.V. (1995) A new technique for recovering heavy oil and tar sands. *Scientia Iranica International Journal of Science and Technology*, **21**, 15–52.

Izumi, J., Tomonaga, N., Yasutake, A., Tsutaya, H., and Oka, N. (2002) Development on high performance gas separation process using gas adsorption. *Technical Review – Mitsubishi Heavy Industries*, **39**, 6.

Jack, T.R., Lee, E., and Mueller, J. (1985) Anaerobic gas production: controlling factors, in *Microbes and Oil Recovery. Proceedings of the International Conference on Microbial Enhancement of Oil Recovery* (eds J.E. Zajic and E.C. Donaldson). Petroleum Bioresources, El Paso, TX, pp. 167–80.

Jack, T.R., Stehmeier, L.G., Ferris, F.G., and Islam, M.R. (1991) Microbial selective plugging to control water channelling, in *Microbial Enhancement of Oil Recovery, Recent Advances* (ed. S. Donaldson). Elsevier Science Publishing Co., New York, pp. 433–40.

Jackson, C. (2002) *Ugrading a Heavy Oil Using Variable Frequency Microwave Energy.* SPE/PS-CIM/CHOA, International Thermal Operations and Heavy Oil Symposium and International Horizontal Well Technology Conference, SPE Paper No. 78082, November 4–7, Alberta, Canada.

Jain, R. and Schultz, J.S. (1982) Numerical technique for solving carrier mediated transport problems. *Journal of Membrane Science*, **11**, 79–106.

Javaherdashti, R. (1999) A review of some characteristics of some MIC caused by sulfate-reducing bacteria: past, present and future. *Anti-Corrosion and Materias*, **46**, 173–80.

Jean-Baptiste, P. and Ducroux, R. (2003) Energy policy and climate change. *Energy Policy*, **31**(2), 155–66.

Jemma, N. and Noble, R.D. (1992) Improved analytical prediction of facilitation factors in facilitated transport. *Journal of Membrane Science*, **70**, 289–93.

Jennings, H.Y. Jr. (1975) A study of caustic solution-crude oil interfacial tensions. *Society of Petroleum Engineers Journal*, **SPE-5049**, 197–202.

Jeong, B.-H., Hasegawa, Y., Kusakabe, K., and Morooka, S. (2002) Separation of benzene and cyclohexane mixtures using an NaY-type zeolite membrane. *Separation Science and Technology*, **37**, 1225.

Jepson, P.D., Arbelo, M., Deaville, R., Patterson, I.A.P., Castro, P., Baker, J.R., Degollada, E., Ross, H.M., Herráez, P., Pocknell, A.M., Rodríguez, F., Howie, F.E., Espinosa, A., Reid, R.J., Jaber, J.R., Martin, J., Cunningham, A.A., and Fernández, A. (2003) Gas-bubble lesions in stranded cetaceans. Was sonar responsible for a spate of whale deaths after an Atlantic military exercise? *Nature*, **425**, 575–6.

Jo, Y.-D. and Ahn, B.J. (2002) Analysis of hazard area associated with high pressure natural-gas pipeline. *Journal Loss Prevention Process Industries*, **15**, 179.

Jo, Y.-D. and Ahn, B.J. (2005) A method of quantitative risk assessment for transmission pipeline carrying natural gas. *Journal of Hazardous Materials*, **123**, 1–12.

John, M., Chris, B., Andrew, P., and Charlotte, T. (2001) *An Assessment of Measures in Use for Gas Pipeline to Mitigate against Damage caused by Third Party Activity.* Printed and published by the health and safety executive, C1 10/01.

Johnsen, S., Frost, T.K., and Hjelsvold, M. [all: Statoil FU – Trondheim] (2000) *The Environmental Impact Factor – A Proposed Tool for Produced Water Impact Reduction, Management and Regulation.* SPE International Conference on Health, Safety, and the Environment in Oil and Gas Exploration and Production, June 26–28, Stavanger, Norway.

Johnson, C.E. Jr. (1976) Status of caustic and emulsion methods. *Journal of Petroleum Technology*, **January**, 85–92.

Johnson, C.R., Greenkorn, R.A., and Woods, E.G. (1966) Pulse-testing: a new method for describing reservoir flow properties between wells. *Journal of Petroleum Technology*, **December**, 1599–1604.

Jones-Meehan, J., Walch, M., Little, B.J., Ray, R.I., and Mansfeld, F.B. (1992) *ESEM/EDS, SEM/EDS and EIS Studies of Coated 4140 Steel Exposed to Marine, Mixed Microbial Communities Including Sulfate-reducing Bacteria.* ASME International Power Generation Conference, Atlanta, GA, USA, Publ. ASME, NY, USA, pp. 1–16.

Judes, U. (2000) Towards a culture of sustainability, in *Communicating Sustainability* (ed. W.L. Leal Filho). Peter Lang, Berlin, pp. 97–121.

K.U. (Khulna University) (1995) Integrated resources development of the Sundarbans reserved forest, Bangladesh. FAO, UNDP. BGD/84/056.

Kaisera, M.J. and Byrd, R.C. (2005) The non-explosive removal market in the Gulf of Mexico. *Ocean & Coastal Management*, **48**, 525–70.

Kakita, Y, Funatsu, M, Miake, F., and Watanabe, K. (1999) Effects of microwave irradiation on bacteria attached to the hospital white coats. *International Journal of Occupational Medicine and Environmental Health*, **12**(2), 123–6.

Karim, A. (1994) Environmental impacts, in *Mangroves of the Sundarbans, Volume II: Bangladesh* (eds Z. Hussain and G. Acharya). IUCN, Bangkok, Thailand, pp. 203–18.

Karnik, S.V., Hatalis, M.K., and Kothare, M.V. (2002) Palladium based micro-membrane hydrogen gas separator-reactor in a miniature fuel processor for micro fuel cells. *Proceedings of the Materials Science of Microelectromechanical Systems (MEMS) Devices IV*, **687**, 243.

Kasevich, R.S., Price, S.L., and Faust, D.L. (1994) *Pilot Testing of a Radio Frequency Heating System for EOR from Diatomaceous Earth.* 69th Annual Technical Conference and Exhibition, SPE Paper No. 28619, New Orleans, LA, USA.

Katebi, M.N.A. and Habib, G. (1989) Sundarbans and forestry, in *National Workshop on Coastal Area Resource Development and Management*, Dhaka, pp. 79–100.

Katsimpiri, M. (2001) *Managing Ecological Risk of Produced Water from Offshore Oil and Gas Drilling.* Graduate Student Project, Marine Affairs Program, Dalhousie University, Unpublished.

Keeney, R.L. (1992) *Value-Focused Thinking.* Harvard University Press, Cambridge, MA.

Kelland, M.A., Svartaas, T.M., and Dybvik, L. (1995) *New Generation of Gas Hydrate Inhibitors.* Proceedings of SPE Annual Technical Conference and Exhibition, October 22–25, 1995, Dallas, TX, USA, pp. 529–37.

Kelland, M.A., Svartaas, T.M., Ovsthus, J., and Namba, T. (2000) A new class of kinetic inhibitors. *Annals of New York Academy of Sciences*, **912**, 281–93.

Kemena, L.L., Noble, R.D., and Kemp, N.J. (1983) Optimal regimes of facilitated transport. *Journal of Membrane Science*, **15**, 259–74.

Kessler-Taylor, I. (1986) An examination of alternative causes of Atlantic salmon decline and surface of water acidification in southwest Nova Scotia. Environment Canada, Working Paper No. 46. Environment Canada. Ottawa, Ontario.

Kesting, R.E. (1985) *Synthetic Polymer Membranes.* Wiley, New York.

Kesting, R.E., Fritzsche, A.K., Murphy, M.K., Handermann, C.A., Cruse, C.A., and Malon, R.F. (1989) Process for forming asymmetric gas separation membranes having graded density skins. US Patent, 871, April.

Ketata, C. and Islam, M.R. (2005a) Expert System Knowledge Management for Laser Drilling in the Oil and Gas Industry, paper presented at *IAWTIC'*2005 in Vienna, Austria, November 28–30.

Ketata, C. and Islam, M.R. (2005b) Knowledge Selection for Laser Drilling in the Oil and Gas Industry, presented at *IAWTIC'*2005 in Vienna, Austria, November 28–30.

Ketata, C. and Islam, M.R. (2005c) Testing an Expert System for Laser Drilling in the Oil and Gas Industry, paper presented at *IAWTIC'*2005 in Vienna, Austria, November 28–30.

Ketata, C., Satish, M.G., and Islam, M.R. (2005a) Stochastic Evaluation of Rock Properties by Sonic-While-Drilling Data Processing in the Oil and Gas Industry, *ICCES '05 Proceedings, International Conference on Computational & Experimental Engineering and Sciences*, Chennai, India, December 1–6.

Ketata, C., Satish, M.G., and Islam, M.R. (2005b) Knowledge-Based Optimization of Rotary Drilling System in the Oil and Gas Industry, *ICCES '05 Proceedings, International Conference on Computational & Experimental Engineering and Sciences*, Chennai, India, December 1–6.

Ketata, C., Satish, M.G., and Islam, M.R. (2006a) Cognitive Work Analysis of Expert Systems Design and Evaluation in the Oil and Gas Industry, *ICCIE '06 Proceedings, 36th International Conference on Computers and Industrial Engineering*, Taipei, Taiwan, R.O.C., June 20–23.

Ketata, C., Satish, M.G., and Islam, M.R. (2006b) Laying the Foundation for Pro-Nature Expert Systems in Petroleum Engineering, *ICCIE '06 Proceedings, 36th International Conference on Computers and Industrial Engineering*, Taipei, Taiwan, R.O.C., June 20–23.

Kevin, J.L., Zlochower, I.A., Casdollar, K.L., Djordjevic, S.M., and Loehr, C.A. (2000) Flammability of gas mixtures containing volatile organic compounds and hydrogen. *J. Loss Prev. Process. Ind.*, **13**, 377–84.

Khalil, C.N. (1995) *New Process for the Chemical De-Waxing of Pipelines Advances.* Multiphase Operations Offshore Conference, November 29–30, 1995, London, UK.

Khan, M.I. (2006a) Development and application of criteria for true sustainability. *Journal of Nature Science and Sustainable Technology*, **1**(1), 1–37.

Khan, M.I. (2006b) *Towards sustainability in offshore oil and gas operations*. Ph.D. Dissertation, Dalhousie University.

Khan, M.I. (1997) Feeding practices and environmental degradation in the brackishwater shrimp culture in Bangladesh. *Shade of Green*, **2**, 136–44.

Khan, M.I. and Islam, M.R. (2002) Oil and Gas in Sundarbans: Effects of Hydrocarbon Operations in the World's Largest Mangrove Ecosystem, Shared Water Management, Coastal Zone Canada 2002, Hamilton, Canada.

Khan, M.I. and Islam, M.R. (2003a) Wastes management in offshore oil and gas: a major challenge in integrated coastal zone management, in *CARICOSTA 2003 – 1st International Conference on Integrated Coastal Zone Management ICZM* (ed. L.G. Luna). University of Oriente, Santiago du Cuba, May 5–7.

Khan, M.I. and Islam, M.R. (2003b) *Ecosystem-Based Approaches to Offshore Oil and Gas Operation: An Alternative Environmental Management Technique*. SPE Annual Technical Conference and Exhibition, Denver, USA. October 6–8.

Khan, M.I. and Islam, M.R. (2003c) *A Restoration Technique for Enclosed Coastal Seas: Creation of Artificial Rigs, Fish Shelters, and Recreation Centers through Utilizing Oil Rigs*. Environmental Management of Enclosed Coastal Seas 2003, Bangkok, November 18–21.

Khan, M.I. and Islam, M.R. (2004a) *Management Based on Environmental Carrying Capacity: An Innovative Approach in Coastal Zone Management*. Coastal Zone Canada, St. John's, Newfoundland, June 27–30.

Khan, M.I. and Islam, M.R. (2004b) Assessing environmental fate and behaviour of oil discharges in marine ecosystem: using fugacity model, in *Offshore Oil and Gas Environmental Effects Monitoring: Approaches and Technologies*, *Armsworthy* (eds S.L.P.J. Cranford and K. Lee). Battelle Press, Ohio, pp. 145–65.

Khan, M.I. and Islam, M.R. (2005a) Sustainable wealth generation: Community-based offshore oil and gas operations. CARICOSTA 2005 – International Conference on Integrated Coastal Zone Management, University of Oriente, Santiago du Cuba, May 11–13.

Khan, M.I. and Islam, M.R. (2005b) *Assessing Sustainability of Technological Developments: An Alternative Approach of Selecting Indicators in the Case of Offshore Operations*. Proceedings of ASME International, Mechanical Engineering Congress and Exposition, Orlando, Florida, November 5–11.

Khan, M.I. and Islam, M.R. (2005c) Assessing Sustainability: A Novel Approach of Selecting Indicators in the Case of Intensive Fish Farming, CARICOSTA 2005 – International Conference on Integrated Coastal Zone Management, University of Oriente, Santiago du Cuba, May 11–13.

Khan, M.I. and Islam, M.R. (2005d) *Environmental Fate and Behaviour of Oil Discharges in Marine Ecosystem: Using Fugacity Model*. Proceedings of Third International Conference on Energy Research and Development (ICERD-3). State of Kuwait, November 21–23.

Khan, M.I. and Islam, M.R. (2005e) *Environmental Modeling of Oil Discharges from Produced Water in the Marine Environment*. Proceedings of Ann. Conf. Canadian Society of Civil Engineers, Toronto, Canada, June 2–4.

Khan, M.I. and Islam, M.R. (2005f) *Sustainable Marine Resources Management: Framework for Environmental Sustainability in Offshore Oil and Gas Operations*. Proceedings of Fifth International Conference on Ecosystems and Sustainable Development, Cadiz, Spain, May 03–05.

Khan, M.I. and Islam, M.R. (2005g) Achieving True technological sustainability: pathway analysis of a sustainable and an unsustainable product, International Congress of Chemistry and Environment, Indore, India, December 24–26.

Khan, M.I. and Islam, M.R. (2006a). Technological analysis and quantitative assessment of oil and gas development on the scotian shelf, Canada. *Int. J. Risk Assessment and Management*, in press.

Khan, M.I. and Islam, M.R. (2006b) *True Sustainability in Technological Development and Natural Resources Management*. Nova Science Publishers, New York. [ISBN: 1-60021-203-4].

Khan, M.I. and Islam, M.R. (2006c) Developing sustainable technologies for offshore seismic operations. *Journal of Petroleum Science and Technology*, in press.

Khan, M.I. and Islam, M.R. (2006d). *Alternative Management and Mitigation for Potential Impacts of Oil and Gas Operations in the World's Largest and Most Productive Ecologically Sensitive Site*. SPE Annual Technical Conference and Exhibition, SPE Paper No. 103269, September 24–27, San Antonio, Texas, USA.

Khan, M.I. and Islam, M.R. (2006e). Sustainable management techniques for offshore oil and gas operations. *Energy Sources*, in press.

Khan, M.I. and Islam, M.R. (2006f) *Handbook of Sustainable Oil and Gas Operations Management.* Gulf Publishing Company, Austin, TX [Scheduled to be published in April, 2007].

Khan, M.M. and Islam, M.R. (2006g) A new downhole water-oil separation technique. *Pet. Sci. Tech.*, **24**(7), 789–805.

Khan, M.M., Prior, D., and Islam, M.R. (2006h) Zero-waste living with inherently sustainable technologies. *Journal of Nature Science and Sustainable Technology*, **1**(2), 271–296.

Khan, M.I., Zatzman, G.M., and Islam, M.R. (2005a) A novel sustainability criterion as applied in developing technologies and management tools, in *Second International Conference on Sustainable Planning and Development*. Bologna, Italy.

Khan, M.I., Zatzman, G., and Islam, M.R. (2005b) *New Sustainability Criterion: Development of Single Sustainability Criterion as Applied in Developing Technologies.* Jordan Proceedings of International Chemical Engineering Conference V, Paper No. JICEC05-BMC-3-12, September 12–14, Amman, Jordan.

Khan, M.M., Prior, D., and Islam, M.R. (2005a) Direct-usage solar refrigeration: from irreversible thermodynamics to sustainable engineering, in *33rd Annual Conference of the Canadian Society for Civil Engineering (CSCE)*, Toronto, ON, Canada.

Khan, M.M., Prior, D., and Islam, M.R. (2005b) Environment-friendly wood panels for healthy homes, in *Int. Chemical Engineering Conference*, September, Jordan.

Khan, M.I., Chhetri, A.B., and Islam, M.R. (2005c) Analyzing sustainability of community-based energy development technologies. *Energy Sources*, in press.

Khan, M.I., Chhetri, A.B., and Islam, M.R. (2006a) Community-based energy model: a novel approach in developing sustainable energy. *Energy Sources*, in press.

Khan, M.I., Chhetri, K.C., and Lakhal, S.L. (2006b) A comparative pathway analysis of a sustainable and an unsustainable product. *Journal of Nature Science and Sustainable Technology*, **1**(2), 233–262.

Khan, M.I., Lakhal, Y.S., Satish, M., and Islam, M.R. (2006c) Towards achieving sustainability: application of green supply chain model in offshore oil and gas operations. *Int. J. Risk Assessment and Management*, in press.

Khan, M.I., Smit, E., Lakhal, S.L., and Islam, M.R. (2006d). *Sustainable Decommissioning of Offshore Oil and Gas Drilling and Production Facilities.* Proceedings of 36th CIE Conference on Computer and Industrial Engineering, Taiwan, July, 2006, pp. 4357–73.

Khan, M.M., Prior, D., and Islam, M.R. (2006e) A novel sustainable combined heating/cooling/refrigeration system. *Journal of Nature Science and Sustainable Technology*, **1**(1), 133–162.

Khan, M.I., Ketata, C., and Islam, M.R. (2006f) Potential impacts and remediation guidelines of hydrocarbon operations on the world's largest sundarbans mangrove ecosystem. *International Journal of Risk Assessment and Management*, in press.

Khatun, B.M.R. and Alam, M.K. (1987) Taxonomic studies on the genus Avicinia Lin. from Bangladesh. *Bangladesh Journal of Botany*, **16**, 39–44.

Khire, A. and Khan, M. (1994). Microbial enhance oil recovery. *Enzyme Micro. Tech.*, **16**, 170–2.

Kikinides, E.S., Sikivitas, V.I., and Yang, R.I. (1995) Natural gas desulfurization by adsorption: feasibility and multiplicity of cystic steadystate. *Industrial and Chemical Engineering Research*, **34**, 255.

Kikkinides, E.S., Sikavitsas, V.I., and Yang, R.T. (1995) Natural gas desulfurization by adsorption: feasibility and multiplicity of cyclic steady states. *Industrial and Chemical Engineering Research*, **34**, 255–62.

Kimber, K., Farouq Ali, S.M., and Puttagunta, V.R. (1988) New Scaling criteria for chemical flooding experiments. *Journal of Canadian Petroleum Technology*, **27**(4), 86–94.

Kingston, H.M. and Jassie, L.B. (eds) (1998) *Introduction to Microwave Sample Preparation.* American Chemical Society, Washington, DC.

Kingston, P.F. (2002) Long-term Environmental Impact of Oil Spills. *Spill Science & Technology Bulletin*, **7**(1), 53–61.

Kirkhorn, S.S. (1995) *Developments in Hydrate Control Employed*, Advances in Multiphase Operations Offshore Conference, November 29–30, London, UK.

Kislenko, N., Aphanasiev, A., Nabokov, S., and Ismailova, H. (1996) *New Treating Processes for Sulfur Containing Natural Gas*, Proceedings, Annual Convention – Gas Processors Association, p. 279.

Kleijnen, J. and Smith, M.T. (2003) Performance metrics in supply chain management. *Journal of Operations Management Research Society*, **54**, 507–27.

Kndudsen, V.O., Alford, R.S., and Emling, J.W. (1944) *Survey Underwater Sound-Ambient Noise (Report No. 3)*. Office of Scientific Research and Development, National Defense Research Committee, Division 6, Section 6.1.

Knott, D. (1998) 20/20 hindsight over Brent spar. *Oil & Gas Journal*, **96**, 31–2.

Kogut, B. and Zander, U. (1993) Knowledge of the firm and the evolutionary theory of the multinational corporation. *Journal of International Business Studies*, **15**, 151–68.

Kokubu, K. and Nashioka, E. (2001) Environmental accounting practices of listed companies in Japan, in *International Forum on Business and the Environment*, September 27–29, Ministry of the Environment, Japan, pp. 13–25.

Koros, W.J., Ma, Y.H., and Shimidzy, T.J. (1996) Terminology for membranes and Membrane processes. *Journal of Membrane Science*, **120**, 149–59.

Kowalski, C., Berruti, F., Chakma, A., Gianetto, A., and Islam, M.R. (1991) A mechanistic model for microbial transport, growth and viscosity reduction in porous media. *American Institute of Chemical Engineers Symp. Series*, **87**(280), 123–33.

KRÜSS (2006) Instruments for Surface Chemistry, Measuring Principle of KRÜSS Tensiometers, KRÜSS GmbH, Wissenschaftliche Laborgeräte, Borsteler Chaussee 85-99a, D-22453 Hamburg, Germany.

Kulkarni, S.S., Funk, E.W., Li, N.N., and Riley, R.L. (1983) Membrane separation processes for acid gases. *American Institute of Chemical Engineers Symposium Series*, **79**(229), 172–8.

Kumaraswamy, P.R. and Datta, S. (2006) Bangladeshi gas misses India's energy drive? *Energy Policy*, **34**(15), 1971–3.

Kunisue, T., Masayoshi, M., Masako, O., Agus, S., Nguyen, H.M., Daisuke, U., Yumi, H., Miyuki, O., Oyuna, T., Satoko, K., et al. (2006) Contamination status of persistent organochlorines in human breast milk from Japan: recent levels and temporal trend. *Chemosphere*, **64**(9), 1601–8.

Labuschange, C., Brent, A.C., and Erck, R.P.G. (2005) Assessing the sustainability performances of industries. *Journal of Cleaner Production*, **13**, 373–85.

Laddha, S.S. and Danckwerts, P.V. (1981) Reaction of CO2 with ethanolamine: kinetics from gas absorption. *Chemical Engineering Science*, **36**, 479–82.

Lakhal, S.Y. and H'Mida, S. (2003) A gap analysis for green supply chain benchmarking, in *32th International Conference on Computers & Industrial Engineering*, Volume 1, August 11–13, Ireland, pp. 44–9.

Lakhal, S.Y. and H'Mida, S. (2004) A model for assessing the greenness effort in a supply chain, in *35th Decision Sciences Institute Annual Meeting*, Decision Science Institute, Boston, USA, pp. 5331–41.

Lakhal, S.Y., H'mida, S., and Islam, R. (2005) A green supply chain for a petroleum company, in *Proceedings of 35th International. Conference on Computer and Industrial Engineering*, Volume 2, June 19–21, Istanbul, Turkey, pp. 273–1280.

Lakhal, S.L., Khan, M.I., and Islam, M.R. (2006a) A framework for a green decommissioning of an offshore platform, in *Proceedings of 36th CIE Conference on Computer and Industrial Engineering*, July, Taiwan, pp. 4345–56.

Lakhal, S.L., Khan, M.I., and Islam, M.R. (2006b) Offshore platform decommissioning: an application of the "Olympic" Green Supply Chain, in *Proceedings of 36th CIE Conference on Computer and Industrial Engineering*, July, Taiwan, pp. 4334–44.

Lamparelli, C.C., Rodirigues, F.O., and Moura, D.O. (1997) Long term assessment of oil spill in a mangrove forest in Sao Paulo, Brazil, in *Mangrove Ecosystem Studies in Latin America and Africa* (eds B. Kjerfve, L.D. de Lacerda and E.H.S. Dipo).UNESCO, Paris, pp. 191–203.

Lang, K. (2000) Managing produced water, in *State of the Art Technology Summary* (PTTC Network News Volume 6, 3rd Quarter. Hart Energy Publications) at <http://www.pttc.org/news/v6n4nn7.htm> [Accessed on June 16, 2005].

Lange, J.-P. (2002) Sustainable development: efficiency and recycling in chemicals manufacturing. *Green Chemistry*, **4**, 546–50.

Larrondo, L.E., Urness, C.M., and Milosz, G.M. (1985) *Laboratory Evaluation of Sodium Hydroxide, Sodium Orthosilicate, and Sodium Metasilicate as Alkaline Flooding*, SPE Paper No. 13577, pp. 307–15.

Lasschuit, W. and Thijssen, N. (2004) Supporting supply chain planning and scheduling decisions in the oil and chemical industry. *Computers & Chemical Engineering*, **28**, 863–70.

Leal Filho, W.L. (1999) Sustainability and university life: some European perspectives, in *Sustainability and University Life: Environmental Education, Communication and Sustainability* (ed. W. Leal Filho). Peter Lang, Berlin, pp. 9–11.

Lecomte du Noüy, P. (1919) A new apparatus for measuring surface tension. *Journal of General Physiology*, **1**, 521.

Lederhos, J.P., Longs, J.P., Sum, A., Christiansen, R.L., and Sloan, E.D. (1996) Effective kinetic inhibitors for natural gas hydrates. *Chem. Eng. Sci.*, **51**(8), 1221–9.

Lee, B.-W. (2001) Environmental accounting in Korea: cases and policy options, in *International Forum on Business and the Environment, Ministry of the Environment, Japan*, September 27–29, Japan, pp. 13–25.

Lee, J. (1982) *Well Testing, First printing*, SPE Textbook series. SPE of AIME, New York, Dallas, USA.

Lee, W.D. and Kamilos, G.N. (1985) Chevron field tests its foam diverter in steam floods with encouraging results. *Pet. Eng. Int.*, **November**, 36–44.

Lehman-McKeeman, L.D. and Gamsky, E.A. (1999) Diethanolamine inhibits choline uptake and phosphatidylcholine synthesis in Chinese hamster ovary cells. *Biochemical and Biophysical Research Communications*, **262**(3), 600–4.

Lems, S., van derKooi, H.J., and deSwaan Arons, J. (2002) The sustainability of resource utilization. *Green Chemistry*, **4**, 308–13.

Lenormand, R., Zarcone, C., and Sarr, A. (1983) Mechanism of the displacement of one fluid by another in a network of capillary ducts. *J. Fluid. Mech.*, **135**, 337–53.

Lenormand, R., Touboul, E., and Zarcone, C. (1988) Numerical models and experiments on immiscible displacements in porous media. *J. Fluid Mech.*, **189**, 165–87.

Leontieff, W. (1973) Structure of the world economy: outline of a simple input-output formulation, Stockholm: Nobel Memorial Lecture, December 11, Lervik, J.K. et al., *Direct Electrical Heating of Pipelines As A Method of Preventing Hydrates and Wax Plugs*, Proceedings of Int. Offshore Polar Eng. Conf., May 24–29, Montreal, Canada, pp. 39–45

Leppin, D. (1995) *Natural Gas Production: Performance of Commercial Technology*, SPE 29743.

Les, W. and Norse, A. (1998) Disturbance of the seabed by mobile fishing: a comparison with forest clear-cutting. *Conservation Biology*, **53**, 239–52.

Lia, X.B., Summersb, D.A., Rupertb, G., and Santib, P. (2001) Experimental investigation on the breakage of hard rock by the PDC cutters with combined action modes. *Tunneling and Underground Space Technology*, **16**, 107–14.

Liao, S.Y., Cheng, Q., Jiang, D.M., and Gao, J. (2005) Experimental study of flammability limits of natural gas – air mixture. *Journal of Hazardous Materials*, **B119**, 81–4.

Lin, Y.-M. and Rei, M.-H. (2001) Separation of hydrogen from the gas mixture out of catalytic reformer by using supported palladium membrane. *Separation and Purification Technology*, **25**, 87.

Lion, M., Skoczylas, F., Lafhaj, Z., and Sersar, M. (2005) Experimental study on a mortar. Temperature effects on porosity and permeability. Residual properties or direct measurements under temperature. *Cement and Concrete Research*, **35**, 1937–1942.

Liu, P. (1993) *Introduction to Energy and the Environment.* Van Nostrand Reinhold, New York.

Liua, Q., Dong, M., Zhou, W., Ayub, M., Zhang, Y.P., and Huang, S. (2004) Improved oil recovery by adsorption – desorption in chemical flooding. *Journal of Petroleum Science and Engineering*, **43**, 75–86.

Livingston, R.J. and Islam, M.R. (1999) Laboratory modeling, field study and numerical simulation of bioremediation of petroleum contaminants. *Energy Sources*, **21**(1/2), 113–30.

Lo, H.Y. and Fiungan, N. (1973) *Effect of Temperature on Water-Oil Relative Permeabilities in Oil-Wet and Water-Wet Systems.* 48th Annual Meeting of SPE, SPE Paper No. 4505, September 30–October 3, Las Vegas.

Lojan, S. (2005) Personal Communication, Senior Petroleum Geologist, Clipper Energy Supply Co., Houston, TX, USA.

Lokkeborg, S. and Soldal, A.V. (1993) The influence of seismic exploration with airguns on cod (Gadus morhua) behavior and catch rates. *ICES Marine Science Symposium*, **196**, 62–7.

Low, K.S., Lee, C.K., and Liew, S.C. (2000) Sorption of cadmium and lead from aqueous solutions by spent grain. *Process Biochemistry*, **36**(1–2), 59.

Low, N.M.P., Fazio, P., and Guite, P. (1984) Development of light-weight insulating clay products from the clay sawdust-glass system. *Ceramics International*, **10**(2), 59–65.

Low, N.M.P., Fazio, P., and Guite, P. (1984) Development of light-weight insulating clay products from the clay sawdust-glass system. *Ceramics International*, **10**(2), 59–65.

Lowe, E.A., Warren, J.L., and Moran, S.R. (1997) *Discovering Industrial Ecology – An Executive Briefing and Sourcebook*. Battelle Press, Columbus.

Lowy, J. (2004) Plastic left holding the bag as environmental plague. Nations around world look at a ban. <http://seattlepi.nwsource.com/national/182949_bags21.html>.

Lozada, D. and Farouq Ali, S.M. (1988) *Experimental Design for Non-equilibrium Immiscible Carbon Dioxide Flood*. The Fourth UNITAR/UNDP International Conference on Heavy Crude and Tar Sands, Paper No. 159, August 7–12, Edmonton.

Lubchenco, J.A., Olson, A.M., Brubaker, L.B., et al. (1991) The sustainable biosphere initiative: an ecological research agenda. *Ecology*, **72**, 371–412.

Lumley, D.E. and Behrens, R.A. (1998) Practical issue of 4D seismic reservoir monitoring: what a engineer needs to know? SPE Reservoir *Evolution and Engineering*, **December**, 528–38.

Lunder, S. and Sharp, R. (2003) *Mother's Milk, Record Levels of Toxic Fire Retardants Found in American Mother's Breast Milk*. Environmental Working Group, Washington, USA.

MacDonald, R.C. and Coats, K.H. (1970) Methods for numerical simulation of water and gas coning. *Society of Petroleum Engineers' Journal*, **10**(4), 425–36.

Mackay, D. and Paterson, S. (1991) Evaluating the regional multimedia fate of organic chemicals: a level III fugacity model. *Environ. Sci. Technol.* **25**, 427–36.

Mackay, D. (1987) *Chemical and Physical Behavior of Hydrocarbons in Freshwater*. Proceedings, Symposium of Oil Pollution in Freshwater, Edmonton, AB, Canada, October 15–19, Pergamon, New York, USA, pp. 10–21.

Mackay, D. and Paterson, S. (2004) Evaluating the multimedia fate of organic chemicals: a level III fugacity model. *Environ. Sci. Technol*, **38**, 1505–12.

Mackay, D. (1987) *Chemical and Physical Behavior of Hydrocarbons in Freshwater*. Proceedings, Symposium of Oil Pollution in Freshwater, Edmonton, AB, Canada, October 15–19, Pergamon, New York, NY, USA, pp. 10–21.

Mackay, D. (2001) *Multimedia Environmental Models: The Fugacity Approach*, 2nd edn. Lewis Publishers, Boca Raton, pp. 201–13.

Mackay, D. and Paterson, S. (1991) Evaluating the multimedia fate of organic chemicals: a level III fugacity model. *Environ. Sci. Technol.*, **25**, 427–36.

Mackay, D. (1985) The fugacity concept in environmental modeling, in *The Handbook of Environmental Chemistry* (ed. O. Hutzinger), Volume 2/Pat C, pp. 120–40.

Mackay, D. (2004) Finding fugacity feasible, fruitful, and fun. *Environmental Toxicology and Chemistry*, **23**(10), 2282–9.

Mackay, D., Joy, M., and Paterson, S. (1983) A quantitative water, air, sediment interaction (QWASI) fugacity model for describing the fate of chemicals in lakes. *Chemosphere*, **12**, 981–97.

Mackay, D., Shiu, W.Y., and Ma, K.C. (2000) *Physical-Chemical Properties and Environmental Fate and Degradation Handbook*. CRCnetBASE 2000, Chapman & Hall CRCnetBASE, CRC Press LLC., Boca Raton, FL. (CD-ROM.).

MacNeil, T. (2002) *Commissioner's Report: Results of the Public Review on the Effects of Oil and Gas Exploration Offshore Cape Breton*. Public Review Commission, Sydney, Nova Scotia.

Mahmood, N., Chowdhury, S.Q., and Sakait, S.Q. (1994) Indiscriminate expansion of coastal aquaculture in Bangladesh, genesis of conflicts: some suggestions, in *Proceedings, Coastal Zone Canada '94. Cooperation in the Coastal Zone Conference Proceedings*, Volume 4 (eds P.G. Wells and P.J. Ricketts). Coastal Zone Canada Association, pp. 1697–706.

Maine Shore Stewards (2001) *Marine Phytoplankton Monitoring Program (MPMP)*. Maine Shore Stewards <http://www.ume.maine.edu/ssteward/phyto.htm> [Accessed on April 17, 2001].

Maini, B.B. and Batycky, J.P. (1983) *Effect of Temperature on Heavy-Oil/Water Relative Permeabilities in Horizontal and Vertically Drilled Core Plugs*. SPE 58th Annual Technical Conference and Exhibition, SPE Paper No. 12115, October 5–8, San Francisco, CA.

Malik, Q.M. and Islam, M.R. (1999) An overview of the potential of greenhouse gas storage and utilization through enhanced oil recovery. *Reviews in Process Chemistry and Engineering*, **2**,129–68.

Malik, Q.M. and Islam, M.R. (2000) *CO2 injection in the Weyburn field of Canada: optimization of enhanced oil recovery and greenhouse gas storage with horizontal wells*. SPE/DOE Improved Oil Recovery Symposium, SPE Paper No. 59327, April 3–5, Tulsa, Oklahoma.

Managia, S., Opaluchb, J.J., Jinc, D., and Grigalunasb, T.A. (2005) Technological change and petroleum explorationin the GulfofMexico. *Energy Policy*, **33**, 619–32.

Market Development Plan (1996) *Market Status Report: Postconsumer Plastics, Business Waste Reduction*. Integrated Waste Development Board, Public Affairs Office, California.

Markiewicz, G.S., Losin, M.A., and Campbell, K.M. (1988) *The Membrane Alternative for Natural Gas Treating: Two Case Studies*. 63rd Annual Technical Conference end Exhibition of the Society of Petroleum Engineers, SPE Paper No. 18230, October 2–5, Houston, TX.

Marsh, R. (2003) A database of archived drilling records of the drill cuttings piles at the North West Hutton oil platform. *Marine Pollution Bulletin*, **46**(5), 587–93.

Martinez, A.B. (2003) The effects of microwaves on the trees and other plants. Valladolid, <Spain: www.stopumts.nl/pdf/onderzoek_bomen_planten.pdf> [Accessed on April 17, 2006].

Marx, K. (1883) *Capital: A Critique of Political Economy II: The Process of Circulation of Capital*. Edited by Frederick Engels, London.

Marzouki, M. (1999) Determining pore size distribution of gas separation membranes from adsorption isotherm data. *Energy Sources*, **21**(1–2), 31–8.

Matsuoka, K., Iriyama, Y., Abe, T., Matsuoka, M., and Ogumi, Z. (2005) Electro-oxidation of methanol and ethylene glycol on platinum in alkaline solution: poisoning effects and product analysis. *Electrochimica Acta*, **51**, 1085–90.

Matsuyama, H., Teramoto, M., and Sakakura, H. (1996) Selective Permeation of CO_2 through poly{2-(*N,N*-dimethyl)aminoethyl methacrylate*}* Membrane Prepared by Plasma-Graft Polymerization Technique. *Journal of Membrane Science*, **114**, 93–200.

Matta, J.R. and Alavalapati, J.R.R. (2006) Perceptions of collective action and its success in community based natural resource management: An empirical analysis. *Forest Policy and Economics*, **9**(3), 274–84.

Matthews, C.S. and Russell, D.S. (1967) *Pressure Buildup and Flow Tests in Well*, Monograph Series. SPE of AIME, New York, Dallas, USA.

Mauer, W.C., Anderson, E.E., Hood, M., Cooper, G., and Cook, N. (1990) *Deep Drilling Basic Research. v. 5 – System Evaluations*, Final Report. GRI-90/0265.5. June 5–1.

Maurer, W.C. and Heilhecker, J.K. (1969) *Hydraulic Jet Drilling*. SPE Paper No. 2434.

Mayer, E.H., Berg, T., Carmichael, R.L., and Weinbrandt, R.M. (1983) Alkaline injection for enhanced oil recovery – a status report. *Journal of Petroleum Technology*, **January**, 209–21.

McAuliffe, C.D. (1973a) Oil-in-water emulsions and theirflowpropertiesinporous media. *Journal of Petroleum Technology*, **June**, 727–33.

McAuliffe, C.D. (1973b) Crude-oil-in-water emulsions to improve fluid flow in an oil reservoir. *Journal of Petroleum Technology*, **June**, 723–26.

McCarthy, B.J. and Greaves, P.H. (1988) Mildew-causes, detection methods and prevention. *Wool Science Review*, **85**, 27–48.

McCauley, et al. (2000) Marine seismic surveys – a study of environmental implications. *Australian Petroleum Production and Exploration Association Journal*, **40**, 692–708.

McCauley, R., Fewtrell, J., and Popper, A. (2002) High intensity anthropogenic sound damages fish ears. *Journal of the Acoustical Society of America*, **113**(1), 638–42.

McCay, D.F. (2003) Development and application of damage assessment modeling: example for the North Cape oil spill. *Marine Pollution Bulletin*, **47**, 341–59.

McCay, D.P.F. and Payne, J.R. (2001) *Modeling of Oil Fate and Water Concentrations with and without Application of Dispersants*. Proceedins of the 24th Arctic and Marine Oilspill (AMOP) Technical Seminar, Edmonton, Alberta, Canada, June 12–14, Environment Canada, pp. 611–45.

McGee, B.C.W. and Vermeulen, F.E. (1996) *Electrical Heating with Horizontal Wells, the Heat Transfer Problem*. SPE International Conference on Horizontal Well Technology, SPE Paper No. 37117, November 18–20, Calgary, Canada.

McKenzie, C., Godley, B.J., Furnessand, R.W., and Wells, D.E. (1999) Concentrations and patterns of organochlorine contaminants in marine turtles from Mediterranean and Atlantic waters. *Marine Environmental Research*, **47**(2), 117–35.

MEA (Millennium Ecosystem Assessment) (2005) The millennium ecosystem assessment, Commissioned by the United Nations, the work is a four-year effort by 1,300 scientists from 95 countries.

Mehedi, M., Khan, M.I., Ketata, C., and Islam, M.R. (2005) A risk management model for the valued ecosystem component in offshore operations. *International Journal of Risk Assessment and Management*. [Accessed on on June, 2005].

Mehedi, M.Y., Chhetri, A.B., Ketata, C., and Islam, M.R. (2007a) An Approach for Conflict Resolution in Oil and Gas Operations, *Int. J. Risk Assessment and Management*, in press.

Mehedi, M., Khan, M.I., Ketata, C., and Islam, M.R. (2007b) A risk management model for the valued ecosystem components in offshore operations, *Int. J. Risk Assessment and Management*, in press.

Meldon, J.H., Kang, Y., and Sung, N. (1985) Analysis of transient permeation through a membrane with immobilizing chemical reaction. *Industrial and Engineering Chemistry Fundamentals*, **24**, 61–4.

Mendoza, G.A. and Prabhu, R. (2003) Qualitative multi-criteria approaches to assessing indicators of sustainable forest resources management. *Forest Ecology and Management*, **174**(1–3), 329–43.

Menzie, C.A. (1982) The environmental implications of offshore oil and gas activities. *Environ. Sci. Technol.*, **16**, 454A–72A.

Michaelson, S.M. (1971) Biological effects of microwave exposure-an overview. *Journal of Microwave Power*, **6**, 259–67.

Miller, G. (1994) *Living in the Environment: Principles, Connections and Solutions*. Wadsworth Publishing, California.

Miller, M.A. and Ramey, H.J. Jr.(1985) Effect of temperature on oil/water relative permeabilities of unconsolidated and consolidated sands. *Society of Petroleum Engineering Journal*, **December**, 945–56, SPE 12116.

Minkkinen, A. and Jonchere, J.P. (1997) *Methanol Simlifies Gas Processing*. Proceedings, Annual Convention – Gas Processors Association, 227 Offshore Sour Gas Treatment, Hydrocarbon Engg., February.

Mishra, D.P. and Rahman, A. (2003), An experimental study of flammability limits of LPG/air mixtures. *Fuel*, **82**, 863–66.

Mittelstaedt, M. (2006a) *Toxic Shock series (May 27–June 1)*, The Globe and Mail, Saturday, May 27.

Mittelstaedt, M. (2006b) Chemical used in water bottles linked to prostate cancer, *The Globe and Mail*, Friday, June 09.

MMS (2002) Minerals Management Service, Department of the Interior SubChapter B – Offshore. Part 250 – Oil and Gas and Sulpher Operations in the Outer Continental Shelf. SubPart Q – Decommissioning Activities. 67 FR 35406, May 17.

Moavenzadeh, F., McGarry, F.J., and Williamson, R.B. (1968) *Use of Laser and Surface Active Agents for Excavation in Hard Rocks*. 43rd Annual Fall Meeting of the SPE, SPE Paper No. 2240, September 29–October 2, Houston, TX.

Mohammadi, S.S. (1989) *Steam-foam pilot project at Dome-Tumbador, Midway Sunset Field: Part 2.* Symposium on Enhanced Oil Recovery, SPE r DOE Paper No. 20201, April 7, Tulsa, Okla.

Molero, C., Lucas, A.D., and Rodrıguez, J.F. (2006) Recovery of polyols from flexible polyurethane foam by "split-phase" glycolysis: glycol influence. *Polymer Degradation and Stability*, **91**, 221–8.

Mollet, C., Touhami, Y., and Hornof, V. (1996) A comparative study of the effect of ready made and in-situ formed surfactants on IFT measured by drop volume tensiometry. *Journal Colloid Interface Science*, **178**, 523.

Morillon, A., Vidalie, J.F., Hamzah, U.S., Suripno, S., and Hadinoto, E.K. (2002) *Drilling and Waste Management.* SPE International Conference on Health, Safety, and the Environment in Oil and Gas Exploration and Production, SPE Paper No. 73931, March 20–22.

Moritis, G. (2006), Special report: EOR/Heavy oil survey. *Oil Gas Journal*, **104**, 2241–51.

Moritis, G. (1998) EOR production up slightly. *Oil & Gas Journal*, **96**(16), 49–56.

Moritis, G. (2000) EOR weathers low oil prices. *Oil & Gas Journal*, **98**(12), 39–61.

Moritis, G. (2004) Point of view: EOR continues to unlock oil resources. *Oil and Gas Journal*, ABI/INFORM Global, **102**(14), 45–9.

Motyka, D. and Mascarenhas, A. (2002) Incineration innovation. *Hydrocarbon Engineering*, **7**(2), 75–7.

MSDS (Material Safety Data Sheet) (2006) Canadian Centre for Occupational Health and Safety, 135 Hunter Street East, Hamilton, ON, Canada L8N 1M5.

MSDS (2005) Ethylene Glycol Material Safety Data Sheet. <www.sciencestuff.com/msds/C1721.html> [Accessed on August 12, 2006].

Mtrinc (2005) <www.mtrinc.com/Pages/NaturalGas/ng.html> [Accessed on June 2, 2006].

Mukherjee, A.K. (1975) The Sundarbans of India and its biota. *J. Bombay Nat. Hist. Soc.*, **72**, 1–20.

Mulcahy, T. and Islam, M.R. (2006) Carbon dioxide sequestration for Enhanced Oil Recovery – A Study of Weyburn Mega Project of Canada. *J. Nat. Sci. Sust. Tech.*, in press.

Mulder, M. (1996) *Basic Principles of Membrane Technology.* Kluwer Academic Publishers, the Netherlands.

Munger, C.G. (1992) Coating requirements for offshore structures. *Material Performance*, **June**, 36–40.

Muntasser, Z.A., AlDarbi, M., and Islam, M.R. (2002), *Prevention of Microbiologically Influenced Corrosion using Zinc Coatings.* Proceedings of NACE Annual Conference, April, Denver, CO.

Murphy, M.L., Heintz, R.A., and Short, J.W. (1999) Recovery of pink salmon spawning areas after the Exxon Valdez oil spill. *Journal of the American Fish Society*, **128**(5), 909–18.

Mustafiz, S. (2002) *A novel method for heavy metal removal from aqueous streams.* MASc Thesis, Dalhousie University.

Mustafiz, S. and Islam, M.R. (2005) Adomian decomposition of two-phase, two-dimensional non-linear PDEs as applied in well testing, in *Proceedings of 4th International Conference on Computational Heat and Mass Transfer* (ed. R. Bennacer), Paris-Cachan, May 17–20, pp. 1353–6.

Mustaifz, S. and Islam, M.R. (2006) State-of-the-Art of Reservoir Simulation. *Journal of Petroleum Science and Technology*, in press.

Mustafiz, S. (2002) *A novel method for heavy metal removal from aqueous streams.* MASc Thesis, Dalhousie University, Canada.

Mustafiz, S., Basu, A., Islam, M.R., Chalaal, O., and Dowaidar, A. (2002) A novel method for heavy metal removal. *Energy Sources*, **24**(11), 1043–52.

Mustafiz, S., Biazar, J., and Islam, M.R. (2005a) *An Adomian Decomposition Solution to the Modified Brinkman Model (MBM) for a 2-Dimensional, 1-Phase Flow of Petroleum Fluids.* Proceeding of the Canadian Society for Civil Engineering (CSCE), 33rd Annual Conference, Toronto, ON, Canada, June 2–4.

Mustafiz, S., Biazar, Z. and Islam, M.R. (2005b) Formulation of Two-Phase, Two-Dimensional Non-Linear PDE's as Applied in Well Testing. Proceeding of the 1st International Conference on Modeling, Simulation and Applied Optimization, Sharjah, UAE, February 1–3.

Mustafiz, S., Bjorndalen, N., and Islam, M.R. (2004) Lasing into the future: potentials of laser drilling in the petroleum industry. *Journal of Petroleum Science and Technology*, **22**(9–10), 1187–98.

Mustafiz, S., Genyk, R., and Islam, M.R. (2006a) Breaking of stable emulsions using ultrasonic irradiation. *Journal of Petroleum Science and Technology*. in press.

Mustafiz, S., Mousavizadegan, H., and Islam, M.R. (2006b) Adomian decomposition of buckley leverett equation with capillary terms. *Journal of Petroleum Science and Technology*, in press.

Mustafiz, S., Mousavizadegan, H., and Islam, M.R. (2006c) The role of linearization in reservoir simulation problems. *Journal of Petroleum Science and Technology*, in press.

Mustafiz, S., Rahaman, M.S., Kelly, D., Tango, M., and Islam, M.R. (2003) The application of fish scales in removing heavy metals from energy-produced waste streams: the role of microbes. *Energy Sources*, **25**(9), 905–16.

Nakornthap, K. and Evans, R.D. (1982) *Temperature Dependent Relative Permeability and Its Effect Upon Oil Displacement by Thermal Methods.* SPE 57th Annual Technical Conference and Exhibition, SPE Paper No. 11217, September 26–29, New Orleans.

Nallinson, R.G. (2004) *Natural Gas Processing and Products.* Encyclopedia of Energy IV, Elsevier Publication.

Narayan, R. (2004) Drivers & rationale for use of biobased materials based on life cycle assessment (LCA). GPEC 2004 Paper.

Natural Environment Research Council (NERC) (1997) *Thematic Research Programme. Plankton Reactivity in the Marine Environment (PRME).* Natural Environment Research Council London, UK <http://www.sos.bangor.ac.uk/prime/intro.html> [Accessed on April 20, 2001].

Natural Resources of Canada (2002). *Statistics Canada, for the Office of Energy Efficiency*, Commercial and Institutional Buildings Energy Use Survey, 2000, Ottawa.

Neff, J.M., Rabalais, N.N., and Boerch, D.F. (1987) Offshore oil and gas development activities potentially causing long-term environmental effects, in *Long-Term Environmental Effects of Offshore Oil and Gas Development* (eds D.F. Boerch and N.N. Rabalais). Elsevier Applied Science, New York, pp. 149–73.

Neiro, S.M.S. and Pinto, J.M. (2004) A general modeling framework for the operational planning of petroleum supply chains. *Computers and Chemical Engineering*, **28**, 871–96.

NETL (National Energy Technology Laboratory) (2004a) NETL's advanced E&P technology research promotes environmental protection. *Advanced Technology*, **9**(1), 1–10.

NETL (2004b) CO_2: Time Is Right For Growth Economics, technology advances, environmental benefits align for 'perfect storm' of opportunity, <http://www.fossil.energy.gov/programs/oilgas/publications/eor_co2/CO2brochure2004.pdf>.

NETL (2005a) *Oil and Natural Gas Supply, Future Supply and Emerging Resources*, <www.netl.doe.gov/technologies/oil-gas/FutureSupply/FutureSupply_main.html>.

NETL (National Energy Technology Lab) (2005b) *Oil and Natural Gas Supply, Exploration and Production Technologies* <http://www.netl.doe.gov/technologies/oil-gas/EP_Technologies/EP_main.html> [Accessed on August 23, 2005].

Newmann-Bennett, M. (2007) Co-optimization of natural gas recovery and CO_2 storage, North Triumph Reservoir, Nova Scotia, MASc Thesis, Department of Civil and Resource Engineering, Dalhousie University.

NGO (2005) <www.naturalgas.org/naturalgas/processing_ng.asp> [Accessed on June 2, 2006].

NHNS (Natural History of Nova Scotia) (2002) Marine Fishes. Natural History of Nova Scotia, Volume I: Topics.

Nichols, C., Anderson, S., and Saltzman, D. (2001) *A Guide to Greening Your Bottom Line Through a Resource-Efficient Office Environment.* City of Portland, Office of Sustainable Development, Portland.

Nikiforuk, A. (1990) Sustainable rhetoric. *Harrowsmith*, **October**, 14–16.

NIOSH (1977) *Criteria Document: Coal Tar Products*, DHEW Pub. NIOSH 78-107.

Nolen, J.S. and Berry, D.W. (1972) Tests of the stability and time step sensitivity of semi-implicit reservoir simulation techniques. *Transactions AIME*, **253**, 253–66.

North Atlantic Fisheries Organization (NAFO) (2002) *Maps. North Atlantic Fisheries Organization.* <http://www.nafo.ca/imap/map.htm> [Accessed on April 02, 2002].

Notz, P.K., Bumgardner, S.B., and Schaneman, B.D. (1996) Application of kinetic inhibitors to gas hydrate problems. *SPE Production and Facilities Journal*, **11**, 256.

Nouri, A., Donald, A., Frempong, P., Butt, S., Vaziri, H., and Islam, M.R. (2003) Using acoustic measurements in evaluating pore collapse occurrence and associated sand production. *Proceeding of Oil and Gas Symposium*, CSCE Annual Conference, refereed proceeding, Moncton, June.

Nouri, A., Vaziri, H., Belhaj, H., and Islam, M.R. (2006a) Sand-Production Prediction: A New Set of Criteria for Modeling Based on Large-Scale Transient Experiments and Numerical Investigation, *SPE Journal*, **11**(2), 227–37.

Nouri, A., Vaziri, H., Kuru, E., and Islam, M.R. (2006b) A Comparison of Two Sanding Criteria in a Physical and Numerical Modeling of Sand Production, *J. Pet. Sci. Eng.*, **50**, 55–70.

Novosad, Z., McCaffery, F.G., Urness, C., and Hodges, J. (1981) *Comparison of Oil Recovery Potential of Sodium Orthosilicate and Sodium Hydroxide for the Wainwright Reservoir, Alberta.* Petroleum Recovery Institute, Alberta, Canada, Report 81-10.

NRC (National Research Council) (2003) *Oil in the Sea III.* National Academy, Washington, DC.

NRCan (Geological Survey of Canada -Atlantic – NRCan) (2001) Interactions between offshore oil and gas operations and the marine environment. *Bedford Institute of Oceanography 2001 in Review* (Dartmouth NS: BIO). <http://www.mar.dfo-mpo.gc.ca/science/review/e/html/2001/BIO-English.html#7> [Accessed on June 16, 2005].

Nultine, P.G. (1927) Soda process for petroleum recovery. *Oil and Gas* **3**, 25.

O'Brien, D.G., Graves, R.M., and O'Brien, E.A. (1999a) StarWars Laser Technology for Gas Drilling and Completions in the 21st Century, SPE 56625. SPE Annual Technical Conference. Houston, Texas, Oct. 3–6.

O'Brien, D.G., Graves, R.M., and O'Brien, E.A. (1999b) Paper on Lasser Technology, International Society for Optical Engineering V3614. San Jose California, Jan 25–29.

O'Brien, D.G., Graves, R.M., and O'Brien, E.A. (1999c) StarWars Technology for Gas and Completions in the 21st Century, paper SPE 56625 presented at the 1999 Annual Technical Conference and Exhibition, Houston, Texas, October 3–6.

OECD (1993) *Organization for Economic Cooperation and Development Core Set of Indicators for Environmental Performance Reviews.* A synthesis report by the Group on State of the Environment, Paris.

OECD (1998) *Towards Sustainable Development: Environmental Indicators.* Organization for Economic Cooperation and Development, Paris.

Ofiara, D.D. (2002) Natural resource damage assessments in the United States: rules and procedures for compensation from spills of hazardous substances and oil in waterways under US jurisdiction. *Marine Pollution Bulletin* **44**, 96–110.

Oil and Gas Journal (1998) Worldwide look at reserves and production. *Oil and Gas Journal*, **96**(52), 38–68.

Oil in the Sea (2003) *Committee on Oil in the Sea: Inputs, Fates, and Effects, Ocean Studies Board and Marine Board, Divisions of Earth and Life Studies and Transportation Research Board, National Research Council.* The National Academies Press, Washington, DC.

OLF (2000) *Physical, Chemical, and Biological Characterisation of Offshore Drill Cuttings Piles.* The Norwegian Oil Industry Association, Lervigsvein, Stavanger, 44pp.

Olivastri, B. and Williamson, M. (2001) *Review of International Initiatives to Accelerate the Reduction of Sulphur in Light and Heavy Fuel Oils.* Environment Canada Contract.

Oram, R.K. (1995) *Advances in Deepwater Pipeline Insulation Techniques and Materials.* Deepwater Pipeline Technology Congress, London, UK.

Oren, P.E., Billiote, J., and Pinczewski, W.V. (1992) Mobilisation of Waterflood Residual Oil by Gas Injection for Water-Wet Conditions, SPE-20185.

OSHA (2005) <http://www.osha.gov/dts/osta/otm/otm_toc.html> [Accessed on June 02, 2006].

Ovalles, C., Fonseca, A., Lara, A., Alvarado, V., Urrecheaga, K., Ranson, A., and Mendoza, H. (2002) *Opportunity of Downhole Dielectric Heating in Venezuela: Three Case Studies Involving, Medium, Heavy and extra-Heavy Crude Oil Reservoirs.* SPE 78980. SPE International Thermal Operations and Heavy Oil Symposium and International Horizontal Well Technology Conference, Alberta, Canada.

OWTG (Offshore Waste Treatment Guidelines) (2002) National Energy Board, Canada-Newfoundland Offshore Petroleum Board, Canada-Nova Scotia Offshore Petroleum Board. ISBN 0-921569-40-8, Released: January 9.

Paez, J.E. (2001). Screening Criteria for the Selection of Low-Dosage Hydrate Inhibitors as Applied in Natural Gas Problems, Dalhousie University, MASc Thesis.

Palermo, T., Argo, C.B., Goodwin, S.P., and Henderson, A. (2000) Flow loop tests on a novel hydrate inhibitor to be deployed in north sea ETAP field. *Annals of the New York Academy of Sciences*, **912**, 355–65.

Pamboris, X. (2004) *Sonar and Seismic Exploration: A Major Headache for Whales.* Vancouver Aquarium Aqua News. <http://www.vanaqua.org/aquanews/features/sonar.html> [Accessed on June 23, 2005].

Panawalage, S., Rahman, M., and Islam, M.R. (2004) Analytical and numerical solution of inverse problem of reservoir permeability. *Far East Journal of Applied Mathematics (FJAM)*, **17**(2), 207–19.

PanCanadian Report (1997) Weyburn unit CO_2 miscible flood EOR. Saskatchewan Energy and Mines, Saskatchewan, Canada.

Pandey, P. and Chauhan, R.S. (2001) Membranes for gas separation. *Progress in Polymer Science*, **26**, 853–93.

Paranjpe, M. (2002) <http://www.ench.ucalgary.ca> [Accessed on July 12, 2005].

Parillo, D.J., Thaeron, C., and Sircar, S. (1997) Separation of bulk hydrogen by selective surface flow membrane. *American Institute of Chemical Engineers Journal*, **43**(13), 2239.

Park, H.B. and Lee, Y.M. (2003) Pyrolytic carbon-silica membrane: a promising membrane material for improved gas separation. *Journal of Membrane Science*, **213**, 263.

Park, M V., Gillett, A.G., and Burema, J.R. (2000) "A GIS-framework for modeling environmental fate of Chernobyl-derived radiocesium". 4th International Conference on Integrating GIS and Environmental Modeling (GIS/EM4): Problems, Prospects and Research Needs. Banff, Alberta, Canada.

Parnell, G.S., Jackson, J.A., Kloeber, J.M. Jr., and Deckro, R.F. (1999) *Improving DOE Environmental Management: Using CERCLA-Based Decision Analysis for Remedial Alternative Evaluation in the RI/FS Process.* VCU-MAS-99-1, M SE Technology Applications, Inc., Butte, Montana.

Parro, D. (1999) Membrane CO_2 proves out at SACROC tertiary recovery project. *Oil & Gas Journal*, 82 (mixtures by selective surface flow membrane, Sep. and Purification tech. **15**, 121–9.

Patin, S. (1999) *Environmental Impact of the Offshore Oil and Gas Industry.* EcoMonitor Publishing, East Northport, New York.

Pearson, H. (2006) Vioxx may trigger heart attacks within days. *News and Nature.*

Penn, J.W. (1983). *An Assessment of Potential Yields from the Offshore Demersal Shrimp and Fish Stock in Bangladesh Waters (Including Comments on the Trawl Fishery 1981/820.* FAO, Rome. FI:DP/BGD/81/034, Field Document 29 p.

Pessier, R.C. and Fear, M.J. (1992) Quantifying common drilling problems with mechanical specific energy and a bit-specific coefficient of sliding friction. Paper 24584 presented at the SPE Annual Technical Conference and Exhibition in Washington, DC, USA, October 4–7.

Peterson, C.H. (1993) Improvement of environmental-impact analysis by application of principles derived from manipulative ecology – lessons from coastal marine case-histories. *Australian Journal of Ecology*, **18**(1), 21–52.

Peterson, C.H. (2001) The "Exxon Valdez" oil spill in Alaska: acute, indirect and chronic effects on the ecosystem. *Advancements in Marine Biology* **39**, 1–103.

Pfahi, U.J., Ross, M.C., and Shepherd, J.E. (2000), Flammability limits, ignition energy, and flame speeds in H_2–CH_4–NH_3–N_2O–O_2–N_2 mixtures, combust. *Flame* **123**, 140–58.

Piipari, R., Tuppurainen, M., Tuomi, T., Mantyla, L., Henriks-Eckerman, M.L., Keskinen, H., and Nordman, H. (1998) Diethanolamine-induced occupational asthma, a case report. *Clinical and Experimental Allergy*, **28**(3), 358–62.

Pinder, D. (2001) Offshore oil and gas: global resource knowledge and technological change. *Ocean & Coastal Management*, **44**, 579–600.

Pizzaro, J.O.S. and Trevisan, O.V. (1990) Electricalheating of oil reservoirs: numerical simulation and field test results. *Journal of Petroleum Technology*, **26**, 1320.

Pizzaro, J.O.S. and Trevisan, O.V. (1989) Electrical heating of oil reservoirs: Numerical simulation and field test results. SPE 19685, proceedings of the SPE annual Technical Conference and exhibition, San Antonio, October 8–11.

Plastic Task Force (1999) *Adverse Health Effects of Plastics.* <http://www.ecologycenter.org/erc/fact_sheets plastichealtheffects.html#plastichealthgrid> [Accessed on October 12, 2006].

Plissner, J.H. and Haig, S.M. (2000) Status of a broadly distributed endangered species: results and implications of the second International Piping Plover Census. *Canadian Journal of Zoology*, **78**(1), 128–39.

Pokharel, G.R., Chhetri, A.B., Devkota, S., and Shrestha, P. (2003) En route to strong sustainability: can decentralized community owned micro hydro energy systems in Nepal Realize the Paradigm? A case study of Thampalkot VDC in Sindhupalchowk District in Nepal. International Conference on Renewable Energy Technology for Rural Development. Kathmandu, Nepal.

Pokharel, G.R., Chhetri, K.C., Khan, M.I., and Islam, M.R. (2006) En route to strong sustainability: can decentralized community owned micro hydro energy systems in Nepal realize the paradigm? *Energy Sources*, in press.

Pollkar, M., Farouq Ali, S.M., and Puttagunta, V.R. (1990) High-Temperature Relative Permeabilities for Athabasca oil sands, SPE Reservoir Engineering, February, pp. 25–32, first presented at the 1988 SPE California Regional Meeting held in Long Beach. March 23–25, SPE 17424.

Pope, D.H. and Morris, E.A., III. (1995) *Microbial Corrosion*. Material Performance, p. 23.

Poruban, S. (2001) Platforms, producers, and the public. *Oil & Gas Journal*, **99**, 15.

Post Note (1995) Oil "Rig" Disposal, P.O.S.T. Note: 65. Parliamentary office of Science and Technology (Extension 2840).

Poston, S.W., Ysrael, S., Hossain, A.K.M.S., Mongomery, E.F., III, and Ramey, H.J. Jr. (1970) The effect of temperature on irreducible water saturation and relative permeability of unconsolidated sands. *Sot. Pet. Eng. J.* **June**, 171–80.

Powell, T. (2004) *The UK Offshore Sector – A Situation Report, Business Briefing: Exploration & Production: the Oil & Gas Review.* Head of Offshore Division, Health & Safety Executive.

Prager, J.C. (1998) *Environmental Contaminant Reference Databook*, Vol. 1–3. John Wiley & Sons, USA.

Preuss, L. (2001) In dirty chains? Purchasing and greener manufacturing. *Journal of Business Ethics*, **34**, 345.

Pulsipher, A.G. and Daniel, W.B., IV. (2000) Onshore disposition of offshore oil and gas platforms: Western politics and international standards. *Ocean & Coastal Management*, **43**, 973–95.

Putin, S. (1999) *Environmental Impact of the Offshore Oil and Gas Industry.* EcoMonitor Publishing, East Northport, New York.

Rahaman, M.S. (2003) Experimental and Numerical Studies of Arsenic Adsorption on Various Natural Adsorbents, Dalhousie University, MASc Thesis.

Rahbar, S. and Islam, M.R. (2005). Making of good plastics. *Journal Nature Science and Sustainable Technology*, in press.

Rahbur, S., Khan, M.M., Satish, M., Ma, F., and Islam, M.R. (2005) Experimental & numerical studies on natural insulation materials, ASME Congress, 2005, Orlando, Florida, November 5–11, IMECE2005-82409.

Rahman, M.H., Wasiuddin, N., and Islam, M.R. (2004) Experimental and Numerical Modeling Studies of Arsenic Removal with Wood Ash from Aqueous Streams, *CSChE J*, 968–77.

Rahman, M.S., Hossain, M.E., and Islam, M.R. (2006) An environment-friendly alkaline solution for enhanced oil recovery. *Journal of Petroleum Science and Technology*, in press.

Rahman, M.A., Mustafiz, S., Biazar, J., Koksal, M., and Islam, M.R. (2006) Investigation of a novel perforation technique in petroleum wells – perforation by drilling. *Journal of Franklin Institute*, in press.

Rahman, M.H. (2002) Experimental and numerical modeling studies of arsenic removal with wood as from aqueous streams. Faculty of Engineering, Dalhousie University, MASc Dissertation.

Ramachandran, S., Breen, P., and Ray, R. (2000) *Chemical Programs Assure Flow and Prevent Corrosion in Deepwater Facilities and Flow lines.* Baker Petrolite, Baker Hughes Inc, TX.

Ramakrishnan, T.S. and Wasan, D.T. (1983) A model for interfacial activity of acidic crude oil-caustic systems for alkaline flooding. *Society of Petroleum Engineers Journal*, **23**, 602–18.

Ramanathan, R. (2001) Comparative risk assessment of energy supply technologies: a data envelopment analysis approach. *Energy*, **26**, 197–203.

Ramsay, C.G., Bolsover, A.J., Jones, R.H., and Medland, W.G. (1994) Quantitative risk assessment applied to offshore process installations. challenges after the piper alpha disaster. *Journal of Loss Prevention in the Process Industries*, **7**(4), 295–304.

Ramskrishnan, T.S. and Wasan, D.T. (1982). A Model for Interracial Activity of Acidic Crude Oil-Caustic Systems for Alkaline Flooding, SPE/DOE 10716. Presented at the SPE/DOE Third Joint Symp. On Enhanced Oil Recovery, Tulsa.

Ransohoff, T.C. and Radke, C.J. Mechanisms of foam generation in glass-bead packs. *SPE Reservoir Engineering*, **3**(2), 573–85.

Rao, M.B. and Sircar, S. (1996) Performance and pore characterization of nanoporous carbon membranes for gas separation. *Journal of Membrane Science*, **3**(7), 109–18.

Rao, M.B. and Sirkar, S. (1993) Liquid-phase adsorption of bulk ethanol-water mixtures by alumina. *Adsorption Science and Technology*, **10**(1–4), 93–104.

Rao, M.B., Sircar, S., and Golden, T.C. (1992) Gas separation by adsorbent membranes. *US Patent*, 5, 104, 425.

Rao, P. (2002) Greening the supply chain: a new initiative in South East Asia. *International Journal of Operations & Production Management*, **22**, 632.

Ray, R., Little, B., Wanger, P., and Hart, K. (1997) Marine coatings used to protect carbon steel. *Scanning*, **19**, 98.

Rees, W. (1989) Sustainable development: myths and realities. Proceedings of the Conference on Sustainable Development Winnipeg, Manitoba: IISD.

Reis, J.C. (1996) *Environmental Control in Petroleum Engineering*. Gulf Publishing Company, Houston, Tex.

Rezende, C.E., Lacerda, L.D., Ovalle, A.R.C., Souza, C.M.M., Gobo, A.A.R., and Santos, D.O. (2002) The effect of an oil drilling operation on the trace metal concentrations in offshore bottom sediments of the Campos Basin oil field, SE Brazil. *Marine Pollution Bulletin*, **44**, 680–4.

Rice, S.A., Kok, A.L., and Neate, C.J. (1992) A test of the electric heating process as a well means for stimulating productivity of an oil well in the Schoonebeek Field. Paper CIM 92-04,1992 Annual Technical Meeting of the Petroleum Society of CIM, Calgary, June.

Rios, P., Stuart, J.A., and Grant, E. (2003) Plastics disassembly versus bulk recycling: Engineering design for end-of-life electronics resource recovery. *Environmental Science & Technology*, **37**, 5463.

Robert, A. (1999) Integrating environmental issues into the mainstream: an agenda for research in operations management. *Journal of Operations Management*, **17**, 575–99.

Robertson, J.K. (2001) A vertical micromachined resistive heater for a micro-gas separation column. *Sensors and Actuators, A: Physical*, **91**, 333.

Robinson, J.G. (1993) The limits to caring: sustainable living and the loss of biodiversity. *Conservation Biology*, **7**, 20–8.

Robinson, R.J., Bursell, C.G., and Restine, J, L. (1977) A Caustic Steam flood Pilot-Kern River Field, paper SPE 6523 presented at SPE AIME 47th Annual California Regional Meeting, 13akerafield, California, April 13–15.

Rojas, G., Farouq Ali, S.M., Zhu, T., and Dyer, S. (1991) Scaled model studies of carbon dioxide flooding. *SPE Reservoir Eng.*, **6**(2), 169–78.

Rojey, A., Jaffret, C., Cornot-Gandolphe, S., Durand, B., Jullian, S., and Valais, M. (1997) Natural gas production processing and transport. Institut Francais du Petrole Publications. Translation (updated and expanded) of *Le gaz naturel. Production. Traitement. Transport*. Editions Technip, Paris.

Rosenbery, J.I., Brashear, J.P., Mercer, J., Morra, F., and O'Shea, P. (1983) A Sensitivity Analysis of the NPC Study of Tight Gas, Soc. Pet. Eng. AIME, Pap., Vol/Issue: SPE/DOE11645; SPE symposium on low permeability; 14 Mar 1983; Denver, CO, USA.

Rushing, S. (1994) CO_2 Recovery improves cogen plant economics. *Power Engineering*, **April**, 46–47.

S.L. Ross Environmental Research Ltd. (2001) *Blowout and Spill Probabilities for Exploration Drilling off Cape Breton, Nova Scotia*. S. L. Ross Environmental Research Ltd., Ottawa, Ontario.

Sable Island Preservation Trust (2001) *Sable Island Preservation Trust Island Operations Business Plan 2001–2002.* Sable Island Preservation Trust. <http://www.sabletrust.ns.ca/documents/BusinessPlan.pdf> [Accessed on April 18, 2002).

Sadiq, R., Khan, F.I., and Veitch, B. (2005) Evaluating offshore technologies for produced water management using GreenPro-I – a risk-based life cycle analysis for green and clean process selection and design, *Computers and Chemical Engineering*, **29**, 1023–39.

Saeed, N.O., Ajijolaiya, L.O., Al-Darbi, M.M., and Islam, M.R. (2003) Mechanical properties of mortar reinforced with hair fibre. In *Proc. Oil and Gas Symposium, CSCE Annual Conference*, Moncton.

Saenger, P. and Siddiqi, N.A. (1993) Land from the sea: the mangrove afforestation program of Bangladesh, *Ocean & Coastal Management*, **20**(1), 23–39.

Saghir, M.Z.H., Vaziri, M.R., and Islam, X.X. (2001) Heat and mass transfer modeling of fractured formations. *International Journal of Computerized Fluid Dynamics*, **15**(4), 279–92.

Saha, M., Sarkar, S.K., and Bhattacharya, B. (2006) Interspecific variation in heavy metal body concentrations in biota of Sunderban mangrove wetland, northeast India, *Environment International*, **32**(2), 203–7.

Saha, S. and Chakma, A. (1992) An energy efficient mixed solvent for the separation of CO_2. *Energy Conversion Management*, **33**, 413.

Sahni A., Kumar, M., and Knapp, R.B. (2000) Electromagnetic Heating Methods for Heavy Oil Reservoirs. *SPE 62550, SPE/AAPG Western Regional Meeting*. California, 19–23, June.

Saito, K., Ogawa, M., Takekuma, M., Ohmura, A., Kawaguchi, M., Ito, R., Inoue, K., Matsuki, Y., and Nakazawa, H. (2005) Systematic analysis and overall toxicity evaluation of dioxins and hexachlorobenzene in human milk, *Chemosphere*, **61**, 1215–20.

SAL (2006) *Soil Acidity and Liming: Internet Inservice Training, Best Management Practices for Wood Ash Used as an Agricultural Soil Amendment.* <http://hubcap. clemson.edu/~blpprt/ bestwoodash.html> [Accessed on June 7, 2006].

Sarkar, A.K., Goursaud, J.C., Sharma, M.M., and Georgiou, G. (1989) A critical evaluation of MEOR processes. *In Situ*, **13**, 207–38.

Sarker, M.S.U. and Sarker, N.J. (1988) *Wildlife of Bangladesh. A Systemitic List with Status, Distribution and Habitat.* Department of Zoology, University of Dhaka, Dhaka.

Sarkis, J. (2003) A strategic decision framework for green supply chain management. *Journal of Cleaner production*, **11**, 397–409.

Sarma, H.K. and Bentsen, R.G. (1987) An experimental verification of a modified instability theory for immiscible displacements in porous media. *J. Cdn. Pet. Tech.*, **35**(7), 47.

Sarwar, M. and Islam, M.R. (1997) A non-fickian surface excess model for chemical transport through fractured porous media. *Chemical Engineering Communication*, **160**, 1–34.

SBOR (2003) *Sustainable Business Online Resources.* <http://www.communityfutures.ca/provincial/SustainableBusiness/English/Main_Pages/home_e.htm> [Accessed on September 04, 2006].

Schembre, J.M., Tang, G.Q., and Kovscek, A.R. (2005) Effect of Temperature on Relative Permeability for Heavy-oil Diatomic reservoirs, paper SPE 93831 presented at the 2005 Western Regional Meeting held in Irvine, CA, USA, 30 March–1 April.

Schlumberger (1991) *Schlumberger Log Interpretation Principles/Applications*, 3rd printing. Schlumberger Educational Services, 5000 Gulf Freeway, Houston, Texas, USA.

Schmidt, J.H., Friar, W.L., Bill, M.L., and Cooper, G.D. (1999) Large-scale injection of north slope drilling cuttings. SPE 52738, Presented at the 1999 SPE/EPA Exploration and Production Environmental Conference, Austin, TX, February 28–March 3.

Schmoker, J.W. and Attanasi, E.D. (1996) *The Importance of Reserve Growth to the Nation's Supply of Natural Gas.* U.S. Geological Survey Fact Sheet FS-202-96, 2 p.

Schroeder, D.M. and Love, M.S. (2004) Ecological and political issues surrounding decommissioning of offshore oil facilities in the Southern California Bight. *Ocean & Coastal Management*, **47**, 21–48.

Schwartz, B. (2000) The crude supply chain. *Transportation and Distribution*, **41**, 49–54.

Schwartz, F.W., and Smith, L. (1988) A continuum approach for modelling mass transport in fractured media. *Water Resources Research*, **24**(8), 1360–72.

Senate Committee on Fisheries (SCF). (1993) *The Atlantic Commercial Inshore Fishery, Committee Report.* <http://www.parl.gc.ca/36/1/parlbus/commbus/ senate/com-e/fish-e/past_rep-e/93repen1.htm> [Accessed on April 17, 2002].

Service, R.F. (2005) Is it time to shoot for the sun? *Science*, **309**, 549–51.

Seuring, S. (2004) Industrial ecology, life cycles, supply chains: differences and interrelations. *Business Strategy and the Environment*, **13**, 306.

Shapiro, R., Zatzman, G.M., and Mohiuddin, Y. (2006) Towards Uniderstanding Disinformation, *Journal of Nature and Sustainable Technology*, in press.

Sharma, M.M. and Yen, T.F. (1983) A thermodynamic model for low interracial tension in alkaline flooding. *Society of Petroleum Engineers' Journal*, Paper number 10590-PA, 125–34.

Sharma, S. (2001) Different strokes: regulatory styles and environmental strategy in the North-American oil and gas industry. *Business Strategy and the Environment*, **10**, 344–64.

Shebeko, Yu.N., Fan, W., Bolodian, I.A., and Navzenya, V. Yu. (2002) An analytical evaluation of flammability limits of gaseous mixtures of combustible-oxidizer-diluent. *Fire Safety Journal*, **37**, 549–68.

SIEAP (Sable Island Environmental Assessment Panel) (1983) *Venture Development Project: Report of the Sable Island Environmental Assessment Panel.* Ministry of Supplies and Services, Ottawa, Ontario.

Sierra, R., Tripathy, B., Bridges, J.E., and Farouq Ali, S.M. (2001) Promising progress in field application of reservoir electrical heating methods. SPE 69709, International Thermal Operations and Heavy Oil Symposium held in Margarita, Venezuela, 12–14 March.

Silva, C.A.R. and A.A. Mozeto, (1977) Relase and retention of phosphorus in mangrove sediments: Septiba Bay, Brazil, in *Mangrove Ecosystem Studies in Latin America and Africa* (eds B. Kjerfve, L.D. de Laverda and E.H.S. Dipo). UNESCO, Paris, pp. 179–90.

Simpson, D.F. and Power, D.J. (2005) Use the supply relationship to develop lean and green suppliers. *Supply Chain Management*, **10**, 60.

Sinnokrot, A.A., Ramey, H.J.Jr., and Marsden, S.S.Jr. (1971) Effect of temperature level upon capillary pressure curves. *Sot. Pet. Eng. J.*, **March**, 13–22.

Skalski, J.R., Hoffmann, A., and Smith, S.G. (1992) Effects of sound from geophysical surveys device on catch-per-unit-effort in a hook-an-line fishery for rockfish (Sebastes spp.). *Canadian Journal of Fisheries and Aquatic Sciences*, **49**, 1357–65.

Sloan, E.D.Jr. (1998) *Clathrate Hydrates of Natural Gases*, 2nd edn. Marcel Dekker Inc., New York.

Smith, P. (2001) How green is my process? A practical guide to green metrics. In: Proceedings of the Conference Green Chemistry on Sustainable Products and Processes; 2001, April 3–6, 2001, *University of Wales, Swansea, U.K.*

Smith, D.R. and Quinn, J.A. (1979) The prediction of facilitation factors for reaction augmented membrane transport. *American Institute of Chemical Engineers Journal*, **25**(1), 197–200.

Smith, D.R., Lander, R.J., and Quinn, J.A. (1977) Carrier mediated transport in synthetic membranes. *Recent Developments in Separation Science*, **3B**, 225–41.

Smith, P. (2001) How green is my process? A practical guide to green metrics. In: Proceedings of the Conference Green Chemistry on Sustainable Products and Processes.

Soliman, M.Y. (1997) Approximate solutions for flow of oil heated using microwaves. *Journal of Petroleum Science and Engineering*, **18**, 93–100.

Song, F.Y. and Islam, M.R. (1994) Effect of salinity and rock type on sorption behaviour of surfactants as applied in cleaning of petroleum contaminants. *Journal of Petroleum Science and Engineering*, **10**(4), 321–36.

Spangenberg, J.H. and Bonniot, O. (1998) Sustainability indicators-a compass on the road towards sustainability. Wuppertal Paper No. 81, ISSN No. 0949-5266.

Spicer, G.W. and Woodward, C. (1991) H_2S control keeps the gas from offshore field on specifications. *Oil & Gas Journal*, **89**(21), 76–80.

Spies, R.B. (1987) The biological effects of petroleum hydrocarbons in the sea, in *Long-Term Environmental Effects of Offshore Oil And Gas Development* (eds D.F. Boesch and N.N. Rabalais). Elsevier Applied Science, London and New York, pp. 287–341.

Sraffa, P. (1960) *Production of Commodities by Means of Commodities*. Cambridge University Press, Cambridge.

Sresty, G.C., Dev, H., Snow, R.H., and Bridges, J.E. (1986) Recovery of bitumen from Tar sand deposits with the radio frequency process. *SPE Reservoir Engineering*, SPE Paper no. 10229-PA, **January**, 85–94.

Srivastava, R.K. and Huang, S.S. (1995) Technical Feasibility of CO_2 Loading in Weyburn Reservoir – A Laboratory Investigation. CIM. paper no. 95–1119 presented at the 6th Saskatchewan Petroleum Conference, Regina, Oct. 16–18.

Srivastava, R.K., Huang, S.S., Dyer, S.B., and Mourits, F.M. (1993) A scaled physical model for Saskatchewan heary oil reserves-design fabrication and preliminary CO_2 flood studies, paper no. 33, presented at the fifth petroleum conference of South Saskatchewan Sction, the Petroleum Society of CIM, Regina, 18–20 October.

Srivastava, R.K., Hutchence, K., and Huang, S.S. (1997) Methane water-alternating-gas injection for heavy oil reservoirs, Year One technical report, Saskatchewan Research Council, Petroleum Branch (March 1997) P 110–373-C-97.

Stagg, R.M., Rusin J., and McPhail, M.E. (2000) Effects of polycyclic aromatic hydrocarbons on expression of CYP1A in salmon (Salmo salar) following experimental exposure and after the Braer oil spill. *Environmental Toxicology and Chemistry*, **19**, 2797–805.

Stanislav, J.F. and Kabir, C.S. (1990) *Pressure Transient Analysis*. Prentice Hall, Englewood Cliffs, New Jersey, USA.

STATOIL. (2000) *The Norne 6608/10 FPSO*. <http://www.offshore-technology.com/projects/statoil/statoil2.html> [Accessed on June 16, 2005].

Stejskal, I.V. (2000) Obtaining approvals for oil and gas projects in shallow water marine areas in Western Australia using an environmental risk assessment framework. *Spill Science & Technology Bulletin*, **6**(1), 69–76.

Stephenson, M.T. (1992) A survey of produced water studies, in *Produced Water: Technological/Environmental Issues and Solutions*, Volume 46 (eds J.P. Ray and F.R. Engelhardt). Plenum Press, New York, pp. 1–12.

Stewart, R.B., Gill, D.S., Lohbeck, C.M., and Baaijens, M.N. (1997) An exampandalbe slotted-tubing, fiber-cement wellbore-lning system. *SPE Drilling and Completion*, **12**(3), 163–7.

Stright, D.H.Jr., Aziz, K., Settari, A., and Starratt, F.E. (1976) Carbon Dioxide Injection Into Bottom-Water, Undersaturated Viscous Oil Reserves. 51st SPE-AIME Annual Fall Technical Conference and Exhibition, New Orleans, Oct. 3–6.

Stuebinger, L.A. and Elphingstone, G.M.Jr. (2000) Multipurpose wells: downhole oil/water separation in the future. SPE 65071. *SPE Production & Facilities*, **15**(31), 191–5.

Sturman, P.J. and Goeres, D.M. (1999) *Control of Hydrogen Sulfide in Oil and Gas Wells With Nitrite Injection*. (SPE 56772).

Subkow, P. (1942) Process for the removal of bitumen from bituminous deposits. *U.S. Patent*, 228857.

Suboor, M.A. and Heller, J.P. (1995) Minipermeameter characteristics critical to its use. *In Situ*, **19**, 225–48.

Sudaryanto, A., Kunisue, T., Kajiwara, N., Iwata, H., Adibroto, T.A., Hartono, P., and Tanabe, S. (2006) Specific accumulation of organochlorines in human breast milk from Indonesia: Levels, distribution, accumulation kinetics and infant health risk. *Environmental Pollution*, **139**(1): 107–17.

Sudicky, E.A. and Frind, E.O. (1982) Contaminant transport in fractured porous media: analytical solution for a single fracture. *Water Resources Research*, **21**, 1677–83.

Sufi, A.S., Ramey, H.J., Jr, and Brigham, W.E. (1982) Temperature Effects on Relative Permeabilities of Oil/water Systems. Presented at the SPE 57th Annual Technical Conference and Exhibition, New Orleans, September 26–29, SPE 11071.

Sundaram, N.S. and Islam, M.R. (1994) Scaled model studies of petroleum contaminant removalfrom soilsusing surfactant solutions. *Journal ofHazardousMaterials*, **38**, 89–103.

Sundaram, N.S., Saleh, S.T., Graves, R.M., and Islam, M.R. (1995) Experimental and numerical modeling of interfacial tension dynamics in oil r water rsurfactant fluid sys-tems. Proceedings of the International Conference on Improved Oil Recovery, Al-Ain, U.A.E., December.

Sundaram, N.S., Sarwar, M., Bang, S.S., and Islam, M.R. (1994) Biodegradation of anionic surfactantsinthepresence ofpetroleum contaminants. *Chemosphere*, **26**(6), 1253–61.

Szokolik, A. (1992) Evaluating single-coat ingorganic zinc silicates for oil and gas production facilities in marine environment. *Journal of Protective Coatings and Linings*, **March**, 24–43.

Szostak-Kotowa, J. (2004) Biodeterioration of textiles. *International Biodeterioration & Biodegradation*, **53**, 165–70.

Tabe-Mohammadi, A. (1999) A review of the applications of membrane separation technology in natural gas treatment. *Separation Science and Technology*, **34**(10), 2095–111.

Taber, J.J. (1994) A Study of Technical Feasibility for the Utilization of CO2 for Enhanced Oil Recovery, The Utilization of Carbon Dioxide from Fossil Fuel Fired Power Stations, IEA Greenhouse Gas R & D program, Cheltenham, England.

Taheri, F. and Hassan, M.A. (2003) A Rational Procedure for Development of Hybrid Composite Masts, *ASTM Journal of Composite Technology and Research*, **25**(2), 13.

Taheri, F., Prior, D., Vaziri, H., and Islam, M.R. (2005) A Novel Petroleum Well Completing Technique, *Proceeding of the Canadian Society for Civil Engineering (CSCE), 33rd Annual Conference*, Toronto, ON, Canada, June 2–4.

Takahashi, A., Urano, Y., Tokuhashi, K., and Kondo, S. (2003) Effect of vessel size and shape on experimental flammability limits of gases. *Journal of Hazardous Materials*, **A105**, 27–37.

Talabani, S. and Islam, M.R. (2000) A New Cementing Technology for Controlling Downhole Corrosion, *J. Pet. Sci. Eng.*, **26**(1–4), 43–8.

Tang, D.H., Frind, E.O., and Sudicky, E.A. (1981) Contaminant transport in fractured porous media: analytical solution for a system of parallel fractures. *Water Resources Research*, **17**(3), 555–64.

Tango, M.S.A. and Islam, M.R. (2002) Potential of extremophiles for biotechnological and petroleum applications. *Energy Sources*, **24**(6), 543–59.

Tanner, R.S. (1991) Microbial enhanced oil recovery from carbonate reservoirs. *Geomicroiol*, **9**, 169–95.

Tarmoon, I.O. (1999) *Middle East Oil Show*. 20–23 February.

Taylor, K.C., Hawkins, B.F., and Islam, M.R. (1990) Dynamic interfacial tension in surfactant-enhanced alkaline flooding. *Journal of Canadian Petroleum Technology*, **291**, 50–5.

Taylor, K.C. and Nasr-E1-Din, H.A. (1996) The effect of synthetic surfactants on the interfacial behavior of crude oil/alkali/polymer systems, colloids and surfaces, A. *Physicochemical and Engineering Aspects*, **108**, 49–72.

Teal, J.M. and Howarth, R.W. (1984) Oil spill studies: a review of ecological effects. *Environmental Management*, **8**(1), 27–44.

Teknowledge Corporation (1993) *M.4 Version 3.0 Release Notes: A Guide to Developing and Delivering Knowledge Systems*. Palo Alto, CA, USA.

Teknowledge Corporation (1995) *User's Guide*. Palo Alto, CA, USA.

Teknowledge Corporation (1996) *M.4 Version 3.01 Release Notes: A Guide to Developing and Delivering Knowledge Systems*. Palo Alto, CA, USA.

Teramoto, M., Nakai, N., Ohnishi, Q., Huang, T., Watari, H., and Matsuyama, H. (1996) Facilitated transport of carbon dioxide through supported liquid membranes of aques amine solutions. *Industrial and Engineering Chemistry Research*, **35**, 538–45.

Teresa, K.L., Wong, W.Y., Wong, Y.S., and Tam, N.F.Y. (2002) Fate of polycyclic aromatic hydrocarbon (PAH) contamination in a mangrove swamp in Hong Kong following an oil spill. *Marine Pollution Bulletin*, **45**(1–12), 339–47.

Thaeron, C.D., Parrillo, J., Clarke, P.F., Paranjape, M., and Pruden, B.B. (1986) Separation of hydrogen sulfide-methane, in *Thomas Graham and Gaseous Diffusion, proceedings of fourth BOC Priestley Conference* (ed. M. Stanly). Leeds University, Royal Society of Chemistry, London, Special Publication No. 62. pp. 1–15.

Thanyamanta, W., Hawboldt, K., Husain, T., Bose, N. and Veitch, B. (2004) Evaluation of offshore drilling cuttings management technologies using multicriteria decision making, in *Proceedings of the Offshore Oil and Gas Environmental Effects Monitoring Workshop: Approaches and Technologies* (eds S.L., Armsworthy, P.J. Cranford and K. Lee). Battelle Press, Columbus, Ohio, pp. 167–79.

Thibodeau, L., Sakanoko, M., and Neale, G.H. (2003) Alkaline flooding processes in porous media in the presence of connate water. *Powder Technology*, **32**, 101–111.

Thomas, E.R., Voorhees, R.J., and Kennelly, K.J. (1991) Paper on Gas processing presented, Proceedings, Annual Convention- Gas Processors Association, p. 45.

Thomas, S. and Farouq Ali, S.M. (1999) Status and assessment of chemical oil recovery methods. *Energy Sources*, **21**, 177–89.

Thompson, D., Sjoberg, M., Bryant, M.E., Lovell, P., and Bjorge, A. (1998) Behavioural and physiological responses of harbour (Phoca vitulina) and grey (Halichoerus grypus) seals to seismic surveys. Report to European Commission of BROMMAD Project. MAS2 C7940098.

Tidwell, V.C., Gutjahr, A.I., and Wilson, J.L. (1999) What does an instrument measure? Empirical spatial weightning functions calculated from permeability data sets measure on multiple sample supports. *Water Resource Research*, **35**, 4354.

Tidwell, V.C. and JWilson, L. (1997) Laboratory method for investigating permeability upscaling. *Water Resource Research*, **33**, 1607–16.

Timashev, S.F. (1991) *Physical Chemistry of Membrane Processes*. Ellis Horwood Ltd., London.

Tipton, T., Johnston, C.T., Trabue, S.L., Erickson, C., and Stone, D.A. (1993) Gravimetric/ FT-IR apparatus for the study of vapor sorption on clay films. *Revue Scientifique Instrumention*, **64**(4), 1091–2.

TNO Purple Book (1999) *Guideline for Quantitative Risk Assessment, Committee for the Prevention of Disasters*. The Netherlands, (Chapter 6).

Toninello, A., Pietrangeli, P., De Marchi, U., Salvi, M., and Mondov, B. (2006) Amine oxidases in apoptosis and cancer. *Biochimica et Biophysica Acta*, **1765**, 1–13.

Torabzadeh, S.J. and Handy, L.L. (1984) The Effect of Temperature and Interracial Tension on Water/Oil Relative, Permeabilities of Consolidated Sands, presented at the SPE/DOE Fourth Symposium on Enhanced oil Recovery held in Tulsa, OK, April 15–18, SPE 12689.

Trujillo, E.M. (1983) The static and dynamic interracial tensions between crude oils and caustic solutions. *Society of Petroleum Engineers' Journal*, SPE Paper no. 10917-PA, **August**, 645–56.

Turksoy, U. and Bagci, S. (2000) Improved oil recovery using alkaline solutions in limestone medium. *Journal of Petroleum Science and Engineering*, **26**(1–4), 105–19.

Twu, C.H, Tassone, V., Sim, W.D., and Watanasiric, S. (2005). Advanced equation of state method for modeling TEG–water for glycol gas dehydration. *Fluid Phase Equilibria*, **228–229**, 213–221.

U.S. Department of Energy (USDEO) (2002) <http://www.fe.doe.gov/oilgas/drilling/laserdrilling.shtml> [Accessed on October 7, 2006].

UKOOA (United Kingdom Offshore Operations Association) (1999) UKOOA Sustainability Strategy, "Striking a Balance", Sustainable development for the UK offshore industry. 232–242 Vauxhall Bridge Road, London.

UKOOA (2005) *Decommissioning Index*. <http://www.ukooa.co.uk/issues/decommissioning/background.htm> [Accessed on April 22, 2006].

UN (2002) *Born to Wild*. The UN work for the Environment. The United Nations, The UN Works Programme, Department of Public Information, Room S-955, New York, NY.

UNCLOS Exclusive Economic Zone, Article55, Specific legal regime of the exclusive economic zone, Rights, jurisdiction and duties of the coastal State in the exclusive economic zone. Oceans and Law of the Sea, Division Ocean Affair and Law of the Sea, UN, <http://www.un.org/depts/los/convention_agreements/texts/unclos/part5.htm>.

UNCSD (United Nations Commission on Sustainable Development) (2001) *Indicators of Sustainable Development: Guidelines and Methodologies*. United Nations, New York.

Union Gas (2006) *Chemical Composition of Natural Gas*, Union Gas, Chatham, Ontario, Canada.

United Kingdom Offshore Operators Association (UKOOA). *Environment: 1999 Environmental Report*. <http://www.oilandgas.org.uk/issues/1999report/ enviro99_water.htm> [Accessed on 12 July, 2001].

United States Coast Guard (1990) *Update of Inputs of Petroleum Hydrocarbons Into the Oceans Due to Marine Transportation Activities*. National Research Council. National Academy Press, Washington, DC.

USEPA (U.S. Environmental Protection Agency) (1982) *Development Document for Effluent Limitations Guidelines and Standards for the Petroleum Refining Point Source Category.* USEPA, Washington, DC.

USEPA (1999) cited in DWMIS (Drilling Waste Management Information System) (2005). *The Drilling Waste Management Information System, Fact Sheet – The First Step: Separation of Mud from Cuttings, Argonne National Laboratory, Natural Gas & Oil Technology Partnership program.* <http://web.ead.anl.gov/dwm/index.cfm.> [Accessed on October 22, 2006].

USGS (2005) The Significance of Field Growth and the Role of Enhanced Oil Recovery Denver Federal Center, Box 25046, MS 939, Denver, CO 80225.

Vancouver Aquarium Marine Science Center (2004) *Canada: Seismic Surveys a Danger to Marine Life, Report Warns. Vancouver Aquarium Aqua News.* <http://www.vanaqua.org/aquanew/fullnews.php?id=1722> [Accessed on June 23, 2005].

Vanderstraeten, B., Tuerlinckx, D., Berghmans, J., Vliegen, S., Oost, E.V, and Smit, B. (1997) Experimental study of the pressure and temperature dependence on the upper flammability limit of methane/air mixtures. *Journal of Hazardous Materials,* **56,** 237–46.

Veil, J.A. (1997) Costs for Offsite Disposal of Nonhazardous Oil Field Wastes: Salt Caverns versus Other Disposal Methods," prepared for DOE's Office of Fossil Energy, April; also published by DOE- National Petroleum Technology Office as DOE/BC/W-31-109-ENG-38-3, DE97008692, September.

Veil, J.A. (1998) *Data Summary of Offshore Drilling Waste Disposal Practices.* Prepared for U.S. Environmental Protection Agency, Engineering and Analysis Division, and U.S. Department of Energy, Office of Fossil Energy.

Veil, J.A. (2002) *Drilling Waste Management: Past, Present and Future.* Annual Technical Conference and Exhibition, San Antonio, Texas.

Verma, M.K., Ulmishek, G.F., and Gilbershtein, A.P. (2000) *Oil and Gas Reserve Growth – A Model for the Volga-Ural Province.* Russia: Proceedings of SPE/AAPG Western Regional Meeting, June 19–23, Long Beach, California, SPE no. 62616, 10 p.

Vermeulen, F. and McGee, B. (2000) In situ electromagnetic heating for hydrocarbon recovery and environmental remediation. *Journal of Canadian Petroleum Technology,* **39**(8), 24–8.

Vignier, V., Vandermeulen, J.H., and Fraser, A.J. (1992) Growth and food conversion by Atlantic salmon parr during 40 days exposure to crude-oil. *Journal of American Fish Society,* **121**(3), 322–32.

Voorhees, R.J., Thomas, E.R., and Kennelley, K.J. (1991) Diaryldisulfide solves sulfur deposition problems at sour gas field. *Oil & Gas Journal,* **89**(35), 85–9.

Wackernagel, M. and Rees, W. (1996) *Our Ecological Footprint.* New Society Publishers, Gabriola Island.

Wadadar, S.S. and Islam, M.R. (1994) Numerical simulation of electromagnetic heating of alaskan tar sands using horizontal wells, *Journal of Cannda Petroleum Technology,* **33**(7), 37–43.

Wadman, M. (2006) How does a painkiller harm the heart? *Nature,* **441,** 262.

Wagner, N., Lobo, R., Doren, D., and Foley, H. (2002) Molecular Transport in Nano Structured Materials. NSF Nanoscale Science and Engineering Grantees Conference, Arlington, VA, December 11–13.

Wang, D., Li, K., and Teo, W.K. (1999) Preparation and characterization of polyvinylidene fluoride (PVDF) hollow fiber membranes. *Journal of Membrane Science,* **163**(2), 211–20.

Wang, D., Li, K., and Teo, W.K. (2000) Highly permeable polyethersulfone hollow fiber gas separation membranes prepared using water as non-solvent additive. *Journal of Membrane Science,* **176**(2), 147–58.

Wang, H., Zhaol, Y., Jusys, Z., and Behm, R.J. (2005) Ethylene glycol electrooxidation on carbon supported Pt, PtRu and Pt_3Sn catalysts: a comparative DEMS study. *Journal of Power Sources,* **115**(1), 33–46.

Ward, W.J. (1970) Analytical and experimental studies of facilitated transport. *American Institute of Chemical Engineers Journal,* **16**(3), 405–10.

Wardle, C.S., Carter, T.J., Urquhat, G.G., Johnstone, A.D.F., Ziolkowski, A.M., Hampson, G., and Mackie, D. (2001) Effects of seismic air guns on marine fish. *Continental Shelf Research,* **21,** 1005–27.

Washburn, E.W. (1921) The dynamics of capillary flow. *Physics Review,* **17,** 273–83.

Wasiuddin, N.M., Tango, M., and Islam, M.R. (2002) A novel method for arsenic removal. *Energy Sources,* **24**(11), 1031–42.

Waste Online (2005) Plastic recycling information sheet. <http://www.wasteonline.org.uk/resources/InformationSheets/Plastics.htm> [Accessed on February 20, 2006].

Watson, R.W. and Ertekin, T. (1988) *The Effect of Steep Temperature Gradient on Relative Permeability Measurements.* Presented the SPE Rocky Mountain Regional Meeting, held in Caspar, WY, May, 11–13, SPE 17505.

WCED (World Commission on Environment and Development) (1987) *Our common future. World Conference on Environment and Development.* Oxford University Press, Oxford.

Webster, E., Mackay, D., Di Guardo, A., Kane, D., and Woodfine, D. (2003) *Regional Differences in Chemical Fate Model Outcomes.* Chemosphere, Accepted for publication, August.

Weinbrandt, R.M., Ramey, H.J.Jr., and Casse, F.J. (1975) The effect of temperature on relative and absolute permeability of sandstones. *Society of Petroleum Engineers' Journal*, **March**, 376–84.

Welford, R. (1995). *Environmental Strategy and Sustainable Development: The Corporate Challenge for the 21st Century.* Routledge, London.

Wells, P.G., Buttler, J.N., and Hughes, J.S. (1995) Introduction, overview, issues, in *Exxon Valdez Oil Spill: Fate and Effects in Alaskan Waters*, ASTM STP 1219 (eds P.G. Wells, J.N. Butler and J.S. Hughes). American Society for Testing and Materials, Philadelphia.

Wenger, L.M., Davis, C.L., Evensen, J.M., Gormly, J.R., and Mankiewicz, P.J. (2004) Impact of modern deepwater drilling and testing fluids on geochemical evaluations. *Organic Geochemistry*, **35**, 1527–36.

Wenz, G.M. (1962) Acoustic ambient noise in the ocean; spectra and sources. *The Journal of the Acoustical Society of America*, **34**, 1936.

West Coast Offshore Exploration Environmental Assessment Panel (WCOEEAP) (1986) *Offshore Hydrocarbon Exploration: Report of the West Coast Offshore Exploration Environmental Assessment Panel.* WCOEEAP, Vancouver British Columbia.

Wierzba, I. and Ale, B.B. (2000), Rich flammability limits of fuel mixtures involving hydrogen at elevated temperatures. *Int. J. Hydrogen Energy*, **25**, 75–80.

Wiese, F.K., Montevecchi, W.A., Davoren, G.K., Hunttmann, F., Daimond, A.W., and Link, J. (2001) Seabirds at Risk around offshore oil platforms in the North-Atlantic. *Marine Pollution Bulletin*, **42**(12), 1285–90.

Wilding, T.A. and Sayer, M.D.J. (2002) Evaluating artificial reef performance: approaches to pre-deployment research. *ICES Journal of Marine Science*, **59**, S222–S230.

Willis, J. (2000) Muddied Water: a survey of offshore oilfield Drilling Waters and Disposal Techniques to Reduce Ecological Impact of Sea Dumping. <http://www.offshoreenvironment.com/ospar.html> [bast accessed June 15, 2005].

Wills, J., Shemaria, M., and Mitariten, M.J. (2004) *Production of Pipeline Quality Natural Gas.* SPE/EPA/DOE Exploration and Production Environmental Conference, SPE Paper No. 87644, March 10–12, San-Antonio, TX.

Winterton, N. (2001) Twelve more green chemistry principles. *Green Chemistry*, **3**, G73–5.

Wood, D.A. (2005) Managing portfolios: the impact of petroleum asset life cycles. *Oil & Gas Financial Journal*, **2**(6), 1–2.

Woodrow, C.K. and Drummond, E. (2001) SPE 67729. SPE/IADC Drilling Conference, February 27–March 1, Amsterdam.

World Bank (1998) *Petroleum Refining Pollution Prevention and Abatement Handbook.* World Bank Group.

World Bank (1996) *Pollution Prevention and Abatement: Petroleum Refining.* World Bank Group.

World Health Organization (WHO) (1994) Brominated diphenyl ethers, in *Environmental Health Criteria*, Volume 162. International Program on Chemical Safety.

Wright, T. (2002) Definitions and frameworks for environmental sustainability in Higher education. *International Journal of Sustainability in Higher Education Policy*, **15**(2), 15–36.

Yasuda, N. and Hoshina, M. (1993) Application of ultrahigh pressure water jet for rock drilling. *International Journal of Rock Mechanics and Mining Science & Geomechanics*, **30**(1), A42.

Young, G.R. (1989) *Determining permeability anisotropy from a core plug using a minipermeameter.* Masters Thesis, Univ. of Tex., Austin.

Zaman, M., Agha, K.R., and Islam, M.R. (2004) Detection of precipitation in pipelines. *Journal of Petroleum Science and Technology*, **22**(9–10), 1119–42.

Zander, B.K.A.U. (2003) Knowledge of the firm and the evolutionary theory of the multinational corporation. *Journal of International Business Studies*, **34**, 516.

Zatzman, G.M. and Islam, M.R. (2006) Natural gas pricing, in *Handbook of Natural Gas Transmission and Processing* (eds S. Mokhatab, J.G. Speight and W.A. Poe). Elsevier Inc., 672pp.

Zatzman, G. and Islam, M.R. (2007) *Economics of Intangibles*. Nova Science Publishers, New York, 393pp.

Zatzman, G.M. and Islam, M.R. (2006) Truth, consequences and intentions: the study of natural and anti-natural starting points and their implications. *Journal of Nature Science and Sustainable Technologies*, **1**(2), 1–42.

Zatzman, G.M. and Islam, M.R. (2007) *Economics of Intangibles*. Nova Science Publisher, New York.

Zatzman, G.M., Mehedi, M.Y., and Islam, M.R. (2005) Comparative evaluation methodology and some implications for produced-water management strategy. *Energy Sources*, in press.

Zekri, A.Y., Shedid, S.A., and Alkashef, H. (2001) *Use of Laser Technology for the Treatment of Asphaltene Deposition in Carbonate Formation*. SPE Annual Technical Conference and Exhibition, SPE Paper No. 71457, September 30–October 3, New Orleans, Louisiana.

Zero, W. (2005) *The Case for Zero Waste*. <http://www.zerowaste.org/> [Accessed on August 12, 2006].

Zhang, C.Y. and Zhang, J.C. (1993) A test pilot of EOR by *in-situ* microorganism fermentation in the Daquang oil field. *Dev. Petr. Sci.*, **39**, 231–44.

Zhang, L., Luo, L., Zhao, S., Xu, Z.-C., An, J.-Y., and Yu, J.-Y. (2004) Effect of different acidic fractions in crude oil on dynamic interfacial tensions in surfactant/alkali/model oil systems. *Journal of Petroleum Science and Engineering*, **41**, 189–198.

Zharikova, A.V., Vitovtovab, V.M., Shmonovb, V.M., and Grafchikov, A.A. (2003) Permeability of the rocks from the Kola superdeep borehole at high temperature and pressure: implication to fluid dynamics in the continental crust. *Tectonophysics*, **370**, 177–191.

Zhong, L. and Islam, M.R. (1995a) Microbial Consolidation Can Improve Sand Control and Fracture Remediation, *Oil and Gas Reporter*, Dec., 83–6.

Zhong, L. and Islam, M.R. (1995b) *A New Microbial Plugging Process and Its Impact on Fracture Remediation*. Proceedings of SPE Annual Conference and Exhibition, SPE Paper No. 30519, October, Dallas, TX.

Zhou, S., Heras, H., and Ackman, R.G. (1997) Role of adipocytes in the muscle tissue of Atlantic salmon (Salmo salar) in the uptake, release and retention of water-soluble fraction of crude oil hydrocarbons. *Marine Biology*, **127**(4), 545–53.

Zhu, Q. and Sarkis, J. (2005) The link between quality management and environmental management in firms of differing size: an analysis of organizations in China. *Environmental Quality Management*, **13**, 53–63.

Zimmerman, J. (1984) *Polyamide in Encyclopedia of Polymer Science and Technology*. John Wiley and Sons, New York.

Zimmerman, R. (1999) *Forecast for the 1999 Brown Shrimping Season in the Western Gulf of Mexico, from the Mississippi River to The U.S. – Mexico Border*. Galveston Laboratory News: Southeast Fisheries Science Center. Galveston, TX. <http://galveston.ssp.nmfs.gov/galv/news/prediction99.htm> [Accessed on April 19, 2002].

Zwanenburg, K.C.T., Bowen, D., Bundy, A., Drinkwater, K., Frank, K., O'Boyle, R., Sameoto, D., and Sinclair, M. (2001) Decadal changes in the Scotian Shelf Large Marine Ecosystem, in *Large Marine Ecosystems of the North Atlantic – Changing states and Sustainability* (eds K. Sherman and H.R. Skjoldal). Elsevier, Amsterdam, pp. 105–50.

Zwicker, S.L, Collins, J., Gilbert, J.T.E., Moore, J.E., and King, R.J. (1983) Technical Guidelines for offshore oil and gas development, in *For the East-West Environment and Policy Institute* (ed. J.T.E. Gilbert). Penn Well Books, Oklahoma City, OK, pp. 49–104.

Author Index

Subject Index